D1480886

The TREE *of* LIFE

A Phylogenetic Classification

Guillaume Lecointre
Hervé Le Guyader

Illustrated by Dominique Visset
Translated by Karen McCoy

HARVARD
UNIVERSITY
PRESS
REFERENCE
LIBRARY

THE BELKNAP PRESS
OF HARVARD UNIVERSITY PRESS
CAMBRIDGE, MASSACHUSETTS
LONDON, ENGLAND
2006

First published as *Classification phylogénétique du vivant,* 3rd ed.,
© Éditions Belin—Paris, 2006

The drawings of male figures on pages 498 and 499 are reprinted by the kind permission of Dargaud Éditeur: page 498, from *Blake et Mortimer: Le piège diabolique,* © Éditions Blake et Mortimer, by E. P. Jacobs; page 499, from *Blake et Mortimer: La marque jaune,* © Éditions Blake et Mortimer, by E. P. Jacobs

Library of Congress Cataloging-in-Publication Data

Lecointre, Guillaume.
[Classification phylogénétique du vivant. English]
The tree of life : a phylogenetic classification / Guillaume Lecointre, Hervé Le Guyader ; illustrated by Dominique Visset ; translated by Karen McCoy.
p. cm.
Includes index.
ISBN-13: 978-0-674-02183-9 (cloth : alk. paper)
ISBN-10: 0-674-02183-5
1. Biology—Classification.   2. Cladistic analysis.   I. Le Guyader, Hervé.   II. Title.

QH83.L42513 2006
578.01'2—dc22   2006047736

Contents

Preface

This book is an attempt to fill a void in scientific and university publishing. Indeed, up to now, no book has provided a complete and coherent classification of the living world based solely on evolutionary relationships. Today these relationships are sister-group relationships and represented by tree diagrams called phylogenies.

Over the last few decades, the general use of cladistics, an analytical method whose basis is outlined in the introduction of this book, has fulfilled a wish that Charles Darwin initially formulated in 1859 in the *Origin of Species:* "difficulties in classification are explained, if I do not greatly deceive myself, on the view that the natural system is founded on descent with modification; that the characters which naturalists consider as showing true affinity between any two or more species, are those which have been inherited from a common parent, and, in so far, all true classification is genealogical; that community of descent is the hidden bond which naturalists have been unconsciously seeking, and not some unknown plan of creation, or the enunciation of general propositions, and the mere putting together and separating objects more or less alike. But I must explain my meaning more fully. I believe that the *arrangement* of the groups within each class, in due subordination and relation to the other groups, must be strictly genealogical in order to be natural." In this sense, constructing a phylogeny, "the history of the descendants of living beings" (cf. P. Darlu and P. Tassy, 1993), is one and the same with establishing a natural classification. This is why recent results presented here follow the cladistic method (or phylogenetic systematics); this is the only method that provides a coherent and logical guideline for understanding the distribution of characters among species and, ultimately, the evolution of these characters.

In addition to the methodological and conceptual advances that followed the advent of cladistics, progress in molecular biology has now given us access to new characters—those contained in the sequences of biological macromolecules—to consider along with the morphological features we have relied on up to this point. These sequences have enabled us to compare organisms that have no obvious *a priori* similarities, like bacteria, plants, fungi and animals, and have allowed us to place them on the same phylogenetic tree. For the first time since the birth of biology, we have the means of broadly tracing the historical structure of all biodiversity within the framework of a single tree of life. The details of this tree can be examined almost infinitely.

The inheritance of contemporary systematics

Traditional classifications, still taught today, are based on the combined influence of Carolus Linnaeus (1707–1778) and Charles Darwin (1809–1882). The heritage from Linnaeus and his period are demonstrated by the persistence of an anthropocentrism originating from the concept of the *Scala Naturae* (all organisms have a fixed place on the scale with humans at the summit) and the continued use of groups constructed on the absence of a character (such as the invertebrates, defined as non-vertebrates, or the fishes, defined as non-tetrapod vertebrates). These apparent aberrations were justified by the central place of humans in Creation and by the philosophy of nature of Linnaeus's time (namely essentialism and fixism—sciences associated with theology), which was largely incompatible with the idea of species transformations. By contrast, nominalism and science politically emancipated from theology favored the rise of evolutionary thought during Darwin's time. However, despite the change in thinking initiated by Darwin, the *Scala Naturae* has survived through time along with pre-cladistic evolutionary classifications. Some paraphyletic groups based on negations were also maintained because they were thought to have given rise to another group deserving of special recognition. For example, fishes are non-tetrapod vertebrates and gave rise to tetrapods via an adaptive jump.

Darwin's legacy has produced many important re-arrangements. After the introduction of the term *phylogeny* in 1866 by Ernst Haeckel (1834–1919), a scientifically valid, natural classification must reflect phylogeny. However, Darwin did not possess the conceptual tools necessary to produce strictly phylogenetic classifications, where groups such as the invertebrates or fishes would have no place. This conceptual revolution would be the work of the German entomologist Willi Hennig (1913–1976), who founded phylogenetic systematics (also called *cladistics*), the only method that produces only monophyletic groups (groups that include an ancestor and all of its descendants). Following the publication of his book in 1950 (in German) and especially after the publication of the English translation in 1966, systematists across the world have progressively changed the way they work. Darwin's objective was finally reached more than a century after his proposal. Nowadays, all serious comparative analyses of anatomical characters follow this way of thinking. As we will see, the situation is less clear for the analysis of molecular characters.

Over the last thirty years, phylogenetic systematics has completely changed biological classifications; it has already permitted a better comprehension of the structure of biodiversity, a structure that can only really be understood in light of phylogenetics, even if much work remains to be done.

Teaching phylogenetic systematics

Despite spectacular progress, and the fact that phylogenetic classifications now dominate research laboratories, traditional classifications (with their "negatively defined groups") are still taught from primary school up to university and are widely diffused in the media. Everyone loves dinosaurs, traditionally defined as all the archosaurs excluding birds. Nowadays, virtually all evolutionary biologists accept that birds lie deep within this group. The group "Dinosauria" is, in the traditional sense, paraphyletic. In the modern classification it is monophyletic and does contain birds. It is well known that the traditional classifications are linked to a cortege of false ideas: a linear vision of the organization of biodiversity; finalism and anthropocentrism, which describe biological evolution as a series of improvements that led to humans (confusion with the nonscientific notion of "progress," already criticized by S. J. Gould); misunderstanding about the link between sister-group relationships and the notion of the ancestor; the persistence of the concept of the living fossil. This archaic image of systematics is fed by an essentialism embellished by false positivism and maintained by a mix of scientific statements and nonscientific representations of life (mythology, morality, ideology) and the combination of popular and specialized vocabulary (think of words like *relatedness, ancestor,* etc). These misconceptions can lead, for example, to a naïve extension of a genealogy (in the strict sense) into a phylogeny.

Today, the situation for high school teachers is difficult. They learned the traditional classification during their studies but must explain how to construct a phylogeny, in the modern sense of the word, in senior science courses. In addition, they must do so quickly, a skill that requires a deep understanding of the concepts. This exercise leads them to observe that the traditional classification has no phylogenetic sense. Moreover, they are not always helped by books: many textbooks in biology, zoology and botany are ambiguous in this respect. The authors provide a chapter about modern thoughts on the historical relationships among organisms and classifications; this chapter is, most of the time, inspired by cladistics. However, they forget about this chapter in the rest of the book, where one can find "invertebrates," "fish," etc. To facilitate the replacement of terms like *reptile, fish, invertebrate* (experience shows that senior students easily see the illogic of these non-groups), teachers need a basic textbook that presents an overview of the phylogenetic systematics of life. We hope to fulfill this need.

This book is therefore aimed at high school teachers who have the difficult task of updating their knowledge without easy access to the abundant scientific literature or to a complete and rigorous summary. This book is equally addressed to all university students, professors and scientists unfamiliar with phylogenetics. Of course, we should not forget the self-taught person with a passion for systematics, zoology, botany or paleontology, or the amateur naturalists who will find in these pages a phylogeny as well as a classification and a better understanding of the fascinating diversity of the living world.

Structure of the book

The tree of life that we present here is composed of a series of nested monophyletic groups (or clades). No living organism is left out: any organism that you can think of should be found within one clade or another. However, certain groups have been examined more closely than others. We wanted to provide a synthesis of the phylogeny of all life while zooming in on the groups that are most important for teaching purposes, such as the metazoans. Given space constraints, certain clades could not be described in their own individual chapter, despite their considerable abundance (for example, the hexapods, angiosperms, teleost fishes, etc.). However, phylogenies of these clades are provided in the appendices, but without any detailed information on the supporting characters.

This type of summary is risky. Indeed, opinions on the phylogenetic position of the groups can differ greatly. If we had taken into account the opinion of every scientist, most of the nodes of the tree of life would be composed of multifurcations, polytomies signalling unresolved relationships. Taking these various opinions into account would therefore not have been of great service to readers. We therefore had to make some decisions, decisions that followed the criteria outlined below.

When it was necessary to decide among several contradictory trees, the tree resulting from a cladistic analysis was preferred to trees based on the opinion of the authors. In the first case, the analysis is transparent because the data employed is available (often morphological characters) and the criterion used to select the tree is provided; these elements follow the logic of this book. When trees based on morphological and anatomical data differed from those of molecular data, decisions were made on a case-by-case basis. If the morphological data contained many internal contradictions, the tree based on molecular data was retained. However, trees constructed from molecular data can also be fraught with artifacts. We have therefore been extremely careful in dealing with this type of data. Most often, molecular phylogenetic results were taken into consideration only when the phylogenetic hypothesis was supported by studies of several independent genes and research teams. This criterion of choice transcends all other more technical criteria (as explained in the Introduction: reconstruction method used, conflicting branch lengths, statistical robustness of inferences, associated homoplasy, risk of a long-branch attraction artifact, or risk of grouping based on symplesiomorphies, etc.). Indeed, we can show that the reliability of a clade is strongly correlated with the independence of the different sources that provided the same result. A clade is a complex inference that cannot be found several times by chance alone. If it is found numerous times, it is because the traces of the common history are found in different independent genes of the species that carry them (except in cases of extremely special artifacts). Polytomies in the tree are shown for two reasons: in the absence of data or, more infrequently, when there are strong phylogenetic conflicts with no measure of the reliability of the contradictory trees. Finally, certain groups have an unresolved phylogenetic position. In these instances, we have placed these groups on the branch that seemed most reasonable to us but have indicated the uncertainty with question marks.

Some clades are found on branch tips, whereas others are found within the internal nodes of the tree. There is no qualitative difference between "tip" and "node" clades. Each type has its own summary page that outlines the exclusive characters that define it. Each clade is illustrated with figures. In the interest of space, we have not included fossil species.

The summary pages of "tip" clades contain seven rubrics: general description, some unique derived features, number of species, oldest known fossils, current distribution, ecology, and examples. The summary pages of the "node" clades contain the same rubrics except general description and ecology; this was done to eliminate any repetition of the information. Finally, we have chosen to adopt a relatively informal style for writing taxon names, without capitals.

Last, it should be noted that the median size is given for the organisms. For example, the common marmoset ranges from 19 to 21 cm in body length and from 25 to 29 cm in tail length. We therefore state a BL (total body length including the head): 20 cm, TL (tail length): 27 cm. For continuously growing organisms, we have indicated the maximum recorded size.

General Description

When a group is defined by its innovations (or synapomorphies), we are interested only in the anatomical features or genes linked to the innovations. However, a synapomorphy is not a general description. Rather, a general description consists of global information on the appearance of the species included in the clade without making the distinction between derived characters (synapomorphies) and ancestral characters (symplesiomorphies).

Some Unique Derived Features

Here, the synapomorphies are described, the features that support the monophyly of the group. More exactly, we describe some of the derived character states. Often, the ancestral state of a character (found outside the clade) is also described in order to better appreciate the innovation. The adjective *unique* means that it is "unique to the clade." However, it can happen that a character state may be used to define a clade, but this state has arisen by convergence in another distant clade. For example, the perissodactyls (tapirs, rhinoceroses, horses) are defined by a supporting axis that passes through the third toes of the posterior limbs, *with a corresponding reduction of the lateral toes*. However, this arrangement has also been acquired by the litopterans, fossil ungulates from South America that are not directly related to the perissodactyls. The words *unique* or *innovation* therefore designate relative notions and must be understood as "locally unique." Acquiring the shape of a mole is an innovation in the eulipotyphlan mammals and defines the talpidae, but it is not really "innovative" among mammals because both afrotherians and marsupials have also independently acquired mole-like forms (chrysochloridae and notoryctidae, respectively). It can also happen that the character state that defines the clade has been secondarily modified in part of the clade. For example, a tetrapod has walking limbs with five digits. However, among the tetrapods, some clades have remodelled this character state. Birds only have four toes, horses have only one, and snakes have lost their limbs entirely. Embryology demonstrates that the origin of these secondary arrangements has been acquired from a pentadactyl limb.

Number of Species

The reader will find the number of currently known and recorded species in this rubric. It is not an estimation of the number of species that truly exist in the clade, including both described species and an estimate of those yet to be described. For clades that include large organisms, these two numbers are more or less the same. For others, like the nematodes and eubacteria, they are very different.

Given the nested nature of the clades, the number of species adds up as we move from the branch tips towards the root of the tree. However, the number of species is well known for some clades, whereas it is only a very approximate number for clades for which no recent inventories have been performed. In certain cases, we were therefore obliged to add very precise numbers to very imprecise numbers. It may seem ridiculous to give the number of metazoan species to within one species. Yet, this is what we have done in order to respect the logic of the nested clades.

Oldest Known Fossil

Indicating the oldest recorded fossil for a clade seems simple. However, we were sometimes obliged to provide several fossils in this section when a clade was closely linked to fossilized sister lineages. For example, in the chapter on current mammals, we should have provided the oldest known fossil for each order. However, the carnivores do not branch off alone from the other mammalian orders. The lineage that branches from the other extant orders is not called "Carnivora" but rather "Ferae" and includes both the carnivores and the creodonts (exclusively fossil species). In this case, we have indicated the oldest known fossil for the clade, strictly speaking, along with that of the expanded clade that includes the fossils. When the fossil from the first clade is the oldest of the entire lineage, only this fossil is provided.

Current Distribution

This refers to the geographic distribution of the clade. In a few cases, one will also find information on the distribution of the fossils.

Ecology

Under this rubric, we have summarized the life style of the organisms, the biotopes that they are found in, their position in the trophic food chain, aspects of their reproduction, and their relationship to humans (domestication, for example). We have also sometimes added interesting anecdotes.

Examples

Some example species are provided. It was impossible to provide an exhaustive list, so these examples have been chosen to illustrate the taxonomic diversity in the group.

Introduction

A brief history of biotic classification

The word *taxonomy*—from the Greek words *taxis*, meaning order or arrangement, and *nomos*, meaning law—was first proposed in 1813 by the Swiss botanist Augustin Pyrame de Candolle (1778–1841) to designate the science of biotic classification. Although it was not until the nineteenth century that it received a formal name, the discipline had preoccupied biologists, and particularly botanists, for centuries. Why were botanists so interested in classification systems? While it was easy for an individual to recognize important animals (those used for breeding, hunting, or fishing) and to memorize their names, knowing the vast number of plants—medicinal plants in particular—demanded a long apprenticeship, typically reserved for professionals. It was thus early on that, for utilitarian reasons, the need to inventory and classify plants was felt.

From Aristotle to the Renaissance

The Greek philosopher Theophrastus (372–287 B.C.), Aristotle's (385–322 B.C.) successor in the leadership of the Lyceum, left behind the first history of plants. We can therefore consider him the founder of botany. Later, in the first century A.D., Dioscorides wrote a botanic treatise in 6 volumes, devoted in succession to aromatic plants, food plants, medicinal plants (two volumes), vinous plants, and poisonous plants. The Latin world was represented primarily by Pliny the Elder (23–79), author of the vast *Historia Naturalis* in 37 volumes, who also proposed a utilitarian classification system.

Until the sixteenth century, it was the works of these three authors in particular that were widely commented upon. With the proliferation of descriptions and illustrations, recognition criteria started to take shape: it was slowly realized that the same plant could be found under several names and that the same name could designate different plants. The need for rules and rigorous principles started to be appreciated. However, it is one thing to point out the errors of a classification system, but something quite different to propose the basic principles of a new system. Eventually, a series of botanists—Fuchs in 1531, Gesner in 1541, Camerarius in 1586—shattered the existing classification system by classifying plants . . . in alphabetical order! This was a useless enterprise *a priori*—those knowing the name of the plant did not require the book, those not knowing the name couldn't use it. This endeavor, however, proved extremely beneficial in the end: botany could start over with a new foundation because the work of antiquity was no longer considered untouchable.

From the Renaissance to Linnaeus

From 1500 on, there were many attempts at classification, or what we call classification systems or methods. During the next two centuries history records at least a hundred. All possible criteria were tried out: size, general shape, root form, leaf type, and, with Andrea Cesalpino (1519–1603), flower and fruit types.

Two broad types of logic were employed. The first, inherited from Aristotle, consists of dividing a group of organisms into subgroups according to predefined criteria and repeating this process until the species level is reached. This logic, termed "divisive," typically results in a dichotomous choice with one term opposing the other (for example, plants with flowers vs. plants without flowers). The second logic involves agglomeration, that is, assembling species into different groups based on similarity criteria. This process is then reiterated, using the newly formed groups as the primary units, until all organisms are united within a single category. One may think that these two logics are equivalent and should lead to the same result, but that is not the case. In the first instance, the criteria must be chosen *a priori,* whereas in the second they can be the result of observations of the organisms themselves.

It was in 1694 that a botanist from Montpellier, Joseph Pitton de Tournefort (1656–1708), finally understood that the fundamental process consisted of reuniting species into genera. Thus we arrive at the notion of hierarchical levels, corresponding to the reiterated process of classification. These levels were coded by Carolus Linnaeus (1707–1778) as *kingdom, class, order, genus, species, variety.* These rankings would later be modified for the animal kingdom into the seven traditional levels *kingdom, phylum, class, order, family, genus, species;* at the time, it was desirable to have seven levels—seven being a supposedly perfect number. We will see in what follows that this coding system was too rigid and divisive. Proceeding by agglomeration, there was no *a priori* reason to end up with seven categories each time. This rule was therefore quickly broken by the utilization of intermediate levels (suborders, superclasses, etc.). In fact, Linnaeus drew on both of the preceding logics: the first, divisive, for taxa superior to the genus; the second, agglomerative, for species and genera. As pointed out by Candolle in 1835, this was why Linnaeus named organisms by the genus name followed by the species name. This system of *binary nomenclature* stemmed directly from Linnaeus's conception of natural groups.

The numerous systems and methods of this period led to a particularly fortunate result: in the large majority of cases, although arrived at via different routes, most of the main plant families (rosaceae, papilionaceae, umbelliferae, gramineae, orchidaceae, and so on) were distinguishable. From this realization grew the impression that there might be a natural classification, one that could be found in any of the artificial classifications. The existence of a unique, natural classification led to the idea that there was an intrinsic order to nature.

According to the thinking of the time, this order was evidently that of divine creation.

The search for a natural method

From this moment on, botanists moved toward a single goal: to find a logic that enabled access to the natural classification. The principle of the subordination of characters, devised by Bernard de Jussieu (1699–1776) and his nephew Antoine Laurent de Jussieu (1748–1836), resulted in a veritable revolution.

The story goes that Bernard de Jussieu had been ordered by Louis XV of France to organize the garden of the Trianon in the most natural manner possible. In doing so, he illustrated an observation that Linnaeus had made in 1751: different plants show relationships similar to those of different territories on a map. Jussieu realized that the different characters used to define a given taxonomic level were not necessarily equal: "Different characters should not be considered as individual units, but each following its relative value, in such a way that a single constant character might be of equal or greater value than several inconsistent characters taken together." In other words, it is better to define a taxon by a few, constant characters that are shared by the species ensemble than by many, labile characters that lead to an ambiguous result. This principle enabled the construction of an effective classification system in botany that, at the time, was also considered to be natural. By the end of the eighteenth century, Georges Cuvier (1769–1832) had already applied this system to the animal kingdom by splitting animals into four groups (Vertebrata, Arthropoda, Mollusca, Radiata), each characterized by its body plan.

From Jussieu to Darwin

With the work of the Jussieus and Cuvier, an important step had been taken; the biological community thought that they now had theoretical tools powerful enough to achieve an acceptable classification—even if the main progress was actually made in what followed. In this way, the basic problem had changed; they were no longer looking for an efficient method to classify organisms in a natural manner, but rather to change the fundamental meaning of classification. Indeed, at the end of the "century of enlightenment," there was no longer any consensus that natural classification corresponded to a divine order. In his work *Philosophie zoologique* (1809), Jean Baptiste

Lamarck (1744–1829) gave substance to an intuition that circulated among many philosophers during the second half of the eighteenth century. Lamarck outlined the first mechanism of change in living organisms, associated with what was later termed the *inheritance of acquired traits*. As Lamarck did not give priority to the variation within species, he was obliged to consider factors from the external environment as the motors of organic change. In fact, to understand the idea of variation and selection later conceived by Darwin and Wallace, Lamarck's attention should have been focused on the polymorphisms of closely related individuals within a single species. However, the naturalists of Lamarck's era observed only absolute forms and considered variation a negligible disorder. The individuals that make up a species were not considered in their own right but instead as holders of an essential property of the species to which they belonged, and this property was defined *a priori* as the fruit of divine creation. Essentialism prevented an interest in the differences between individuals.

It was Charles Darwin (1809–1882) who turned his back on essentialism. Before seeing the species, he saw the variation among individuals, with the average and variance of traits providing a certain impression about what the species was. Darwin's nominalism enabled him to discover the hidden value of the variation that had previously been considered negligible. From this moment on, the question was no longer "Why are living organisms different from each other?" but rather "Why, despite all this variation, are organisms still similar?" Darwin postulated that it was not because of divine or environmental instruction but, instead, because of natural selection. Under the particular environmental conditions of a given moment in time, certain variants are favored and become more numerous because they leave behind more descendants than do competing variants. A change of milieu can modify this equilibrium: the population evolves over the course of generations and thus over the course of its genealogy. Here Darwin arrived at the key idea of descent with modification proposed in the *Origin of Species* (1859). This idea would henceforth structure phylogenetic thinking.

According to Darwin, the key was to study the variability not of acquired characters but only that of inherited characters, those transmitted to future generations. Resemblance between species is thus due to characters inherited from an ancestral species. It follows that the community of descendants of a given ancestral species is the

hidden link that explains the similarity of species and therefore their arrangement into a single taxon. As this reasoning can be used in the reverse sense, to go back in time, it can be used to build a classification system based on species relatedness. When we compare extant species, certain characters are pertinent for identifying common ancestry. These characters, inherited from a common ancestor, are called *homologues*. As long as we are able to identify them with the appropriate methods, these elements allow us to reconstruct history. It is for this reason Darwin stressed the necessity of classifying organisms following their genealogical development (or phylogeny); the current distribution of observed biological structures is nothing more than a reflection of life's long history. This is what led to Theodosius Dobzhansky's (1900–1975) most famous quote: "Nothing in biology makes sense except in the light of evolution."

We had to wait until Darwin to understand that nature's order is a reflection of the evolutionary history of Earth's organisms: finding a natural classification system and uncovering this history are the same endeavor. Nonetheless, after this fundamental conceptual breakthrough, it took almost a century for the method to become fully operational.

From Darwin to Hennig

For a hundred years scientists endeavored, without an explicit method, to link their classifications to their ideas about the genealogy (who descended from whom?) and the phylogeny (who is more closely related to whom?) of different taxa. Operationally, the two notions were confounded. For example, branching bubbles (fig. 1), a favorite tool of American paleontologist Alfred S. Romer (1884–1973), confounded ancestral relationships (genealogy) and relatedness (phylogeny) because it showed only the ancestral relationships of high-ranking taxa (mammals descend from reptiles, reptiles from amphibians, etc.). The branching points between two groups frequently included ancestors identified from the fossil record. To be completely thorough, criteria other than strict relatedness were frequently incorporated into the classifications of this period, criteria that took into account adaptive or ecological considerations or that integrated relative complexity and global similarity.

The classic example is the reptile group. By the end of the nineteenth century, it was recognized

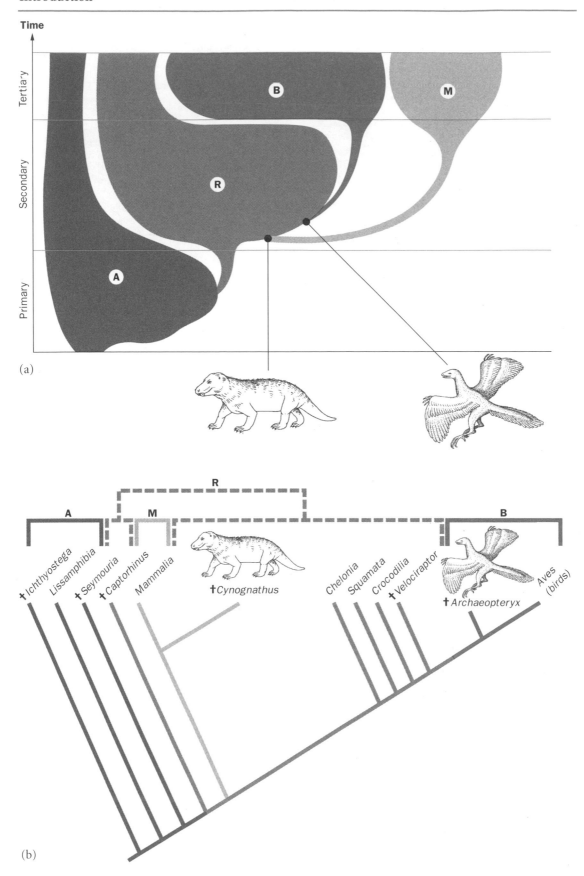

Figure 1. Two ways of representing the relationship among tetrapod vertebrates, with the inclusion of certain fossils. (a) The classic representation of the relationship among taxa as conceived by the eclectic systematics of Alfred Romer. The fossils are *Cynognathus* and *Archaeopteryx*. (b) The same relationship displayed in terms of phylogenetic systematics (a cladistic analysis). Amphibians and reptiles are indicated to show the absence of monophyly in the groups. Note that the fossils are placed at the end of the branches, as though they were extant species. A: amphibians, M: mammals, B: birds, R: reptiles.

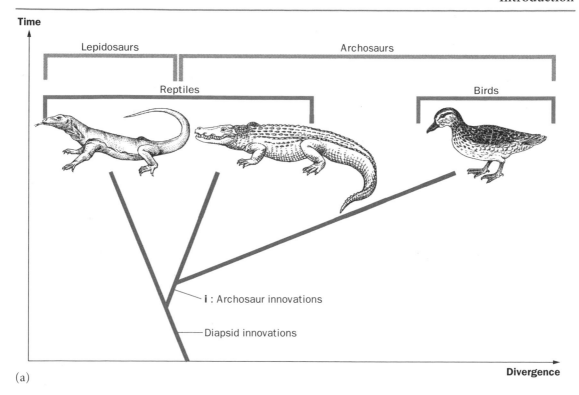

(a)

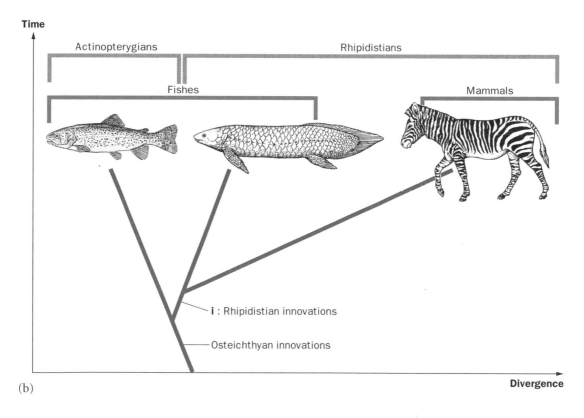

(b)

Figure 2. Classification by relatedness as determined by phylogenetic systematics (cladistics), as opposed to classification based on overall similarity (phenetics) or on adaptations (eclectic systematics). (a) The Komodo dragon (*Varanus komodensis*) globally resembles a crocodile (*Crocodylus nitolicus*), while the duck (*Anas querquedula*) does not resemble either one. Yet the bird shares exclusive characteristics with the crocodile that it does not share with the Komodo dragon: these are innovations or synapomorphies (labeled i). A classification based on overall resemblance, or attempting to underline the "adaptive leaps" of the birds, would bring the Komodo dragon and the crocodile together within the reptiles and would leave the birds outside. A classification based on common ascent (phylogenetic systematics, shown in color) would classify the crocodile and the bird within the archosaurs and would leave the dragon with the lepidosaurs. (b) The same sets of relationships displayed above, but using a different group of organisms: a trout (*Salmo trutta*), a lungfish (*Neoceratodus forsteri*), and a zebra (*Equus zebra*). A phenetic classification would distinguish between fishes and mammals, while a classification based on adaptations (shown in color) would distinguish between actinopterygians and rhipidistians.

15

that, given the anatomy of their toes and pelvis, birds were archosaurs and that their origin should therefore be linked to the theropod dinosaurs. However, this relationship remained hidden by the separation of reptiles and birds under the pretext that the birds, with their ability to fly, had undergone an "adaptive leap" that merited putting them in a separate class (fig. 2). Also, with the disappearance of the dinosaurs (in the classic sense), the only extant archosaurs are crocodiles. Despite numerous anatomical structures exclusive to extant birds and crocodiles (mandibular fenestra, gizzard, etc.), their degree of overall similarity and their ecological situations do not readily group these taxa together. A purely phylogenetic classification would give priority to relatedness despite adaptive or ecological considerations and would therefore require an independent group to underline this relationship. In order to avoid this problem, we no longer speak of a class of reptiles. The birds are archosaurs; they are dinosaurs; they are theropods. The reptiles themselves constitute not a *clade* but rather a *grade,* a classification meant to show a degree of complexity inferior to that of the birds. A grade is defined not by a unique character but by adaptation and, sometimes, by fate: it therefore corresponds to a level of complexity that precedes that of the group to which it gives rise. It possesses nonunique characters and is defined by the absence of characters found in the "next evolutionary step." Reptiles exist because birds and mammals are descended from them; they are amniotes without hair or feathers. From a cladistic point of view, it is easy to define a grade as *paraphyletic:* it possesses members that are more closely related to organisms outside the grade than to other members within the grade. Dromaeosaurids (theropod dinosaurs) are closer to birds than to any other reptile. Considering only extant fauna, crocodiles are closer to birds than to any other reptile (turtles, lizards, snakes, tuataras).

Phylogenetic systematics—or cladistics—was founded by Willi Hennig (1913–1976) in 1950 and has greatly developed over the last thirty years. It attempts to recognize the evolutionary relationships among species by assembling them into *monophyletic groups,* that is, into groups that are composed of the common ancestor and all of its descendants. To do this, it rigorously exploits the Darwinian concept of descent with modification.

Phylogenetic systematics is best understood by focusing on taxonomic sampling. Systematists classify only subsamples of the biodiversity. Taxonomic sampling is choosing a number of

representative individuals about whom we ask "Who is related to whom?" If there is inherited variation for a character within a studied sample, then the character is present in at least two states. Hennig understood that observing only one of these states would fail to indicate relatedness among members of the sample because it is already present outside the sample; this is the ancestral state. States that result from a more recent transformation are therefore considered derived. A shared ancestral trait, or *symplesiomorphy,* should never be taken into account in cladistics because, as it is already present outside the studied group, it does not provide any information about the relationships among species within the group. For example, if we study primates, we observe that certain primates have two frontal bones. Although this character is inherited from a common ancestor, the presence of two frontal bones does not permit one to define a sub-group of primates, because it is also present in all other mammals. The hypothetical ancestor that displayed two frontal bones for the first time is clearly one of the ancestors of the primates, but it is also the ancestor of many other mammals. Therefore, in a sample of primates, the presence of two frontal bones is a shared ancestral trait, a *symplesiomorphy.* Only the sharing of a derived trait, the result of a unique evolutionary event, by several species within the group indicates a relationship among sample members. This shared derived character state, or *synapomorphy,* identifies a monophyletic group within the studied group. Therefore, within the primates, the fusion of the two frontal bones into a single bone is a novelty unique to the simiiformes (true monkeys).

Cladistics can run into difficulties in its application because not all character states are necessarily homologous. Certain resemblances are convergent—that is, the result of independent evolution. We cannot always detect these convergences immediately, and their presence may contradict other similarities, "true homologies" yet to be recognized. Thus, we are obliged to assume at first that, for each character, similar states are homologous, despite knowing that there may be convergences among them. These assumptions are always explicit and are most often coded into the data matrix describing all the characters at hand. In cladistics, we build all possible trees, incorporating the fewest number of evolutionary events required by the data matrix. In the end, this amounts to maximizing the proximity of identical character states in the tree. We keep only the most parsimonious tree—the one with the fewest number of evolutionary steps—and then reexamine the assumption of homology. The result is the most simple representation, the tree with the fewest number of transformation assumptions. This is not to say that with another dataset the same tree would necessarily be constructed. After other characters or species are added to the data matrix, the original tree may be supported or refuted, totally or partially, by the new result. A phylogenetic tree may be considered a working hypothesis, one that is openly linked to a clear procedure. It is only after having numerous concordant results that one may conclude that a proposed clade is reliable.

The clarity of the cladistic approach implied a considerable change in the mentality of the systematists of the twentieth century. Earlier approaches expressed the author's opinion more than the methodology. Today, it is easy to test the effect of a given assumption on the result when one has at one's disposal the phylogenetic analyses of other researchers. This was impossible before systematics had a clear methodology.

Philosophical foundations of classification

The history of classification cannot be presented without evoking the concept of the Great Chain of Being, the *scala naturae*. This concept was developed in detail by Gottfried Leibnitz (1646–1716), but its basis was set out by Plato and Aristotle. Indeed, it was Aristotle who proposed the linear gradation principle, whereby living organisms are classified according to their degree of perfection—either in the degree of development attained by the living organism at birth or in the power of its soul. In this way, a gradient of enriched souls can be established—from plants up to rational man and those beings superior to him. Each being possesses the powers of those below it, along with its own. The Great Chain of Being is a linear ordering, and this order is fixed.

The Great Chain of Being is, in fact, firmly ingrained in our culture and spirits. It leads to certain grave errors that are commonly acknowledged but difficult for teachers to correct. For example:

Anthropocentrism. This is the view that man is the measure of all things, that if we exclude the angels humans are at the top of the Great Chain, the most evolved organisms on Earth. It is difficult for people to understand that all extant organisms are equally *evolved*, in the sense that the lineages to which they belong have all existed over the same

evolutionary time. Indeed, the Great Chain of Being is a view that considers human uniqueness qualitatively superior to that of other organisms, an idea with no scientific value. In addition, the influence of anthropocentrism helps to maintain grades in classification. The prokaryotes are cells without a nucleus (whose cells have a nucleus? the eukaryotes, which include humans); the invertebrates are metazoans without a backbone (who has a backbone? the vertebrates, which include humans); the fishes are craniates without legs (who has legs? the tetrapods, which include humans); the reptiles are amniotes without hair (who has hair? the mammals, which includes humans), etc.

Finalism. According to this view, evolution is mysteriously driven toward the emergence of man. Yet, our current understanding of evolutionary mechanisms does not assign a goal to evolution; natural selection sorts variants as a function of the conditions at the time. In other words, as in history, if evolution could be started over, the probability that everything would happen in the same manner is almost zero.

Essentialism. The idea that evolution has a goal requires as a corollary belief that, from a philosophical perspective, the intimate nature of beings, or their essence, precedes their existence. Essentialism gives to living beings an absolute that, in science, should reside only in nature's laws and not in objects that are subject to these laws. It constructs entities *a priori* and then forces the reality of living beings into this form. Living beings are thus only seen by that which links them to this essence. Essentialism led Linneaus and his contemporaries to neglect the variation within species and pushed the paleontologist Richard Owen (1804–1892) to argue with Darwin. Essentialism amounts to denying the fundamental role of the conditional processes taking place on Earth. It is incompatible with progress in phylogenetic classification, in that changing a classification because a taxon is one day found to be polyphyletic might contradict a classification based on essence (for example, the classification of elephants, rhinoceroses, and hippopotamuses as pachyderms because of their thick skins no longer exists). Moreover, in phylogenetic classifications organisms are seen not as instances of a single essence but as character mosaics in which character states are sometimes primitive, sometimes derived, sometimes convergent, depending on the scale of the taxonomic sampling of the study. Finally and most importantly, most phylogeneticists that use cladistic analyses have a nominalistic attitude toward living beings; this idea that abstract principles exist in nature but only in our words for them, held by Darwin, is the polar opposite of essentialism. The reality of the living world is one of individual organisms on which we superimpose the conventions of language. The species is nothing more than a collection of individuals, defined at best by a synapomorphy or at worst by the mean and variance of some measured parameters.

The ideas underlying the concept of the Great Chain of Being were overwhelmed by the concept and consequences of descent with modification. However, this idea of a linear ordering to nature explains why, among the essential concepts of modern science, those linked with phylogenetics are the least assimilated by the general public. Indeed, if the biological basis of a concept is difficult to grasp, it is often easier on our self-esteem to avoid learning it and to replace it by some nonscientific icon from the past.

Systematics today

Definition and importance of systematics

In the natural sciences, the science of classification is called systematics. Its first task is the identification, description, and inventory of nature's living organisms, past and present. The second is their classification, that which makes their immense diversity intelligible. This is in fact the objective of all science: to understand the complexity of nature in terms of categories that can give rise to the most coherent language and reasoning possible and that may serve as a basis for other studies. However, these categories do not rest on just any logic. Individuals, populations, and then species are all linked through their genealogies, inheritances they pass on, with changes, to their descendants. Species therefore have an evolutionary history that alone explains the natural order of life. In biology, the most pertinent logic for classifying species is that of evolutionary relatedness. Classifications should reflect the succession of branches of the unique tree of life (and should not be confounded with an identification key used to recognize species).

In addition to responding to the simple desire to know and understand, systematics is vital because of its many applications: medical, pharmaceutical, agronomical, ecological, geological, etc. The environmental and medical sciences illustrate the importance of systematics particularly well. For example, the species that inhabit an area and their respective abundances are excellent descriptors of the degree of local environmental degradation. This is the case for pollution indicators like marine annelids and freshwater fishes. Any decision related to the protection of the environment relies on a good understanding of its biodiversity. Likewise, the control of large disease epidemics, responsible for millions of deaths each year, would not progress without a clear understanding of the systematics of the parasites and the vectors involved.

Systematics and biodiversity

The diversity of living organisms, or biodiversity, can be viewed in two ways. The first results from a habitat-based (ecosystem) approach, that of ecological biodiversity. Species are classified in relation to the habitat they occupy, their position in the trophic chain, or their reproductive strategy—a classification based on functional, relational data that is linked to biotic space and an ecological time scale. For example, burrowing animals include those taxa that have in common behaviors typically linked to similar morphological adaptations; moles and mole crickets have anterior appendages that are short, wide, and flattened. Such functional groups are useful in ecology and environmental sciences, the process sciences.

The second, more abstract, perception of biodiversity relies on history to explain its structure. Species are classified on the basis of comparative, structural data and are considered on a paleontological time scale. This integrative classification responds to the question "Where does a species come from?" It is this question that interests us here. The first perception of biodiversity (horizontal) does not necessarily cover the second (vertical). For example, the constraints of a given environment will often give rise to similar, convergent structures in species that are phylogenetically quite distant; recall, for example, the hydrodynamic profile of the ichthyosaurs, the shark, and the dolphin. Even if ecological groups help us to analyze how a biotope functions, we still cannot understand it completely without integrating the history of its species. However, this history could not be reconstructed without phylogenetic tools, without reference to monophyletic groups. For those who want to explain biodiversity, the historical dimension is unavoidable.

Character and homology

What is a character?

A character is an observable attribute of an organism. Constructing a tree requires that characters be comparable and compared. When two structures are similar, but not identical, they are distinguished by a minimum of two states for the same character. In other words, the *character* is a designation of that which we observe (eye color or site number 177 of the hemoglobin β gene sequence) and its *state* is the way we discriminate it within a sample of organisms (brown vs. blue for eye color, G vs. T for site 177). The character and the character state differ in their designation, but not in their nature. In a comparative context, a character is thus a structure identified as being similar in two or more organisms. *Similar* means that (1) the structure can either be identical or differ to various degrees among the observed organisms and (2) this similarity, as analyzed phylogenetically, constitutes a hypothesis of homology (we call this *primary homology*). In a comparative approach, the notion of the character is tightly associated with the notion of similarity, and thus homology. *The complete definition of a character in systematics is therefore: any set of observable organismal attributes for which we can formulate at least one hypothesis of homology.*

What is homology?

In many natural science courses, two different definitions of homology may be presented without relating one definition to the other from an operational point of view.

First, we say structures are homologous when they are inherited from a common ancestor. This is homology by descent, otherwise termed *secondary homology*. Expressed in this way, the notion is of no operational interest because it does not tell us how we know that it is inherited from a common ancestor, nor does it specify the taxonomic context in which the structures were observed. Indeed, homology by descent only makes sense within a given taxonomic framework. Given that the tree of life is unique, everything would be homologous in the end without a taxonomic framework for reference. For example, a parrot's wing is homologous to an eagle's wing when one is considering wings (homology at the taxonomic level of birds) and also when one is considering anterior limbs (at the taxonomic level of the tetrapods). However, the parrot's wing and the bat's wing, even if they remain homologous at the level of the tetrapods as anterior limbs, are not homologous as wings (at this same taxonomic level).

Second, we use the term *homologous* when a structure, compared among organisms with the same body plan, has the same arrangement with neighboring structures, regardless of its form or function. This recalls the definition that Richard Owen (1804–1892) formulated in 1843 based on Étienne Geoffroy Saint-Hilaire's (1772–1844) principle of connections between parts.

One must place these two definitions in time in order to understand them better. Initially, we formulate a hypothesis of homology (second definition), which may then be confirmed by the tree that is found for the sample organisms (first definition). Owen's definition corresponds to the initial hypothesis of homology, the primary homology. This is what we do when first we declare that the radius bone of a dolphin can be compared (or is similar) to that of an opossum or a bat (fig. 3a), and when we then code their differences as diverse states under the "radius" character. None of these radii have the same form. If they are perceived as homologues, it is because they are each part of the stable organizational plan of the tetrapod forelimb, found between the stylopodium (humerus) and the distal basipodium (carpal bones). The radius is one of the components of the intermediate segment: the zeugopodium.

We do the same thing when we align nucleic acid or amino acid sequences. When similar sequences are written out one below the other, we initiate a plan of interpretation. In each column (or site), the nucleotides or amino acids are assumed to be homologous because their position in the sequence is the same. It is the same principle of connections, but applied to molecules.

To formulate a hypothesis of homology, we therefore use a plan, a type of abstract chart of the relationships among structures. We can also use data from the ontogeny (the development of the organism), however, to show that the opposing structures derive from one another. For example, using this type of information, and in particular by examining marsupial embryology, we know that two of the three bones of the mammalian middle ear (malleus and incus) are homologous to certain bones involved in the articulation between the jaw and the skull of other vertebrates (respectively, the articular and the quadrate, fig. 3b). The articulation between these two bones was conserved in the mammals despite their transfer to an auditory function. Remarkably, this transfer, recognized through ontogenesis, is equally documented in the Permo-Triassic fossil record.

Homology by descent can only be demonstrated by constructing a tree. Indeed, the cladistic method uses the most parsimonious tree obtained from the starting data, that is, the tree

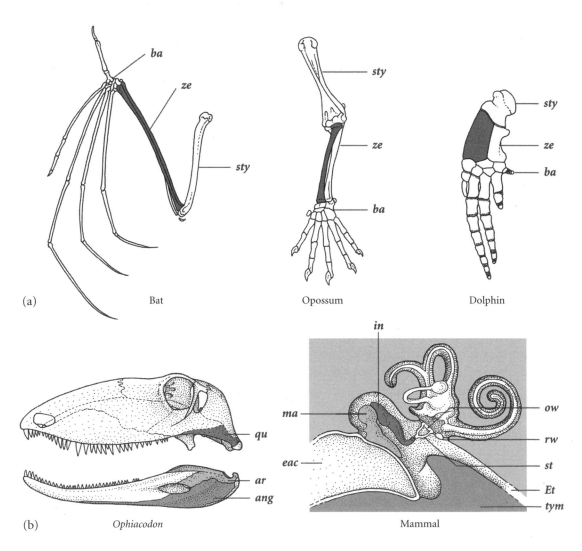

Figure 3. Primary homology based on body plan (a) and ontogeny (b). (a) The forelimbs of three mammals illustrate the notion of body plan. The radius is colored. *ba*: basipodium, *sty*: stylopodium, *ze*: zeugopodium. (b) Ontogeny (and more specifically marsupial embryogenesis) shows us that two of the three bones of the mammalian middle ear originate from bones previously involved in jaw articulation and correspond to two bones of the articulation between the jaw and the skull in other amniotes (here a synapsid fossil, *Ophiacodon*). The articular bone (*ar*) is homologous to the malleus (*ma*), the quadrate bone (*qu*) is homologous to the incus (*in*), the angular (*ang*) is homologous to the tympanic bone (*tym*), the stapes (*st*) is homologous to the hyomandibula (not shown). *eac*: external auditory canal, *ow*: oval window, *rw*: round window, *Et*: Eustachian tube.

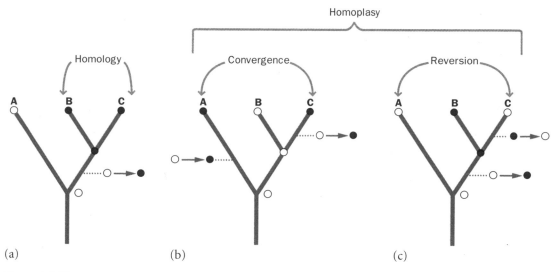

Figure 4. Different categories of resemblance. (a) Resemblance due to secondary homology; (b) Resemblance due to convergence; (c) Resemblance due to reversion. A white circle represents the primitive state of the character, a black circle the evolved state, and the arrows show the direction of the transformation.

containing the fewest number of evolutionary events in its branches. Once a tree is chosen, the position of each character change in the tree is determined. If a character state is distributed in an aggregated manner in the tree, such that it appears only on a single branch, it can be considered a homologous structure because it was inherited from a single hypothetical common ancestor in whom the structure appeared for the first time. This homologous structure is a synapomorphy, or secondary homology. Conversely, if the structure of a different character appears in several places in the tree, it is homoplastic; the resemblance was not inherited from a common ancestor. Finding this arrangement in a tree reveals that the initial hypothesis of homology was false (examples of homology and homoplasy are given later, fig. 11).

To conclude this question, we reiterate that the concept of similarity can be expressed as homology (similarity inherited from a common ancestor) or as homoplasy (similarity not inherited from a common ancestor: convergences, reversions) (fig. 4). For homologies within a given sample of species, there are similarities among the organisms that are also present outside the sample (symplesiomorphies) and there are similarities that are restricted to only part of the species sample and originate from the common ancestor of this subgroup (synapomorphies). Symplesiomorphies are synapomorphies at a higher taxonomic level

than the level of the sample. The discovery of a similarity within a given sample translates into a hypothesis of homology, or primary homology; we postulate that this similarity was inherited from a common ancestor. The most parsimonious tree will provide the response to our hypothesis. Homology by descent is called secondary homology because it is the result of an analysis: all confirmed primary homologies are secondary homologies or synapomorphies. These are resemblances inherited from a hypothetical common ancestor of the sample. All refuted primary homologies are homoplasies, that is, resemblances that are not inherited from a common ancestor (convergences, reversions).

What is a tree?

A description of the term *tree* simultaneously uses terms from mathematics (graph theory) and metaphors from the evolutionist tradition. By its mathematical definition, a tree is a noncyclic, connected graph. It comprises links (or arches, branches) that join up the tips. The graph is noncyclic because only a single pathway links two tips (the tree is not a network). The internal points are nodes (the meeting point of three links); the external tips are leaves (or terminal taxa, or evolved units). The internal links are also called internodal branches (or segments); the external links, terminal

21

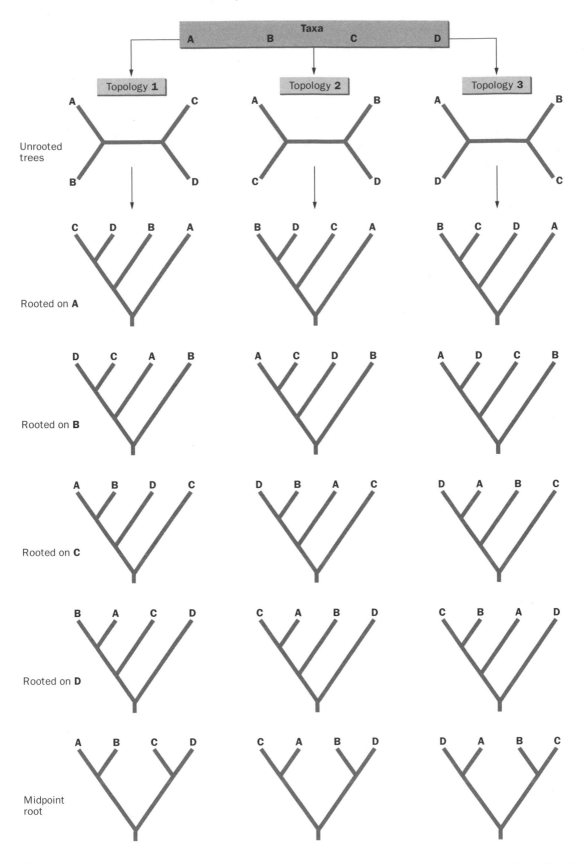

Figure 5. For 4 taxa, there are 3 corresponding topologies, 3 unrooted representations, and 15 possible rooted trees. None of the 15 trees retrace the same phylogeny. When one knows the outgroup *a priori,* one knows the root in advance and in that case there are only 3 possible topologies, as shown on each of the 5 horizontal lines. Note that tree-construction programs find the optimal solution without relying on a preestablished root (first line). The user then subsequently roots the tree with the help of one or several outgroups that were added to the matrix beforehand.

branches. A tree in which each node is the meeting point for three links is said to be totally resolved. When the tree is rooted, each internal branch gives rise to two daughter branches. These terms correspond to the amount of information contained in the tree, which in this case would be maximal: all relationships are supported. In contrast, a tree with nodes where more than three links meet is said to be unresolved (or partially unresolved). Multifurcation (or polytomy) signifies ambiguity in the tree concerning the relationships among taxa. Whether a tree is fully dichotomous or not implies nothing about the speciation modes. Details about the type and tempo of speciation remain unknown. A tree can also be unrooted. To become a phylogeny, however, it must be rooted (although this is not the only condition); a point of origin, or root, is defined *a priori* and determines how the tree will be read or, rather, how it will be oriented in time. A given tree can correspond to several different phylogenies depending on the position of the root. Fig. 5 shows the three possible topologies (unrooted trees) for the simplest case involving four taxa. To classify four taxa is to separate them two by two, as shown by the three topologies: (AB, CD); (AC, BD); (AD, BC). In this type of presentation, the nodes can pivot around the median segment, but the time of origin cannot be represented. To indicate the root of the tree is to indicate the time of origin. In fig. 5, all possible roots are given for each topology. How should one choose the root? According to the classical Hennigian approach, a tree must be rooted on the outgroup that was chosen to orient the characters (that is, to determine the direction of the character state transformations). In computerized analyses, characters from one or more outgroups are carefully coded and are subsequently used to root the resulting tree. Tree-constructing programs find the optimal solution without relying on a preestablished root; it is the user that identifies the root in the end.

We give specific names to trees according to the method that was used for their construction (see below). The term *dendrogram* is neutral and simply refers to a tree. A *cladogram* is a dendrogram constructed by cladistic analysis; it expresses the phylogenetic relationships among taxa and its nodes are defined by synapomorphies. A *phenogram* is a dendrogram constructed on the basis of phenetic (observable) characteristics of the taxa. It expresses the degree of global similarity among taxa. A *phylogram* is a cladogram in which the length of the branches is proportional to the

divergence of the taxa—in other words, to the number of character changes (synapomorphies, autapomorphies, homoplasies).

What is a taxon?

A taxon is a group of organisms that is recognized as a formal unit. *Homo sapiens* is a taxon at the species level; *Homo* is a taxon at the genus level; the hominidae form a taxon at the family level; the primates form a taxon at the level of the order, etc. In phylogenetic systematics, the cladogram itself expresses a nested series of taxa (fig. 6). Indeed, a taxon corresponds to each node and tip in a phylogenetic tree. It is defined by at least one unique, derived character (or synapomorphy). It is monophyletic: it includes a hypothetical common ancestor (in whom the synapomorphy appeared for the first time) and all of its known descendants. Two sister taxa have the same rank.

Two remarks should be made concerning this book. When presenting cladograms, we do not specify the rank of the described taxa (class, infradivision, suborder, superfamily). The rank is given by the position of the branch in the tree. A tree shows us that mammals are a taxon of the same level as the sauropsids, a level inferior to that of the tetrapods and superior to that of the primates. It is better to remember the logic behind the rankings than the ranks themselves. In other words, it is better to retain the relative rank (the position in the tree) than the absolute rank (which becomes useless when changes are made). For example, the former logic of the rankings has been turned upside-down by phylogenetic systematics. The division of extant vertebrates into five classes (fishes, amphibians, reptiles, mammals, birds) no longer holds, not only because the first three are not monophyletic but also because a phylogenetic classification does not put them all at the class level. According to the tree presented in Chapter 12, if we are to conserve the class level for mammals, we must also have a class for sauropsids, a subclass for chelonians and for diapsids, an infraclass for lepidosaurians and for archosaurians, a superorder for crocodiles and for birds—all this without taking into consideration the many fossil branches.

The second remark concerns the terminal taxa of the trees presented in the pages that follow. These phylogenetic trees represent a series of nested taxa that could be used to describe all groups to infinity, or at least as far as knowledge allows. It was thus necessary to cut the tree at a given height for each large branch. These choices—always arbitrary—were limited by time and space constraints. The fact that we do not include the insect phylogeny or that of the chelonians reflects neither their taxonomic rank (they do not have the same rank) nor the phylogenetic knowledge we have of these taxa.

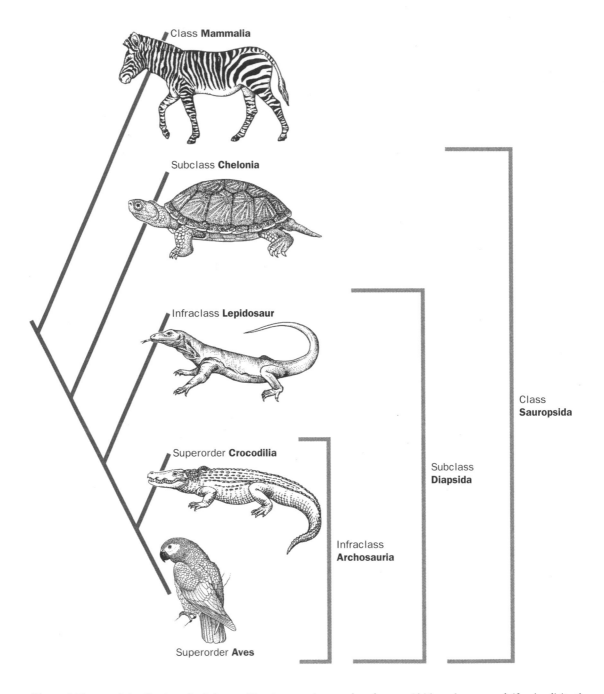

Figure 6. The nested classification of a cladogram. Two sister taxa (mammals and sauropsids) have the same rank (for simplicity, the fossils have been omitted).

Current Methods

Systematics today incorporates tools and concepts from outside the Hennigian approach. For historical and practical reasons, algorithms for tree construction also exist that specifically exploit overall similarity or use probabilistic approaches. Phenetic algorithms, based on overall similarity, originated outside of evolutionism and are effective only insofar as similarity remains proportional to the degree of relatedness (which is not always the case). Probabilistic algorithms are generally employed for sequence data and use the probability that one nucleotide or one amino acid will be replaced by another to calculate the likelihood of the observed sequence for a given tree. Briefly stated, three large families of algorithms are used today: phylogenetic (*sensu* Hennig: cladistic), phenetic, and probabilistic (maximum likelihood). Of course, they are all employed with the help of computers, which could lead one to question the potential problems associated with each type (for example, algorithms that search for the optimal tree without examining all possible trees, tree exploration techniques used, etc.). Here, we will simply recall the foundations of each approach.

Phylogenetic systematics (cladistics)

A phylogenetic analysis *sensu* Hennig aims to reconstruct the phylogeny of a sample of organisms by distinguishing between the primitive state (plesiomorphy) and the derived state(s) (apomorphies) of each character. These qualities are only valid within the framework of a given sample. Hennig was the first to understand that one could not group species on the basis of a shared primitive character because this character was already present outside the studied sample. Within a taxon, only shared derived character states (synapomorphies) are signs of relatedness; grouping based on shared derived states thus leads to the creation of monophyletic groups. In this book, we refer to unique derived characters to describe the unique derived character states of the considered group that are shared by all group members (unless secondarily modified). These are synapomorphies or innovations of the group.

Elements for reading trees

Before discussing phylogenetic analyses in more detail, we must first review some points of vocabulary. In order for a biological classification to be strictly phylogenetic, the classification must contain only monophyletic groups. A monophyletic group includes a hypothetical common ancestor and all of its descendants (fig. 7a). Of course, of this ancestor we know only the derived characters that define the group. For example, in extant fauna, birds are monophyletic. There are no sister groups to some birds that are not already included in the group. The most ancient common ancestor of all birds possessed the exclusive characters of the birds (new at the

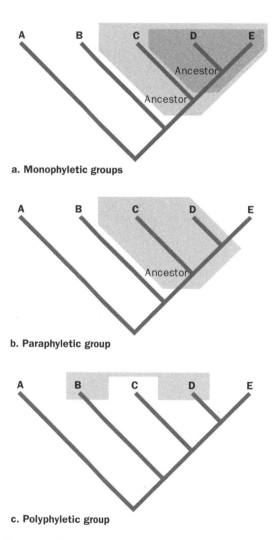

a. Monophyletic groups

b. Paraphyletic group

c. Polyphyletic group

Figure 7. (a) A monophyletic group includes an ancestor and all of its descendants. DE and CDE are both monophyletic groups. (b) A paraphyletic group includes an ancestor, but only part of its descendants. One of the group members (D) is more closely related to a taxon outside the group (E) than to its fellow group members (C). (c) A polyphyletic group includes members (BD) with no common ancestor.

time), such as flight feathers or a backward-facing first toe.

Paraphyletic and polyphyletic groups oppose monophyletic groups in that they are missing something. Certain descendants are missing from paraphyletic group, while the common ancestor is missing from polyphyletic groups.

A paraphyletic group includes the ancestor but only part of its descendants (fig. 7b). For example, the reptiles (in the classic sense: turtles, rhynchocephalians, squamates, crocodiles) all have the same ancestor (that of the amniotes), but within the reptiles certain sister groups are missing (such as the birds, a sister group of crocodiles, fig. 1b). Founding groups on the basis of ecological or adaptive characteristics, or those linked to progression or an increase in complexity, often leads to the construction of paraphyletic groups.

A polyphyletic group does not contain the common ancestor of all group members (fig. 7c). For example, the vultures are polyphyletic in the sense that Old World vultures are Falconiformes, whereas the Cathartidae (New World vultures) are Ciconiiformes. Within the vulture group, there is no ancestor common to all species. This ancestor does in fact exist because there is only a single tree of life, but to find it one must go back in time well beyond the scope of the vultures; the last common ancestor is also the ancestor of a large number of other Neornithes (modern birds). It is therefore linked to a taxonomic scope beyond that of the vultures alone.

What is evolutionary relatedness?

To look for relatedness is to look for sister groups, not the ancestor. It means searching for a group (or species) that shares a unique common feature with another group (or species), a feature not shared with any other group (or species) in the sample. The presence of the unique feature attests to an inheritance from an exclusive hypothetical common ancestor. This does not amount to looking directly for the ancestor (in genetic terms), because the ancestor will always be inaccessible. Rather, the process of tree reconstruction leads to the deduction of certain characters that the ancestor should have possessed based on the derived characters that define the group, the synapomorphies. (Note that the final tree can also invalidate the inheritance of the feature from a single hypothetical ancestor.)

How do we determine evolutionary relatedness?

In the classic Hennigian approach, the first step in the construction of a phylogeny is to polarize

character changes, that is, to distinguish the primitive state (or plesiomorphy) from the derived state(s) (or apomorphy). The second step is to deduce the relationships (that is, to construct groups) from the shared derived states.

How does one give direction to a character change? We apply two criteria, the outgroup criterion and the ontogenetic criterion, to a collection of species that have been sampled in order to determine their evolutionary relationships. This species sample constitutes the study group or *ingroup;* it is assumed to be monophyletic, but this does not imply that we know the relationships among its members in advance. Within the context of the study group, the goal is to differentiate between the primitive and the derived state(s) of each character.

The outgroup criterion (fig. 8), the most frequently used, consists of finding a species (or group of species) that we are sure is external to the study group; that is, it will not be found within the internal branches of the tree. This species is called the *outgroup* and serves to root the tree. Its definition uses no *a priori* information on the possible relationships within the study group and must be considered as part of the starting assumptions. Next, we compare each of the two (or more) states present in the study group with the state of the same character in the outgroup. If, in

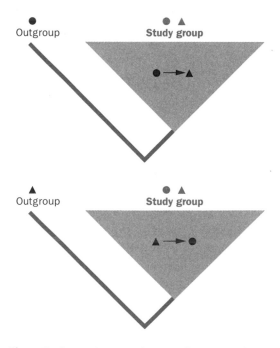

Figure 8. The transformation direction of a two-state character (circle, triangle) as determined by the outgroup criterion.

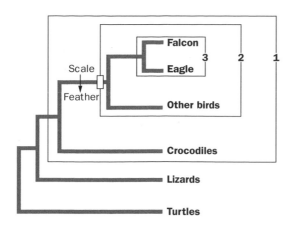

Figure 9. A character state is primitive or derived according to the taxonomic framework of the study. Box 1: archosaurs, box 2: birds, box 3: falconiformes. Within the archosaurs, the feather is a derived state shared by all birds; it signifies their relatedness. Within the falconiformes, the feather is a shared primitive state because it is already present outside of this group.

one or several species of the study group (for example, a crocodile and an alligator), the character state (scale) is the same as in the outgroup (scales on a turtle), this state will be considered primitive (or plesiomorphic). If, in other species (chicken, ostrich), the state is different (feather) from that in the outgroup (scale), this state will be considered derived (or apomorphic).

The second criterion for defining the direction of character transformations is the ontogenetic criterion. It is founded on Haeckel's law, according to which ontogeny recapitulates phylogeny. During embryonic development, general characters appear before more specific characters. If, within the species or study group, it is possible to observe the development of the character of interest, we can determine the order of appearance of the two states. The state that appears first will be considered the primitive state and then, if it is modified, the modified state will be considered the derived state.

Recall that the terms *primitive* and *derived* are relative to the taxonomic framework within which we are working (fig. 9). The feather is the derived state of the epidermal character within the archosauromorpha (the group including extant birds and crocodiles, box 1). However, within the falconiformes (the order including diurnal predators, box 3), the feather becomes a primitive state because it is already present outside the order, within numerous other avian orders. For this reason, the simple presence of feathers cannot be used to construct a group within the falconiformes.

Recall also that it is when giving directionality to character changes that the concept of descent with modification intervenes. Indeed, the direction signifies that the derived state (feather) is the result of a transformation of the primitive state (scale) and that this product was transmitted from generation to generation. This evolutionary event is evidence of the relatedness of the species (birds) that carry the derived state (feather).

Using the parsimony method

A simple biological example will allow us to place the preceding elements in context one by one. The most simple case of phylogenetic reconstruction involves a sample of three species and an outgroup of a single species.

Let us consider the following species: human (primate), the African grey parrot (bird), and the European free-tailed bat (bat). The question is to determine which two species of the three are the most closely related. There are only three possibilities: bird and bat; human and bat; human and bird. For choosing among them, a collection of observed characters are available. We can now examine these characters to determine which of the three trees they support.

First, we must polarize character states. Using information independent of the analysis, we are able to bring in a fourth species, the trout, which we know is external to the sample (study group: human, bird, bat). For example, the fact that the three species are amniotes (the embryo develops in the amniotic sac) excludes the trout because it does not have an amnion. The trout is therefore the outgroup and will serve to root the tree. Note that this choice uses no information about the anticipated relationships among human, bat, and bird. Contrary to a common misconception, the selection of an outgroup is not circular.

In figure 10, we examine six characters: (1) Jaw; two states: present, absent; (2) Type of paired appendage; two states: fins, walking limbs; (3) Teeth; two states: present, absent; (4) Composition of the lower jaw; two states: a single bone (the dentary), several bones; (5) Vitellus reserves of egg; two states: alecithal egg with little vitellus, telolecithal egg with a large quantity of vitellus separate from the cytoplasm; (6) Anterior paired appendages; two states: transformed into wings, absence of wings. The first thing to note is that all characters are not equally informative.

Noninformative characters. Character 1, the jaw, is present in the trout. By convention, we code the character state found in the outgroup as 0. We find jaws in all four species. The presence of jaws will therefore be coded as 0 for all four. The presence of jaws cannot discriminate a closer relatedness for two of the three species in the study group and therefore constitutes a noninformative character for this species sample. Character 2, the type of paired appendage, is coded 0 for the trout, whose paired appendage is in the form of fins (primitive state). The human, bat, and bird all have paired appendages in the form of walking limbs, and this state is coded 1 (derived state). Nevertheless, this character is not informative because again it does not allow us to separate two from the three species of the group being studied. Character 3, teeth, is coded as 0 for the trout. By definition, the presence of teeth is therefore primitive. For humans and bats, which have teeth, this state is also coded 0 (primitive state). The bird has no teeth, so this state is coded 1. We now have a character that can discriminate two of the three species in the study group. It therefore seems to be informative. It will not allow us to form a group within the studied group, however, because the cladistic method does not permit groupings based on shared primitive characters. Humans and bats cannot be grouped

together because they both have teeth because teeth are already found in the trout. The acquisition of a derived character, the absence of teeth, is observed only in the bird. This character is therefore not informative from the *point of view of the method used*: it tells us that the bird is a bird.

Informative characters. Character 4, the composition of the lower jaw, is coded 0 for the trout and the bird, in which we find a mosaic of bones (primitive state), and 1 for the human and the bat, in which we find a single bone on each side of the jaw, the dentary bone (derived state). This character enables us to group the two latter species because they share the same derived character. The situation is the same for character 5, the relative amount of vitellus reserves in the egg: the trout and the bird have vitellus-rich eggs in which the vitellus is separated from the cellular cytoplasm (the primitive state, because it is found in the trout), while the human and the bat have eggs that are practically devoid of vitellus (derived state). Character 6, anterior limbs transformed into wings, also permits us to form a group within the study group, but not the same one: (bat, bird). Two remarks are required here: (1) only characters that are present in the derived state at least twice within the study group are informative for the method used; and (2) characters can be contradictory (evolutionary convergences and reversions may both exist).

Now superimpose our characters on each of the three possible trees of relatedness (the cladograms in fig. 11). Recall that a group can be founded only on a shared derived character, that is, a common black circle. At the level of the black circle, we pass from state 0 to state 1. Tree 11a shows that the node (human, bird) is not based on a shared black circle. Tree 11b shows that the node (bat, human) is supported by characters 4 and 5: both the jaw consisting of a single dentary bone and an alecithal egg are common derived states, or synapomorphies, indicating a possible relationship between humans and bats. Tree 11c presents the grouping of the bird with the bat based on the presence of wings. Therefore, two characters defend the group [bird, (human, bat)], one defends [human, (bird, bat)], and none defends [bat, (human, bird)].

One must now calculate the number of character transformations that make up each tree. Character 6 contradicts characters 4 and 5. There are two possibilities to consider when faced with homoplasy: we can assume that the homoplastic character is the result of either a convergence

	Human	Bat	Bird	Trout
1 **Jaw**	0	0	0	0
2 **Paired appendage**	1	1	1	0
3 **Teeth**	0	0	1	0
4 **Composition of the lower jaw**	1	1	0	0
5 **Vitellus reserves of the egg**	1	1	0	0
6 **Wings**	0	1	1	0

Figure 10. Character matrix. 1: jaw (presence/absence), 2: type of paired appendage (fin/walking limb), 3: teeth (present/absent), 4: composition of the lower jaw (several bones, including the dentary/only the dentary), 5: vitellus reserves of the egg (enormous/very little), 6: anterior paired appendage (wings/no wings).

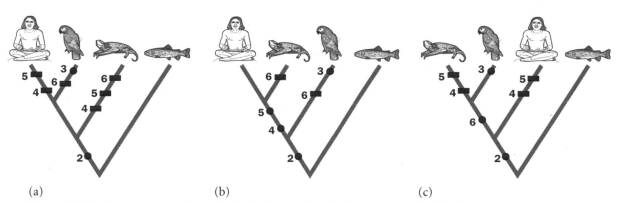

Figure 11. The three trees that may be produced when the transformation hypotheses are placed in the most parsimonious manner and homoplasies are interpreted as convergences. Circles: synapomorphies, rectangles: convergences.

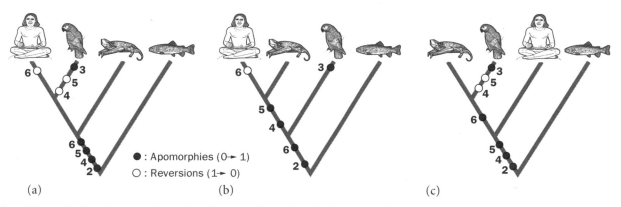

Figure 12. The three trees (as in fig. 11) that may be produced when the transformation hypotheses are placed in the most parsimonious manner and homoplasies are interpreted as reversions. Black circles: synapomorphies, white circles: reversions.

(fig. 11) or a reversion (fig. 12: these are the same three trees). These two options end up being identical in terms of the number of events occurring in the tree. They differ simply in their position. Let us take the tree that is the most economical in terms of evolutionary events, figure 11b. The black circle symbolizes a synapomorphy, that is, a secondary homology in the descendants. On the segment of the branch that unites human, bat and bird, character 2 passes from state 0 to state 1 (this transformation will be the same on all three trees). This already counts as one step. On the segment of the branch that unites human and bat, characters 4 and 5 change state (0 to 1); here, we count two additional steps. Character 3 transforms from 0 to 1 on the branch that leads to the bird (autapomorphy). We therefore count a fourth step. With regard to character 6, wings, we can assume a convergence, that is, two independent transformations from 0 to 1, one on the branch leading to the bird and a second on the branch

leading to the bat (marked by the rectangles). The convergence therefore costs two steps, making a total of six steps for this tree. We could also assume a reversion that would also require two steps (see the white circles in tree 12b, the same tree as 11b): a transformation from 0 to 1 on the branch uniting human, bat, and bird (black circle), then a reversion from 1 to 0 (white circle) on the branch leading to the human. In this case, humans would have lost their wings! The reversion also brings the total number of steps to six for tree 12b.

Let us now consider tree 11a. One must assume one synapomorphy (black circle for character 2), one autapomorphy (black circle for character 3), and three homoplasies (rectangles for characters 4, 5, and 6). Each homoplastic event costs two steps whether it is considered as a convergence (fig. 11a) or a reversion (fig. 12a): this brings the total number of steps to eight. Finally, let us consider tree 11c. One must assume two synapomorphies (characters 2 and 6), one autapomorphy (character 3), and two homoplasies (characters 4 and 5). As each homoplasy costs two steps, the total number of steps is seven for this tree. We now use the minimum number of evolutionary steps as the criterion for tree selection. This criterion conforms to the principle of parsimony or the principle of fewest hypotheses. We therefore choose tree 11b (or the "reversion" form 12b), as it is the most parsimonious.

The chosen relationship is that between the human and the bat. The classification that follows is [bird, (human, bat)]. The most parsimonious tree tells us that hair and teats were acquired by a common ancestor exclusive to humans and bats. *Exclusive* here means they are shared by the human and the bat and not the bird. In terms of homoplasy, wings were either acquired independently twice (11b) or underwent a secondary disappearance in humans (12b). A tree containing more species would show us that the first of the two hypotheses is retained.

Choosing the most parsimonious tree has nothing to do with the so-called parsimony of evolution. We will never know the complete sequence of events that have led to the present state of nature. Choosing the most parsimonious tree is simply choosing the most consistent tree, the one that involves the fewest number of *ad hoc* hypotheses of multiple changes to resolve character conflicts. In other words, tree 11c involves one fully consistent character and four *ad hoc* hypotheses of homoplasy, while the tree 11b involves two fully consistent characters and only two *ad hoc* hypotheses of homoplasy.

Summary of the steps for constructing a phylogenetic classification

The reconstruction of evolutionary relationships is the result not of fortuitous discovery but of well-considered decisions that can be considered postulates:

1. Pose an initial question. This step defines the phylogenetic problem or delimits the group of interest (*ingroup*).
2. Choose the organisms to be sampled to answer the question: the taxonomic sampling. This step also consists of choosing one or several external reference points, or outgroups. These organisms are considered outside the group of interest, that is, they are not included in the group for which we wish to define relationships. The outgroup will serve to root the future tree(s).

3. Choose or identify the characters that will be informative for the question posed: the character sampling.
4. Code the data and construct the character matrices. Coding includes establishing primary homologies (by the principle of connections, sequence alignment) and polarizing character transformations (outgroup or ontogenetic criteria). The results of this analysis are then arranged in a character matrix.
5. Explore all possible trees. For a fixed number of taxa in the matrix, there is a corresponding number of possible trees. For four taxa, of which one is defined *a priori* as the outgroup, three rooted trees are possible. For five taxa, 15 rooted trees are possible, etc. It is best is to evaluate all trees with regard to the information contained in the matrix.
6. Evaluate each tree with regard to the character matrix.
7. Choose the tree-selection criterion: in phylogenetic systematics, this criterion is parsimony. One selects the tree that is the most parsimonious, that is, the tree that requires the fewest number of evolutionary events to explain the character states of the organisms. However, one should be aware that, outside phylogenetic systematics, other selection criteria exist (in the probabilistic approach it is the least-squares method).
8. Choose a tree. This will be shortest tree, the one requiring the fewest number of character transformation hypotheses. This tree enables:

- the identification of character states that define each node (= clades) of the resulting tree (synapomorphies = homologies by descent or secondary homologies).
- the identification of homoplastic states and potentially the establishment of conclusions about the mode of evolution of the characters.
- the creation of a phylogenetic classification, that is, the allocation of a group name, or *ranking,* to each supported node (= clade) of the tree.

Note: The order of steps 2, 3, and 4 are sometimes rearranged. Indeed, the variability of the characters may lead to the modification of the taxonomic sampling. In molecular phylogenies, steps 4 through 8 are sometimes combined. For example, certain computer programs directly calculate the alignment of the sequence data that will produce the shortest tree for the sampled species.

Species			Aligned sequences			→
Common carp	SLSDKDKAAV	KIAWAKISPK	ADDIGAEALG	RMLTVYPQTK	TYFAHWADLS	PGSGPVKHGK
Human	V--PA--TN-	-A--G-VGAH	-GEY-----E	--FLSF-T--	---P-F*---	H--AQ--GHG
Fruit bat	V--SA--TNI	-A--D-VGGN	-GEY-----E	--FLSF-T--	---P-F*---	H--AQ--GHG
Chicken	V--AA--NN-	-GIFT--AGH	-EEY---T-E	--F-T--P--	---P-F*---	H--AQI-GHG

		Aligned sequences			→
KVIMGAVGDA	VSKIDDLVGG	LASLSELHAS	KLRVDPANFK	ILANHIVVGI	MFYLPGDFPP
-KVAD-LTN-	-AHV--MPNA	-SA--D---H	------V---	L-SHCLL-TL	AAH--AE-T-
-KVGD-LTN-	-GHL---P-A	-SA--D---Y	------V---	L-SHCLL-TL	ANH--S--T-
-KVVA-LIE-	ANH---IA-T	-SK--D---H	------V---	L-GQCFL-VV	AIHH-AALT-

—	Aligned sequences	—
EVHMSVDKFF	QNLALALSEK	YR
A--A-L---L	ASVSTV-TS-	--
A--A-L---L	ASVSTV-TS-	--
---A-L---L	CAVGTV-TA-	--

Figure 13. Aligned sequences from a portion of hemoglobin α. The letters correspond to the amino acid codes. A hyphen signifies that the amino acid is identical to that of the first line; an asterisk indicates the absence of an amino acid.

Phenetics

Phenetic methods analyze data in another manner. In contrast to phylogenetic systematics, phenetics attempts to quantify the general similarity between organisms. An index is calculated of overall similarity between two taxa, that is, a distance for each pair of taxa. In the case of aligned sequences, this distance is the number of nucleotides that differ between the two species (or amino acids in the case of protein sequences), divided by the number of examined sites. In the end, this number amounts to the percentage of sequence differences between the two species. These distances are recorded in a matrix (a double-entry table, fig. 14). It should be pointed out that these distances can be corrected to incorporate evolutionary information. For example, they can take into account the fact that certain types of mutations occur more often than others, but this goes beyond the limited scope of this chapter.

There are several techniques for tree construction using a distance matrix. The most simple is the UPGMA (Unweighted Pair Group Method using Averages), which starts from the matrix and results in a single tree by successively grouping together the species, from the closest to the most distant. This method has the advantage of being able to be done by "hand" if the matrix is not too large. However, it has the inconvenience of requiring the assumption that all sequences evolve at the same rate throughout the tree. This is a real disadvantage because this assumption is rarely verified. If the assumption is not valid, the tree is false.

A second commonly used method is that of W. Fitch and E. Margoliash. It requires no particular assumptions about the rate of molecular evolution. Given the homoplasy present in all biological data sets, it is impossible to construct a tree in which the distance between species, obtained by adding the branch lengths, is strictly equal to the distance represented in the starting matrix. As a consequence, tree distances always deviate around (larger or smaller) the initial matrix distances. With this method, the criterion for selecting a tree is to minimize this deviation: we choose the tree in which the distances between the species are as close as possible to the distances in the initial matrix.

A third method, for which the algorithm is more elaborate, is called neighbor-joining. The originality of this distance method is that it introduces a criterion to minimize the total tree length. It constructs a single tree, but not necessarily by initially grouping the closest species (as does UPGMA). Rather, at each step, it groups those species that will minimize the total length of the tree.

Two remarks should follow these generalities. First, distance trees express the degree of similarity among taxa. This similarity is not always proportional to the relatedness: if a taxon is extremely diverged or if the phylogenetic signal is insufficient to resolve the relationships, distance methods can form groups based on symplesiomorphies (that we detect by analyzing the same data using a cladistic analysis). Second, distance methods are more frequently used for molecular data today than for morphological data. Contrary to the false idea that is widespread among the informed public,

31

however, analysis of sequence data need not be based on distance but can quite easily be done using a cladistic approach (also called parsimony analysis).

Using the UPGMA method

The algorithm can be summarized as follows:
1. In the distance matrix, find taxa i and j for which the distance d_{ij} is the smallest.
2. Place the root at an equal distance between i and j, or at $d_{ij}/2$.
3. Create at new unit U that includes i and j. If i and j are the last units, the tree is finished.
4. Recalculate a distance matrix from U to each of the other taxa k by taking the average of the distances d_{ki} and d_{kj}: $dU_k = (d_{ki} + d_{kj})/2$.

5. From this new matrix (in which you have one less entry because i and j have been grouped together in unit U), return to step 1.

Remark: Steps 2 and 4 require that the amount of evolution (the distance) be the same on two sister branches. If the data do not conform to this property, the tree will be incorrect.

Let us look at the aligned protein sequences of hemoglobin α (fig. 13) for humans (H), the Egyptian fruit bat (a bat, Bt), the domestic chicken (a bird, B), and the carp (a fish, F). It is possible to calculate the percentage of differences between each pair of sequences. We can find these values in the matrices of figure 14. The smallest distance is that linking the human and the bat: 9.93%. The root of the human and the bat will be put at 9.93/2,

	Common carp	Human	Egyptian fruit bat	Chicken
Common carp	0.00			
Human	52.11	0.00		
Egyptian fruit bat	50.70	9.93	0.00	
Chicken	52.82	29.79	31.21	0.00

- Shortest distance :
  d (Human, fruit bat) ≈ 10 %
  Rooted at 10/2 = 5 %

- d (**M**, carp) = (52.11 + 50.70)/2 = 51.40 %
  d (**M**, chicken) = (29.79 + 31.21)/2 = 30.5 %

	M	Carp	Chicken
M	0.00		
Carp	51.4	0.00	
Chicken	30.5	52.82	0.00

- Shortest distance :
  d (**M**, chicken) = 30.5 %
  Rooted at 30.5/2 = 15.25 %

- d (**A**, carp) = (52.82 + 51.4)/2 = 52.11 %
  Rooted at 52.11/2 = 26.055%

	A	Carp
A	0.00	
Carp	52.11	0.00

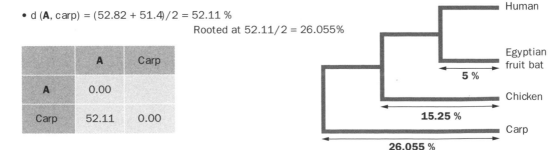

Figure 14. Tree calculation using the UPGMA method. The distance matrices are drawn from sequence data (fig. 13). The values refer to the percentage of differences in the amino acids of two sequences.

or approximately 5% (fig. 14). Consider M as the new unit (human and bat). Now, we calculate a new matrix with dMB = (dHB + dBtB)/2 = 30.2%; and dMF = (dHF + dBtF)/2 = 51.4%.

In the new matrix, dMB is the shortest distance. The bird is now grouped together with M to form a new unit A. The root is positioned at a distance from B equal to dMB/2 = 15.25%. There now remains only the carp to group; dFA = (dFB + dFM/2) = 52.11%. The root is positioned at a distance from the carp of dFA/2 = 26.055%. The resulting tree is found at the bottom of figure 14.

Remark: This method does not allow one to choose an outgroup: the tree is rooted by the principle of steps 2 and 4, at the median point. The risk of basing group formation on shared primitive characters is therefore great.

Other methods of tree construction

The methods described above are those most commonly used today. The Hennigian method, also called phylogenetic systematics or the cladistic method, and its computerized offshoots (the parsimony methods utilizing programs like PAUP, HENNIG86, DNAPARS, NONA, etc.) are universally employed. Distance methods (Fitch, neighbor-joining) are especially used for nucleic acid or amino acid sequence data. However, other methods also exist. The compatibility method is similar to the parsimony method, but rather than retaining the shortest tree it selects the tree in which the number of simultaneous characters that appear without homoplasy (compatible characters) is the greatest.

Probabilistic methods establish a model made up of a set of parameters expressing some hypotheses or assumptions about the process of character evolution. These assumptions concern, in particular, the probabilities of character state evolution. For example, a model will stipulate that the probability of an adenine changing to a cytosine is 0.3, whereas the probability of it changing to a guanine is 0.5. As all trees have character state changes within their branches, all of the associated transformation probabilities for a given tree will be multiplied together to provide a global likelihood value for the data in relation to the tree. Among the possible trees, the selected tree will be the one for which the likelihood of the data is maximal under a given model. This method functions well for molecular characters, for which we can readily establish models of protein or nucleic acid evolution. It is much more difficult to apply to morphological characters because, given their highly integrated nature, we have no idea about the probabilities of one state transforming into another or, more generally, about the type of background model that should be constructed.

Some properties of the methods

In situations where the true phylogeny is known—that is, when carrying out computer simulations or experimental phylogenetic work (for example, rapid phylogenesis of controlled viruses in the laboratory)—it has been shown that all methods result in the same phylogeny if the data contain little homoplasy and enough information. When the data contain little discriminating information, all methods provide trees with short internal branches. In these instances, it is not unusual to find contradictory topologies among methods, notably for distance methods that construct groups based on shared primitive characters (symplesiomorphies). All methods are prone to an artifact caused by long-branch attraction (see below, fig. 19). This artifact results from a difference in the evolutionary rates of characters in the analyzed lineages. Species that evolve more quickly than others have longer branches and introduce homoplasy. It has been shown, both theoretically and experimentally, that beyond a certain difference in evolutionary rates, tree analyses tend to group taxa with similar rates of change rather than similar degrees of relatedness. This is what we call long-branch attraction.

When is a tree a phylogeny?

Starting with historical definitions (those of Haeckel, 1866, and Darwin, 1872), Darlu and Tassy (1993) define a phylogeny as: "the historical pattern of descent of organized beings." They insist—correctly—on the operational success of the tree as a representation of the phylogeny. Although a phylogeny can be efficiently represented by a tree, a tree is not necessarily a phylogeny. This is because a tree, a noncyclical connecting graph, can be used outside of a phylogenetic context anytime it is necessary to symbolize a hierarchy. For example, a key for identifying species may summarize the identification criteria in a hierarchy that takes the form of a tree. This will never result in a classification. A tree can also serve to summarize the selection results of a sporting event, to describe the hierarchical relationships among company personnel, or, more generally, to illustrate any kind of successive sorting (organizational chart).

How does a tree become a phylogeny? To make this point, we find it convenient to give both a general definition and a strict, more technical definition. To classify is to create logical links among objects according to a predefined purpose. According to Darwin, the purpose of classification is to describe the degree of evolutionary relatedness among organisms. Modern tree-construction methods are classifying procedures that aim to do this. According to P. Darlu, however, no construction method alone has the capacity to produce an evolutionary model; that is, none will automatically make a tree a phylogeny. This point is generally agreed upon. A tree becomes phylogenetic in the context of the particular evolutionary hypotheses made by the biologist when applying the classification method. In other words, phenograms and cladograms are phylogenies if the author constructed them and speaks of them in this way. This last point is debated by those who support a stricter definition. The strict definition has two conditions:

1. The idea of descent with modification must be implicit in the method of tree construction. Indeed, living organisms have the particular property of having evolved through the transmission of selected, heritable modifications to their offspring. This unique property must be an integral part of any tree that claims to be a phylogeny (although this point is still debated among the experts). Technically, this concept is used in the classical Hennigian method when characters are oriented *a priori* in time. In computerized parsimony methods, this idea is produced by the *a priori*

choice of one or several outgroups, the incorporation of outgroups into the matrix, and the *a posteriori* rooting of the tree based on this information.

2. A tree is phylogenetic if it enables the researcher to detect *a posteriori* the similarities inherited from a common ancestor (homologies) and the false similarities (homoplasies). In particular, a cladogram allows us to identify our successes (supported homologies) and our errors (homoplasies) among those characters we initially assumed homologous. This is possible because character modifications are directly mapped onto the branches of the cladogram (fig. 11). If we had assumed homology between a bat's wing and a bird's wing, the most parsimonious tree obtained for the vertebrates using a dozen other characters would indicate two independent appearances of wings: once on the branch leading to bats (fig. 11b), and once on the branch leading to birds. These two appearances are separated by other branches. This is therefore a clear case of homoplasy: a similarity that is not inherited from a common ancestor. If at the same time, we had assumed that the mandibles of humans and bats were homologous, we would discover in this same tree a single appearance of the mandible made up exclusively of the dentary bone at the point joining these two species (fig. 11b), the position of their hypothetical common ancestor. The mandible composed exclusively of the dentary bone is therefore a homology assumption that is supported by the tree, a similarity inherited from a common ancestor. Cladograms therefore permit the identification of homologies and homplasies. Distance trees alone do not have this advantage because they do not work directly with the characters; once a primary homology is assumed, the character states are transformed into distances.

Managing phylogenetic conflicts

The molecular data available to the biologist provide a large number of characters that are expensive to obtain but that are, in general, structurally simple; nucleic acids, the most frequently analyzed, have only five possible states: A, G, C, T, insertion/deletion. In contrast, morphological and anatomical data describe characters that are intrinsically richer in information but fewer in number. Despite these differences, we can reasonably say that conflicts among different morphological or molecular trees are at least as frequent as those between morphological and molecular trees.

Phylogenetic conflicts between morphological and molecular data can be due to multiple factors: differences among lineages in the evolutionary rates of molecular characters, sample sizes, the number of characters considered, unsuitable tree-construction methods for the mode of character evolution, and, for molecular phylogenies, the treachery of genes as markers for a species history; a gene found in a given species may be a copy that originated in another species and arrived in the genome by horizontal transfer.

These conflicts can initially be illustrated by comparing trees produced by multiple data sets, and then resolved by employing consensus techniques. However, to understand the basis of the conflict—that is, to measure the force and potentially detect the artifacts that are at its origin—we must examine different trees and compare their robustness. We can even go a step further and measure the force of the apparent conflict using statistical tests. Finally, we can produce a simultaneous phylogenetic analysis using all the existing data, that is, construct a tree with all available data in a single matrix. This last approach has received the name *total evidence,* but it can be shown that this term is inappropriate. If the source of the conflict can be reliably identified, however, we can carry out this last approach by replacing the character states that are the result of an artifact by question marks and in this way clean up the matrix.

In this book, we manage conflicts by comparing trees and their robustness and by evaluating the number of times a clade was found independently of the number of times that its competitor was found. If the conflict remains insoluble, the strict consensus tree is used (see fig. 16a).

Analyzing a tree's robustness

A phylogenetic tree cannot be published these days without an analysis of its robustness. The robustness of a phylogenetic tree is conceived differently depending on the culture of the biologist who constructed it. Statistical biologists refer to the *bootstrap* value of a tree node as a measure of its statistical robustness. The bootstrap method, a technique based on sampling characters with replacement, will be explained later. Morphologists, anatomists, and paleontologists prefer to consider the number and quality of synapomorphies that support a clade. When a

cladogram is presented as a phylogram, the branch lengths are proportional to the number of evolutionary events that produced them. These events can be synapomorphies (the acquisition of exclusive derived states) or homoplasies (the acquisition of nonexclusive derived states). The longer an internal branch—and more specifically, the greater the proportion of synapomorphic events—the more solid or robust the branch is. A long branch means that many convergence hypotheses would be required to eliminate a clade member. For morphological characters, the complexity of a character that changes at a given node and the number of times that this character changes at other places in the tree are used as indices of the robustness of the node. Another way of evaluating the strength of a clade is to examine trees that are less parsimonious than the most parsimonious tree. Recall that the most parsimonious tree is the one with the fewest number of evolutionary transformations, or steps, in its branches. For a given clade, we determine how many evolutionary steps must be added back to less parsimonious trees to find a contradicting clade; the more steps that must be added, the stronger the support for the clade of interest. If the addition of a single step to the minimal number results in a contradictory clade, little is required to break the clade of interest. This number of steps is called the *Bremer index.*

Biologists who construct molecular phylogenies use a large number of characters that are structurally simple. The fact that a given site of a sequence mutates at one or another node is of little interest to the character itself. It is therefore not shocking for these biologists to apply a statistical treatment to their data. It is certainly for this reason that *bootstrapping* is commonly used in molecular phylogenies.

In 1985, J. Felsenstein proposed the *bootstrap* method as a tool to measure the robustness of phylogenies. It measures the degree to which a data set (most frequently sequence data) supports a grouping without using external information, and it can be used with any technique of tree construction. Bootstrapping involves repeated random samplings of the characters with replacement (the characters corresponding to the sites of a sequence) until the number of sampled sites is identical to the number of sites in the initial data set (see fig. 15). After the sampling of a given site, this site is available for subsequent sampling. During the sampling of a second site, the first can easily be sampled a second time. This means that when *n* sites have been sampled among *n* sites, the

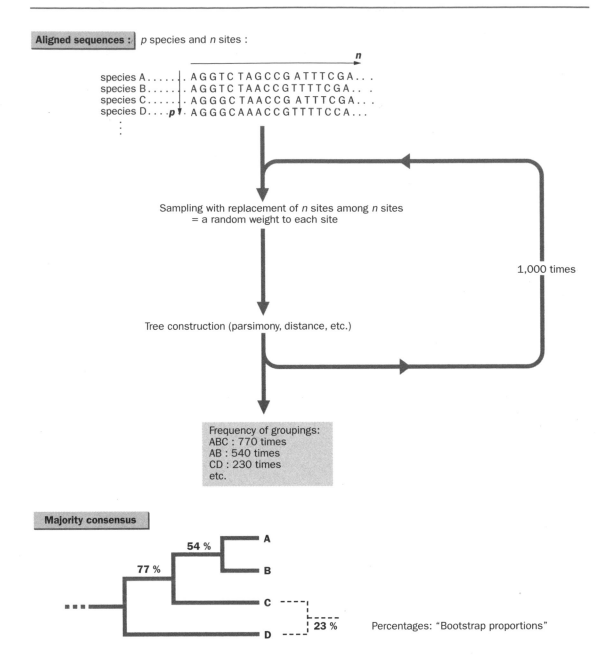

Figure 15. The procedure for sampling with replacement (the *bootstrap* method).

artificial data set will be made up of certain sites that have been sampled numerous times and other sites that are absent. From this artificial data set, a tree is constructed (all construction techniques are possible: parsimony, distance, etc.). This process of sampling with replacement *n* sites among *n* sites and the construction of a tree is repeated a large number of times (generally 1,000 times). We thus obtain a large number of trees that represent a large number of groupings, certain of which are, of course, contradictory. The overall result is typically represented as a consensus tree in which all the major groups appear (fig. 15). Each group (or node) possesses a percentage that indicates the proportion of sampled trees in which it is present. This is the *bootstrap* value or the *bootstrap* proportion of the node. This value signifies that all contradictory topologies appeared with a lower percentage. The consensus tree is not an obligatory representation. One can also present the most parsimonious tree that follows from a standard cladistic analysis and place *bootstrap* values on the clades.

Using this tool, we can therefore evaluate the resistance of a node to perturbations in the data: the smaller the number of sites that support a given group, the less chance there is that these sites will be selected during random sampling. As a result, the proportion of trees that present the group will be lower and the *bootstrap* value will be weaker. On the other hand, if there is a strong signal in the sequence data (many sites) for a given species group, these sites will have a high probability of being selected at each sampling and the *bootstrap* value will be strong. In this way, bootstrapping reflects the characteristics of the phylogenetic signals in the data set, as interpreted by a given tree-construction technique. A node "supported" at 95% indicates that there is a strong signal in the sequence data to group the species together. The node is therefore considered to be robust. However, this robustness should not be thought of as the probability that a group is real—a frequently made error. This value simply indicates the degree of support for a group in *the analyzed data*.

Measuring robustness permits one to be more objective when managing phylogenetic conflicts. Indeed, no importance is given to a phylogenetic conflict if the clades involved are not robust. Similarly, we expect that any novel (or surprising) phylogenetic result will be robust. But robustness is not everything. Indices of robustness can also fall victim to artifacts. Robustness is not equal to reliability. In order for a new phylogenetic result to be considered reliable, it must be found independently several times, by different researchers and with independent data sets. For example, in the last few years it was discovered multiple times—with several very different molecular data sets (numerous independent genes,

retrotranspositions at unique sites in the genome)—that the hippopotamuses are a sister group of the cetaceans. Likewise, since the 1970s it was thought that the pinnipeds (sea lions, seals, walrus) were polyphyletic, sea lions being closer to bears and seals being closer to weasels. Recently, however, several independent molecular phylogenies of the carnivores have found that the pinnipeds are in fact monophyletic; this result is now starting to be taken seriously.

Consensus techniques

When one wants to summarize the taxonomic information coming from equally parsimonious but conflicting trees, one uses consensus trees (fig. 16). These techniques emphasize the stability of a given node and the common information found in numerous trees. The most frequently used consensus techniques are those of strict consensus (fig. 16a) and majority consensus (fig. 16b, used to represent the *bootstrap* results). In the first case, only those nodes found in all source trees are kept. The other nodes are reduced to multifurcations. In the second case, only those branches that are found in the majority of the source trees are conserved. Another type of consensus technique is the "semi-strict" or "combinable component" method. It also entails showing the compatible clades and reducing the incompatible clades to multifurcations (fig. 16c), but if one of two considered trees is unresolved (fig. 16c, left), the unresolved tree is not considered to be fundamentally incompatible with the resolved tree (fig. 16c, right). Strict consensus would not recognize the internal clades of BCD because it retains only those clades that are present in all trees. In an analysis utilizing the semi-strict consensus method, these two trees would not be interpreted as contradicting the sister-group status of B and C and thus would support this relationship.

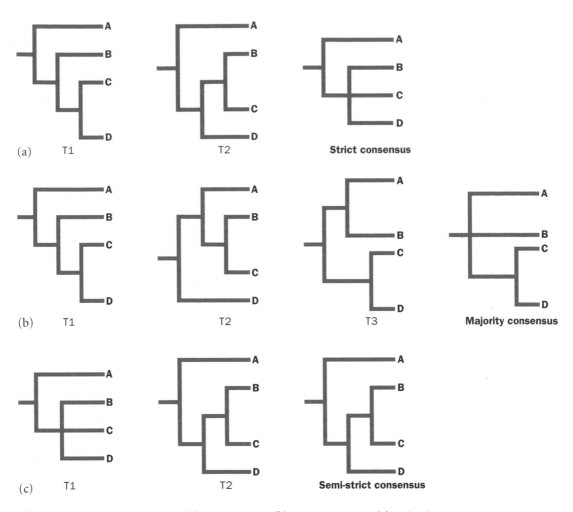

Figure 16. Consensus tree techniques: (a) strict consensus. (b) majority consensus. (c) semi-strict consensus.

Upheavals in our classifications

The upheavals in our classifications that have occurred over the last thirty years have had two sources. The first is purely technical, a result of the application of biotechnological tools to the field of systematics while the same fundamental concepts were retained. The second source runs much deeper: it relates to our way of thinking about how we order biodiversity—or, rather, the algorithm we select to analyze the structures of living organisms so that we obtain a purely phylogenetic classification. Molecular systematics only appears to be modern because it uses impressive tools that are highly valued by our commercial societies. This technological modernization is praised because it is manufactured, but it hides the true modernism,

which is the change in ideas that has occurred over time. After a delay of 100 years, the Hennigian revolution has enabled us to fulfill Darwin's wish that classifications should follow genealogies (we say today the phylogenies) as much as possible. The true modernism is thus the renunciation of the paraphyletic group.

A new method and the use of some new characters

The ensemble of trees illustrated in this book exposes the upheavals that have occurred in our classifications by this conceptual and methodological revolution, initially started by the cladistic analysis of classic anatomical characters. For the record, we can cite the most important results, such as the abandonment of classical groups like the protozoans, the algae, the bryophytes, the gymnosperms, the invertebrates, the diploblasts, the acoelomates, the agnatha, the fish, the reptiles, the prosimians (see the appendices).

Hennigian systematics (like its historical competitors) is powerless to resolve questions if there are no comparable characters, however. Access to DNA and protein sequences, initially limited between 1967 and 1985, has been largely facilitated by the technique of *in vitro* gene amplification, or PCR (the polymerase chain reaction) and the widespread use of automatic sequencers. These tools do not offer a new method to systematists; rather, they give us a new class of available characters, the basis for molecular phylogenies. It should be noted that, early on, biochemistry made it possible to base phenetic classifications on molecular characters: enzymatic polymorphisms by the end of the 1950s, the strength of heteroduplex interactions between antibodies and antigens starting in 1963, and the

strength of DNA-DNA heteroduplex interactions starting in 1970. However, the use of these techniques remained marginal in systematics because they were restricted to lower taxonomic levels—with the exception of DNA-DNA hybridizations, which have been used at a large scale to classify 168 of the 171 extant families of birds.

In our times, the great majority of results produced in molecular systematics involve sequence data. Although the pioneering experimental work of W. Fitch and E. Margoliash in 1967 and that of M. Goodman's team (starting in 1963 and using protein sequences) were of great importance, the true cultural shock took place in 1977, when C. Woese and G. E. Fox undertook the classification of the bacteria using ribosomal sequences. They discovered that the living world was not, in fact, divided into prokaryotes and eukaryotes but rather into three domains—the archaea, the eubacteria, and the eukaryotes—and that the genetic gulf between eukaryotes and archaea was as wide as that separating the eubacteria and the eukaryotes. The prokaryote world was thus revealed to contain unsuspected levels of divergence. In 1986,

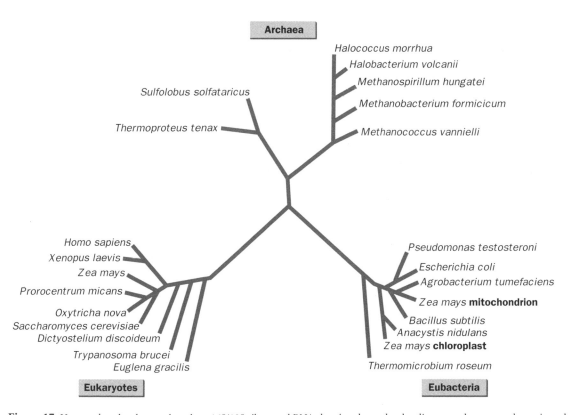

Figure 17. Unrooted molecular tree based on 16S/18S ribosomal RNA showing the molecular divergence between eubacteria and archaea and the endosymbiotic origin of mitochondria and chloroplasts.

C. Woese and G. Olsen had the idea to incorporate homologous 16S mitochondrial and chloroplastic sequences into the data set. Most surprisingly, the mitochondrial and chloroplastic sequences were not closest to the homologous nuclear sequence of their eukaryote host in the tree, but instead were most similar to sequences well within the interior of the eubacteria (fig. 17). Many see the definitive proof for the endosymbiotic origin of mitochondria and chloroplasts in this still famous tree.

At the end of the 1980s, a debate about the paraphyly of the archaea started because certain archaea were found to be more closely related to eukaryotes than to other archaea, the tree of life being rooted on the eubacteria. Today, discussion about the position of the root of life is ongoing and even includes arguments in favor of a eukaryote rooting. Likewise, gene amplifications carried out on unidentified microorganisms from a diversity of different environments suggest an incredible level of genetic divergence. We may therefore have seen only the tip of the genetic iceberg in the world of the bacteria: numerous species may exist whose divergence from currently described species is comparable to the maximum divergence found among the described species themselves.

Today, the phylogeny (and therefore the classification) of the eukaryotes is in the midst of a reconstruction phase. Indeed, once the artifacts of phylogenetic reconstruction are cleared up, molecular phylogenies show us completely unexpected relationships, relationships that will be found in the pages that follow. We can cite as examples:

- the separation of the eukaryotes into two clades, the bikonts and unikonts;
- within the bikonts, the construction of the following clades: the green eukaryotes with a close relatedness between the rhodobionts and chlorobionts; chromoalveolates that bring together the alveolates (ciliates, dinophytes, apicomplexa), the stramenopiles, the cryptophytes and the haptophytes; the rhizarians that include the actinopods, forams and chlorarachniophytes; the excavobionts that take in the euglenobionts, the parabasalids, the metamonads and the percolozoans;
- within the unikonts, the construction of the opisthokont clade, including the fungi and the metazoans, and the amoebozoan clade that bring together the rhizopods and the mycetozoans;

- the relatedness of the eumycetes (or true fungi) and the microsporidia;
- the construction of the opisthokont clade, including the eumycetes, the microsporidia, the choanoflagellates, and the metazoans;
- the relatedness between the chlorobionts and the rhodobionts;
- the construction of the alveolate clade, including the ciliates, dinophytes, and apicomplexa (the apicomplexa group together the sporozoans and the hematozoans);
- the construction of the stramenopiles clade, including the brown algae, the diatoms, the chrysophytes, the xanthophytes, the eustigmatophytes, the oomycetes, the hypochytridiomycetes;
- the paraphyly of the green algae;
- the paraphyly of the bryophytes;
- the paraphyly of the diploblastic animals;
- the division of the protostomians into lophotrochozoans and ecdysozoans, which renders obsolete the former subdivision of the triploblasts into acoelomates, pseudocoelomates, and coelomates;
- the paraphyly of the crustaceans (or, in other words, hexapods being viewed as terrestrial crustaceans);
- the paraphyly of the fish and the reptiles;
- the new relationship among the orders of placental mammals.

The role of fossils in phylogenetic reconstruction

The phylogenetic position of the birds, strongly debated during the 1980s, illustrates the importance of fossils in phylogenetic reconstruction.

The classic theories

Paleontologists generally consider that birds are archosaurs. From here, however, there are two opposing theories. According to the first, birds originated within a paraphyletic group of fossil archosaurs, the thecodonts. This latter group also includes the ancestors of the crocodiles and the dinosaurs. According to the second theory, which dates from the end of the nineteenth century, birds originate from another paraphyletic group of fossil archosaurs, the dinosaurs. Currently, most paleontologists consider the Dromaeosauridae (theropod dinosaurs, including *Deinonychus* and *Velociraptor*) as a sister group to the birds. If the dinosaurs are to constitute a monophyletic group, therefore, the birds must be included. We are thus justified in our claim that the dinosaurs are still among us. Thecodonts and dinosaurs (in the classic sense) have all disappeared, however. The only archosaurs still alive are the birds and the crocodiles. Today, the majority of paleontologists therefore consider that the closest living relatives of the birds are the crocodiles.

Debate on the origin of birds and the role of fossils

The above-mentioned debate started from a relatively inoffensive statement by English paleoichthyologist C. Patterson. In 1981, he expressed the idea that fossils played only a limited role in phylogenetic reconstruction: "Instances of fossils overturning theories of relationships based

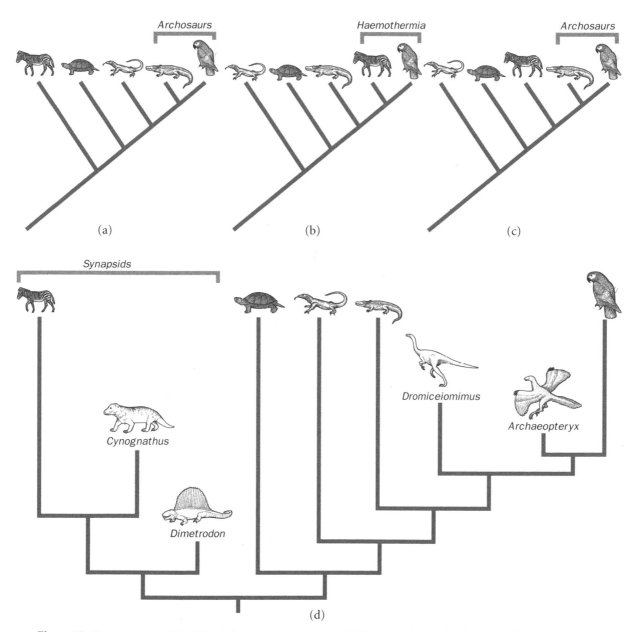

Figure 18. The importance of fossils in phylogenetic reconstruction. (a) The classic theory of evolutionary relationships among extant amniotes. (b) The tree obtained by Gardiner based on characters from the soft parts of living animals, then on mixed characters incorporating fossils. (c) The relationships obtained by Gauthier and his colleagues based on 109 characters from soft and bony body parts (including those of Gardiner) of living amniotes only. (d) The relationships obtained by Gauthier and his colleagues when they added fossils to the previous matrix.

on recent organisms are very rare and may be nonexistent." Another English researcher, B. Gardiner, took this statement to heart and, in 1982, published a now famous study on the phylogeny of the amniotes. The five extant groups of amniotes were classified according to 37 derived characters, 25 of which were anatomical or physiological characters and thus unavailable from fossils. Without a single fossil in the analysis, the most parsimonious tree, based on 17 shared derived characters (fig. 18b), showed the birds as a sister group to the mammals, rather than following the classical grouping of birds and crocodiles. Subsequently, Gardiner introduced fossils into his study and maintained this point of view. The introduction of the fossils did not change the conclusion: birds and mammals constituted the group Haemothermia. The reasoning behind and the implications of Gardiner's work concerned not only the role of fossils but also the theoretical foundations and the practice of phylogenetic analyses. In other words, this study on the origin of birds was also (and maybe especially) the occasion to underline certain work methods. It . defended cladistic analyses. In 1985, S. Løvtrup added other characters that supported Gardiner's cladogram of the amniotes.

In 1988, however, J. Gauthier and his colleagues convincingly demonstrated—using a cladistic analysis (the same method that Gardiner used)—that the addition of fossils to the characters used by Gardiner (1982) and Løvtrup (1985) could indeed modify the most parsimonious tree. In other words, they contradicted Patterson by underlining the importance of fossils. By reanalyzing the data of Gardiner and Løvtrup, to which they added some original data taken from living animals (109 characters), Gauthier found the cladogram of figure 18c, illustrating that birds are the sister group of the crocodiles. In this case, the mammals are the sister group of the ensemble (bird + crocodile), an arrangement that does not correspond to the classic phylogeny of the amniotes. Gauthier then introduced numerous reptile fossils to the analysis, increasing the number of characters to 207. This provoked a change in the position of the mammals and a return to the classic cladogram of the amniotes (fig. 18d). Among the

added fossils, the authors then looked for the ones that were responsible for the change in the the tree: it was the synapsid fossils (from a paraphyletic group, the therapsids or mammal-like reptiles). Indeed, these animals had a combination of characters that are absent in extant fauna: they possessed some of the evolved characteristics of mammals, mixed with certain primitive characteristics that have long since been lost in extant diapsid lineages (lepidosaurs, crocodiles, and birds). Better yet, the presence of a single synapsid (any one) could provoke this change in the tree; these fossils hold character combinations that profoundly influence the inferences made about the evolutionary history of the amniotes. In fact, it is the particular property of all fossils that they are able to present combinations of apomorphic and plesiomorphic characters that have disappeared from present-day nature. Fossils are not ancestors. They are not primitive. They are no greater proof of evolution than present-day organisms. They are simply new mosaics for us.

The key of the morphological characters

In terms of morphology, one can count 28 characters common to birds and mammals. Most of these characters are linked to an evolutionary convergence toward endothermy (elevated and constant body temperature), whereas others have been criticized and rejected. We can cite, by way of example, some simple common characters, such as the loop of Henle in the kidney, a single aortic trunk, and pancreatic islet cells. There are also 77 characters common to birds and crocodiles, such as the mandibular fenestrae, the presence of a gizzard, a membrane on the eye that forms a sort of second eyelid, the reduction or disappearance of certain cranial bones, such as the postfrontal and the postparietal. The current consensus is toward the relatedness of birds and crocodiles. Many of the characteristics that are common to birds and mammals are in fact due to the convergent evolution of endothermy in these two groups. Indeed, endothermy has emerged at least twice in the amniotes, in two distinct lineages: a first time in the fossil synapsids of the Permian, giving rise to this character in the mammals; then a second time in certain dinosaurs, giving rise to it in the birds.

Important concepts, vague concepts

The notion of homology and similarity

Resemblance is the raw material of homology. It is resemblance that will provoke a bet about the common ascendance of a structure that is similar in two or more organisms. The difference between a homology and a similarity is that a homology is a similarity inherited from a common ancestor, whereas all similarities are not necessarily inherited from a common ancestor (fig. 4): resemblances that we detect among organisms are a mosaic of similarities made up of symplesiomorphies, synapomorphies, and evolutionary convergences. This is because resemblances are due not only to inheritance from shared common ancestors but also to adaptations to diverse environments. Using global similarity to determine the evolutionary relationships among organisms does not guarantee success because (1) global dissimilarity (the inverse of resemblance) can hide subtle anatomical traits that may signal a relationship that is nonintuitive at first glance, such as the sister-group relationship between birds and crocodiles, and (2) certain structures that make organisms resemble each other may be the result of spectacular evolutionary convergences. We can cite, for example, the marsupial mole *Notoryctes* and the placental mole *Talpa,* whose common ancestor was not a mole. The marsupials in Australia and the placental mammals in Eurasia each gave rise to animals we call "moles."

The notion of the molecular clock

The concept of the molecular clock arose from the work of M. Kimura: selectively neutral mutations are fixed and replaced in a population at a regular rhythm through time by genetic drift. In molecular phylogenetics, the notion of the clock implies that the global rate of nucleotide (in DNA) or amino acid (in proteins) replacement is roughly equal in all reconstructed evolutionary lineages. A constant rate is not a prerequisite but rather a null hypothesis to be tested. Indeed, today we easily construct trees without necessarily evoking the clock, whether the tree is based on a cladistic analysis from sequence data or from a distance matrix. Irregularities in the molecular clock can also be seen in a tree but, when they are extreme,

they can provoke serious artifacts in tree construction. Indeed, one should understand that when the sequences of a taxonomic group (truly related or not) evolve more quickly than the sequences present in the sample, all tree-construction methods will tend to group sequences with similar evolutionary rates, regardless of their relatedness. A classic artifact is the exclusion of long branches (that is, of species with accelerated evolution at a considered gene) toward the base of the phylogenetic tree, near the outgroup. The outgroup was chosen because it is more distant from the members of the study group than they are from each other. As such, it often carries a greater number of differences than the members of the study group and thus possesses a long branch. As it is the outgroup that defines the base of the tree, all of the long branches are reunited there (fig. 19).

Our methods are becoming more and more efficient at detecting and managing irregularities in the clock. We now know of a range of genes that, depending on the taxonomic sample, vary in their ability to act as clocks. The molecular clock is observed in certain restricted taxonomic cases, roughly observed in other cases, and not at all in others. All depends on the analyzed gene and on the study group. For example, in certain families of teleost fishes, the cytochrome b gene behaves as a good molecular clock and enables the reconstruction of a strongly reliable phylogeny. On the other hand, the cytochrome b gene does not evolve at a homogeneous rate among the mammals, and not even among the rodents. We therefore now talk about a "local molecular clock." It is because of this variability that the clock is considered as a null hypothesis that can be supported or rejected (as is recognized by Kimura himself) or as a way to speak about homoplasy in molecular characters. When sequences have substitution rates that seem to reflect a clock, it is easy to deduce the time since the divergence of two species by using the known rate of evolution of the molecule and the number of differences found between the two species. This is the way that we use molecular data to date the last common ancestor of two species.

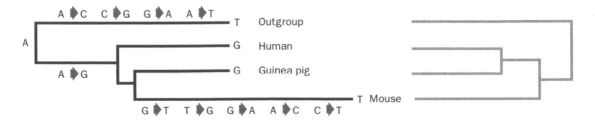

Figure 19. The attraction of long branches is a major artifact in molecular phylogenies. Presented here is a simple phylogeny of four taxa, including two rodents: the guinea pig and the mouse. On the left, we see what actually happens: the nucleotides symbolize the mutations that occurred at particular positions in gene X. On the right, we see the inferred molecular phylogeny. Assume that gene X, the gene that we would like to use to build our phylogeny for these four organisms, has an accelerated rate of evolution in the mouse. The outgroup underwent many transformations along its branch because it is very distant (the time since it diverged from the three others is longer than the divergence time among the three). The very large divergence between the outgroup and the mouse means that, by convergence, they end up with identical nucleotides at several positions on the gene. The mouse is thus "pulled" toward the outgroup and we are brought to the conclusion that rodents are paraphyletic (to the right). The increase in the number of mutations in the mouse is a typical problem for testing the null hypothesis of the molecular clock. If genes behaved as true clocks, such artifacts would not occur.

The notion of the living fossil

The term *living fossil,* as it used in the popular media, is unclear. A living fossil is simply an extant species that is morphologically identical to a known fossil. For example, the coelacanth *Latimeria chalumnae* possesses a body structure that is identical to certain fossils from the upper Cretaceous period. One should keep in mind that nothing enables us to say that this coelacanth and the fossils are the same species. Overall genetic divergence can occur, even if the morphology remains stable. Indeed, less than five percent of the vertebrate genome controls the morphology of the animal.

The status of the ancestor

A genealogy shows us who descended from whom, the ancestors being individually identified. A phylogeny shows us who is closest to whom. The ancestors are not identified but rebuilt, piece by piece, as an unfinished puzzle. In your family's genealogy, the ancestors are identified thanks to the local registry offices. From these records, we know who gave birth to whom. The links in a genealogical tree symbolize the genetic relationships from the ancestors to the descendants. There is, of course, a genealogy for all living organisms, but there is no general registry office: not all living organisms leave behind documents. We have thus lost all traces of

exact individual ancestors. When a fossil is discovered, it is impossible to know whose ancestor it is, in a genetic sense. It is therefore necessary to work with this fossil as we do with living species.

When constructing phylogenies, Hennigian systematists must clarify the status of the ancestor on two fronts. On the one hand, they must clearly state that an identified ancestor is unavailable, otherwise we return to the confusion between a genealogy and a phylogeny and end up constructing paraphyletic groups. On the other hand, they cannot deny the notion of the ancestor—far from it: the ancestor underlies all logic in phylogenetic reconstruction. Hennig provided the methodological tools that enable us to construct monophyletic groups— that is, those containing the ancestor and all of its descendants—but we often forget to say that this ancestor is always inferred and not identified. A cladogram shows each node and the most parsimonious state for each character. At a given node in the tree, we thus reconstitute the missing ancestor. All we know about it are the innovations, the synapomorphies, that it had when alive and that it passed on to the descendants we study. The ancestor of all tetrapods passed on its four legs to us (along with some other characters). It is only through these characters that the ancestor is known. We should note that not all character states at a given node were necessarily found in the same organism. It is possible that these different states appeared in several organisms and that we are simply unable to distinguish among them.

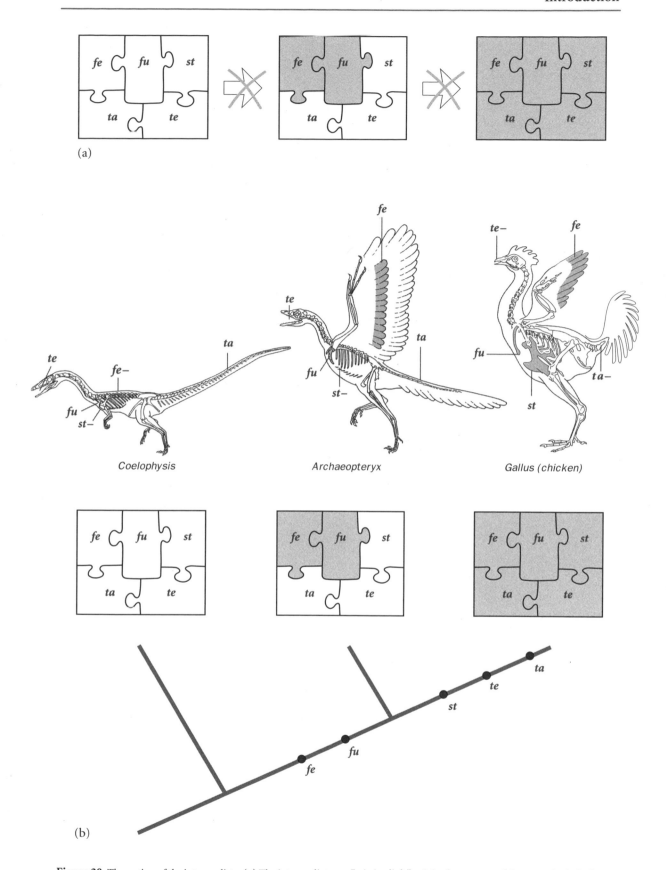

Figure 20. The notion of the intermediate. (a) The intermediate as a "missing link," as it is often portrayed, in a genealogical schema linking ancestor to descendant (to be banished). (b) The structural intermediate, as illustrated by a schema of shared structures, is more compatible with phylogenetic reasoning and methods. *st:* "expanded sternum" character, *te:* "teeth" character, *fu:* "furcula" character, *fe:* "feather" character, *ta:* "tail" character. White puzzle: tail present, teeth present, feathers absent, furcula absent, expanded sternum absent. Colored puzzle: tail absent, teeth absent, feathers present, furcula present, expanded sternum present.

The notion of the intermediate

In popular views of phylogenetics, there are many incorrect ideas about the intermediate, or missing link. We know that a phylogeny does not express the notion of ancestry (who descended from whom) but rather the notion of relatedness (who is closest to whom). A real organism (living or fossil) that is considered as an intermediate is therefore not an ancestor. An organism (living or fossil) qualifies as an intermediate only when it carries a collection of characters unknown from living beings. Let us imagine a real organism (living or fossil) that is like a white-colored puzzle and another organism that has puzzle pieces arranged in the same way except that they are colored black. An intermediate would be a puzzle with the same pieces, but some would be colored white and others would be colored black (fig. 20).

For example, *Archaeopteryx* is an intermediate in the sense that fauna known from before its discovery consisted of the theropod dinosaurs on one side (white puzzle: no feathers, no furcula (or wishbone), no expanded sternum, a tail, teeth), and modern birds on the other (colored puzzle: feathers, furcula, expanded sternum, no tail, no teeth). *Archaeopteryx* carries an original mosaic of traits, some elements from one and some from the other: feathers (color), furcula (color), no expanded sternum (white), tail (white), teeth (white). However, this intermediate is not the genealogical ancestor. To eliminate any confusion about the term *intermediate,* we will speak of a structural intermediate. An intermediate is placed not at a node in a phylogenetic tree but at the end of its own branch, positioned between two previously known branches. Those who pretend that there are no intermediates between defined classes of organisms create a false problem. There are no intermediate ancestors, but not for the reasons that they hope: the reason is purely methodological. There are, however, structural intermediates. Feathered dinosaurs, the birds of the Secondary era, are structural intermediates, as are the therapsids (a paraphyletic group of mammal-like reptiles) and many other such examples. Opponents to evolutionary theory who argue that genealogical intermediates between the large classes of organisms do not exist fail to take into account the difference between genealogical and structural intermediates. They simply negate the existence of any kind of intermediate. Objections of this type could easily go on forever. Even if structural intermediates truly exist, one can always look for structural intermediates of the known intermediates, because the probability of recovering the exact genealogical chain of all living species, generation by generation, is infinitely small!

References

Darlu, P., and Tassy, P. *Reconstruction phylogénétique. Concepts et méthodes.* Collection "Biologie théorique," no. 7. Paris: Masson, 1993.

Darwin, C. *On the Origin of Species.* A facsimile of the 1st edition [1859]. Cambridge, MA: Harvard University Press, 1964.

———— *Autobiographie.* Collection "Un savant, une époque." Paris: Belin/Pour la Science, 1985.

De Bonis, L. *La famille de l'Homme.* Collection "Bibliothèque scientifique." Paris: Belin/Pour la Science, 2000.

Dupuis, C. Permanence et actualité de la Systématique: la "Systématique phylogénétique de W. Hennig (historique, discussion, choix de références)." In *Cahiers des Naturalistes: Bulletin des Naturalistes Parisiens,* n.s. 34 (1978):1–69.

Forey, P. L., et al. *Cladistics: A Practical Course in Systematics.* A Systematic Association publication, no. 10. Oxford: Clarendon Press, 1992.

Futuyama, D. J. *Evolutionary Biology,* 2nd ed. Sunderland, MA: Sinauer Associates, 1986.

Gardiner, B. G. "Tetrapod classification." *Zoological Journal of the Linnean Society* 74 (1982):207–232.

Gauthier, J., Kluge, A. G., and Rowe, T. "Amniote phylogeny and the importance of fossils." *Cladistics* 4 (1988):105–209.

Hennig, W. *Phylogenetic Systematics,* 3d ed. Trans. D. D. Davis and R. Zangerl. Urbana and Chicago: University of Illinois Press, 1979.

Hillis, D., Moritz, C., and Mable, B. (eds.) *Molecular Systematics,* 2nd ed. Sunderland, MA: Sinauer Associates, 1996.

Kimura, M. *Théorie neutraliste de l'évolution.* Collection "Nouvelle Bibliothèque scientifique." Paris: Flammarion, 1990.

Le Guyader, H. (Coord.) *L'évolution.* Collection "Nouvelle Bibliothèque scientifique." Paris: Belin/Pour la Science, 1997.

Lecointre, G. (Coord.) *Comprendre et enseigner la classification du vivant.* Belin. 2005.

Ridley, M. *Evolution biologique.* Paris: De Boeck University, 1997.

Solignac, M., et al. *Génétique et évolution,* 2 vols. Collection "Méthodes." Paris: Hermann, 1995.

Tassy, P. (Coord.) *L'ordre et la diversité du Vivant.* Paris: Fondation Diderot, Fayard, 1986.

———— *L'arbre à remonter le temps.* Collection "Epistémé." Paris: Christian Bourgois, 1991.

Tassy, P., and Lelièvre, H. (Coord.), *Systématique et phylogénie; modèles d'évaluation biologique,* 2d ed. Collection "Biosystema," no. 11. Paris: Société française de systématique, 1999.

Tassy, P. *Le paléontologue et l'évolution.* Collection "Quatre à quatre." Paris: Editions Le Pommier, 2000.

Tillier, S., et al. *Systèmatique: Ordonner la diversité du vivant.* Collection "Rapports sur la Science et la Technologie," no. 11. Paris: Editions Tec & Doc, 2000.

Tort, P. (Coord.), *Dictionnaire du Darwinisme et de l'Evolution,* 3 vols. Paris: Presses Universitaires de France, 1996.

———— *Darwin et la darwinisme.* Collection "Quadrige." Paris: Presses Universitaires de France, 1997.

Wiley, E. *Phylogenetics: The Theory and Practice of Phylogenetic Systematics.* New York: Wiley-Liss, 1981.

Chapter 1

See chap. 2, p. 58

Representative species: *Rhizobium leguminosarum*.
Size: 2 μm.

See chap. 3, p. 80

Representative species: *Desulfurococcus mobilis*.
Size: 1 μm.

See chap. 4, p. 98

Representative species: Nordmann fir,
Abies nordmanniana.
Size: 50 m.

LIFE

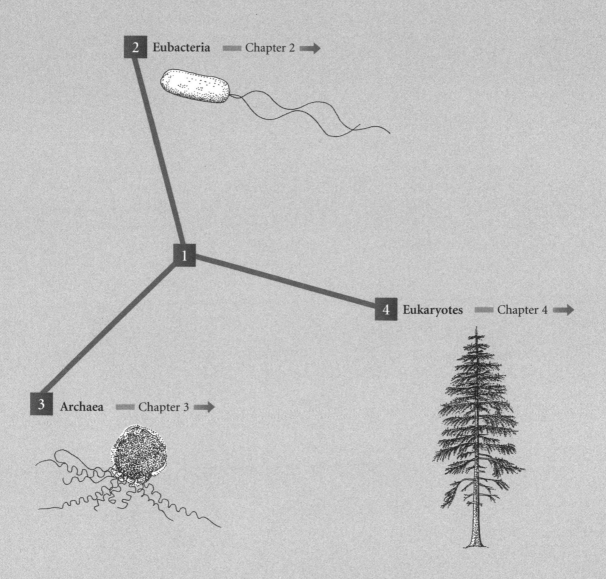

Living organisms have a cellular structure. The cell is the fundamental unit of life, and even the smallest organisms are made up of at least one cell.

The 1,749,577 described extant organisms all possess two exclusive properties. They are all able to make copies of their own DNA sequences and, in this way, are able to propagate their genetic information in time and space. They are also capable of translating this genetic information into enzymatic proteins or components. These autonomous properties define life and oppose the inclusion of viruses as living beings, because viruses cannot independently reproduce themselves. Viruses are not autonomous and must exploit the cellular machinery of the host cell for their replication and synthesis. It is for this reason that they are excluded from this book.

Traditionally, the structure of the prokaryote organism (fig. 1a; *chr:* chromosome, *cmb:* cytoplasmic membrane, *emb:* external membrane, *cw:* cell wall) was viewed as the ancestral structure, and the structure of the eukaryote organism (fig. 1b; *N:* nucleus, *G:* Golgi apparatus, *mi:*

mitochondrion, *ret:* endoplasmic reticulum) as the derived structure. However, this view all depends on how the tree of life is rooted.

Because of the particular metabolism of the archaea, certain microbiologists thought that these organisms were the first to emerge from the tree of life and could therefore be used as the outgroup for all other organisms (eubacteria + eukaryotes). It is for this reason that they received their name. This idea has since been strongly contested and all possible scenarios have been proposed. Certain biologists have good arguments in support of the hypothesis that the eukaryotes were the first group to emerge. This scenario makes the prokaryotes monophyletic and reverses the traditional view that the "simple" prokaryotes gave rise to the "complex" eukaryotes.

If the eukaryotes were indeed the first organisms to emerge, this obviously calls into question the classic gradualist vision of things! The prokaryote structure, viewed as the most simple, may have resulted from a secondary simplification of a more complex structure.

The fundamental problem for situating the root is that we have no outgroup when we

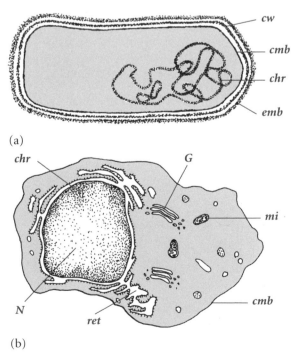

(a)

(b)

Figure 1

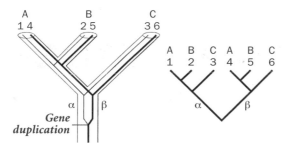

Figure 2

consider all life. We are therefore obliged to use a different procedure: we search for genes that are found in multiple copies in the genome and for which the duplications predate the divergence of the three groups, archaea, eubacteria, and eukaryotes. Let us consider an example (fig. 2) of an ancestral gene that was duplicated into α and β before the separation of the lineages. The phylogeny of the two genes will show a tree with two subtrees, one giving the phylogeny of gene α and the other giving the phylogeny of gene β. As the duplication is the oldest event, it enables us to root the tree; in other words, each subtree can serve as an outgroup to the other. This rooting method has contributed to the proposition that the eubacteria are the most ancient branch, an idea that has resulted in the view that the archaea are a sister group of the eukaryotes. Results remain difficult to interpret for the moment, however. Do not forget that these events are at least a billion years old! The fact that their traces are partially, if not totally, erased is not surprising.

Finally, there is another alternative solution that should not be overlooked, even if it is less likely: maybe some of these lineages are not monophyletic. In this case, evolutionary history could be even more complicated than we have imagined.

Living organisms are present from the atmosphere to the furthest depths of the soil and thus define the biosphere. The physical and chemical conditions that favor active life can vary—and incredibly so, if we consider dormant life (resistant forms: cysts, spores, seeds, etc.). Life can develop in both aerobic and anaerobic conditions. We find halophilic archaea in aquatic environments with salt (NaCl) concentrations higher than 2 moles per liter. Some archaea can grow in acidic milieus with pH less than 1, while others can survive in temperatures greater than 110°C. Numerous life forms develop at −1.8°C in the polar seas. Using a complex system to regulate their internal temperature, some metazoans—for example, mammals like the yak or musk ox—are able to tolerate external temperature as low as −40°C. Some life forms, such as eukaryote spores or tardigrades, are able to resist freezing in liquid nitrogen (−196°C) or liquid helium (−268.5°C), while others are able to resist desiccation: plant seeds, spores, mosses, mites, tardigrades. In such cases, the cells greatly reduce their metabolism but never lose their ability to reproduce.

Energetic resources and metabolic pathways are incredibly variable in the living world. Certain cells are chemoautotrophic: in the absence of all organic matter and light, they are able to extract energy by oxidizing chemical compounds (nitrites, ammonia, sulfides, etc.). These cells are of vital importance in biospheric nitrogen, sulfur, and carbon cycles because they convert gases and mineral salts, useless to most living organisms, into usable compounds. Other cells are autotrophic and photosynthetic: they obtain their energy from the sun. Depending on the group, photosynthesis can be either aerobic or anaerobic. Other living organisms are heterotrophic and obtain their chemical energy by absorbing previously constituted organic compounds. They are thus dependent on other autotrophic or heterotrophic organisms; they either consume these organisms directly or eat their by-products.

One of the consequences of the properties of life is that, because of mutations and selection, organismal diversification is an ongoing and continual process. If we take extinctions into account, extant living organisms represent only a small fraction of the ensemble of organisms that have existed over the course of life's history.

➡ *See chap. 2, p. 58*

Eubacteria

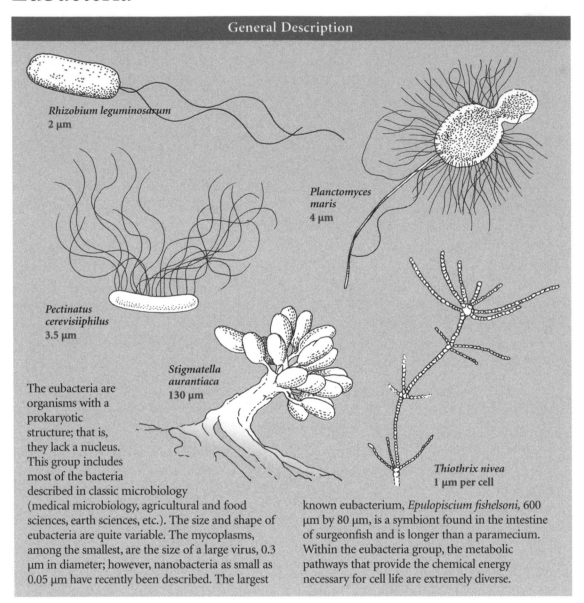

General Description

Rhizobium leguminosarum
2 μm

Planctomyces maris
4 μm

Pectinatus cerevisiiphilus
3.5 μm

Stigmatella aurantiaca
130 μm

Thiothrix nivea
1 μm per cell

The eubacteria are organisms with a prokaryotic structure; that is, they lack a nucleus. This group includes most of the bacteria described in classic microbiology (medical microbiology, agricultural and food sciences, earth sciences, etc.). The size and shape of eubacteria are quite variable. The mycoplasms, among the smallest, are the size of a large virus, 0.3 μm in diameter; however, nanobacteria as small as 0.05 μm have recently been described. The largest known eubacterium, *Epulopiscium fishelsoni*, 600 μm by 80 μm, is a symbiont found in the intestine of surgeonfish and is longer than a paramecium. Within the eubacteria group, the metabolic pathways that provide the chemical energy necessary for cell life are extremely diverse.

Ecology

The diversity of energetic resources and metabolic pathways used by the eubacteria is incomparably larger than that of the eukaryotes, despite the fact that the variety of forms is modest in comparison. The chemoautotrophic eubacteria obtain their energy by oxidizing mineral compounds in the absence of organic matter or sunlight. In the presence of simple molecules like nitrogen salts, oxygen, or carbon dioxide, certain eubacteria oxidize nitrites or nitrates and even convert ammonia to nitrite. Others oxidize methane into methanol, and again others convert sulfites, sulfides, or thiosulfates into sulfates. These eubacterian cells are of great importance for biospheric nitrogen, sulfur, and carbon cycles because they convert gases and minerals that are useless to most living organisms into usable compounds. The eubacteria represent a range of life conditions: autotrophic, heterotrophic, strictly or facultatively aerobic, strictly or facultatively anaerobic, photosynthetic. This diversity is amplified again by the multitude of chemical compounds that can serve as substrates (electron donors or acceptors). Photosynthetic eubacteria draw their energy from solar photons. This energy is then transformed into the chemical energy contained in the bonds of the synthesized organic molecules. Depending on the group, photosynthesis can take place with or without the production of oxygen. The cyanobacteria, incredibly numerous in the Precambrian era (between 2.6 and 0.6 billion years ago), increased the level of oxygen of the terrestrial atmosphere from 1% to the current 20%. Certain eubacteria are fermenting, anaerobic heterotrophes. The heterotrophic state signifies that the cells take their energy from previously constituted organic molecules.

The eubacteria are extremely important organisms because—owing to their metabolic diversity and their ability to live and grow in the most hostile environments, without air, without light—they are at the base of the ecological pyramid. They are involved in the initial steps of soil formation, well before any plant could colonize. They are a food source for many unicellular eukaryotes and small metazoans. They are symbionts in the digestive tube of vertebrates and even in the cytoplasm of unicellular eukaryotes; the mitochondria and chloroplasts of eukaryote cells are eubacteria. They are parasites, and sometimes pathogens, of plants and animals (in humans: cholera, typhus, the plague, etc.). The eubacteria are of vital importance in the recycling of organic material (degradation of dead organisms and excrement, etc.).

Some unique derived features

■ The eubacteria possess a cell wall of peptidoglycan containing muramic acid. Those that do not possess this structure have lost it secondarily. Fig. 1 shows the structural arrangement of a peptidoglycan molecule (*ma:* muramic acid; *pch:* peptide chain; *N-ac:* N-acetylglucosamine; *peb:* pentaglycin bridge).

■ Transfer RNA that initiates translation carries an N-formylmethionine molecule.

■ Molecular phylogenies based on several different molecules (RNA polymerases, ribosomal RNA, elongation factors: EF-TU, EF1-α, ATPase subunit β) all support the monophyly of the eubacteria.

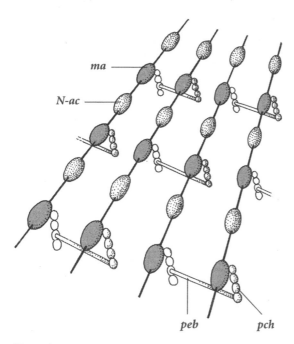

Figure 1

Examples

Escherichia coli; Desulfovibrio desulfuricans; Rhodomicrobium vanielii; Azotobacter vinelandii; Pseudomonas multivorans; Nitrobacter winogradskyi; Micrococcus radiodurans; Bacillus anthracis (causal agent of anthrax); *Vibrio cholerae* (causal agent of cholera); *Mycobacterium leprae* (causal agent of leprosy); *Pectinatus cerevisiiphilus; Planctomyces maris; Rhizobium leguminosarum; Stigmatella aurantiaca; Thiothrix nivea; Treponema pallidum* (causal agent of syphilis); *Clostridium tetani* (causal agent of tetanus); *Mycobacterium tuberculosis* (causal agent of pulmonary tuberculosis).

Number of species: 10,593 classified on the basis of RNA 16S (but this number is greatly underestimated).

Oldest known fossils: the most ancient stromatolites date back to 3.5 billion years ago. These are layered sedimentary rocks created by the metabolic activity of filamentous, photosynthetic cyanobacteria. Fig. 2 shows sections of fossils from western Australia estimated to be 3.5 billion years old. They are thought to be filaments of cyanobacteria 2 μm in length.

Current distribution: the eubacteria are present throughout the biosphere (from the atmosphere to the furthest limits of the soils and marine floors) and in the digestive tract of a great many animals.

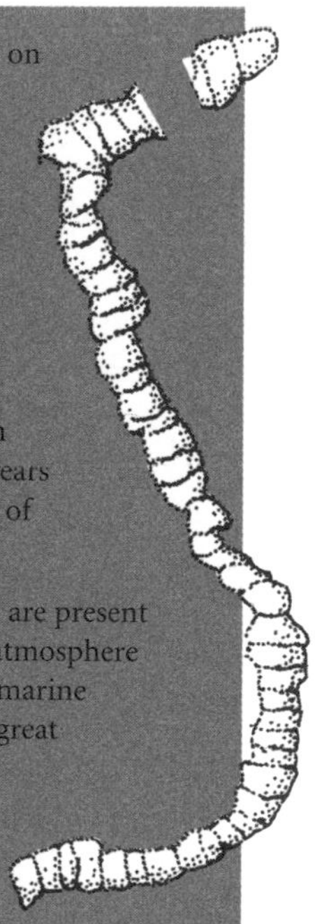

Figure 2

➡ *See chap. 3, p. 80*

Archaea

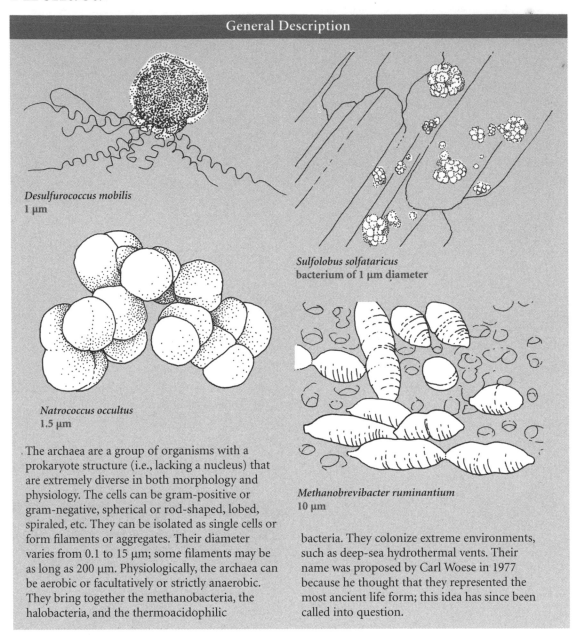

General Description

Desulfurococcus mobilis
1 μm

Sulfolobus solfataricus
bacterium of 1 μm diameter

Natrococcus occultus
1.5 μm

Methanobrevibacter ruminantium
10 μm

The archaea are a group of organisms with a prokaryote structure (i.e., lacking a nucleus) that are extremely diverse in both morphology and physiology. The cells can be gram-positive or gram-negative, spherical or rod-shaped, lobed, spiraled, etc. They can be isolated as single cells or form filaments or aggregates. Their diameter varies from 0.1 to 15 μm; some filaments may be as long as 200 μm. Physiologically, the archaea can be aerobic or facultatively or strictly anaerobic. They bring together the methanobacteria, the halobacteria, and the thermoacidophilic bacteria. They colonize extreme environments, such as deep-sea hydrothermal vents. Their name was proposed by Carl Woese in 1977 because he thought that they represented the most ancient life form; this idea has since been called into question.

Ecology

The archaea are often found in extreme aquatic and terrestrial environments, habitats that are anaerobic, highly saline, high in temperature, or at great depths. Recently, they have also been found in cold environments; they constitute 34% of the prokaryote biomass found at the surface of the Antarctic coastal waters. The archaea prosper in the digestive tract of certain vertebrates, like the rumen of the ruminants.

Examples

Desulfurococcus mobilis; Halobacterium halobium; H. marismortui; Methanobacillus amelianski; Methanobacterium ruminantium; Methanobrevibacter ruminantium; Methanococcus vannielii; Natrococcus occultus; Sulfolobus solfataricus.

Some unique derived features

■ The cellular membrane contains very specific lipids; in contrast to the lipids in the eubacteria and the eukaryotes, the fatty acid is not bound to the alcohol via an ester bond but rather via an ether bond. Fig. 1a shows a normal lipid with an ester bond (*esb*) between the glycerol (*gly*) and the fatty acid (here, stearic acid: *st ac*), and fig. 1b shows a lipid from an archaea with an ether bond (*etb*) between the glycerol and the fatty acid (here, phytanol: *phy*). Tetraethers, made up of two glycerol molecules linked by two fatty acids, may also form.

■ The membrane lipids may form a bilayer as they do in the eubacteria and the eukaryotes (fig. 2a: bilayer formed of glycerol esters; *est*: ester; *prot*: protein), but they may also form rigid monolayers, made up of tetraethers (fig. 2b: monolayer formed by glycerol tetraethers; *tet*: tetraether).

■ The thymine of the TψC loop of transfer RNA has been replaced by a pseudo-uracil or a 1-methylpseudo-uracil.

■ The ribosomes of the archaea have a characteristic shape.

■ Molecular phylogenies based on different molecules (RNA polymerases, ribosomal RNA, elongation factors: EF-TU, EF1-α, ATPase subunit β) all support the monophyly of the archaea.

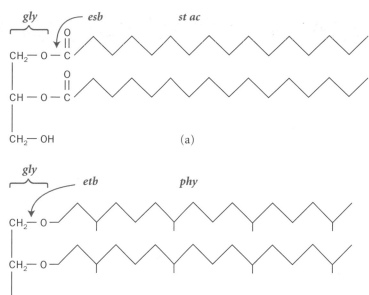

Figure 1

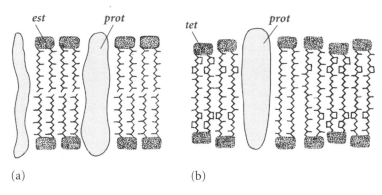

Figure 2

Number of species: 409, classified on the basis of RNA 16S (but this number is greatly underestimated).
Oldest known fossils: none.
Current distribution: worldwide, terrestrial and aquatic. Certain methanogenic archaea live in the rumen of the ruminants (mammals), which are found across the globe. With this exception, the archaea tend to be associated with seas, saline lakes, oxygen-free sediments, and deep-sea hydrothermal vents.

➡ *See chap. 4, p. 98*

Eukaryotes

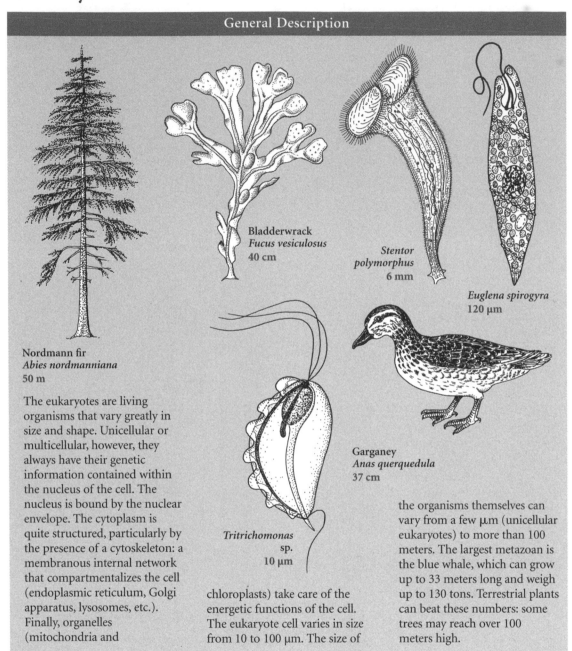

General Description

Nordmann fir
Abies nordmanniana
50 m

Bladderwrack
Fucus vesiculosus
40 cm

*Stentor
polymorphus*
6 mm

Euglena spirogyra
120 μm

Tritrichomonas
sp.
10 μm

Garganey
Anas querquedula
37 cm

The eukaryotes are living organisms that vary greatly in size and shape. Unicellular or multicellular, however, they always have their genetic information contained within the nucleus of the cell. The nucleus is bound by the nuclear envelope. The cytoplasm is quite structured, particularly by the presence of a cytoskeleton: a membranous internal network that compartmentalizes the cell (endoplasmic reticulum, Golgi apparatus, lysosomes, etc.). Finally, organelles (mitochondria and chloroplasts) take care of the energetic functions of the cell. The eukaryote cell varies in size from 10 to 100 μm. The size of the organisms themselves can vary from a few μm (unicellular eukaryotes) to more than 100 meters. The largest metazoan is the blue whale, which can grow up to 33 meters long and weigh up to 130 tons. Terrestrial plants can beat these numbers: some trees may reach over 100 meters high.

Ecology

The eukaryotes are present in all environments, at all altitudes and depths, and at all latitudes. They represent a considerable biomass in all ecosystems. They are fundamentally aerobic—that is, their metabolism requires the presence of oxygen. The only exceptions to this are secondary modifications that took place in some ancient aerobic eukaryotes. The eukaryotes have a relatively homogeneous type of oxidative metabolism (glycolysis, Krebs cycle, oxidative phosphorylation involving a chain of cytochromes, the electron transport chain, etc.).

Photosynthetic species have photosynthetic enzymes contained within the membrane-bound chloroplasts and always produce oxygen as a metabolic by-product. This contrasts with photosynthesis in the prokaryotes, in which different photosynthetic pathways result in the production of sulfur, sulfates, or oxygen. Eukaryotic cell division is initially made up of a diploid phase and a haploid phase. Chromosomes go from the diploid phase to the haploid phase via meiosis, and from the haploid phase to the diploid phase by fertilization.

Some unique derived features

- The DNA is contained within a nucleus that is surrounded by a nuclear envelope.

- Microtubules are tubulin polymers and make up the major part of the cell's cytoskeleton.

- The flagellum has a unique structure: 9 pairs (or triplets) of microtubules surround a central pair (fig. 1: cross-section of the flagellum of *Chlamydomonas* with its central pair of microtubules; *dyn*: dynein arm, *mic*: microtubule).

- Extant eukaryotes possess mitochondria that assure cell respiration. The absence of mitochondria is always the result of a secondary loss.

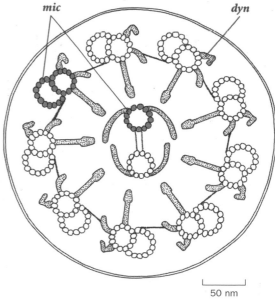

50 nm

Figure 1

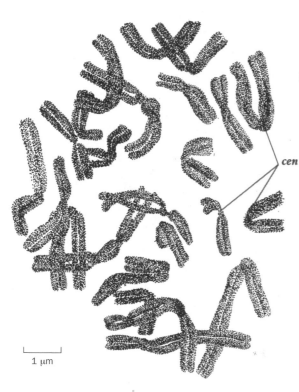

1 μm

Figure 2

- During cell division, DNA is divided and packaged into chromosomes (fig. 2: metaphase plate of a *Triturus* newt showing the duplicated chromosomes; *cen*: centromere).

- Cell division occurs through mitosis and involves centrioles and spindle fibers, both made up of cytoplasmic microtubules; in other living organisms, cell division is achieved by binary fission.

- There are true sexes, with each sexual type contributing an equal proportion of the genetic material to the next generation.

- Molecular phylogenies based on different molecules (RNA polymerases, ribosomal RNA, elongation factors: EF-TU, EF1-α, ATPase subunit β) all support the monophyly of the eukaryotes.

Examples

Gonyaulax tamarensis; Tetrahymena pyriformis; Tritrichomonas sp.; euglena: *Euglena spirogyra;* bladderwrack: *Fucus vesiculosus;* Nordmann fir: *Abies nordmanniana;* corn: *Zea mays; Stentor polymorphus;* large amoeba: *Amoeba proteus;* baker's yeast: *Saccharomyces cerevisiae;* red cracking bolete: *Boletus chrysenteron;* gregarious desert locust: *Schistocerca gregaria;* cat: *Felis catus;* blue whale: *Balaenoptera musculus.*

Number of species: 1,738,575.
Oldest known fossils: certain fossils, estimated to be one billion years old, are probably those of unicellular eukaryotes. Spheres resembling characteristic eukaryotic spores are numerous in the Precambrian.
Current distribution: worldwide.

Chapter 2

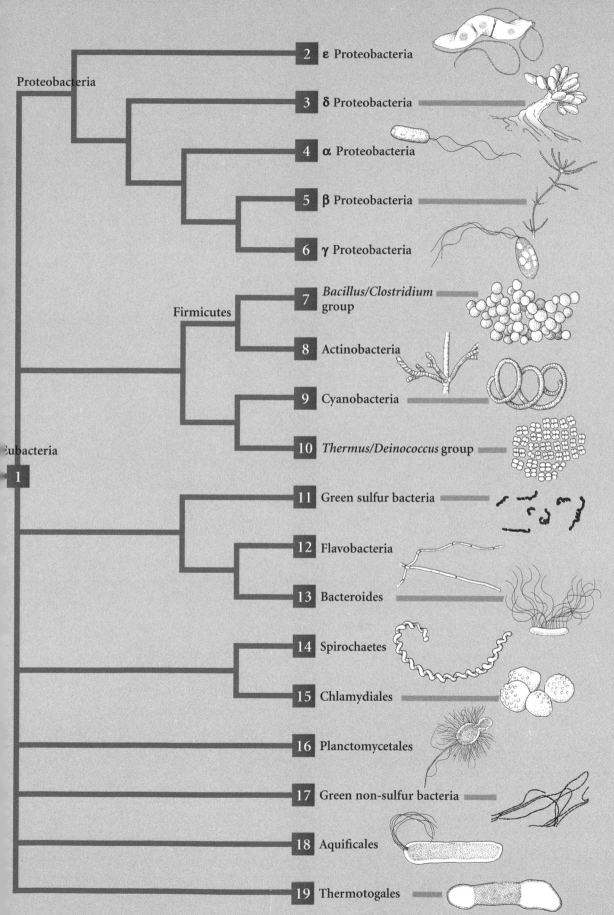

EUBacteria

Molecular phylogenies based on the 16S ribosomal RNA gene have had an extraordinary impact on the classifications of the bacterial world and on our perception of this world. Indeed, until recently, bacteria were classified by microbiologists by their cellular (cell shape, type of cell wall, etc.) and metabolic characteristics. However, useful characters are rare and the difficulties of employing a phylogenetic approach with available characters are insurmountable:

- the morphological simplicity of bacteria limit the number of available morphological characters.

- most "good" characters are autapomorphies that are used to define the phyla and are not useful for determining the interrelationships among phyla.

- certain types of characters that were classically used by microbiologists have been found to be extremely labile and therefore homoplastic.

- in order to identify bacteria, it is necessary to cultivate them; therefore only those bacteria that microbiologists can grow in cultures are available for study.

With the advent of cladistics as the major method for classifying organisms, microbiologists realised that little could be done with the classic characters used for bacteria. For the most part, they decided to retain a strictly utilitarian classification (or rather, an identification key) that was not phylogenetic but was indispensable because of the importance of the eubacteria in the medical, agricultural and agro-food industries. It is then that molecular phylogenies appeared and enabled them to:

- propose a reliable phylogeny of the major lineages;

- show the homoplastic nature of metabolic characters;

- construct molecular probes to detect eubacteria in the environment that were previously inaccessible due to our inability to cultivate them.

The new edition of *Bergey's*, the microbiologist's "bible," is based entirely on molecular phylogenies, and particularly those of the 16S ribosomal RNA gene (see the website: *www.cme.msu.edu/Bergeys/*). Now, all new eubacteria are sequenced at the 16S RNA gene in order to be classified. This is why we have given only the number of species that have been sequenced at this gene under the section "number of species." In the most complete database (for example: *The European ribosomal RNA data base: http://www.psb. ugent.be/rRNA/index.html; NCBI: The National Center for Biotechnology Information: http://www.ncbi.nlm.nih.gov/*), sequences are included that have been obtained from environmental samples, that is, from eubacteria that have not yet been cultured. Of the 10,593 described eubacteria, 1,514 sequences come from this type of sample (as of January 2006). It is certain that these numbers will increase; some microbiologists think that we have, at present, described only 10% of the eubacteria. These numbers should therefore be taken as indications only. This is even more so given that the concept of a species is difficult to apply to the bacterial world.

In the proposed phylogeny, based entirely on 16S ribosomal RNA gene sequences, certain nodes are robust, such as that of the proteobacteria group and the taxon comprising [green sulphur bacteria + (flavobacteria + bacteroides)]. The monophyly of the gram-positive bacteria is now generally accepted by the scientific community, as is the grouping of the cyanobacteria and the *Thermus/Deinococcus* group, and its placement as a sister-group of the gram-positive bacteria. A clade comprising the spirochaetes and the chlamydiales has recently been proposed.

There are now a substantial number of genomes that have been entirely sequenced. An exhaustive list of sequenced genomes is available for each phylum (inventory from January 2006) on the JGI website (*Joint Genome Institute, Integrated Microbial Genomes, http://img.jgi.doe. gov/cgi-bin/pub/main.cgi*).

ε Proteobacteria

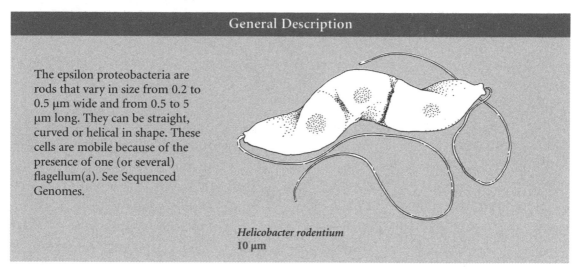

General Description

The epsilon proteobacteria are rods that vary in size from 0.2 to 0.5 μm wide and from 0.5 to 5 μm long. They can be straight, curved or helical in shape. These cells are mobile because of the presence of one (or several) flagellum(a). See Sequenced Genomes.

Helicobacter rodentium
10 μm

Some unique derived features

■ The only unique derived characters that we know of are those that exist in the 16S ribosomal RNA gene sequence.

Ecology

The epsilon protobacteria are weakly aerobic bacteria that we find in the oral cavity, the digestive tube or the genital tract of mammals. Many are pathogenic and provoke abortions in cows and sheep, enteritis in humans and sheep, and gastritis in humans.

Number of species: 232, classified according to 16S RNA. **Oldest known fossils:** none. **Current distribution:** worldwide.

Examples

Campylobacter fetus (causes abortion in cows and sheep); *C. jejuni* (causal agent of human enteritis or food poisoning); *Helicobacter pylori* (causal agent of human gastritis); *H. rodentium.*

δ Proteobacteria

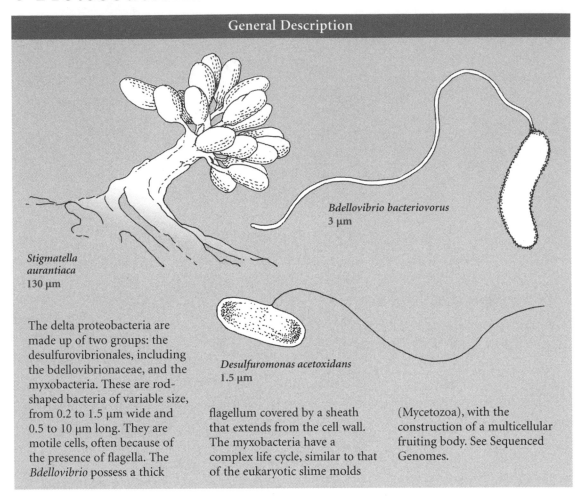

General Description

Stigmatella aurantiaca
130 µm

Bdellovibrio bacteriovorus
3 µm

Desulfuromonas acetoxidans
1.5 µm

The delta proteobacteria are made up of two groups: the desulfurovibrionales, including the bdellovibrionaceae, and the myxobacteria. These are rod-shaped bacteria of variable size, from 0.2 to 1.5 µm wide and 0.5 to 10 µm long. They are motile cells, often because of the presence of flagella. The *Bdellovibrio* possess a thick flagellum covered by a sheath that extends from the cell wall. The myxobacteria have a complex life cycle, similar to that of the eukaryotic slime molds (Mycetozoa), with the construction of a multicellular fruiting body. See Sequenced Genomes.

Some unique derived features

■ The only unique derived characters that we know of are those that exist in the 16S ribosomal RNA gene sequence.

Ecology

The desulfurovibrionales are of considerable ecological importance. They are strictly anaerobic bacteria that reduce sulfates and sulfur during anaerobic fermentation. As they are present in all terrestrial and aquatic habitats, they play a major role in sulfur cycling. *Bdellovibrio* is a predator that feeds itself by attacking other bacteria. Its life cycle is similar to that of a bacteriophage. The myxobacteria produce proteolytic enzymes that they use as predators or detritivores. The fruiting body carries sporangia.

Examples

Bdellovibrio bacteriovorus;
Chondromyces apiculatus;
Desulfovibrio desulfuricans;
Desulfuromonas acetoxidans;
Stigmatella aurantiaca.

Number of species: 332, classified according to 16S RNA.
Oldest known fossils: none.
Current distribution: worldwide.

α Proteobacteria

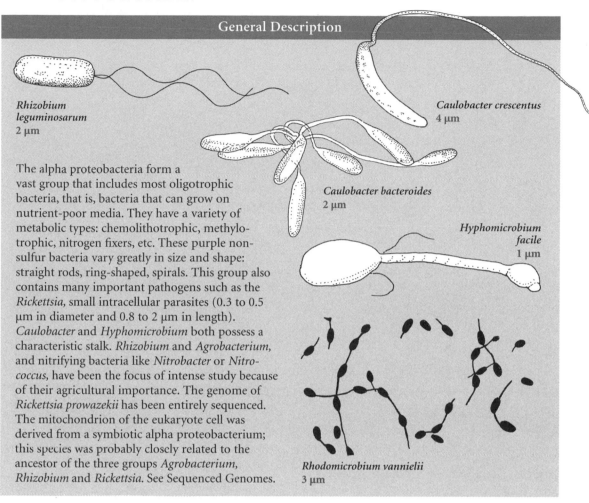

General Description

Rhizobium leguminosarum
2 μm

Caulobacter crescentus
4 μm

Caulobacter bacteroides
2 μm

Hyphomicrobium facile
1 μm

The alpha proteobacteria form a vast group that includes most oligotrophic bacteria, that is, bacteria that can grow on nutrient-poor media. They have a variety of metabolic types: chemolithotrophic, methylotrophic, nitrogen fixers, etc. These purple non-sulfur bacteria vary greatly in size and shape: straight rods, ring-shaped, spirals. This group also contains many important pathogens such as the *Rickettsia*, small intracellular parasites (0.3 to 0.5 μm in diameter and 0.8 to 2 μm in length). *Caulobacter* and *Hyphomicrobium* both possess a characteristic stalk. *Rhizobium* and *Agrobacterium*, and nitrifying bacteria like *Nitrobacter* or *Nitrococcus*, have been the focus of intense study because of their agricultural importance. The genome of *Rickettsia prowazekii* has been entirely sequenced. The mitochondrion of the eukaryote cell was derived from a symbiotic alpha proteobacterium; this species was probably closely related to the ancestor of the three groups *Agrobacterium*, *Rhizobium* and *Rickettsia*. See Sequenced Genomes.

Rhodomicrobium vannielii
3 μm

Some unique derived features

■ The only unique derived characters that we know of are those that exist in the 16S ribosomal RNA gene sequence.

Ecology

The purple non-sulfur bacteria are frequently found in mud and lakes rich in organic matter. There are also marine species. Like other purple bacteria, they possess bacteriochlorophylls and carry out anaerobic photosynthesis. Most *Rickettsia* are intracellular parasites that cause serious disease such as typhus. The *Wolbachia*, bacteria related to the *Rickettsia*, are associated with arthropods, and in particular with insects and isopods. Without being pathogenic, they induce sexual incompatibilities in their hosts. *Hyphomicrobium* are budding bacteria that are able to attach themselves to substrates in their aquatic habitats. *Caulobacter* have a complex life cycle during which they form asymmetrical pre-divided cells. *Rhizobium leguminosarum* form nitrogen-fixing nodules. *Agrobacterium tumefaciens* is responsible for crown gall disease, tumor-like swellings found in certain plants. This species is used as a vector to insert genetic material into the genome of certain plants. The nitrifying bacteria like *Nitrobacter* or *Nitrococcus* are found throughout the soils and oxidize nitrites into nitrates.

Number of species: 1,488, classified according to 16S RNA.
Oldest known fossils: none.
Current distribution: worldwide.

Examples

Agrobacterium tumefaciens; Caulobacter bacteriodes; C. crescentus; Hyphomicrobium facile; Nitrococcus mobilis; Nitrobacter winogradskyi; Rhizobium leguminosarum; Rhodomicrobium vannielii; Rickettsia prowazekii; R. typhi (causal agent of typhus).

β Proteobacteria

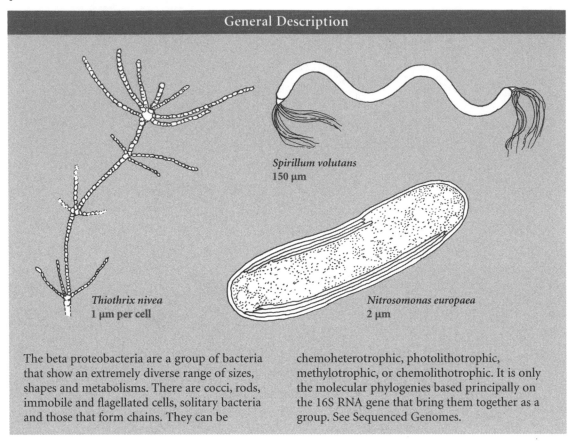

General Description

Spirillum volutans
150 μm

Thiothrix nivea
1 μm per cell

Nitrosomonas europaea
2 μm

The beta proteobacteria are a group of bacteria that show an extremely diverse range of sizes, shapes and metabolisms. There are cocci, rods, immobile and flagellated cells, solitary bacteria and those that form chains. They can be chemoheterotrophic, photolithotrophic, methylotrophic, or chemolithotrophic. It is only the molecular phylogenies based principally on the 16S RNA gene that bring them together as a group. See Sequenced Genomes.

Some unique derived features

■ The only unique derived characters that we know of are those that exist in the 16S ribosomal RNA gene sequence.

Ecology

The nitrosomonadaceae (*Nitrosomonas, Nitrosococcus*) are nitrifying soil bacteria that oxidize ammonia to nitrite. The colorless sulfur bacteria, such as *Thiobacillus* or *Thiothrix*, are aerobic and oxidize sulfur compounds to sulfates. These species are present in all terrestrial and aquatic habitats. The bacteria of the genus *Bordetella* are found in the respiratory tract of mammals and birds. Certain *Neisseria* are human pathogens (gonorrhea, meningitis).

Examples

Bordetella bronchiseptica (causal agent of bronchial infections in mammals and birds); *Neisseria gonorrhoeae* (causal agent of gonorrhea); *N. meningitidis* (causal agent of meningitis); *Nitrosococcus oceanus*; *Nitrosomonas europaea*; *Spirillum volutans*; *Thiobacillus novellus*; *Thiothrix nivea*.

Number of species: 682, classified according to 16S RNA.
Oldest known fossils: none.
Current distribution: worldwide.

γ Proteobacteria

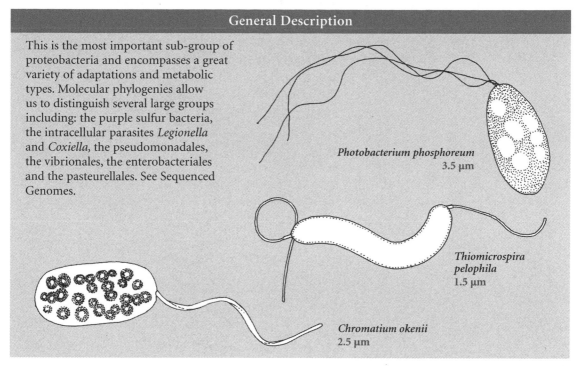

General Description

This is the most important sub-group of proteobacteria and encompasses a great variety of adaptations and metabolic types. Molecular phylogenies allow us to distinguish several large groups including: the purple sulfur bacteria, the intracellular parasites *Legionella* and *Coxiella*, the pseudomonadales, the vibrionales, the enterobacteriales and the pasteurellales. See Sequenced Genomes.

Photobacterium phosphoreum
3.5 μm

Thiomicrospira pelophila
1.5 μm

Chromatium okenii
2.5 μm

Some unique derived features

■ The only unique derived characters that we know of are those that exist in the 16S ribosomal RNA gene sequence.

Ecology

The purple sulfur bacteria are strictly anaerobic and photolithoautotrophic. They have bacteriochlorophyll as photosynthetic pigments. Sulfur is temporarily accumulated in the cell as granules and often acts as an electron donor. *Legionella* are frequently found in thermally polluted waters and are human pathogens. *Methylococcus* uses methane as its unique source of carbon and energy under weakly aerobic conditions. *Pseudomonas* species play an important role in the decomposition of organic matter. *Azotobacter*, widely found in soils and waters, is a nitrogen fixer. The vibrionales are primarily aquatic bacteria; some are pathogenic like the causal agent of cholera. Certain *Photobacterium* are bioluminescent and live in the light organs of fish. Enterobacteria, such as *Escherichia coli*, are largely limited to the mammalian intestine. The group includes many important pathogens that are responsible for a variety of diseases: typhoid fever, gastroenteritis, dysentery, pneumonia. The pasteurellales are vertebrate parasites that are responsible for diseases such as avian cholera and bovine and ovine pneumonia.

Examples

Chromatium okenii; Escherichia coli; Haemophilus influenzae; Legionella pneumophila (causal agent of Legionnaires' disease); *Methylococcus capsulatus; Photobacterium phosphoreum; Pseudomonas aeruginosa; Salmonella typhi* (causal agent of typhoid fever); *Thiomicrospira pelophila; Vibrio cholerae* (causal agent of cholera); *Yersinia pestis* (causal agent of the plague).

Number of species: 1,924, classified according to 16S RNA.
Oldest known fossils: none.
Current distribution: world-wide.

Bacillus / *Clostridium* group
Low G+C gram-positive bacteria

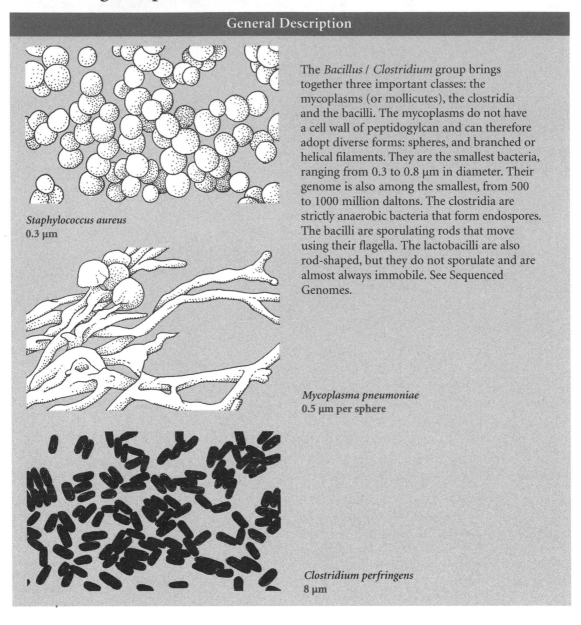

General Description

The *Bacillus* / *Clostridium* group brings together three important classes: the mycoplasms (or mollicutes), the clostridia and the bacilli. The mycoplasms do not have a cell wall of peptidogylcan and can therefore adopt diverse forms: spheres, and branched or helical filaments. They are the smallest bacteria, ranging from 0.3 to 0.8 μm in diameter. Their genome is also among the smallest, from 500 to 1000 million daltons. The clostridia are strictly anaerobic bacteria that form endospores. The bacilli are sporulating rods that move using their flagella. The lactobacilli are also rod-shaped, but they do not sporulate and are almost always immobile. See Sequenced Genomes.

Staphylococcus aureus
0.3 μm

Mycoplasma pneumoniae
0.5 μm per sphere

Clostridium perfringens
8 μm

Ecology

The mycoplasms are extremely widespread in nature: soils, animals, plants. At least 10% of eukaryote cell cultures are contaminated by mycoplasms that are difficult to detect and eliminate. In animals, mycoplasms are often associated with diseases of the respiratory and uro-genital tracts (for example, pneumonia in humans, cattle and swine). Spiroplasms have been found infecting numerous insect species that transmit them to plants (lemon, cauliflower, corn, etc.).

The anaerobic clostridia have temperature resistant spores; it is for this reason that they are responsible for the spoiling of food preserves, causing botulism. They are also major pathogens (tetanus, gangrene).

The bacilli are equally important for humans. Many are pathogenic: streptococci, staphylococci, *Listeria*. Many *Bacillus* species produce antibiotics (bacitracin, gramicin, polymyxin). Among the pathogens, *B. anthracis* causes anthrax; *B. thuringiensis* and *B. sphaericus* are lethal to many insects and are used for biological control. The lactobacilli are heavily used in the food industry for the production of fermented vegetables (sauerkraut, for example), beverages (beer and wine), and milk products (cheeses, yogurts).

Some unique derived features

- Some unique derived characters exist in the 16S ribosomal RNA gene sequence.

- Gram +: bacteria that respond positively to Gram staining (gram +) because of the composition of the cell wall (fig. 1).

Examples

Bacillus subtilis; B. anthracis (causal agent of anthrax); *B. thuringiensis; Clostridium perfringens* (causal agent of gas gangrene); *C. botulinum* (causal agent of botulism); *C. tetani* (causal agent of tetanus); *Lactococcus lactis; Mycoplasma pneumoniae* and *Streptococcus pneumoniae* (causal agents of human pneumonia); *Mycoplasma genitalium; Ureaplasma urealyticum; Streptococcus thermophilus* and *Lactobacillus bulgaricus* (used in yogurt production); *Listeria monocytogenes* (causal agent of listeriosis); yellow staphylococcus: *Staphylococcus aureus*.

Number of species: 2,513, classified according to 16S RNA.
Oldest known fossils: none.
Current distribution: worldwide.

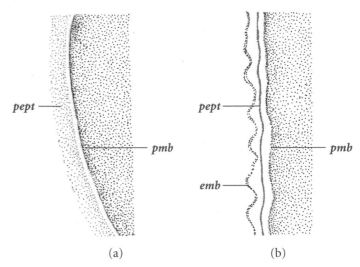

(a)   (b)

Fig. 1: cell wall structure in gram + bacteria (a) and gram – bacteria (b); the gram + bacteria possess a thick cell wall of either peptidoglycan or murein; the gram – bacteria have a thin cell wall that is covered by an external membrane (*emb:* external membrane; *pmb:* plasma membrane; *pept:* peptidogylcan).

Gram staining:

- cover completely with crystal violet for 30 s, then rinse;
- cover with Lugol's solution (iodine) for 1 min, then rinse;
- wash in ethanol or acetone for 10 to 30 s, then rinse;
- cover with safranin, then rinse;

Gram-positive cells retain the crystal violet color; gram-negative cells lose this color when washed with ethanol (or acetone) and are then colored red by the safranin.

Actinobacteria
High G+C gram-positive bacteria

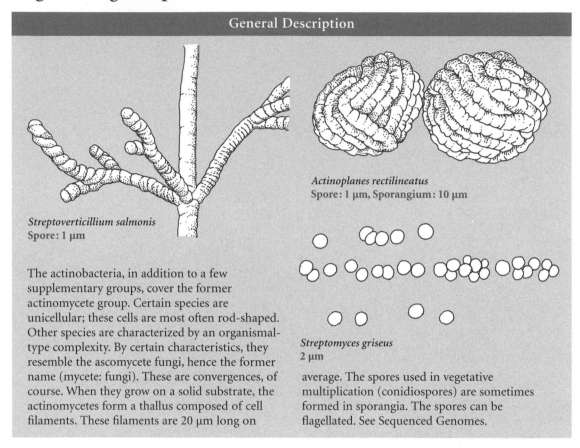

General Description

Streptoverticillium salmonis
Spore: 1 μm

Actinoplanes rectilineatus
Spore: 1 μm, Sporangium: 10 μm

Streptomyces griseus
2 μm

The actinobacteria, in addition to a few supplementary groups, cover the former actinomycete group. Certain species are unicellular; these cells are most often rod-shaped. Other species are characterized by an organismal-type complexity. By certain characteristics, they resemble the ascomycete fungi, hence the former name (mycete: fungi). These are convergences, of course. When they grow on a solid substrate, the actinomycetes form a thallus composed of cell filaments. These filaments are 20 μm long on average. The spores used in vegetative multiplication (conidiospores) are sometimes formed in sporangia. The spores can be flagellated. See Sequenced Genomes.

Some unique derived features

- Some unique derived characters exist in the 16S ribosomal RNA gene sequence.

- These bacteria respond positively to gram staining (gram +). It is not certain that this is an apomorphy shared with the *Bacillus / Clostridium* group. However, for the moment, this similarity is used to justify the firmicutes clade.

Ecology

The actinobacteria are of enormous ecological, agro-food and medical importance. We find them in the soils where, like the *Arthrobacter,* they play a significant role in the decomposition of organic matter. They are also present in the digestive tract of many animals; *Bifidobacterium bifidus* is the first bacterium to colonize the intestine of breast-fed babies. They are important for the production of cheese; species of the *Proprionibacterium* genus are used to make a gruyere. Finally, many species cause important diseases (diphtheria, leprosy, tuberculosis). However, other species can be the source of antibiotics like streptomycin.

Examples

*Actinoplanes rectilineatus;
Bifidobacterium bifidus;
Corynebacterium diphtheriae* (causal agent of diphtheria); *Mycobacterium leprae* (causal agent of leprosy); *M. tuberculosis (*causal agent of human tuberculosis); *Streptomyces griseus* (source of streptomycin); *S. salmonis.*

Number of species: 2,134, classified according to 16S RNA. **Oldest known fossils:** none. **Current distribution:** world-wide.

Cyanobacteria

General Description

Arthrospira platenis
1 μm per cell

Spirulina labyrinthiformis
1.5 μm per cell

Anabaena variabilis
4 μm per cell

This is the most important group of photosynthetic bacteria. The cyanobacteria contain a wide variety of sizes and shapes. Cells can vary in size from 1 to 10 μm. They can be isolated, form colonies of various shapes, or filaments called trichomes. In a trichome, the cells are in tight contact, as in *Oscillatoria*. When phycocyanin is abundant in the cell, it is blue-green in color.

When phycoerythrin is dominant, the color is more brown or deep red. The peptidoglycan layer is often very thick, up to 200 nm. The cells often contain gas vesicles that enable them to move vertically. Many filamentous colonies are mobile. The photosynthetic pigments and the electron transport chain are found in structures attached to the membranes called phycobilisomes. Biological,

ultrastructural and phylogenetic characteristics show that the chloroplasts of photosynthetic eukaryotes are symbiotic cyanobacteria. It is very likely that there was a single event of primary endosymbiosis in the hypothetical common ancestor of the green plants, a clade that includes the chlorobionts, the rhodobionts and the glaucophytes. See Sequenced Genomes.

Some unique derived features

- The photosynthetic pigments and the electron transport chain are found in phycobilisomes.

- The photosynthetic system includes photosystem II (PSII).

- 16S ribosomal RNA gene: as for the other eubacteria, some unique derived characters are found in the 16S ribosomal RNA gene sequence.

Ecology

The cyanobacteria practice oxygenic photosynthesis. The electron donor is water and oxygen is produced as a by-product. The cyanobacteria are very likely the original source of Earth's atmospheric oxygen. They are very tolerant to a range of environmental conditions and are found in almost all waters and soils. Certain species are thermophilic, others can grow in the fissures of desert rocks. The dumping of organic waste into water sources causes the development of large quantities of cyanobacteria that adopt a type of chemoheterotrophic

metabolism. By consuming dissolved oxygen, they kill off fish and other aquatic life forms. Finally, the cyanobacteria are implicated in many symbiotic relationships: lichens, protozoa, etc. The nitrogen fixing species are associated with numerous plant types (liverworts, mosses, conifers, angiosperms).

Examples

Anabaena variabilis; Arthrospira platenis; Mastigocladus laminosus; Oscillatoria agardhii, Prochloron didemni; Spirulina labyrinthiformis; Synechocystis spp.

Number of species: 295, classified according to 16S RNA.
Oldest known fossils: the most ancient stromatolites are 3.4 billion years old. Stromatolites are sedimentary rocks made up of layered mats produced by the metabolic activity of filamentous photosynthetic cyanobacteria.
Current distribution: worldwide.

Thermus / Deinococcus group

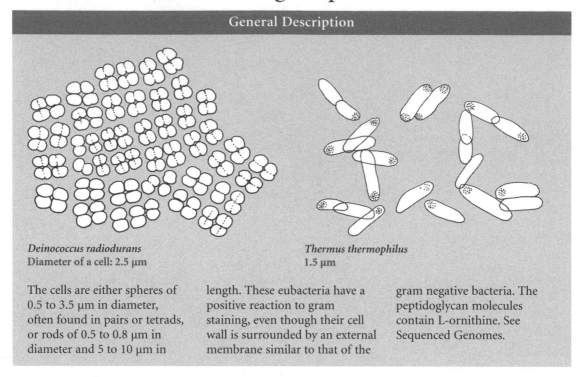

General Description

Deinococcus radiodurans
Diameter of a cell: 2.5 μm

Thermus thermophilus
1.5 μm

The cells are either spheres of 0.5 to 3.5 μm in diameter, often found in pairs or tetrads, or rods of 0.5 to 0.8 μm in diameter and 5 to 10 μm in length. These eubacteria have a positive reaction to gram staining, even though their cell wall is surrounded by an external membrane similar to that of the gram negative bacteria. The peptidoglycan molecules contain L-ornithine. See Sequenced Genomes.

Some unique derived features

- The peptidoglycan is very rich in ornithine.
- 16S ribosomal RNA gene: as for the other eubacteria, some unique derived characters are found in the 16S ribosomal RNA gene sequence.

Ecology

These bacteria are aerobic mesophiles. Species of *Deinococcus* can be isolated from the air, freshwater, or feces, without ever truly knowing its natural habitat. They are extremely resistant to desiccation and to radiation; they can survive an exposure of 3 to 5 million rad, whereas an exposure of 100 rad is lethal for humans. This tolerance comes from their ability to repair their chromosomes. The chromosomes are found in several copies and have a very efficient repair system (involving the protein RecA). *Thermus* is found in hot springs, as well as in freshwater that has been thermally polluted. *Thermus* forms colonies that range in size from 10 to 20 μm in diameter.

Number of species: 52, classified according to 16S RNA.
Oldest known fossils: none.
Current distribution: worldwide.

Examples

Deinococcus radiodurans; Meiothermus ruber; Thermus aquaticus; T. thermophilus.

Green sulfur bacteria

General Description

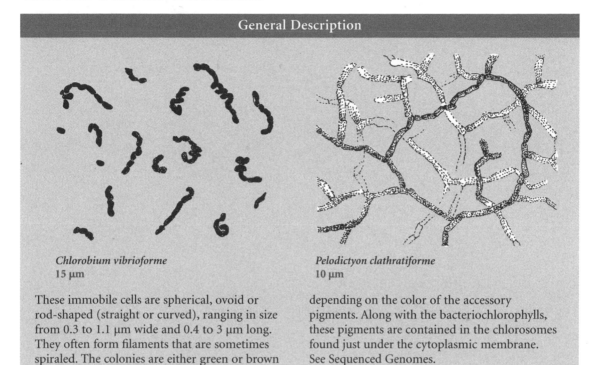

Chlorobium vibrioforme
15 μm

Pelodictyon clathratiforme
10 μm

These immobile cells are spherical, ovoid or rod-shaped (straight or curved), ranging in size from 0.3 to 1.1 μm wide and 0.4 to 3 μm long. They often form filaments that are sometimes spiraled. The colonies are either green or brown depending on the color of the accessory pigments. Along with the bacteriochlorophylls, these pigments are contained in the chlorosomes found just under the cytoplasmic membrane. See Sequenced Genomes.

Some unique derived features

- The only unique derived characters that we know of are those that exist in the 16S ribosomal RNA gene sequence.

Ecology

The green sulfur bacteria are strictly anaerobic phototrophes. Sulfides or sulfur acts as an electron donor. During sulfide oxidation, sulfur globules form on the external surface of the cell, hence the name of the group. This sulfur can then be oxidized into sulfate. The green sulfur bacteria are able to grow at very low light intensities. This is why we often find them in nature under layers of other photosynthetic organisms, in muddy waters or in marine sediment.

Number of species: 28, classified according to 16S RNA.
Oldest known fossils: none.
Current distribution: worldwide.

Examples

Chlorobium vibrioforme; Pelodictyon clathratiforme; Prosthecochloris aestuarii.

Flavobacteria (cytophagales)

General Description

The flavobacteria are rods with rounded extremities of 0.5 μm wide and from 1 to 3 μm long. The cells are immobile. Cultures on solid milieu are often yellow or orange in color, hence the name of the genus (*flavus* = yellow in Latin). See Sequenced Genomes.

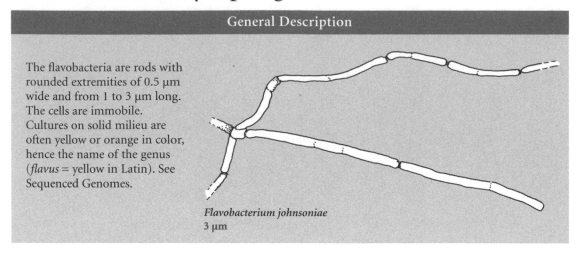

Flavobacterium johnsoniae
3 μm

Some unique derived features

- The only unique derived characters that we know of are those that exist in the 16S ribosomal RNA gene sequence.

Ecology

The flavobacteria are strictly aerobic and are widely distributed in soil and water. However, they can also be found in uncooked foods, and in milk and milk products. Because of the large number of strains that are resistant to antibiotics, we can also find this group in hospital environments. Certain species are pathogenic.

Number of species: 152, classified according to 16S RNA.
Oldest known fossils: none.
Current distribution: worldwide.

Examples

Flavobacterium aquatile; F. johnsoniae; F. meningosepticum (causal agent of meningitis in newborns).

Bacteroides

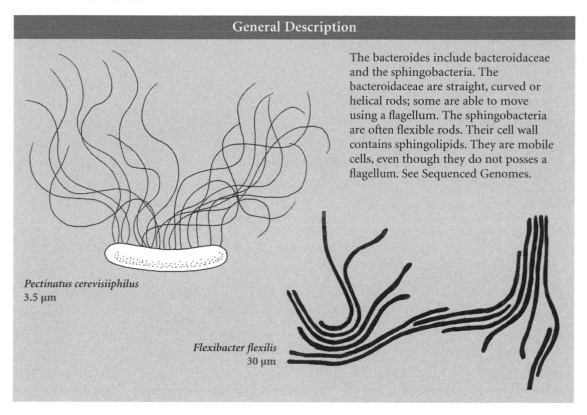

General Description

The bacteroides include bacteroidaceae and the sphingobacteria. The bacteroidaceae are straight, curved or helical rods; some are able to move using a flagellum. The sphingobacteria are often flexible rods. Their cell wall contains sphingolipids. They are mobile cells, even though they do not posses a flagellum. See Sequenced Genomes.

Pectinatus cerevisiiphilus
3.5 μm

Flexibacter flexilis
30 μm

Some unique derived features

■ The only unique derived characters that we know of are those that exist in the 16S ribosomal RNA gene sequence.

Ecology

The bacteroidaceae are strictly anaerobic and never sporulate. They develop in the buccal cavity and digestive tract of numerous animals, including humans and ruminants. *Bacteroides ruminicola* is the most abundant species of the rumen flora; 30% of the human intestinal flora is composed of bacteroidaceae. Some species are pathogenic. The sphingobacteria are aerobic and many of these species degrade complex polysaccharides like cellulose. They play an important role in the decomposition of organic matter. We find them in soils, but also in marine sediments. Certain species are pathogenic.

Number of species: 181, classified according to 16S RNA.
Oldest known fossils: none.
Current distribution: worldwide.

Examples

Bacteroides ruminicola; Flexibacter elegans; F. flexilis; Pectinatus cerevisiiphilus; Sporocytophaga myxococcoides.

Spirochaetes

General Description

Spirochaete cells are characterized by their shape and type of locomotion. They are long, helical-shaped bacteria (0.1 to 3 µm in diameter and 5 to 250 µm in length). Many species are so thin that they cannot be distinguished using phase-contrast microscopy. Because of their particular cell structure, spirochaetes are motile, even in very viscous environments. The central cylinder, containing the protoplasm, has a typical membrane and cell wall. A hundred or so flagella are fixed at each of the cell's extremities. These flagella run parallel to the cell and overlap near the center. They are covered in a flexible tunic. It is the movement of the flagella inside this tunic that allows the cell to move. However, the detailed mechanism is not yet completely understood. See Sequenced Genomes.

Leptospira interrogans
7 µm

Cristispira pectinis
100 µm

Some unique derived features

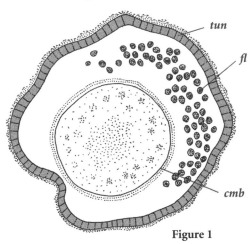

Figure 1

- The flagella are covered by a flexible external tunic that allows the cell to move (fig. 1: cross-section of *Clevelandina*, symbiont of the termite *Reticulitermes flavipes*; *fl*: flagella; *cmb*: cytoplasmic membrane; *tun*: tunic).

- 16S ribosomal RNA gene: as for the other eubacteria, some unique derived characters are found in the 16S ribosomal RNA gene sequence.

Number of species: 341, classified according to 16S RNA.
Oldest known fossils: none.
Current distribution: worldwide.

Ecology

Spirochaetes are often anaerobic or weakly aerobic. Common organic molecules (sugar, amino acids, fatty acids, etc.) can serve as sources of carbon and energy. The ecological niches used by these bacteria are very diverse: free-living in mud and aquatic or marine sediments; symbiotic in the digestive tract of insects (termites and cockroaches, in particular), molluscs, and mammals; pathogenic in the genital tracts and blood of vertebrates, including humans. The external tunic is extremely important for pathogenic spirochaetes; because there are few outer surface proteins, the tunic protects the cell from attack by the host's antibodies. *Borrelia* are transmitted to humans by ticks and fleas.

Examples

Borrelia burgdorferi (causal agent of Lyme disease); *Cristispira pectinis*; *Leptospira interrogans*; *Spirochaeta litoralis*; *Treponema pallidum* (causal agent of syphilis).

Chlamydiales

General Description

The infectious particles are immobile shells called elementary bodies that range in size from 0.2 to 1.5 μm in diameter. Their cell walls have neither muramic acid nor peptidoglycan. Their genome is among the smallest of the prokaryotes, ranging from 400 to 600 million daltons. See Sequenced Genomes.

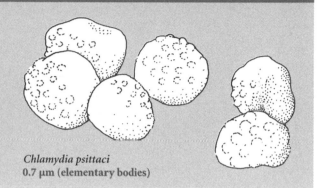

Chlamydia psittaci
0.7 μm (elementary bodies)

Some unique derived features

- Infection occurs by elementary bodies.
- There is a loss of cytochromes.
- There is a loss of both muramic acid and pepitoglycans.

- 16S ribosomal RNA gene: as for the other eubacteria, some unique derived characters are found in the 16S ribosomal RNA gene sequence.

Ecology

The chlamydiales are obligate intracellular parasites of mammals and birds. An elementary body (0.2 to 0.4 μm in diameter) attaches to the membrane of the host cell. It is phagocytized by the cell. The lysosomes of the host cell are inhibited and the *Chlamydia* transforms into a reticulate body that multiplies in the vacuole. The reticulate bodies give rise to the elementary bodies that are then liberated during cell lysis, 72 hours after infection. The chlamydiales are the causal agents of serious diseases (psittacosis, trachoma, pneumonitis).

Number of species: 107, classified according to 16S RNA.
Oldest known fossils: none.
Current distribution: worldwide.

Examples

Chlamydia psittaci (causal agent of psittacosis); *C. trachomatis* (causal agent of trachoma); *C. pneumoniae* (causal agent of pneumonitis).

Planctomycetales

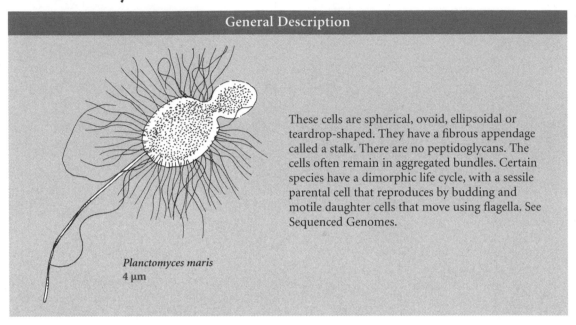

General Description

These cells are spherical, ovoid, ellipsoidal or teardrop-shaped. They have a fibrous appendage called a stalk. There are no peptidoglycans. The cells often remain in aggregated bundles. Certain species have a dimorphic life cycle, with a sessile parental cell that reproduces by budding and motile daughter cells that move using flagella. See Sequenced Genomes.

Planctomyces maris
4 μm

Some unique derived features

- 16S ribosomal RNA gene: as for the other eubacteria, some unique derived characters are found in the 16S ribosomal RNA gene sequence.

- Certain species possess a membrane-bound nucleoid (a single for the *Pirellula*, and two for *Gemmata obscuriglobus*) that resembles the nucleus of the eukaryote cell. Fig. 1 (*Nd*: nucleoid) shows the double membrane system of *Gemmata obscuriglobus*, giving the appearance of a nucleus. This characteristic might have appeared at the origin of the group.

Number of species: 82, classified according to 16S RNA.
Oldest known fossils: none.
Current distribution: worldwide.

Examples

Gemmata obscuriglobus; Pirellula spp.; *Planctomyces maris.*

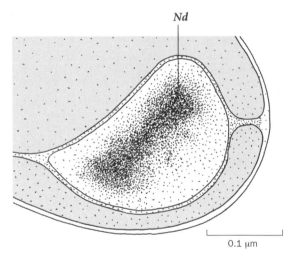

Nd

0.1 μm

Figure 1

Ecology

Planctomyces are found in freshwater bodies high in organic compounds, and in marine waters and estuaries.

Green non-sulfur bacteria

General Description

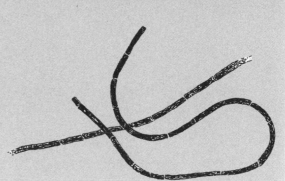

Chloroflexus aurantiacus
10 μm

Heliothrix oregonensis
15 μm

The cells of the green non-sulfur bacteria are arranged in single, multicellular filaments of undefined length (up to several tens of μm) that can vary in diameter from 0.5 to 5.5 μm and are motile. With the exception of *Heliothrix* and

Herpetosiphon, all genera possess chlorosomes, structural elements attached to the interior surface of the cytoplasmic membrane containing the photosynthetic pigments, bacteriochlorophylls. See Sequenced Genomes.

Some unique derived features

- The cells form long, motile multicellular filaments of single-cell width.

- 16S ribosomal RNA gene: as for the other eubacteria, some unique derived characters are found in the 16S ribosomal RNA gene sequence.

Ecology

These bacteria are largely anaerobic, but some are facultatively aerobic. Almost all species practice anoxygenic photosynthesis. They use organic molecules (and not water) as electron sources. As a consequence, they do not release oxygen. The maximum absorption of the bacteriochlorophylls occurs in the infrared range. Certain species are thermophilic. We find *Chloroflexus* and *Heliothrix* living in association with cyanobacteria in natural hot springs of neutral or alkaline pH.

Number of species: 16, classified according to 16S RNA.
Oldest known fossils: none.
Current distribution: worldwide.

Examples

Chloroflexus aurantiacus; Heliothrix oregonensis; Herpetosiphon aurantiacus.

Aquificales

General Description

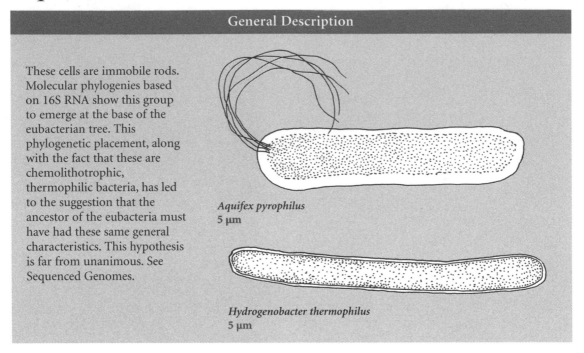

These cells are immobile rods. Molecular phylogenies based on 16S RNA show this group to emerge at the base of the eubacterian tree. This phylogenetic placement, along with the fact that these are chemolithotrophic, thermophilic bacteria, has led to the suggestion that the ancestor of the eubacteria must have had these same general characteristics. This hypothesis is far from unanimous. See Sequenced Genomes.

Aquifex pyrophilus
5 μm

Hydrogenobacter thermophilus
5 μm

Some unique derived features

- The only unique derived characters that we know of are those that exist in the 16S ribosomal RNA gene sequence.

Ecology

The aquificales are hyper-thermophilic bacteria with optimal temperatures of around 80°C found in volcanic hot springs. *Aquifex* uses molecular hydrogen, thiosulfate and sulfur as electron donors and oxygen as an electron acceptor. Carbon dioxide is the carbon source.

Examples

Aquifex aeolicus; A. pyrophilus; Calderobacterium hydrogenophilum; Hydrogenobacter thermophilus.

Number of species: 9, classified according to 16S RNA.
Oldest known fossils: none.
Current distribution: world-wide.

Thermotogales

General Description

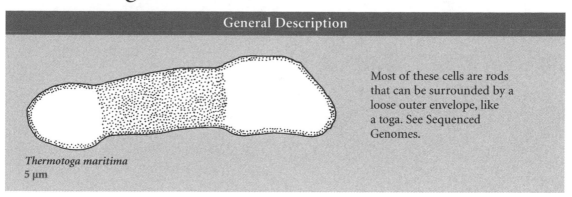

Most of these cells are rods that can be surrounded by a loose outer envelope, like a toga. See Sequenced Genomes.

Thermotoga maritima
5 μm

Some unique derived features

■ An envelope surrounds the cell (fig. 1: cross-section of *Thermotoga maritima*; *cmb*: cytoplasmic membrane, *env*: envelope).

■ 16S ribosomal RNA gene: as for the other eubacteria, some unique derived characters are found in the 16S ribosomal RNA gene sequence.

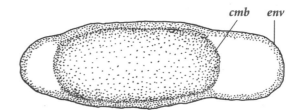

cmb   *env*

Figure 1

Ecology

The thermotogales are hyperthermophilic bacteria with an optimal temperature of around 80°C. They can be found in marine hydrothermal springs and terrestrial sulfur springs. They are chemoheterotrophes that can grow anaerobically using sugars or proteins as energy sources.

Number of species: 25, classified according to 16S RNA.
Oldest known fossils: none.
Current distribution: world-wide.

Examples

Geotoga petraea; Fervidobacterium gondwanense; Thermotoga maritima.

Chapter 3

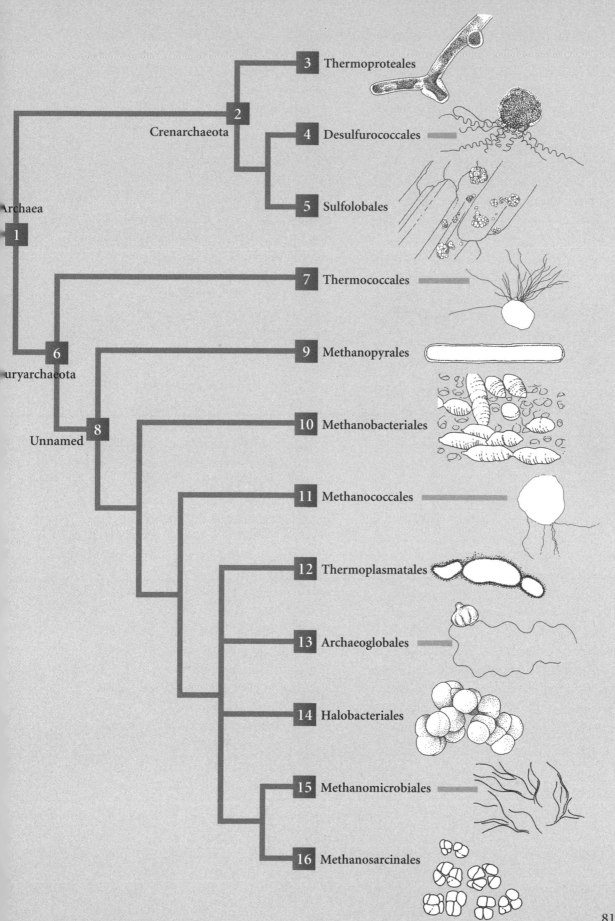

Archaea

1

Crenarchaeota

2

3 Thermoproteales

4 Desulfurococcales

5 Sulfolobales

7 Thermococcales

Euryarchaeota

6

9 Methanopyrales

Unnamed

8

10 Methanobacteriales

11 Methanococcales

12 Thermoplasmatales

13 Archaeoglobales

14 Halobacteriales

15 Methanomicrobiales

16 Methanosarcinales

Archaea

The existence of the archaea as a group was uncovered using molecular phylogenies based on the 16S ribosomal RNA gene. In contrast to the eubacteria, there was no previous rigid classification to call into question. Certain metabolic properties had simply been used to distinguish among the halophilic archaea living in high-salt environments, the thermophilic archaea that support high temperatures, and the methanogenic archaea that produce methane as a byproduct of metabolism. The archaea were therefore quickly classified on the basis of the 16S rRNA gene, the characters linked with metabolism often being revealed as homoplastic.

The archaea have been relatively well studied because of their importance from perspectives both fundamental (how do such organisms resist high temperatures?) and applied (methanogenic archaea are abundant in the digestive tract of humans and economically important animals like ruminants). Moreover, with the development of molecular probes and their use in particular environments (deep-sea hydrothermal vents, geysers, marshes, etc.), it has been shown that the biodiversity of this group is much greater than previously thought.

The new edition of *Bergey's Manual of Systematic Bacteriology,* the microbiologist's "bible," is based entirely on molecular phylogenies, and particularly that of the 16S ribosomal RNA gene (see the website: *www.cme.msu.edu/Bergeys/*). Now, all new bacteria are sequenced at the 16S RNA gene in order to be classified. This is why we have given only the number of species that have been sequenced at this gene under the section "Number of species." In the most complete databases (for example: *The European* *ribosomal RNA database: http://www.psb.ugent.be/rRNA/index.html; NCBI: The National Center for Biotechnology Information: http://www.ncbi.nlm.nih.gov/*), sequences are also provided that have been obtained from environmental samples, that is, sequences from bacteria that have not yet been cultured. Of the 409 currently described archaea, 181 come from such samples (as of January 2006). This leaves one with the impression that this number is going to increase exponentially over the next few years and that the overall percentage of described archaea is currently very low (a few percent).

There are now numerous sequenced genomes from this group. An exhaustive list has been composed for each phylum (inventory as of January 2006). This list is available on the JGI website (*Joint Genome Institute, http://img.jgi.doe.gov/cgi-bin/pub/main.cgi*).

Recently, two new phyla of archaea have been proposed. The nanoarchaeota seem to correspond to a basal branch of the tree and are represented for the moment by *Nanoarchaeum equitans,* a parasite of the genus *Ignicoccus* (Desulfurococcales). This small cell (400 nm in diameter) possesses the smallest genome currently known, only 490,885 base pairs. This species has lost many genes linked to metabolism, such as that for lipids, and is therefore an obligate parasite. Its emergence at the base of the archaea tree seems to be an artefact due to long-branch attraction. More precise phylogenetic studies place *Nanoarchaeum equitans* within the thermococcales. The korarchaeota are exclusively described on the basis of DNA from environmental samples. However, it is still too soon to accept definitively a phylum that does not yet possess any organisms.

Crenarchaeota

A few representatives

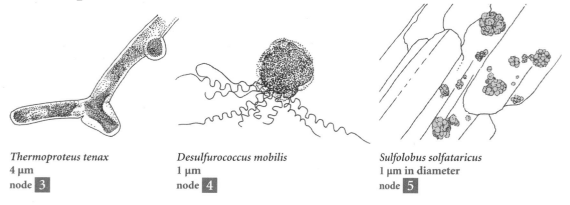

Thermoproteus tenax
4 μm
node 3

Desulfurococcus mobilis
1 μm
node 4

Sulfolobus solfataricus
1 μm in diameter
node 5

Some unique derived features

Some microbiologists think that the crenarchaeota (from the Greek word *crene:* source) have retained many characteristics of the common ancestor of the archaea. They are almost all extreme thermophilic bacteria with strictly anaerobic metabolism. Many are acidophilic. For most, their metabolism depends on sulfur, either as an electron acceptor for anaerobic respiration or as an electron donor for lithotrophic bacteria. We find these bacteria in geothermally heated waters or in sulfur-rich soils. Let us cite a few of their most famous habitats, like the hot springs of Yellowstone National Park (Wyoming, USA), the solfataras of volcanoes such as Vulcano or Vulcanello (Lipari Islands, Italy), or hydrothermal vents on the ocean floor.

■ The crenarchaeota, and the clades within it, are the result of molecular phylogenetic analyses. The most frequently used is the 16S ribosomal RNA gene, along with proteins linked to the transcription and translation machinery.

Examples

THERMOPROTEALES: *Thermoproteus tenax.*
DESULFUROCOCCALES: *Desulfurococcus mobilis; Pyrodictium abyssi.*
SULFOLOBALES: *Sulfolobus solfataricus.*

Number of species: 55, classified according to 16S RNA.
Oldest known fossils: none.
Current distribution: worldwide.

Thermoproteales

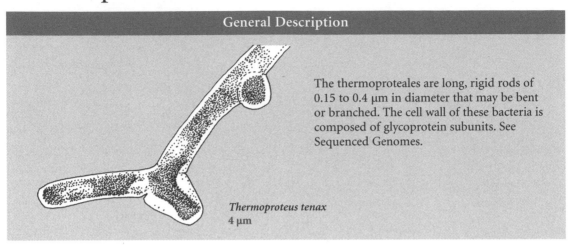

General Description

The thermoproteales are long, rigid rods of 0.15 to 0.4 μm in diameter that may be bent or branched. The cell wall of these bacteria is composed of glycoprotein subunits. See Sequenced Genomes.

Thermoproteus tenax
4 μm

Some unique derived features

■ The unique derived characters are molecular; they are based on sequences of the 16S ribosomal RNA gene, along with genes that code for proteins of the cell's transcription and translation machinery.

Ecology

We find *Thermoproteus* in sulfur-rich hot springs. It is a strictly anaerobic bacterium that grows at temperatures ranging from 70 to 97°C and at pHs of between 2.5 and 6.5. It has a type of organotrophic metabolism; it can carry out anaerobic respiration by oxidizing glucose, amino acids or other organic molecules using sulfur as an electron acceptor. The reduced molecule produced by this reaction is hydrogen sulfide (H_2S). *Thermoproteus* can also undergo chemolithotrophic metabolism, using molecular hydrogen (H_2) and sulfur as electron donors. Carbon monoxide and dioxide are possible carbon sources. Glycogen serves as a storage molecule.

Examples

Hyperthermus butylicus; Thermofilum pendens; Thermoproteus tenax.

Number of species: 16, classified according to 16S RNA.
Oldest known fossils: none.
Current distribution: worldwide.

Desulfurococcales

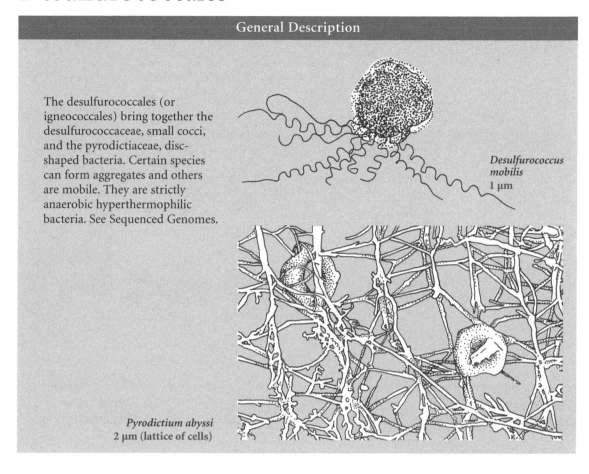

General Description

The desulfurococcales (or igneococcales) bring together the desulfurococcaceae, small cocci, and the pyrodictiaceae, disc-shaped bacteria. Certain species can form aggregates and others are mobile. They are strictly anaerobic hyperthermophilic bacteria. See Sequenced Genomes.

Desulfurococcus mobilis
1 μm

Pyrodictium abyssi
2 μm (lattice of cells)

Some unique derived features

■ The unique derived characters are molecular; they are based primarily on sequences of the 16S ribosomal RNA gene, along with genes that code for proteins of the cell's transcription and translation machinery.

Ecology

The desulfurococcaceae use proteins and sugars for fermentation or anaerobic respiration. *Pyrodictium abyssi*, a pyrodictiaceae isolated from geothermally heated areas of the ocean floor, holds the record for hyperthermophilic bacteria; its growing temperature ranges from 82 to 110°C, with an optimum at 105°C.

Examples

Aeropyrum pernix; Desulfurococcus mobilis; Igneococcus islandicus; Pyrodictium abyssi; Staphylothermus marinus.

Number of species: 14, classified according to 16S RNA.
Oldest known fossils: none.
Current distribution: worldwide.

Sulfolobales

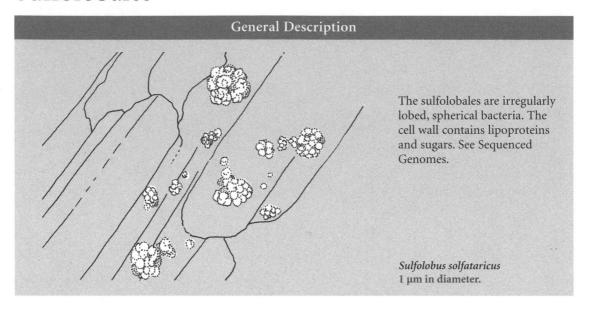

General Description

The sulfolobales are irregularly lobed, spherical bacteria. The cell wall contains lipoproteins and sugars. See Sequenced Genomes.

Sulfolobus solfataricus
1 μm in diameter.

Some unique derived features

■ The unique derived characters are molecular; they are based primarily on sequences of the 16S ribosomal RNA gene, along with genes that code for proteins of the cell's transcription and translation machinery.

Ecology

The sulfolobales are extremely thermoacidophilic with optimal growth at a pH of 2. They are strictly or facultatively aerobic. They are chemolithotrophic, using sulfur as an electron source. These bacteria grow on sulfide crystals in acidic hot springs. The genus *Sulfolobus* is the most well-known group. It has been isolated from mud pots (solfataras) in Yellowstone park (Wyoming, USA) and in Vesuvius park (Italy). Its optimal growth temperature is between 70 and 80°C. These archaea oxidize hydrogen sulfide (H_2S) into sulfur, and sulfur into sulfuric acid. Oxygen is the normal electron acceptor, but ferric ion can also be used. Sugars and amino acids can also serve as sources of carbon and energy.

Examples

Acidianus infernus; Metallosphaera sedula; M. prunae; Sulfolobus solfataricus.

Number of species: 25, classed according to 16S RNA.
Oldest known fossils: none.
Current distribution: worldwide.

Euryarchaeota

A few representatives

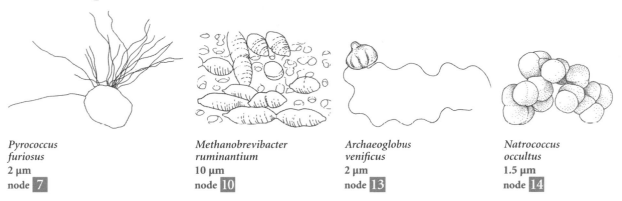

Pyrococcus
furiosus
2 μm
node 7

Methanobrevibacter
ruminantium
10 μm
node 10

Archaeoglobus
venificus
2 μm
node 13

Natrococcus
occultus
1.5 μm
node 14

Some unique derived features

■ The monophyly of the euryarchaeota and the proposed tree are the result of molecular phylogenies based principally on the 16S ribosomal RNA gene, along with genes that code for proteins of the cell's transcription and translation machinery. The euryarchaeota (from the Greek word *eury:* large) represent a wide range of different metabolisms and are adapted to diverse environments. We generally make the distinction among the methanogenic, the extreme halophilic and the thermophilic bacteria, but it is possible to find species that hold several of these adaptations. The prefix of the group names (methano, thermo, halo) indicates the principal

type of specialization. A quick look at the tree shows that these specializations are not synapomorphies. In four of the groups that emerge at the base of the tree (nodes 7 to 11), we find thermophilic species; thus, certain methanococcales can be both methanogenic and thermophilic. The last three groups (nodes 14 to 16) tend to be mesophilic. In the euryarchaeota, we therefore find evolutionary tendencies to go from thermophilic to mesophilic and from non-halophilic to halophilic without any real synapomorphies because the groups are not homogeneous in their adaptations.

Number of species: 354, classified according to 16S RNA.
Oldest known fossils: none.
Current distribution: worldwide.

Examples

THERMOCOCCALES: *Pyrococcus furiosus*
CLADE 8: *Methanopyrus kandleri; Methanobrevibacter ruminantium; Methanococcus jannaschii; Thermoplasma acidophilum; Archaeoglobus venificus; Natrococcus occultus; Methanomicrobium mobile; Methanosarcina barkeri.*

Thermococcales

General Description

The thermococcales are spherical or elongated bacteria of approximately 1 μm in diameter that are well-equipped with flagella. We often observe them in the diplococcus form in culture, that is, as dividing cells. See Sequenced Genomes.

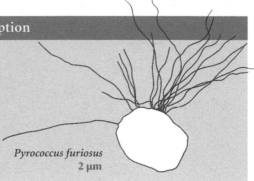

Pyrococcus furiosus
2 μm

Some unique derived features

■ The unique derived characters are molecular; they are based primarily on sequences of the 16S ribosomal RNA gene, along with genes that code for proteins of the cell's transcription and translation machinery.

Number of species: 39, classified according to 16S RNA.
Oldest known fossils: none.
Current distribution: most likely worldwide, across the ocean floor.

Ecology

The thermococcales are strictly anaerobic bacteria that we find in marine sulfur vents of neutral pH and with temperatures of between 80 and 103°C. They can use peptides or sugars as carbon sources. They have anaerobic respiration and produce hydrogen sulfide (H_2S) from sulfur.

Examples

Pyrococcus abyssi; P. furiosus; P. horikoshii; Thermococcus celer.

Unnamed clade

A few representatives

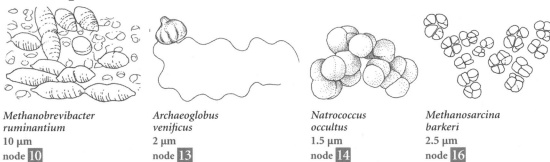

*Methanobrevibacter
ruminantium*
10 μm
node 10

*Archaeoglobus
venificus*
2 μm
node 13

*Natrococcus
occultus*
1.5 μm
node 14

*Methanosarcina
barkeri*
2.5 μm
node 16

Some unique derived features

■ The unique derived characters are molecular; they are based primarily on sequences of the 16S ribosomal RNA gene, along with genes that code for proteins of the cell's transcription and translation machinery.

■ Sub-unit B of the RNA polymerase is coded by two genes, called *rpoB'* and *rpoB''*. In the crenarchaeota and the thermococcales, this sub-unit is coded by a single gene, *rpoB*, and corresponds to the ancestral state of the character.

> **Number of species:** 315, classified according to 16S RNA.
> **Oldest known fossils:** none.
> **Current distribution:** worldwide.

Examples

Methanopyrus kandleri; Methanobrevibacter ruminantium; Methanococcus jannaschii; Thermoplasma acidophilum; Archaeoglobus venificus; Natrococcus occultus; Methanomicrobium mobile; Methanosarcina barkeri.

Methanopyrales

General Description

Methanopyrus kandleri
10 μm

The methanopyrales are represented by a single genus, *Methanopyrus*. They are rod-shaped bacteria that are methanogenic and extreme thermophiles. For a long time, this group was placed at the base of the euryarchaeota tree, suggesting that methanogenic archaea may have been among the first living organisms. However, this position was in fact due to the long-branch attraction phenomenon and recent phylogenies based on proteins of the cell's transcription and translation machinery have revealed the correct phylogenetic position of this group. See Sequenced Genomes.

Some unique derived features

■ The unique derived characters are molecular; they are based primarily on sequences of the 16S ribosomal RNA gene, along with genes that code for proteins of the cell's transcription and translation machinery.

Number of species: 1, classified according to 16S RNA.
Oldest known fossils: none.
Current distribution: most likely worldwide, across the ocean floor.

Ecology

Methanopyrus kandleri was isolated from a deep-sea hydrothermal vent. It grows at temperatures that vary from 84 to 110°C, with an optimum at 98°C.

Examples

Methanopyrus kandleri.

Methanobacteriales

General Description

The methanobacteriales are strictly anaerobic, methanogenic bacteria. Their shape is variable: rods or filaments for *Methanobacterium,* curved rods for *Methanothermus.* The dominant peptidoglycan polymer of their cell wall is pseudomurein. See Sequenced Genomes.

Methanobrevibacter ruminantium
10 µm

Some unique derived features

■ The unique derived characters are molecular; they are based primarily on sequences of the 16S ribosomal RNA gene, along with genes that code for proteins of the cell's transcription and translation machinery.

Number of species: 109, classed according to 16S RNA.
Oldest known fossils: none.
Current distribution: worldwide.

Ecology

The methanobacteriales are widely distributed in nature because we find them in many types of anaerobic habitats: places that accumulate the molecular hydrogen produced by geothermal processes, aquatic sediments, soils, manure, and animal intestines, especially those of ruminants. They grow by oxidizing molecular hydrogen. Carbon dioxide is the normal electron acceptor. These bacteria do not degrade sugars, proteins or any other organic compound other than formate or carbon monoxide.

Examples

Methanobacterium formicicum;
M. thermoautotrophicum;
Methanobrevibacter ruminantium;
Methanothermus fervidus.

Methanococcales

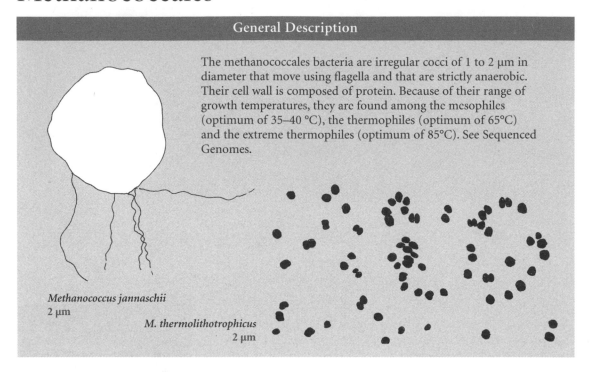

General Description

The methanococcales bacteria are irregular cocci of 1 to 2 μm in diameter that move using flagella and that are strictly anaerobic. Their cell wall is composed of protein. Because of their range of growth temperatures, they are found among the mesophiles (optimum of 35–40 °C), the thermophiles (optimum of 65°C) and the extreme thermophiles (optimum of 85°C). See Sequenced Genomes.

Methanococcus jannaschii
2 μm

M. thermolithotrophicus
2 μm

Some unique derived features

■ The unique derived characters are molecular; they are based primarily on sequences of the 16S ribosomal RNA gene, along with genes that code for proteins of the cell's transcription and translation machinery.

Ecology

The methanococcales are strict methanogenic bacteria that are found in salt marshes, estuaries and marine sediments. Molecular hydrogen and formate serve as electron donors. With the rare exception, they all develop autotrophically on mineral substrates. The nitrogen source can be ammonia, nitrogen gas or alanine. Glycogen is used to store glucose.

Number of species: 20, classified according to 16S RNA.
Oldest known fossils: none.
Current distribution: worldwide, in sediments.

Examples

Methanococcus jannaschii; M. thermolithotrophicus; M. vannielii.

Thermoplasmatales

General Description

The thermoplasmatales are obligate thermoacidophilic bacteria that lack a cell wall. Their shape is variable: spheres of 0.3 to 2 μm or filaments, depending on environmental conditions. The DNA of *Thermoplasma* is stabilized by proteins that are analogous to histones and that enable the formation of structures that resemble eukaryotic nucleosomes. See Sequenced Genomes.

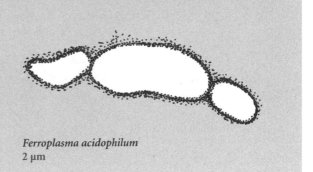

Ferroplasma acidophilum
2 μm

Some unique derived features

■ The unique derived characters are molecular; they are based primarily on sequences of the 16S ribosomal RNA gene, along with genes that code for proteins of the cell's transcription and translation machinery.

■ Sub-unit B of the RNA polymerase is coded by a single gene, *rpoB,* and corresponds to a reversion, or return to the ancestral state.

Ecology

Thermoplasma acidophilum develops in the waste piles of coal mines. These wastes are rich in pyrite (FeS) that is oxidized into sulfuric acid by bacteria. The piles therefore become warm and acidic, making an ideal environment for *Th. acidophilum* whose optimal conditions range in temperature from 55 to 59°C and in pH from 1 to 2. *Picrophilus* has been found in mud pots in Japan. Its optimum temperature is 60°C. However, for the pH, *Picrophilus* holds the record; its optimum pH is 0.7, but it can grow at a pH as low as 0!

Number of species: 5, classified according to 16S RNA.
Oldest known fossils: none.
Current distribution: periodically found in coal mine wastes in India and Pennsylvania, or in certain mud pots.

Examples

Ferroplasma acidophilum; Picrophilus oshimae; Thermoplasma acidophilum.

Archaeoglobales

General Description

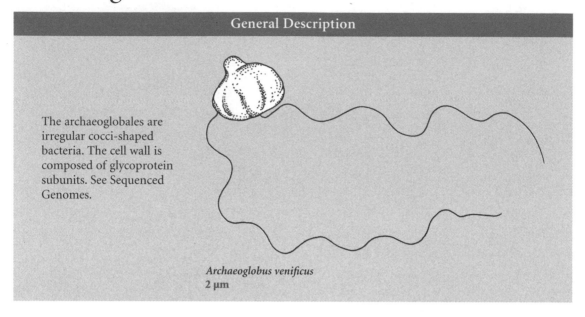

The archaeoglobales are irregular cocci-shaped bacteria. The cell wall is composed of glycoprotein subunits. See Sequenced Genomes.

Archaeoglobus venificus
2 μm

Some unique derived features

■ The unique derived characters are molecular; they are based primarily on sequences of the 16S ribosomal RNA gene, along with genes that code for proteins of the cell's transcription and translation machinery.

Ecology

Archaeoglobus is an extreme thermophile (optimum temperature of 83°C) and has been isolated from deep-sea hydrothermal vents. It can use a variety of electron donors, such as molecular hydrogen, glucose, or lactate. It can reduce sulfate, sulfite and thiosulfate ions.

Examples

Archaeoglobus fulgidus;
A. venificus.

Number of species: 6, classified according to 16S RNA.
Oldest known fossils: none.
Current distribution: likely worldwide, across the ocean floor.

Halobacteriales

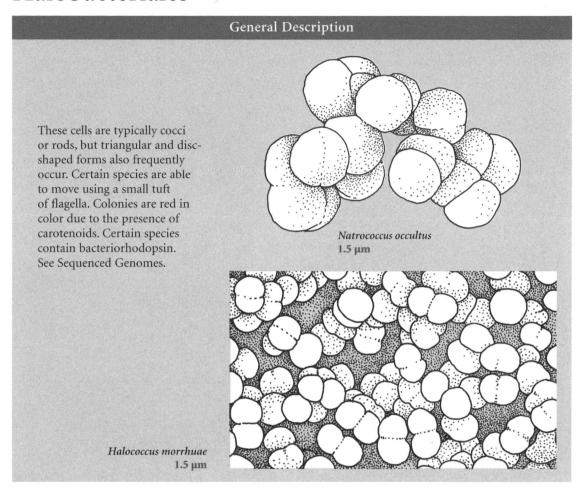

General Description

These cells are typically cocci or rods, but triangular and disc-shaped forms also frequently occur. Certain species are able to move using a small tuft of flagella. Colonies are red in color due to the presence of carotenoids. Certain species contain bacteriorhodopsin. See Sequenced Genomes.

Natrococcus occultus
1.5 μm

Halococcus morrhuae
1.5 μm

Some unique derived features

■ The unique derived characters are molecular; they are based primarily on sequences of the 16S ribosomal RNA gene, along with genes that code for proteins of the cell's transcription and translation machinery.

Ecology

The halobacteriales are found anywhere the salt concentration is usually high, that is, in salty lakes, like the Dead Sea, or salt marshes. These bacteria can also develop in heavily salted protein preserves, particularly fishes; *Halococcus morrhuae* grows on salted cod (*Gadus morrhua*). They are aerobic or facultatively anaerobic and require at least 1.5M of sodium chloride to grow. To compensate for the high salt concentration of the environment, the cells accumulate up to 5M of potassium chloride.

Examples

Halococcus morrhuae; Haloarcula vallismortis; Halobacterium halobium; Haloferax volcanii; Natrobacterium magadii; Natrococcus occultus.

Number of species: 105, classified according to 16S RNA.
Oldest known fossils: none.
Current distribution: worldwide.

Methanomicrobiales

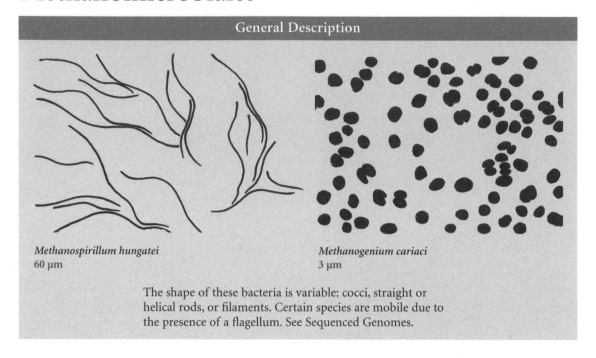

General Description

Methanospirillum hungatei
60 μm

Methanogenium cariaci
3 μm

The shape of these bacteria is variable: cocci, straight or helical rods, or filaments. Certain species are mobile due to the presence of a flagellum. See Sequenced Genomes.

Some unique derived features

■ The unique derived characters are molecular; they are based primarily on sequences of the 16S ribosomal RNA gene, along with genes that code for proteins of the cell's transcription and translation machinery.

Ecology

The methanomicrobiales are strictly anaerobic bacteria that are common in aquatic sediments and in animal intestines. They oxidize formate or molecular hydrogen to reduce carbon dioxide and release methane as a byproduct. They are mesophiles; their optimal temperature range is between 30 and 50 °C.

Examples

Methanogenium cariaci;
Methanomicrobium mobile;
Methanosaeta thermophila;
Methanospirillum hungatei.

Number of species: 33, classified according to 16S RNA.
Oldest known fossils: none.
Current distribution: worldwide.

Methanosarcinales

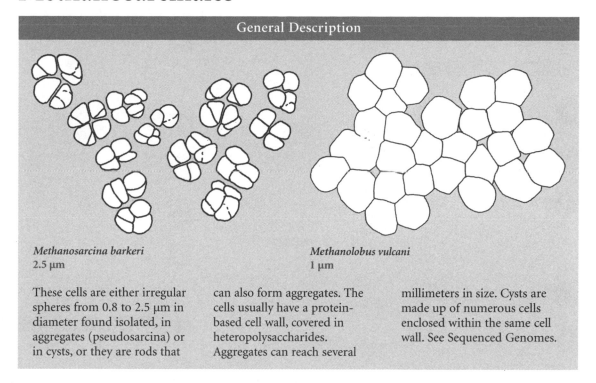

General Description

Methanosarcina barkeri
2.5 μm

Methanolobus vulcani
1 μm

These cells are either irregular spheres from 0.8 to 2.5 μm in diameter found isolated, in aggregates (pseudosarcina) or in cysts, or they are rods that can also form aggregates. The cells usually have a protein-based cell wall, covered in heteropolysaccharides. Aggregates can reach several millimeters in size. Cysts are made up of numerous cells enclosed within the same cell wall. See Sequenced Genomes.

Some unique derived features

- The unique derived characters are molecular; they are based primarily on sequences of the 16S ribosomal RNA gene, along with genes that code for proteins of the cell's transcription and translation machinery.

Ecology

The methanosarcinales are mesophiles or thermophiles that are strictly anaerobic. They live in aquatic sediments or in the digestive tract of animals, particularly in the rumen of the ruminants. They oxidize methanol, amines or molecular hydrogen to reduce carbon dioxide and release methane as a byproduct. Many species require high salt concentrations.

Examples

Methanococcoides methylutens;
Methanolobus vulcani;
Methanosarcina barkeri; M. mazei.

Number of species: 36, classified according to 16S RNA.
Oldest known fossils: none.
Current distribution: worldwide.

Chapter 4

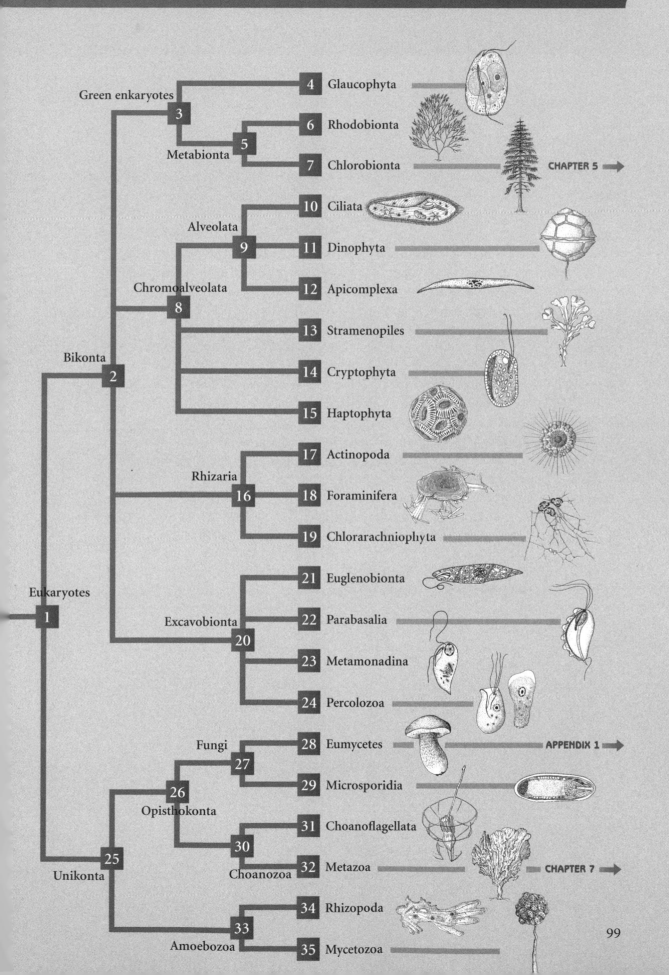

Green enkaryotes

3

4 Glaucophyta

Metabionta

5

6 Rhodobionta

7 Chlorobionta — CHAPTER 5 →

Alveolata

9

10 Ciliata

11 Dinophyta

12 Apicomplexa

Chromoalveolata

8

13 Stramenopiles

14 Cryptophyta

15 Haptophyta

Bikonta

2

Rhizaria

16

17 Actinopoda

18 Foraminifera

19 Chlorarachniophyta

Eukaryotes

1

21 Euglenobionta

22 Parabasalia

Excavobionta

20

23 Metamonadina

24 Percolozoa

Fungi

27

28 Eumycetes — APPENDIX 1 →

29 Microsporidia

26

Opisthokonta

31 Choanoflagellata

30

32 Metazoa — CHAPTER 7 →

Choanozoa

25

Unikonta

34 Rhizopoda

33

Amoebozoa

35 Mycetozoa

Most currently described living organisms are eukaryotes. They vary greatly in size and complexity, ranging from microscopic, unicellular organisms to large, highly organized, multicellular species, such as the sequoia or whale.

The eukaryote cell is unique: a chimera in the true sense of the word. Indeed, via endo-symbiosis, the eukaryote cell *sensu stricto* obtained very efficient organelles that control cellular energy. The mitochondrion, site of cellular respiration, is an α-proteobacterium that has been integrated into the cytoplasm. The chloroplast, site of photosynthesis, is a cyanobacterium. Thus, a photosynthetic eukaryote possesses at least three genomes of independent origin: a nuclear genome that is primarily eukaryotic *sensu stricto,* and two genomes of eubacterian origin, those of the mitochondria and chloroplasts.

Of course, things are more complicated in reality. After an endosymbiotic event, the symbionts' genes (for example, those of the future mitochondrion) have 3 possible fates. Some genes are lost, others remain in the organelle's genome, and still others are exported and integrated into the nuclear genome, which is also therefore a chimera. This genetic integration is of great practical utility as it enables one to verify experimentally if a eukaryote cell lacking mitochondria lost them secondarily or if the species is in fact a living fossil, a direct descendant of an ancestor that existed prior to the endosymbiotic event. Recent research seems to indicate that all eukaryote cells currently lacking mitochondria lost them secondarily; mitochondrial-like genes (such as *Hsp*) have been found in the nuclear genome in all cases. This result, in tandem with molecular phylogenies based on ribosomal RNA genes 16S/18S, supports the hypothesis that all current eukaryotes have a single common ancestor which possessed mitochondria and that only a single mitochondrial endosymbiotic event took place.

The evolutionary events are more complex for the chloroplasts, as one must distinguish between primary and secondary endo-symbioses. The primary endosymbiosis refers to the unique event when a cyanobacterium was integrated into the cell, following a process comparable to that of the mitochondrial endosymbiosis. However, secondary endo-symbioses frequently occurred where non-photosynthetic eukaryotes incorporated photosynthetic eukaryotes into their cells. In certain cases, we can find relics of the diverse membranes involved in this process. Sometimes, we even see nucleomorphs, the vestigial remains of the symbiont nucleus, as we find in the cryptophytes, for example. Indeed, the chloroplast of brown algae comes from a secondary endosymbiotic event involving a photosynthetic eukaryote of the rhodobionta lineage.

A recent book details no less than 13 occurrences of secondary endosymbiosis. The dinophyta are perhaps the most malleable organisms in this sense given that at least 3 independent endosymbiotic events with other eukaryotes have been recorded.

Bikonta

A few representatives

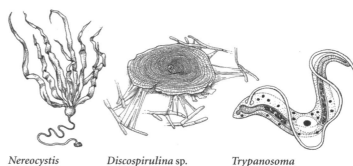

Nordmann fir
*Abies
nordmanniana*
50 m
node 7

Paramecium
*Paramecium
tetraurelia*
100 μm
node 10

*Nereocystis
luetkeana*
10 m
node 13

Discospirulina sp.
2 mm
node 18

*Trypanosoma
brucei*
30 μm
node 21

Some unique derived features

■ The unicellular organisms and free-living multicellular organisms of this clade all have two flagella (fig. 1a: *Bodo caudatus*; fig. 1b: *Glaucocystis*; *fl*: flagella). As the unikonts are the sister group, we cannot say whether this character is ancestral or derived.

■ The dihydrofolate reductase (DHFR) and thymidylate synthase (TS) genes are fused in such a way that they now code for a single protein, DHFR-TS, with two enzymatic activities.

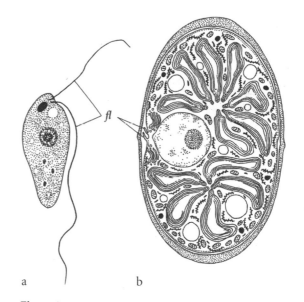

Figure 1

Number of species: 425,311.
Oldest known fossils: cysts similar to those of the genus *Pterosperma* (Prasinophyta) are known from the start of the Precambrian (1200 MYA).
Current distribution: worldwide.

Examples

■ GREEN EUKARYOTES: *Chlamydomonas reinhardtii*; sea lettuce: *Ulva lactuca*; Nordmann fir: *Abies nordmanniana*.

■ CHROMOALVEOLATA: *Plasmodium falciparum*; paramecium: *Paramecium tetraurelia*; bladderwrack: *Fucus vesiculosus*; *Emiliana huxleyi*.

■ RHIZARIA: *Challengeron wyvillei*; *Discospirulina* sp.; *Globigerina pachyderma*; *Chlorarachnion reptans*.

■ EXCAVOBIONTA: euglena: *Euglena gracilis*; *Trichomonas vaginalis*; *Giardia lamblia*; *Tetramitus rostratus*.

Green eukaryotes

A few representatives

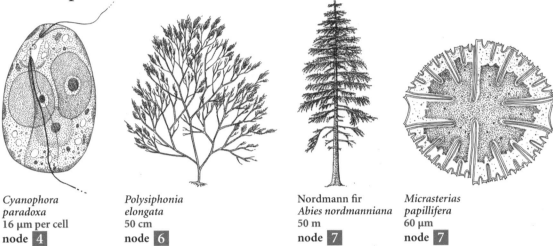

Cyanophora
paradoxa
16 µm per cell
node 4

Polysiphonia
elongata
50 cm
node 6

Nordmann fir
Abies nordmanniana
50 m
node 7

Micrasterias
papillifera
60 µm
node 7

Some unique derived features

Certain authors hypothesize that the green eukaryotes and the chromoalveolates are sister groups.

■ Chloroplast: the green eukaryotes bring together the glaucophyta, the rhodobionta, and the chlorobionta principally on the characteristics of the chloroplasts. They are direct descendants of the organism in which the first chloroplastic endosymbiotic event occurred. Secondary endosymbiosis then followed, with the integration of photosynthetic eukaryote symbionts into the host cells. The chloroplasts of the glaucophyta have such marked ancestral characteristics that it took some time before they were recognized as "true" chloroplasts. Previously called "cyanelles," they seem to correspond to an early stage of endosymbiosis. There are two possible interpretations for this observation: either the endosymbiosis is recent and still being finalized, or the endosymbiosis is ancient and the

glaucophyte chloroplasts are more weakly incorporated into the cell and have remained at an intermediate stage of integration. A detailed study has shown that the latter interpretation is the correct one.

■ Gene structure: recent molecular phylogenetic results using mitochondrial, chloroplastic and nuclear markers all support the monophyly of the group. However, the emergence order of the three taxa remains unclear, even though the monophyly of the rhodobionta/chlorobionta group is very likely. If this hypothesis is correct, the chloroplastic characteristics of the glaucophytes are ancestral.

■ The chloroplast is found within its own compartment and has a thin cell wall of peptidoglycans. Fig. 1 presents the commonly accepted interpretation of the ancestral incorporation and transformation of a

Figure 1

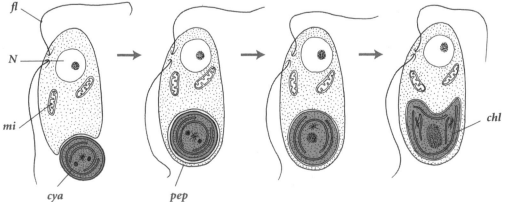

cyanobacterium into the chloroplast (*chl:* chloroplast, *cya:* cyanobacterium, *fl:* flagellum, *mi:* mitochondrion, *N:* nucleus, *pep:* peptidoglycan cell wall).

■ The thylakoids are not lumped together, but are evenly spaced apart in a stack. The chloroplast ultrastructure of *Ceramium* (fig. 2; *dmb:* double membrane, *thy,* thylakoid) illustrates the double membrane and isolated thylakoids.

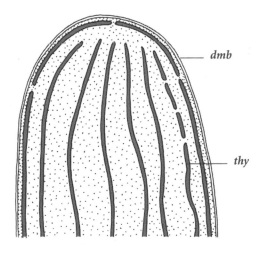

Figure 2

■ The accessory pigments are found in phycobilisomes attached to the thylakoids, as in the cyanobacteria.

■ Chloroplasts contain chlorophyll a.

■ Sugars are stored as starch grains outside the chloroplasts.

■ Chloroplastic DNA is circular and found at the center of the chloroplast, as in the cyanobacteria.

■ Chloroplastic DNA has two inverted repetitions that contain ribosomal RNA.

■ The two ribulose-1,5-biphosphate carboxylase (RuBisCO) sub-units are coded for by the chloroplastic DNA.

■ The flagellated cells are isokonts, that is, the flagella of a cell are structurally identical even if they are of unequal length.

■ The flagella possess two rows of filamentous expansions resembling those found in chlorobionts like *Chlamydomonas*. Fig. 3 shows the tip of a flagellum of *Chlamydomonas;* the glaucophytes possess identical lateral expansions.

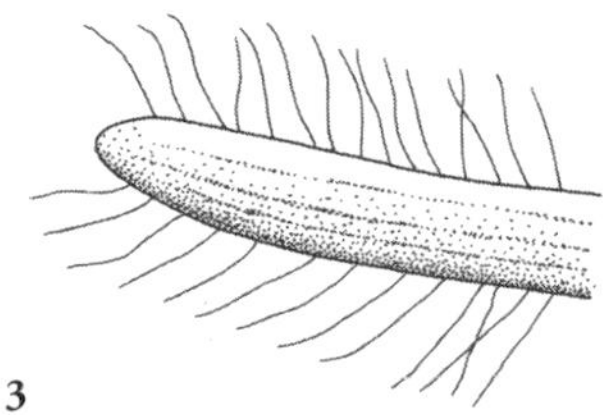

Figure 3

■ A star-like structure marks the transition between the flagellum and the basal body. Fig. 4 shows a longitudinal section and three cross-sections of the flagellum of *Chlamydomonas reinhardtii.* Cross-section A, at the basal body, shows the classic structure in 9 triplets of microtubules; the star-like structure occurs at the level of cross-section B. Cross-section C shows the structure of the flagellum with 9 peripheral pairs of microtubules and one central pair. The star-like structure of the glaucophyta is somewhat different from that shown and calls this synapomorphy into question.

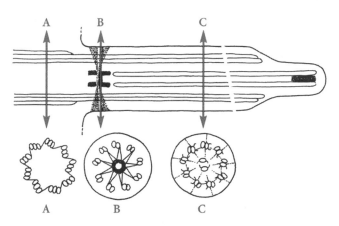

Figure 4

> **Number of species:** 283,415.
> **Oldest known fossils:** cysts similar to those of the genus *Pterosperma* (prasinophyta) are known from the start of the Precambrian (1200 MYA). There may be unicellular and filamentous fossil chlorophycean algae in the rocks of the Bitter Springs formation of central Australia that date from the Precambrian (900 MYA).
> **Current distribution:** worldwide.

Examples

■ GLAUCOPHYTA: *Cyanophora paradoxa.*
■ METABIONTA: *Bathycoccus prasinos; Chlamydomonas reinhardtii;* sea lettuce: *Ulva lactuca; Cladophora vagabunda; Micrasterias papillifera; Spirogyra fluviatillis; Chaetosphaeridium minus; Chara vulgaris; Coleochaete scutata;* bracken fern: *Pteridium aquilinum;* corn: *Zea mays;* European crab apple: *Malus sylvestris; Porphyra umbilicalis; Palmaria palmata; Corallina officinalis; Chondrus crispus; Lithothamnion roseum.*

Glaucophyta

General Description

The glaucophytes – or glaucocystophytes – were for a long time integrated into the rhodobionta. They have been recently separated because the characteristics of their chloroplasts (or cyanelles) are ancestral, not derived. They are unicellular, dorso-ventrally structured organisms with a rounded "back," a flattened "belly," and two flagella of unequal length. Though they represent a phylum of relatively low diversity, they are still interesting from an evolutionary point of view given the unique nature of their chloroplasts. For a long time, they were considered to be "intermediates" between the prokaryotes and the eukaryotes. In reality, they are eukaryotes that have retained certain striking ancestral characteristics.

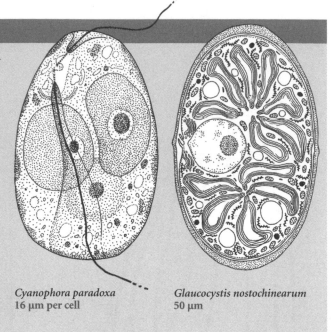

Cyanophora paradoxa
16 µm per cell

Glaucocystis nostochinearum
50 µm

Some unique derived features

- The blue-green color of the chloroplasts is due to the presence of accessory pigments in the phycobilisomes, phycocyanin and allophycocyanin.

- Below the cytoplasmic membrane, the cells contain alveoli supported by microtubules. As in the dinophytes, these alveoli can contain fibrillar material. This character brings the glaucophytes closer to the alveolates.

Number of species: 13.
Oldest known fossils: none.
Current distribution: we find them in the plankton of temperate freshwater ponds.

Ecology

The glaucophytes are rare organisms that swim among the submerged plants of lakes, marshes, bogs and acidic lakes. The cyanelles of *Cyanophora* have been selectively eliminated under laboratory conditions, resulting in the rapid death of the host cell. The metabolic link between the two partners is therefore strong and corresponds to an intimate symbiosis.

Examples

Cyanophora paradoxa, Glaucocystis nostochinearum, Gloeochaete wittrockiana.

Metabionta

A few representatives

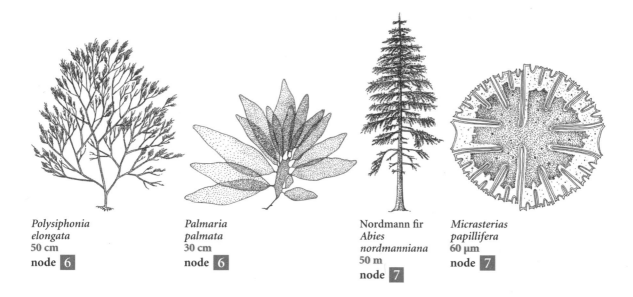

*Polysiphonia
elongata*
50 cm
node 6

*Palmaria
palmata*
30 cm
node 6

Nordmann fir
*Abies
nordmanniana*
50 m
node 7

*Micrasterias
papillifera*
60 µm
node 7

Some unique derived features

■ The peptidoglycan cell wall of the chloroplast has disappeared; the chloroplast now contains a double membrane. The internal membrane is considered to be that of the cyanobacterium from the original endosymbiosis and the external membrane that of the endocytotic vacuole of the host.

■ Chloroplastic DNA: the circular chloroplastic DNA molecules are concentrated into numerous small nucleoids of 1 to 2 µm in diameter and are distributed throughout the chloroplast.

■ There is the possibility of forming multicellular organisms.

■ The EF-2 elongation gene contains unique derived characteristics.

Number of species: 283,402.
Oldest known fossils: cysts similar to those of the genus *Pterosperma* (prasinophyta) are known from the start of the Precambrian (1200 MYA). There may be unicellular and filamentous fossil chlorophycean algae in the rocks of the Bitter Springs formation of central Australia that date from the Precambrian (900 MYA).
Current distribution: worldwide.

Examples

■ RHODOBIONTA: *Chondrus crispus; Lithothamnion roseum; Nemalion multifidum; Polysiphonia elongata; Porphyra umbilicalis; Palmaria palmata.*

■ CHLOROBIONTA: sea lettuce: *Ulva lactuca; Caulerpa taxifolia; Mougeotia tenuis; Chara vulgaris; Coleochaete scutata;* corn: *Zea mays;* European crab apple: *Malus sylvestris.*

Rhodobionta

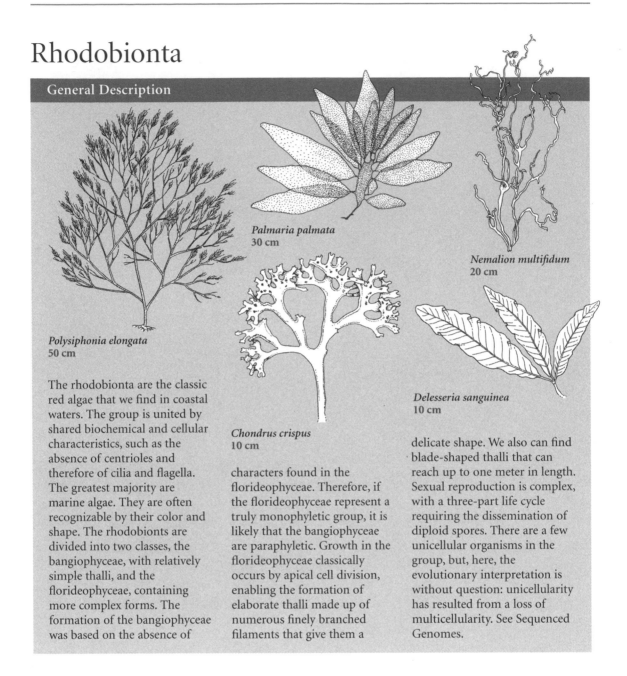

General Description

Polysiphonia elongata
50 cm

Palmaria palmata
30 cm

Nemalion multifidum
20 cm

Chondrus crispus
10 cm

Delesseria sanguinea
10 cm

The rhodobionta are the classic red algae that we find in coastal waters. The group is united by shared biochemical and cellular characteristics, such as the absence of centrioles and therefore of cilia and flagella. The greatest majority are marine algae. They are often recognizable by their color and shape. The rhodobionts are divided into two classes, the bangiophyceae, with relatively simple thalli, and the florideophyceae, containing more complex forms. The formation of the bangiophyceae was based on the absence of characters found in the florideophyceae. Therefore, if the florideophyceae represent a truly monophyletic group, it is likely that the bangiophyceae are paraphyletic. Growth in the florideophyceae classically occurs by apical cell division, enabling the formation of elaborate thalli made up of numerous finely branched filaments that give them a delicate shape. We also can find blade-shaped thalli that can reach up to one meter in length. Sexual reproduction is complex, with a three-part life cycle requiring the dissemination of diploid spores. There are a few unicellular organisms in the group, but, here, the evolutionary interpretation is without question: unicellularity has resulted from a loss of multicellularity. See Sequenced Genomes.

Some unique derived features

- There has been a loss of centrioles and their derivatives.

- The gametes are protoplasts, that is, naked cells with no skeletal cell wall or flagella. Fertilization is always through oogamey (the female gamete is sessile and larger than the male gamete).

- Mitosis is closed (fig. 1, example of *Polysiphonia*); the nuclear membrane persists during mitosis (1a: metaphase; *pb:* polar body, *chr:* chromosome, *sp:* spindle, *nmb:* nuclear membrane) and the spindle remains during telophase (1b); the polar bodies play the role of the centrioles (1a).

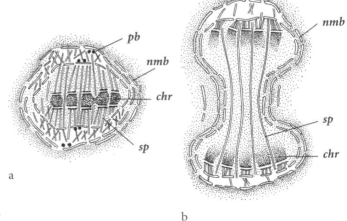

a

b

Figure 1

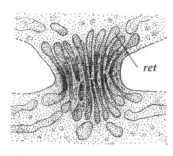
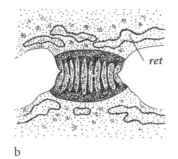

a   b

Figure 2

- Proteinaceous plug: in most cases, cell separation is incomplete. If, at the end of mitosis, the cytoplasms of two daughter cells are still connected, a proteinaceous (or pit) plug is quickly synthetized. Fig. 2 (*ret*: reticulum) shows the formation of the plug: the connection between the cells at the end of metaphase (2a) and the replacement of the reticulum cisterns by the proteinaceous plug (2b).

- Phycobilisomes: the green color of chlorophyll a is masked by two accessory pigments found in the phycobilisomes, a predominant red, phycoerythrin, and a less predominant blue, phycocyanin.

- The chloroplastic DNA is never circular.

- Floridean starch is the most abundant storage molecule. These starch granules are always situated in the cytoplasm. On the cell ultrastructure diagram (fig. 3) of *Porphyridium purpureum*, we note the starch granules (*sg*) outside the double membrane of the chloroplast (*chlm*) and the isolated thylakoids (*cmb*: cytoplasmic membrane, *N*: nucleus, *cw*: cell wall, *py*: pyrenoid, *thy*: thylakoid).

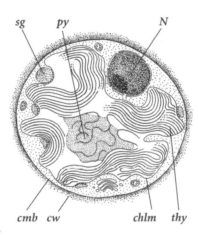

Figure 3

Ecology

The majority of the rhodobionts live anchored to rocks in sea water. However, some may prefer other substrates such as brown algae, eelgrass, bivalve shells or gastropods. The vital habitat of the rhodobionta is restricted to a thin belt around the continents. Nonetheless, their accessory pigments enable these organisms to live at depths greater than those attainable by the green algae (chlorobionta) or brown algae (phaeophyceae). As water filters out low-energy red light, only blue light, that absorbed by red pigments, occurs at great depths. The maximum depth attainable by these organisms can vary from one place to another depending on the transparency of the water. In the Baltic, it is at a few meters, whereas in the Mediterranean they can live at depths of up to 100 meters. The photosynthetic organism found at the greatest ocean depth is a rhodobiont living off the coast of the Bahamas at a depth of 268 m, a depth where only 0.001% of the surface incident light is present. Some rhodobionts are symbionts such as *Porphyridium*, an endosymbiont of tropical benthic foramins. Others have lost their chlorophyll and are obligate parasites of other rhodobionts. By the nature of their cell walls, certain rhodobionts are of considerable economic importance. These organisms are used as sources of agar and carrageen moss used in biochemistry and in the production of candies, ice cream, jams and other foods. The annual harvest is in the order of one million tons fresh weight, half of which comes from aquaculture production in the Far East. Finally, rhodobionts with calcified cell walls have been used as enriching agents in acidic soils. *Lithothamnion roseum* makes up the maerl beds in Brittany, France.

Number of species: 5,500.
Oldest known fossils: Cambrian fossils (590 MYA) have been classified in the family Solenoporaceae, a family related to the coralline algae; these are florideophyceae with calcified cell walls. Some authors have suggested that ancient Precambrian fossils such as *Eosphaera* and *Huroniospora* (1900 MYA) are rhodobionts. This strongly debated diagnostic is based on the hypothesized group age rather than on any real characters.
Current distribution: worldwide.

Examples

Chondrus crispus; Corallina officinalis; Delesseria sanguinea; Furcellaria fastigiata; Gracilaria compressa; Lithothamnion roseum; Nemalion multifidum; Plumaria elegans; Polysiphonia elongata; Porphyra umbilicalis; Palmaria palmata; Scinaia furcellata.

➡ *See chap. 5, p. 148*

Chlorobionta

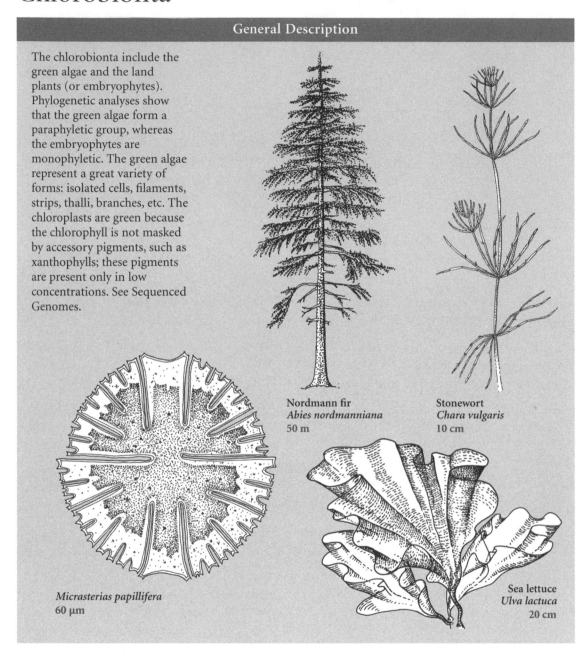

General Description

The chlorobionta include the green algae and the land plants (or embryophytes). Phylogenetic analyses show that the green algae form a paraphyletic group, whereas the embryophytes are monophyletic. The green algae represent a great variety of forms: isolated cells, filaments, strips, thalli, branches, etc. The chloroplasts are green because the chlorophyll is not masked by accessory pigments, such as xanthophylls; these pigments are present only in low concentrations. See Sequenced Genomes.

Nordmann fir
Abies nordmanniana
50 m

Stonewort
Chara vulgaris
10 cm

Micrasterias papillifera
60 μm

Sea lettuce
Ulva lactuca
20 cm

Ecology

The ecology of the green algae is very diverse. We find these organisms in both fresh and salt waters. Unicellular forms are often planktonic, but some form rock films. Multicellular forms are often fixed to a substrate, but filamentous forms can detach and form masses that float with the current. Some stalked algae can be invasive, as is currently the case with *Caulerpa taxifolia* in the Mediterranean. We find green algae in aerial environments, on tree trunks, in soil or on rocks. Finally, some species of the genera *Trebouxia* and *Trentepohlia* are symbiotic with ascomycete fungi (rarely with basidiomycetes) to form lichen. Curiously, the green alga *Chlamydomonas nivalis* that lives in permanent snow, colors the snow . . . red; this time, carotenoids mask the green color of the chlorophyll.

The embryophytes have conquered the terrestrial environment and make up the world's forests, prairies, and steppes.

Some unique derived features

- The chloroplasts are green because the color of the chlorophyll is not masked by accessory pigments, such as xanthophylls, even though these pigments are still present in low quantities. There are never any phycobilins.

- Chlorophyll: in addition to chlorophyll a, the chloroplast also contains chlorophyll b, and, in a single species *Mantoniella squamata,* chlorophyll c. The photosynthetic apparatus, ancestrally based on chlorophyll a and the phycobilins, is transformed into an apparatus based on chlorophylls a and b.

- The thylakoids of the chloroplast are grouped (from 2 to 6) to form lamellae. On fig. 1, the detailed ultrastructure of a chloroplast from *Chlamydomonas reinhardtii* shows the thylakoids *(thy),* the pyrenoid *(py)* and starch granules *(sg).*

- The pyrenoid, when present, is contained within the chloroplast and surrounded by starch granules (fig. 1).

- The carbohydrate reserves are preferentially made up of starch granules (α-1,4 glucane) which surround the pyrenoids or are found in the stroma of the chloroplast.

- The gene that codes for ribosomal sub-unit 1,5-biphosphate carboxylase (RuBisCO) has been transferred from the chloroplastic DNA to the nucleus.

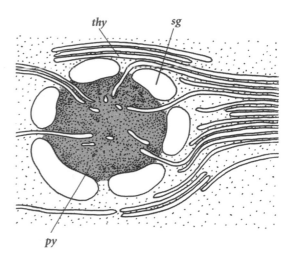

Figure 1

Number of species: 277,902.
Oldest known fossils: cysts similar to those of the genus *Pterosperma* (prasinophytes) are known from the start of the Precambrian (1200 MYA). There may be fossilized unicellular and filamentous chlorophycean algae in the rocks of the Bitter Springs formation of central Australia that date from the Precambrian (900 MYA).
Current distribution: worldwide.

Examples

- PRASINOPHYTA: *Bathycoccus prasino; Pyramimonas lunata.*
- ULVOPHYTA: *Chlamydomonas reinhardtii; Volvox globator;* sea lettuce: *Ulva lactuca; Cladophora vagabunda; Caulerpa taxifolia; Acetabularia acetabulum.*
- STREPTOPHYTA: *Chlorokybus atmophyticus; Klebsormidium flaccidum; Desmidium schwartzii; Micrasterias papillifera; Spirogyra fluviatillis; Mougeotia tenuis; Zygnema peliosporum; Chaetosphaeridium minus;* stonewort: *Chara vulgaris; Nitella gracilis; Coleochaete scutata;* bracken fern: *Pteridium aquilinum;* corn: *Zea mays;* European crab apple: *Malus sylvestris.*

Chromoalveolata

A few representatives

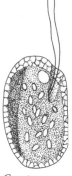

Paramecium
*Paramecium
tetraurelia*
100 μm
node 10

Nereocystis luetkeana
10 m
node 13

*Cryptomonas
erosa*
30 μm
node 14

Emiliana huxleyi
5 μm
node 15

Some unique derived features

This clade brings together the "brown algae" or chromophytes (the stramenopiles and the haptophytes) and the alveolates, two lineages that have only recently been united. Within this clade, we find both photosynthetic organisms, like the phaeophyceae, and non-photosynthetic organisms, such as the ciliates and the apicomplexans. However, some of the non-photosynthetic members still contain residual chloroplasts, relics of a photosynthetic ancestor. The ancestor of the clade had therefore developed a secondary endosymbiosis with a rhodobiont and the non-photosynthetic organisms of this group have experienced either a total or partial secondary loss of this chloroplast. The phylogeny of the chloroplasts indicates a close relationship between the alveolates, for which the monophyly is already well established, and the brown algae, for which the monophyly is questionable. Certain authors have suggested that the green eukaryotes and the chromoalveolates are sister groups.

- There is a periplastidal reticulum situated between the two chloroplastic envelopes and the two external membranes.

- The chloroplasts contain chlorophylls a and c.

- The chloroplast is surrounded by four membranes. There are the two typical membranes, plus two supplementary membranes; the most exterior is often a continuation of the nuclear envelope and is derived from the cell's endoplasmic reticulum (fig. 1: phaeophycean cell; *chl:* chloroplast, *G:* Golgi apparatus, *nmb:* nuclear membrane, *mi:* mitochondrion, *N:* nucleus, *py:* pyrenoid, *ret:* endoplasmic reticulum). This structure is the result of a secondary endosymbiotic event that involved the ingestion of a unicellular, eukaryote cell of the rhodobionta group by another eukaryote cell. The interpretation of the two supplementary membranes is as follows: the external membrane corresponds to that of the phagocytic vacuole, the internal membrane to the cytoplasmic membrane of the integrated eukaryote.

- The chloroplastic form of the glyceraldehyde-3-phosphate dehydrogenase (GAPDH) gene, homologue of a cyanobacterial gene, has been replaced by the cytosolic form of the gene, characteristic of the eukaryotes.

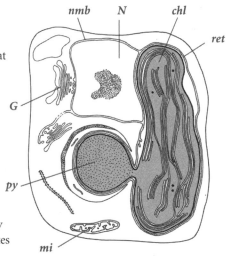

Figure 1

Number of species: 121,822.
Oldest known fossils: spores that might have belonged to the chrysophyceae date from the Precambrian (700 MYA).
Current distribution: worldwide.

Examples

- ALVEOLATA: *Gonyaulax tamarensis; Plasmodium falciparum;* paramecium: *Paramecium tetraurelia.*
- STRAMENOPILES: bladderwrack: *Fucus vesiculosus; Ochromonas danica; Ophiocytium arbuscula;* grapevine downy mildew: *Plasmopara viticola.*
- CRYPTOPHYTA: *Chilomonas paramecium; Cryptomonas erosa.*
- HAPTOPHYTA: *Emiliana huxleyi.*

Alveolata

A few representatives

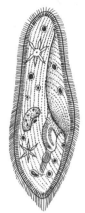

Paramecium
Paramecium tetraurelia
100 μm
node 10

Stentor polymorphus
2 mm
node 10

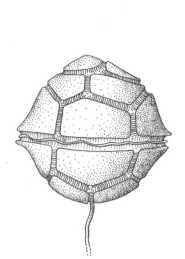

Peridinium cinctum
30 μm
node 11

Monospora
polyacantha
150 μm
node 12

Some unique derived features

■ These unicellular eukaryotes are sometimes photosynthetic and are extremely variable in size and lifestyle. They show a range of morphologies and often have flagella or cilia. The monophyly of this group that includes the ciliata, the dinophyta and the apicomplexa has been suggested by molecular phylogenies based on the 18S rRNA gene. It is probable that the dinophytes and the apicomplexa are sister groups.

■ Sub-membranous vesicles: under the cellular membrane is a widely distributed series of flattened vesicles, the alveoli, that may be used for calcium storage (fig. 1a: cortical structure of a paramecium; *al*: alveola, *fl*: flagellum, *mi*: mitochondrion, *tri*: trichocyst). The alveoli can also contain dense material, like the thecal plates of the dinoflagellates (fig. 1b; cortical structure of *Peridinium bipes*; *mit*: microtubule, *tpl*: thecal plate).

> **Number of species:** 15,200.
> **Oldest known fossils:** the oldest indubitable dinophyte cysts date from the Silurian (420 MYA). However, rigid theca, thought to be those of dinophytes, have been found from the lower Cambrian (540 MYA).
> **Current distribution:** worldwide.

Examples

■ DINOPHYTA: *Gonyaulax tamarensis; Noctiluca miliaris.*
■ APICOMPLEXA: *Plasmodium falciparum; Eimeria falciformis.*
■ CILIATA: *Paramecium tetraurelia; Stentor coeruleus.*

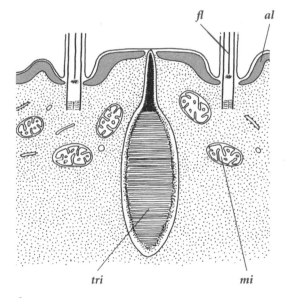

a

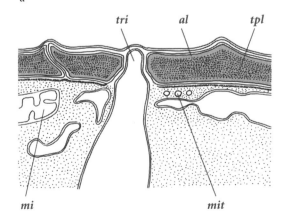

b

Figure 1

111

Ciliata

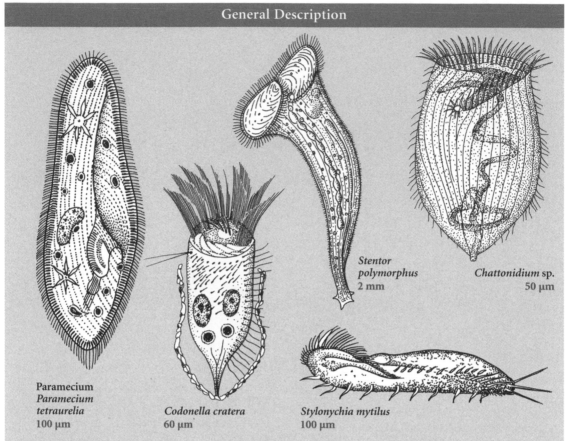

General Description

Paramecium
*Paramecium
tetraurelia*
100 µm

Codonella cratera
60 µm

*Stentor
polymorphus*
2 mm

Chattonidium sp.
50 µm

Stylonychia mytilus
100 µm

These are unicellular eukaryotes characterized by the presence of numerous vibratile cilia on the surface of their cells, at least at some stage of their life cycle. In the majority of documented cases, these cilia or kinetes are anchored at their extremity by a complex fibrous system of microtubules and other microfibers. The morphological diversity of this group is large, not only in terms of the size of the cells (from 30 to 300 µm), but also in the distribution of the cilia. The cilia are often specialized to provide certain locomotory systems or structures for capturing prey. Many species have a functional "mouth" or cytostome. The ciliates are characterized by their heterokaryotic condition; they always possess a reproductive micronucleus and a vegetative macronucleus. We find free-swimming species, those anchored to a substrate by a stalk, colonial species and parasitic species. The ciliates are not photosynthetic, but some possess endosymbionts filled with chloroplasts (in particular, green algae and dinophytes). See Sequenced Genomes.

Some unique derived features

- The nuclear dimorphism is always present (fig. 1: *Tetrahymena pyriformis; mo:* mouth, *cil:* cilia, *kin:* kinete, *cyt:* cytoproctus, *mN:* micronucleus, *MN:* macronucleus, *vp:* contractile vacuole pore). The diploid micronucleus used in sexual reproduction has chromosomes and undergoes mitosis. The polyploid macronucleus is responsible for all basic cellular functions. It forms from the dividing micronucleus during sexual conjugation, a process unique to ciliates. During this process, a complex series of DNA synthesis is followed by the selective elimination of part of the genetic material. The macronucleus does not have chromosomes, but the loose chromatin contains thousands of copies of a small number of genes.

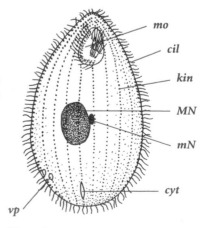

mo
cil
kin
MN
mN
cyt
vp

Figure 1

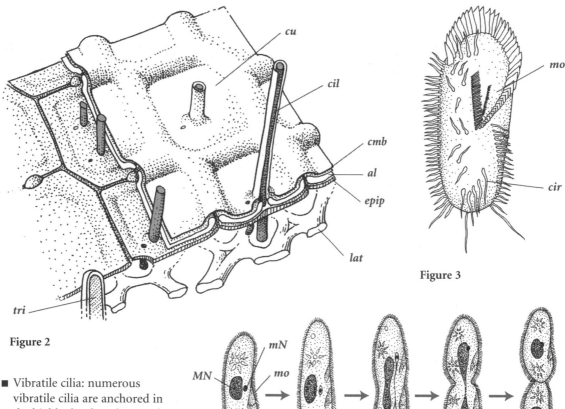

Figure 2

Figure 3

Figure 4

- Vibratile cilia: numerous vibratile cilia are anchored in the highly developed cortical complex of membranes, microtubules and microfibers (fig. 2: overview of the cortical structure of *Paramecium tetraurelia*; *al*: cortical alveola, *epip*: epiplasm, *lat*: infraciliary lattice, *cmb*: cytoplasmic membrane, *cu*: cortical unit, *tri*: trichocyst).

- The somatic ciliature assures locomotion and the oral ciliature is used for capturing prey. This is shown in the ventral view of *Stylonychia mytilus* (fig. 3; *cir*: cirri).

- Asexual division occurs by cell fission perpendicular to the anterior-posterior axis of the cell (fig. 4: cell division of *Paramecium; cv*: contractile vacuole).

Number of species: 8,000.
Oldest known fossils: ciliates of the order Tintinnida secrete loricae that can be fossilized. We find these fossils starting in the Ordovician (450 MYA).
Current distribution: worldwide.

Ecology

The presence of ciliates is linked to water. In the terrestrial environment, we find them in the interstitial waters of mosses and soil. Their ecological niches are quite varied; we can find them in fresh or brackish water, in fast-moving or stagnant pools, as free swimmers or stalked, sessile forms. Other species are symbiotic or non-pathogenic parasites of diverse organisms. The ciliates are heterotrophic, most often phagotrophic, feeding on organic particles, bacteria, and other ciliated or flagellated small animals. Different diets are accompanied by different specializations of the oral structures. Parasitic life can lead to some major adaptations, from the growth of attachment organs to the occurrence of complex developmental cycles. When environmental conditions become unfavorable, many species are capable of forming resistant cysts. Sexual reproduction is by conjugation; cells of different mating types meet and reciprocally exchange haploid nuclei. There are no true gametes in terms of specialized cells. Conjugation is a complex phenomenon during which the genome is reconstructed.

Examples

Blepharisma japonicum; Codonella cratera; Chattonidium sp.; *Discocephalus* sp.; *Euplotes aediculatus;* paramecium: *Paramecium tetraurelia; Stentor coeruleus; Stylonychia mytilus; Tetrahymena pyriformis; Vorticella campanula.*

113

Dinophyta

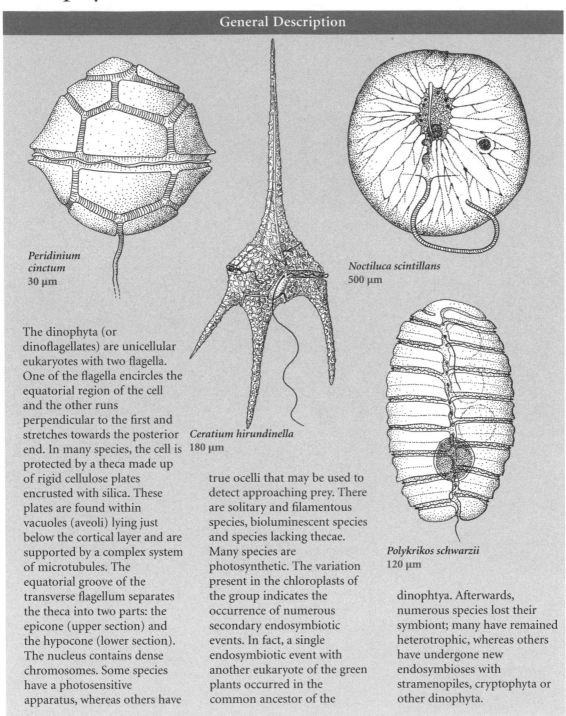

General Description

Peridinium cinctum
30 µm

Noctiluca scintillans
500 µm

Ceratium hirundinella
180 µm

Polykrikos schwarzii
120 µm

The dinophyta (or dinoflagellates) are unicellular eukaryotes with two flagella. One of the flagella encircles the equatorial region of the cell and the other runs perpendicular to the first and stretches towards the posterior end. In many species, the cell is protected by a theca made up of rigid cellulose plates encrusted with silica. These plates are found within vacuoles (aveoli) lying just below the cortical layer and are supported by a complex system of microtubules. The equatorial groove of the transverse flagellum separates the theca into two parts: the epicone (upper section) and the hypocone (lower section). The nucleus contains dense chromosomes. Some species have a photosensitive apparatus, whereas others have true ocelli that may be used to detect approaching prey. There are solitary and filamentous species, bioluminescent species and species lacking thecae. Many species are photosynthetic. The variation present in the chloroplasts of the group indicates the occurrence of numerous secondary endosymbiotic events. In fact, a single endosymbiotic event with another eukaryote of the green plants occurred in the common ancestor of the dinophtya. Afterwards, numerous species lost their symbiont; many have remained heterotrophic, whereas others have undergone new endosymbioses with stramenopiles, cryptophyta or other dinophyta.

Ecology

The dinophytes are principally marine, planktonic, unicellular organisms. The greatest diversity of species is found in the plankton of warm seas. However, some species live in fresh water and others are parasites of marine animals (jellyfish, copepods, fishes). Only half of the species are photosynthetic, or sometimes mixotrophic. Others are carnivorous and feed by capturing their prey with a posterior tentacle or filament. For example, *Noctiluca miliaris* is a carnivorous, bioluminescent dinophyte with a long fishing flagellum. Certain species are toxic and cause a large amount of damage when they swarm during "red tides." Because of their atypical nuclear organization, cellular division is specialized. The different steps of mitosis are absent and the chromatin always remains condensed into chromosomes. Despite this, genes are continually expressed. Sex occurs in certain species, but only the zygote is diploid. Certain species can form resistant cysts.

Some unique derived features

- Flagella: the cell has two flagella that run perpendicularly to one another, each situated in a cellular groove (fig. 1: *Didinium* with its two flagella; *epi:* epicone, *tfl:* transverse flagellum, *lfl:* longitudinal flagellum, *hyp:* hypocone).

- The sub-cortical alveoli contain the cellulose plates that form the theca.

- Dinokaryon: the nucleus, called a dinokaryon, has an unusual organization. In other eukaryotes, histones are associated with the DNA at interphase. In the dinophytes, the DNA is associated with a basic protein and the chromosomes are condensed into helices (fig. 2; a and b: condensing of a chromosome; c: microscope view).

- Mitosis is closed.

- The cell carries characteristic trichocysts.

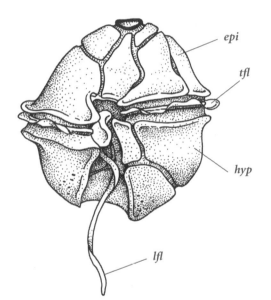

Figure 1

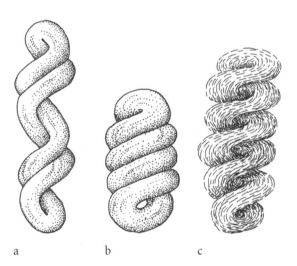

a   b   c

Figure 2

Number of species: 2,200.
Oldest known fossils: the oldest indubitable cysts date from the Silurian (420 MYA). However, rigid thecae, thought to be those of dinophytes, have been found from the lower Cambrian (540 MYA).
Current distribution: worldwide.

Examples

Ceratium hirundinella; Erythropsidinium pavillardii; Gonyaulax tamarensis; Gymnodinium microadriaticum; Nematodinium lebourae; Noctiluca miliaris; N. scintillans; Syndinium rostratum; Peridinium cinctum; Polykrikos schwartzii.

Apicomplexa

General Description

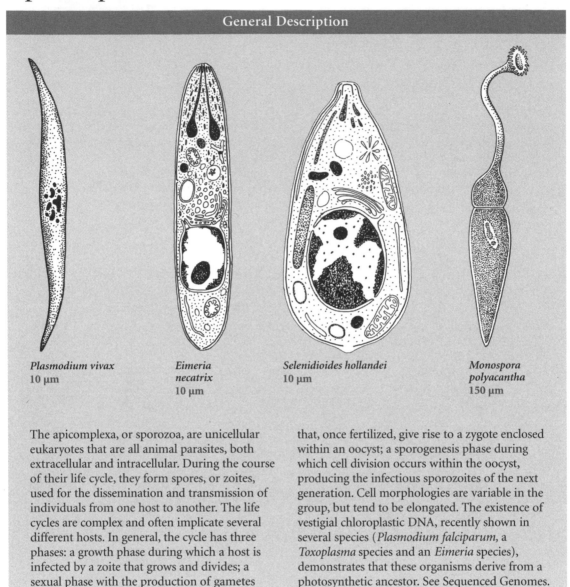

Plasmodium vivax
10 μm

**Eimeria
necatrix**
10 μm

Selenidioides hollandei
10 μm

**Monospora
polyacantha**
150 μm

The apicomplexa, or sporozoa, are unicellular eukaryotes that are all animal parasites, both extracellular and intracellular. During the course of their life cycle, they form spores, or zoites, used for the dissemination and transmission of individuals from one host to another. The life cycles are complex and often implicate several different hosts. In general, the cycle has three phases: a growth phase during which a host is infected by a zoite that grows and divides; a sexual phase with the production of gametes that, once fertilized, give rise to a zygote enclosed within an oocyst; a sporogenesis phase during which cell division occurs within the oocyst, producing the infectious sporozoites of the next generation. Cell morphologies are variable in the group, but tend to be elongated. The existence of vestigial chloroplastic DNA, recently shown in several species (*Plasmodium falciparum,* a *Toxoplasma* species and an *Eimeria* species), demonstrates that these organisms derive from a photosynthetic ancestor. See Sequenced Genomes.

Ecology

The apicomplexa are highly diversified and present in all environments. They are intracellular parasites or endosymbionts of a wide variety of hosts. There is an alternation of haploid and diploid generations. Both forms can multiply by schizogony. During this process, there is rapid multiplication of the nuclei by mitosis, using intranuclear mitotic spindles, without the cell growing in size.

This results in a multinucleated form that will undergo cytoplasmic division around each nucleus to give rise to small infectious cells, the spores. These spores are the dispersive forms. Sexual reproduction is different in this group. The small, sometimes flagellated, microgametic male fuses with the macrogametic female. The generally mobile zygote transforms into an oocyst, a

resistant cell form with a thick cell wall. Then, under certain conditions, meiosis occurs within the oocyst. Sporogony therefore takes place: rapid nuclear division that gives rise to small infectious cells, the haploid sporozoites. After transmission to a new host, these cells can then give rise to male or female gametes. This cycle takes place within several successive hosts. Certain species pose serious health threats at a worldwide scale. Let us examine, for example, the life cycle of *Plasmodium falciparum,* the causal agent of malaria. The vector is a mosquito of the genus *Anopheles.* Fertilization takes place in the digestive tract of the mosquito host. The mobile egg transforms into a resistant oocyst that, after meiosis and multiplication by sporogony, gives rise to sporozoites, infectious haploid cells. The sporozoites migrate to the mosquito's salivary glands. The mosquito then feeds on a human and contaminates him/her. The sporozoites develop in the human blood; their growth requires the heme of the hemoglobin. The parasite enters into the red blood cell, reaches the trophozoite stage, undergoes rapid mitosis and becomes a merozoite. The merozoites break open the red blood cell and are released into the blood stream, where they infect other red blood cells. A new cycle of multiplication can then occur. After several infectious cycles, the merozoites differentiate into male and female gametes. These gametes are then ingested by a feeding mosquito. The cyclic emergence of new merozoite generations from the red blood cells provokes a fever in the host.

Some unique derived features

- The sporozoite form always occurs. The infectious cell starts the cycle and new sporozoites are produced by sporogony in the oocyst.

- Specialized organelles form a unique apical complex at the cell's extremity (fig. 1: anterior end of a zoite; *al:* alveola, *pr:* polar rings, *con:* conoid, *G:* Golgi apparatus, *mn:* microneme, *N:* nucleus, *rh:* rhoptries). One of these structures is the conoid, a spiraled band of microtubules that forms the conical tip. The apical complex also includes fibrils, microtubules, vacuoles, polar rings and micronemes.

- When centrioles are present, they are made up of nine single microtubules and not nine triplets as usual (ancestral state).

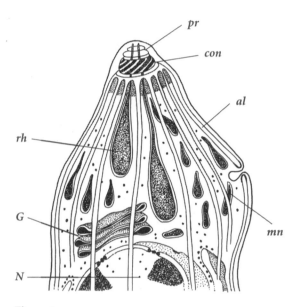

Figure 1

Number of species: 5,000.
Oldest known fossils: none.
Current distribution: worldwide.

Examples

Babesia canis (causal agent of piroplasmosis); a coccidium: *Eimeria falciformis*; gregarines: *Gregarina cuneata, Stylocephalus longicollis*; *Plasmodium falciparum* (causal agent of malaria); *Toxoplasma gondii* (causal agent of toxoplasmosis).

Stramenopiles

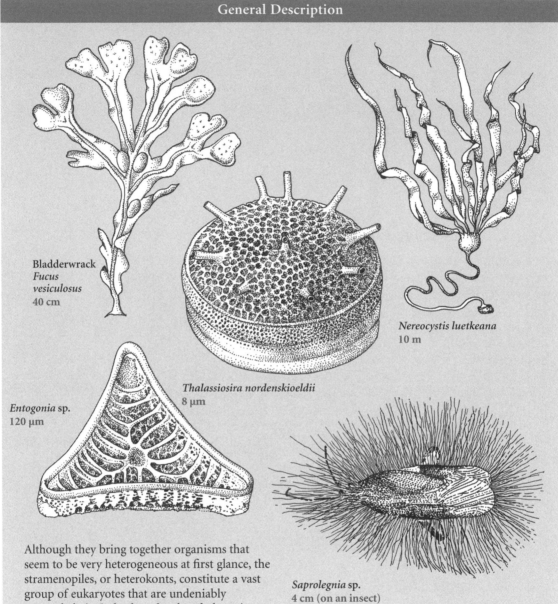

General Description

Bladderwrack
Fucus vesiculosus
40 cm

Thalassiosira nordenskioeldii
8 μm

Entogonia sp.
120 μm

Nereocystis luetkeana
10 m

Saprolegnia sp.
4 cm (on an insect)

Although they bring together organisms that seem to be very heterogeneous at first glance, the stramenopiles, or heterokonts, constitute a vast group of eukaryotes that are undeniably monophyletic. Indeed, molecular phylogenies are unanimous and a clear cellular synapomorphy, involving the presence of small tubular hairs on the flagella, unite them. "Stramenopile" comes from the Latin words *stramen:* straw or hollow stem, and *pilus:* hair.

In this group we find both unicellular and multicellular organisms that may or may not be photosynthetic (using chlorophylls a and c). The most well known stramenopiles groups are the phaeophyceae (the brown algae) and the bacillariophyceae (the diatoms). However, we can also mention groups of lesser ecological importance such as the chrysophyceae (golden algae: unicellular or colonial), the xanthophyceae (yellow-green algae: unicellular, colonial or filamentous), the eustigmatophyceae (unicellular

algae with a typical heterokont photoreceptor apparatus), the raphidophyceae and the dictyochophyceae (flagellated, unicellular algae). Non-photosynthetic organisms were only recently included in this group, having been previously part of the vast polyphyletic group of "lower fungi." These organisms include the oomycetes, typically parasitic organisms such as the grapevine downy mildew (*Plasmopara viticola*) or the pathogenic fish watermold (*Saprolegnia ferax*); the hyphochytridiomycetes, parasites of algae and animals; the labyrinthulomycetes, saprophytes or parasites of algae or fungi; finally the opalines, animal parasites that principally infect anuran amphibians. See Sequenced Genomes.

Some unique derived features

■ The flagella are heterokonts, that is, they differ in their structure, orientation (anterior or posterior), and function (beating rhythm, etc). Of these two unequal flagella (anisokonts), only the anteriorly oriented flagellum carries mastigonemes, tubular, tripartite hairs, as shown on the chrysophycean *Ochromonas danica* (fig. 1; *tfi*: terminal filament, *mast*: mastigoneme, *N*: nucleus).

■ At the base of the flagella, there is an intracellular transition region that includes a small coiled fiber, the transitional helix (fig. 2; *chl*: chloroplast, *fl*: flagella, *th*: transitional helix, *ret*: peripheral endoplasmic reticulum, *sti*: stigma, a photoreceptor).

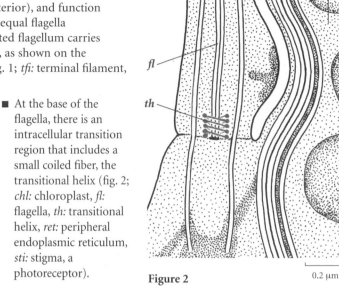

Figure 2   0.2 µm

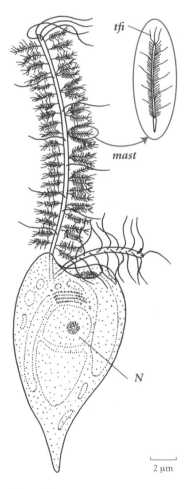

Figure 1

Ecology

The majority of stramenopiles are photosynthetic and make up a large part of the marine biomass. Many species are part of the marine phytoplankton, whereas others are macroscopic primary producers (submerged forests of brown algae, or species found in marine tidal zones in temperate regions). The stramenopiles are at the base of the marine ecological food chain because they produce organic matter and are prey to a large number of other organisms, including zooplankton. They are equally important in fresh (chrysophyceae, eustigmatophyceae) and brackish waters (xanthophyceae). These species can be parasites of plants or vertebrates (oomycetes) or be free-living in moist humus (oomycetes or saprophytic labyrinthulomycetes that feed from decomposing organic matter). Sex can be present or absent and more or less developed depending on the group. It can involve spores and biflagellated gametes with unequal flagella. In the brown algae (phaeophyceae), the life cycle alternates between a diploid sporophyte and a haploid gametophyte (with some exceptions such as *Fucus*, for example).

Examples

■ BACILLARIOPHYCEAE: *Entogonia* sp.; *Navicula incerta; Thalassiosira nordenskioeldii.*
■ PHAEOPHYCEAE: *Ascophyllum nodosum;* bladderwrack: *Fucus vesiculosus; Himanthalia elongata;* kelp: *Laminaria hyperborea; Nereocystis luetkeanea;* rockweed: *Pelvetia fastigiata; Sphacelaria rigidula.*
■ CHRYSOPHYCEAE: *Ochromonas danica; Synura petersenii.*
■ XANTHOPHYCEAE: *Ophiocytium arbuscula.*
■ EUSTIGMATOPHYCEAE: *Eustigmatos magnus.*
■ OOMYCETA: grapevine downy mildew: *Plasmopara viticola;* watermold: *Saprolegnia ferax.*
■ OPALINES: *Opalina* sp.

Number of species: 105,922.
Oldest known fossils: spores that may be related to the chrysophyceae date from the Precambrian (700 MYA). The chrysophyceae and the bacillariophyceae (the diatoms) have left numerous fossils because of the siliceous parts or cell walls that are synthesized by the cells.
Current distribution: present in all seas, oceans, fresh and brackish waters of the world.

Cryptophyta

General Description

The cryptophyta are unicellular, photosynthetic eukaryotes. They possess chlorophylls a and c, along with a variable and characteristic composition of accessory pigments. Certain species like *Chilomonas paramecium* are not photosynthetic, but the presence of leucoplasts in some show that this is a secondary loss. The cryptophytes have two anterior flagella inserted into an oral groove. This type of oral pocket contains trichocysts, special structures used to neutralize prey. The crypto- phytes are interesting from any evolutionary perspective because their structure proves the existence of secondary endosymbiotic events. Indeed, an organelle called a *nucleomorph* can be found connected to the chloroplast and represents the vestigial nucleus of the integrated eukaryote. If we examine the 18S rRNA gene of the nucleomorph genome, the molecular phylogeny suggests that the symbiont is related to the rhodobionta. The presence of mastigonemes bring the cryptophyta closer to the stramenopiles. See Sequenced Genomes.

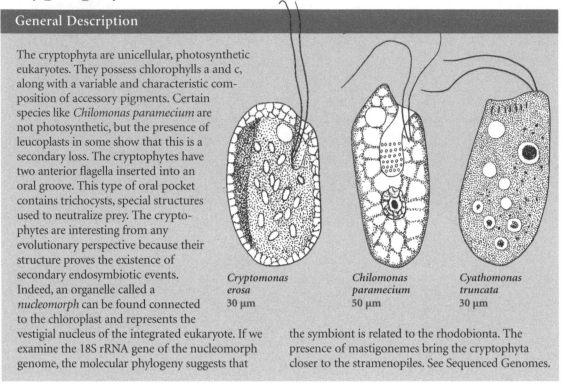

Cryptomonas erosa
30 µm

Chilomonas paramecium
50 µm

Cyathomonas truncata
30 µm

Some unique derived features

■ The ventral side of the cell has a groove that ends in a depression. At the anterior end, there is an oral groove covered in trichocysts (fig. 1: *Cryptomonas; chl:* chloroplast, *Mc:* Maupas

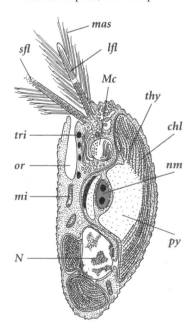

Figure 1

corpuscles, *lfl:* large flagellum, *sfl:* small flagellum, *or:* oral groove, *mas:* mastigoneme, *nm:* nucleomorph, *mi:* mitochon- drion, *N:* nucleus, *py:* pyrenoid, *thy:* thylakoid, *tri:* trichocyst).

■ The trichocysts, or ejectosomes, are unique to this group; they do not resemble the trichocysts of the dinophytes or those of the ciliates.

■ Flagellum: the large flagellum has two rows of mastigonemes (fig. 1).

■ Maupas corpuscles: two structures of unknown function lie close to the oral groove (fig. 1).

■ Mitosis is open and shows characteristics unique to the group; there are no centrioles.

■ The thylakoids of the chloroplasts are grouped into pairs (fig. 1).

■ A unique organelle, the nucleomorph, represents the vestigial nucleus of a symbiotic eukaryote related to the rhodobionta (fig. 1).

Ecology

The cryptophyta are found in all seas, in fresh water and in the interstitial waters of humid terrestrial environments. They make up a significant proportion of the nanoplankton. We can find them in Antarctic fresh water. Some species are intestinal parasites of domestic animals. Others have become endosymbionts of dinophytes.

Number of species: 200.
Oldest known fossils: none.
Current distribution: worldwide.

Examples

Chilomonas paramecium; Cryptomonas erosa; Goniomonas truncata.

Haptophyta

General Description

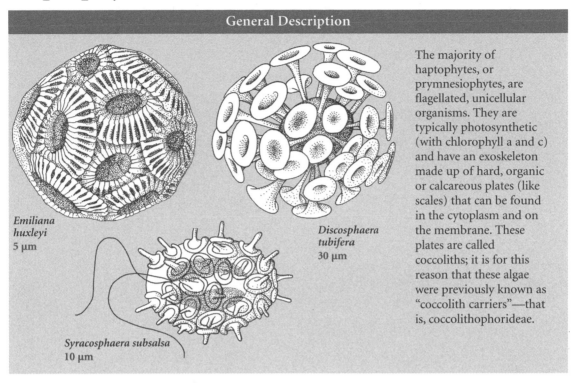

Emiliana huxleyi
5 μm

Discosphaera tubifera
30 μm

Syracosphaera subsalsa
10 μm

The majority of haptophytes, or prymnesiophytes, are flagellated, unicellular organisms. They are typically photosynthetic (with chlorophyll a and c) and have an exoskeleton made up of hard, organic or calcareous plates (like scales) that can be found in the cytoplasm and on the membrane. These plates are called coccoliths; it is for this reason that these algae were previously known as "coccolith carriers"—that is, coccolithophorideae.

Some unique derived features

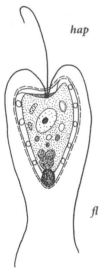

Figure 1

- Haptonema: in addition to two flagella, the cell carries a filamentous appendage called a haptonema. The *haptonema* differs in structure from the flagella having 6 or 7 micro-tubules. The haptonema of *Chrysochromulina polylepis*, only half the length of the flagella, is about the same length as the cell, 10 μm (fig. 1; *hap:* haptonema, *fl:* flagellum).

- The surface of the cell is covered with scales or plates of organic material. These can be calcified and are formed in the Golgi apparatus. *Emiliana huxleyi*, shown in cross-section in fig. 2a (*chl:* chloroplast, *coc:* coccolith; *G:* Golgi apparatus, *mi:* mitochondrion, *N:* nucleus, *cocv:* coccolith vacuole), has coccoliths (fig. 2b) with elementary segments (*es*) that are formed so that they overlap at the cell surface (fig. 2c).

- Mitosis is open with a distinctive metaphase plate.

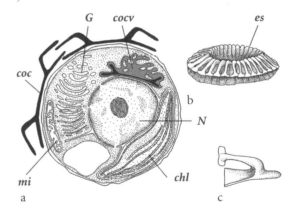

Figure 2

Ecology

The haptophytes make up the majority of photo-synthetic marine nanoplankton. The coccoliths are responsible for a large quantity of sediments. The large chalk deposits of the Cretaceous were principally formed by coccoliths. There are also a few freshwater species.

Number of species: 500.
Oldest known fossils: some coccoliths have been found in rocks of the Carboniferous (300 MYA). The haptophytes diversified and became abundant during the Jurassic (180 MYA).
Current distribution: in all the oceans of the world.

Examples

Chrysochromulina polylepis; Discosphaera tubifera; Emiliana huxleyi; Prymnesium patellifera; Syracosphaera subsalsa.

Rhizaria

A few representatives

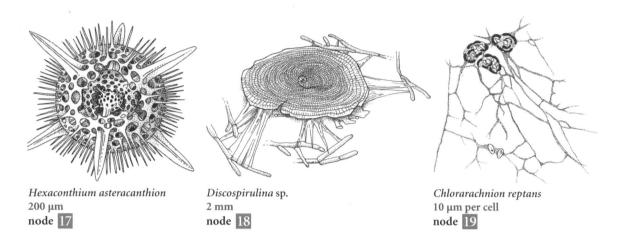

Hexaconthium asteracanthion
200 μm
node 17

Discospirulina sp.
2 mm
node 18

Chlorarachnion reptans
10 μm per cell
node 19

Some unique derived features

■ The 18S ribosomal RNA has unique derived characteristics.

Remark: The actinopods are not monophyletic; the helozoans are likely polyphyletic with their branches spread among the rhizarians. The radiolarians (other actinopods) seem to be the sister-group of the other rhizarians, a clade that some authors refer to as cercozoa. However, for the moment, there is no reliable phylogeny for the rhizarians.

> **Number of species:** 18,006.
> **Oldest known fossils:** the radiolarians and the forams have been recorded since the end of the Precambrian (540 MYA).
> **Current distribution:** worldwide.

Examples

■ ACTINOPODA: *Challengeron wyvillei; Clathrulina elegans; Amphilonche elongata.*
■ FORMINIFERA: *Astrorhiza limicola; Discospirulina* sp.; *Globigerina pachyderma.*
■ CHLORARACHNIOPHYTA: *Chlorarachnion reptans.*

Actinopoda

General Description

The actinopoda bring together the radiolarians, the heliozoans and the acantharians. These are relatively large, unicellular eukaryotes with a complex architecture. They most often show spherical symmetry and are covered in spicules and cytoplasmic extensions (axopodia and filopodia). The spicules originate from an elaborate, compound endoskeleton that, depending on the species, can contain strontium sulfate, silica or a mixture of silica and organic compounds. The axopodia are supported by microtubule beams that attach together at the center of the cell. The spicules are covered

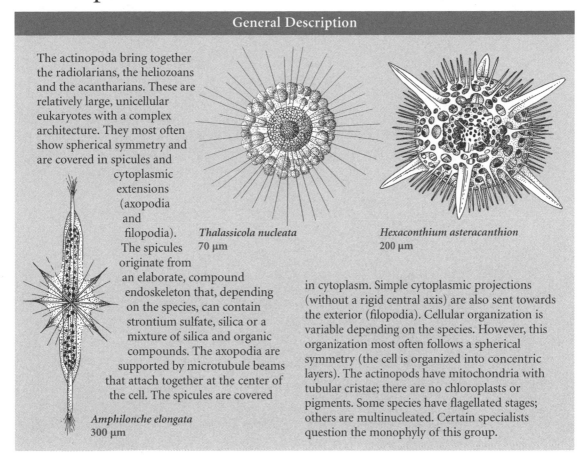

Thalassicola nucleata
70 μm

Hexaconthium asteracanthion
200 μm

Amphilonche elongata
300 μm

in cytoplasm. Simple cytoplasmic projections (without a rigid central axis) are also sent towards the exterior (filopodia). Cellular organization is variable depending on the species. However, this organization most often follows a spherical symmetry (the cell is organized into concentric layers). The actinopods have mitochondria with tubular cristae; there are no chloroplasts or pigments. Some species have flagellated stages; others are multinucleated. Certain specialists question the monophyly of this group.

Some unique derived features

- The cell is armed with axopodia and spicules covered in cytoplasm (fig. 1: *axo:* axopodium, *ccp:* central capsule, *N:* nucleus, *spi:* spicule, *sym:* photosynthetic symbiont).

- The axopodium has a particular structure. It has a rigid center called the axoneme made up of a microtubule bundle. However, this character may have been acquired independently several times.

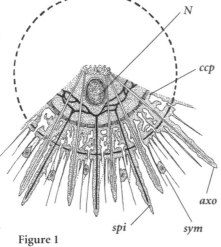

Figure 1

Examples

- RADIOLARIA: *Challengeron wyvillei; Hexaconthium asteracanthion; Thalassicola pellucida.*
- HELIOZOA: *Actinophrys pontica; Clathrulina elegans; Sticholonche zanclea.*
- ACANTHARIA: *Amphilonche elongata; Acanthochiasma fusiforme.*

Ecology

The actinopods are mostly marine, with a small minority living in fresh water. They are planktonic predators whose prey, unicellular animals such as larval crustaceans, stick to the axopodia and are phagocytosed. Cytoplasmic streaming then brings the prey into the central area of the cell where digestion occurs. Certain species carry symbiotic haptophyte algae in the cortical shell. Asexual reproduction is frequent. Sexual reproduction is known in many species and involves the formation of cysts (flagellated, dispersive forms called zoospores). The radiolarians are of great geological importance.

Number of species: 12,000.
Oldest known fossils: radiolarians are recorded from the lower Cambrian (540 MYA).
Current distribution: worldwide.

Foraminifera

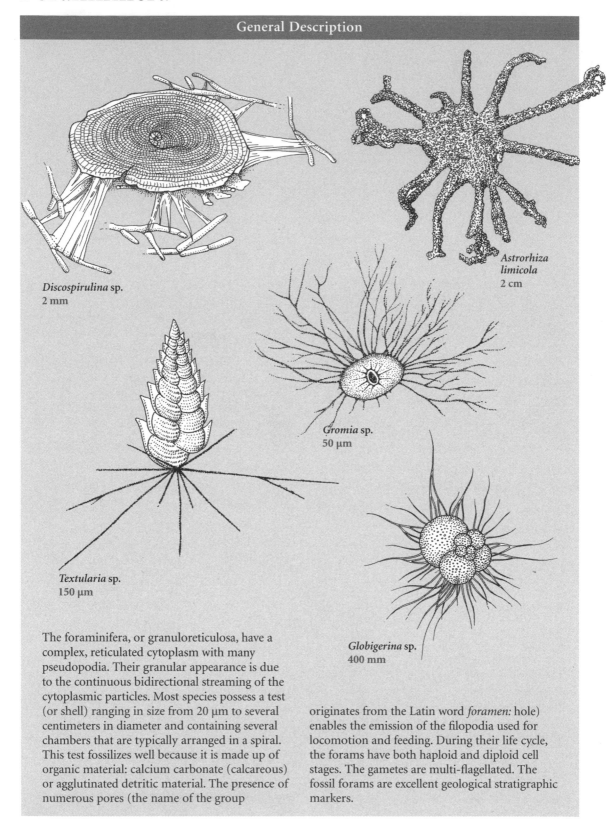

Discospirulina sp.
2 mm

Astrorhiza limicola
2 cm

Gromia sp.
50 μm

Textularia sp.
150 μm

Globigerina sp.
400 mm

The foraminifera, or granuloreticulosa, have a complex, reticulated cytoplasm with many pseudopodia. Their granular appearance is due to the continuous bidirectional streaming of the cytoplasmic particles. Most species possess a test (or shell) ranging in size from 20 μm to several centimeters in diameter and containing several chambers that are typically arranged in a spiral. This test fossilizes well because it is made up of organic material: calcium carbonate (calcareous) or agglutinated detritic material. The presence of numerous pores (the name of the group originates from the Latin word *foramen:* hole) enables the emission of the filopodia used for locomotion and feeding. During their life cycle, the forams have both haploid and diploid cell stages. The gametes are multi-flagellated. The fossil forams are excellent geological stratigraphic markers.

Some unique derived features

■ The reticulated filopodia (reticulopodia) are supported by a dynamic cytoskeleton (fig. 1: *Polystomella; fil:* filopodium, *tch:* test chamber).

■ The test is initially formed of glycoprotein and grows by the addition of new chambers (fig. 1). Secondarily, calcium carbonate or agglutinated detritic material may be added.

Ecology

The forams are exclusively marine and typically benthic, although there are some planktonic species. These single-celled organisms use their filopodia, stretched out through pores in the test, for feeding, locomotion and collecting the mineral particles used for test construction. As predators, they feed on ciliates, actinopods, algae and sometimes nematodes or larval crustaceans. Their complex form of sexual reproduction involves the alternation of haploid and diploid generations. Different associated forms can be taken for distinct species, an error that paleontologists have a difficult time avoiding. Sexual reproduction is complicated by the presence of a nuclear dimorphism. In certain groups, the gametes are flagellated. Some forams harbor photosynthetic symbionts: dinophytes, chrysophytes or diatoms.

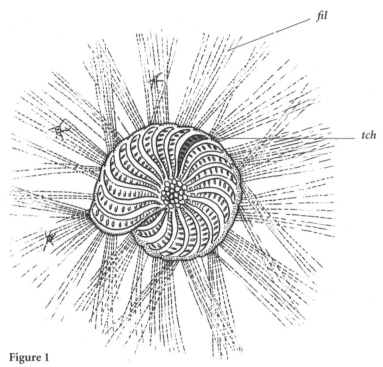

fil

tch

Figure 1

Number of species: 6,000.
Oldest known fossils: the most ancient foraminifera (*Chitinodendron, Archaeochitosa*) date from the lower Cambrian (540 MYA). They become very abundant starting in the Triassic (230 MYA), after which time they are the principal group responsible for the formation of the marine sediment. Their abundance (30,000 described fossil species) and the detailed architecture of their tests make them ideal geological stratigraphic markers.
Current distribution: worldwide.

Examples

Allogromia laticollaris; Astrorhiza limicloa; Discospirulina sp.; *Globigerina pachyderma; Globigerinoides ruber; Iridia serialis; Myxotheca arenilega; Rotaliella roscoffensis; Rosalina leei; Textularia* sp.

Chlorarachniophyta

General Description

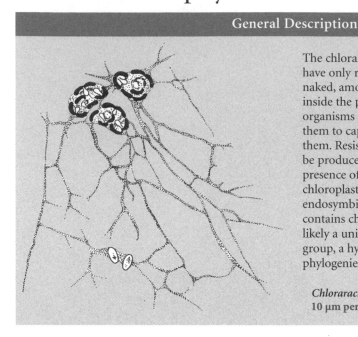

The chlorarachniophyta are a restricted group and have only recently been described. The cells are naked, amoeboid shaped and linked by filopodia inside the plasmodium. Even though these organisms are photosynthetic, the filopodia enable them to capture unicellular prey and phagocytize them. Resistant cysts and flagellated zoospores can be produced. As in the cryptophyta, we note the presence of a nucleomorph close to the chloroplast that indicates a past secondary endosymbiotic event. Given that the chloroplast contains chlorophylls a and b, this symbiont was likely a unicellular chlorobiont of the ulvophyta group, a hypothesis supported by molecular phylogenies.

Chlorarachnion reptans
10 μm per cell

Some unique derived features

- Zoospore: has a single flagellum that spirals around the cell when swimming (fig. 1: zoospore of *Chlorarachnion reptans; chl:* chloroplast, *fl:* flagellum, *N:* nucleus, *py:* pyrenoid).

- Chloroplast: is bilobed and surrounded by four membranes, as in the photosynthetic stramenopiles and the cryptophyta.

The thylakoids are grouped in twos or threes to form lamellae.

- The nucleomorph is found in a depression at the surface of the pyrenoid (fig. 2: *ecyt:* endosymbiont cytoplasm, *dmb:* double membrane, *chlmb:* chloroplastic membrane, *nm:* nucleomorph, *thy:* thylakoid, *rv:* reserve vesicle in the host's

cytoplasm). We should note the presence of chloroplastic membranes and the two external membranes. This double membrane represents the cytoplasmic membrane of the endosymbiotic eukaryote and the phagocytotic membrane. The vestigial cytoplasm of the symbiont is also seen.

Ecology

Two species have been found living in association with green tropical siphon algae. They penetrate into dead filaments of these algae and develop there.

Number of species: 6.
Oldest known fossils: none.
Current distribution: tropical and temperate seas.

Examples

Chlorarachnion reptans; Cryptochlora perforans; Lotharella globosa; Lotharella amoeboformis; Gymnochlora stellata; Bigelowiella natans.

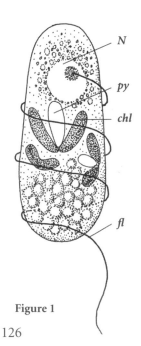

Figure 1

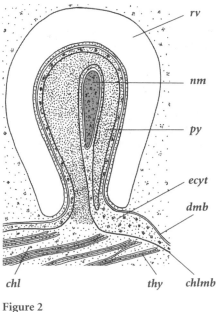

Figure 2

Excavobionta

A few representatives

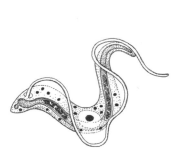

Trypanosoma brucei
30 μm
node 21

Tritrichomonas sp.
10 μm
node 22

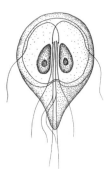

Giardia lamblia
20 μm
node 23

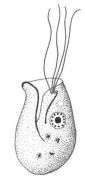

Tetramitus rostratus
50 μm
node 24

Some unique derived features

- The 18S ribosomal RNA has unique derived characteristics.

- These unicellular organisms have an "excavation," a cytopharynx; the flagella emerge from this ampulla or groove (fig. 1: *Euglena spirogyra; amp:* ampulla; *fl:* flagellum).

> Number of species: 2,068.
> Oldest known fossils: none.
> Current distribution: worldwide.

Examples

- EUGLENOBIONTA: euglena: *Euglena gracilis;* trypanosome: *Trypanosoma brucei.*
- PARABASALIA: *Calonympha grassii; Trichomonas vaginalis; Tritrichomonas* sp.
- METAMONADINA: *Giardia lamblia; Retortamonas* sp.
- PERCOLOZOA: *Naegleria gruberi; Tetramitus rostratus.*

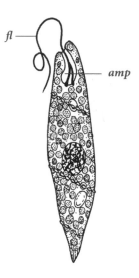

fl — *amp*

Figure 1

Euglenobionta

General Description

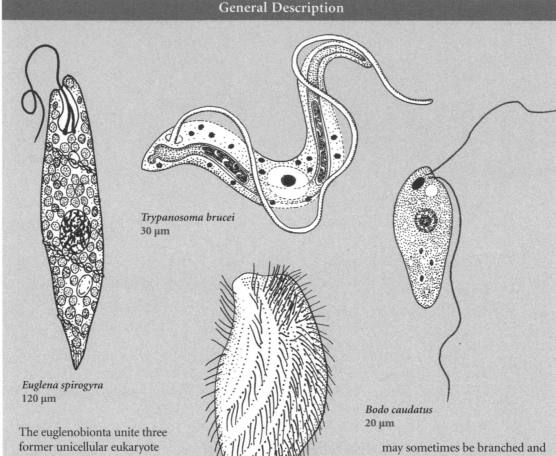

Euglena spirogyra
120 μm

Trypanosoma brucei
30 μm

Stephanopogon apogon
40 μm

Bodo caudatus
20 μm

The euglenobionta unite three former unicellular eukaryote branches. Molecular phylogenies, based primarily on the 18S rRNA gene, have confirmed the grouping of this taxon, a grouping that was initially proposed 30 years ago. This group is biologically surprising because it includes unicellular, photosynthetic algae (the euglena), human blood parasites and their relatives (the trypanosomes), and unique organisms of the genus *Stephanopogon*, organisms originally considered as "primitive ciliates"!

The euglenophyta are highly mobile cells, typically with two flagella (one small and one large) inserted into an anterior canal or ampulla (gullet). In photosynthetic species, a photosensitive organelle is situated close to the ampulla. The longer flagellum has a unilateral line of fine hairs. When present, the chloroplasts have three membranes and contain chlorophylls a and b, β-carotene and xanthophylls. These characteristics are similar to those of the chlorobionta, the lineage from which the plastids were derived by secondary endosymbiosis. The cell is covered in a rigid pellicle composed of spiral protein strips and supported by microtubules. The mitotic spindle lies within the nuclear envelope (mitosis is closed) and the chromosomes remain condensed at interphase.

The kinetoplastida are flagellated parasites lacking pigment. Each cell contains a single, large mitochondrion, which is typically tubular but may sometimes be branched and runs almost the entire length of the cell. The cell contains a large amount of DNA organized into a unique organelle that characterizes the group, the kinetoplast. The kinetoplast is found in close proximity to the kinetosomes (the flagellar basal bodies). The flagella typically arise from a depression in the naked cell surface. One of the flagella may become recurrent and form an undulating membrane. The two kinetosomes are parallel and linked together by filaments. The pseudociliata are non-photosynthetic, single-celled organisms with flagella arranged into several rows. The apical mouth is surrounded by outgrowths that serve for feeding. The multinucleated cell can have 2 to 16 identical nuclei. The mitotic spindle is intranuclear. See Sequenced Genomes.

Some unique derived features

- The cristae of the mitochondria are discoid, that is, they are flattened with a narrow base much like a ping-pong racket (fig. 1: *Trypanosoma congolense, dc:* discoid crista, *fl:* flagellum, *G:* Golgi apparatus, *mi:* mitochondrion, *mit:* microtubules, *N:* nucleus).

- The cellular membrane is supported by characteristic sheets of cortical microtubules. On fig. 1, these microtubules are shown only at the posterior end of the cell.

- Molecular phylogenies based on cytochrome C, 5S and 18S rRNA all support the monophyly of this group.

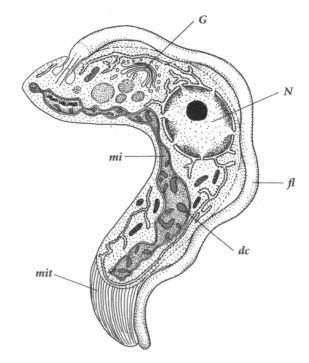

Figure 1

Ecology

The euglenophyta are free-swimming unicellular organisms found in fresh water, particularly those rich in organic compounds. They feed via a cytosome, a type of cellular mouth. Photosynthetic species move towards light using a photosensitive organ. Sexual reproduction is not known in this group.

The kinetoplastida are either free-swimming biflagellated organisms of fresh or salt water or parasitic organisms. Species of the genus *Leishmania* provoke leishmaniasis in humans; the parasite is transmitted from one host individual to another by certain mosquito species. For example, *Leishmania tropica* infests epidermal macrophages and causes cutaneous ulcers; *Leishmania donovani* is a deadly parasite that infests macrophages in the liver, spleen and bone marrow. Species of the genus *Trypanosoma* cause sleeping sickness and Chagas disease. The vectors are biting insects like the tsetse fly of the genus *Glossina*. Trypanosomiasis is a serious chronic disease of livestock responsible for large economic losses in Africa. In humans, the trypanosome life cycle takes place entirely within the blood and lymphatic fluid. When the parasite reaches the central nervous system, it provokes drowsiness and then a coma.

The pseudociliata of the genus *Stephanopogon* are marine benthic unicellular organisms that live in the fissures of marine sediment and feed on bacteria, diatoms and flagellated organisms.

Number of species: 1,398.
Oldest known fossils: none.
Current distribution: worldwide.

Examples

- EUGLENOPHYTA: euglena: *Euglena gracilis; E. spirogyra.*
- KINETOPLASTIDA: *Bodo caudatus; Crithidia fasciculata; Leishmania donovani; L. tropica;* trypanosome: *Trypanosoma brucei; T. cruzi; T. gambiense; T. rhodesiense.*
- PSEUDOCILIATA: *Stephanopogon apogon.*

Parabasalia

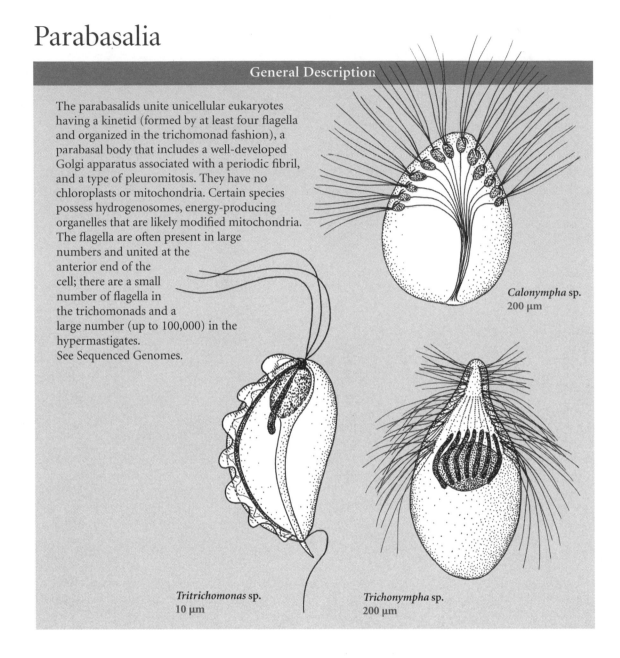

General Description

The parabasalids unite unicellular eukaryotes having a kinetid (formed by at least four flagella and organized in the trichomonad fashion), a parabasal body that includes a well-developed Golgi apparatus associated with a periodic fibril, and a type of pleuromitosis. They have no chloroplasts or mitochondria. Certain species possess hydrogenosomes, energy-producing organelles that are likely modified mitochondria. The flagella are often present in large numbers and united at the anterior end of the cell; there are a small number of flagella in the trichomonads and a large number (up to 100,000) in the hypermastigates.
See Sequenced Genomes.

Calonympha sp.
200 μm

Tritrichomonas sp.
10 μm

Trichonympha sp.
200 μm

Ecology

The unicellular parabasalids feed by phagocytosis. They store their energy in the form of glycogen. Almost all species are endosymbionts or endoparasites of insects or vertebrates. For example, we find them in the digestive tract of cockroaches and termites. In xylophagic termites, some parabasalid species actively participate in wood digestion, a process that the insect could not carry out alone. Other species are parasites of the digestive tube, the urogenital tract or buccal cavity of vertebrates. For example, *Trichomonas vaginalis* is a parasite of the human urogenital tract. Sexual reproduction exists in this group, but is poorly described.

Some unique derived features

- Kinetid: in its simplest form, the kinetid includes four flagella, three anterior and one recurrent (fig. 1: kinetid of *Tritrichomonas*; *Ax:* axostyle, *pc:* parabasal costa, *afl:* anterior flagella, *rfl:* recurrent flagellum, *umb:* undulating membrane, *pel:* pelta). The kinetosomes have a unique organization in this group and are associated with characteristic cytoskeletal structures: the pre-axostylar fiber, the pelta, the parabasal fibers.

- The parabasal body, a structure unique to this group, is composed of striated filamentous fibers and is always associated with a Golgi apparatus (fig. 2: *Joenia annectens; perr:* periaxostylar rings, *bact:* symbiotic bacteria, *parb:* parabasal body, *fl:* flagella, *N:* nucleus).

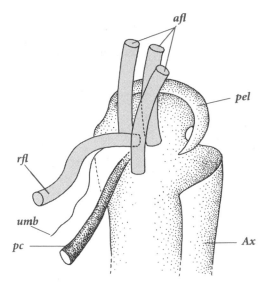

Figure 1

- The mitotic spindle that transports the chromosomes during cell division is extranuclear. The centriole structures that orient the spindle fibers during division are complex. They are the microtubule organizing centers (MTOC), fibrous extensions of certain basal bodies (fig. 3: *MTOC:* microtubule organizing center, *chr:* chromosome, *sp:* spindle, *nmb:* nuclear membrane).

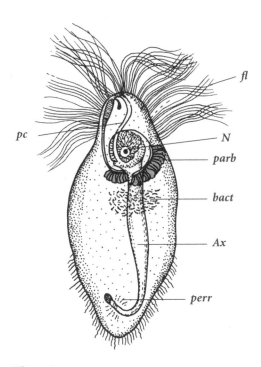

Figure 2

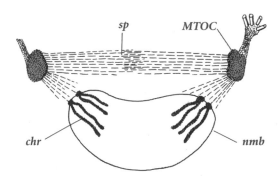

Figure 3

Number of species: 350.
Oldest known fossils: none.
Current distribution: worldwide.

Examples

Calonympha grassii; Joenia annectens; Trichonympha agilis; Trichomonas vaginalis; Tritrichomonas sp.

Metamonadina

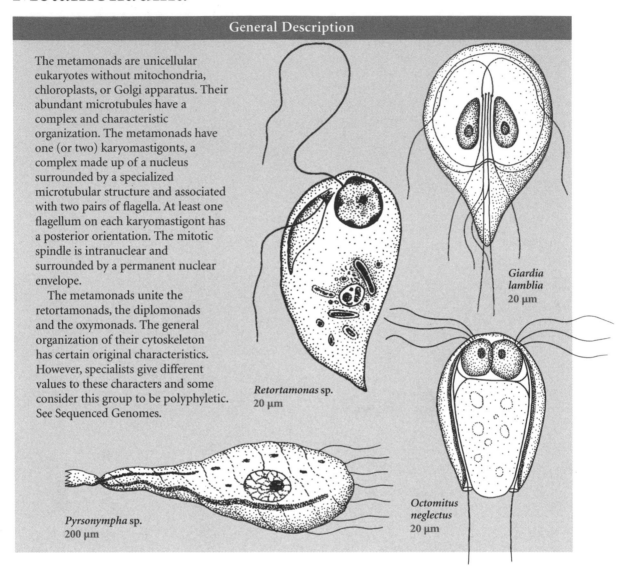

General Description

The metamonads are unicellular eukaryotes without mitochondria, chloroplasts, or Golgi apparatus. Their abundant microtubules have a complex and characteristic organization. The metamonads have one (or two) karyomastigonts, a complex made up of a nucleus surrounded by a specialized microtubular structure and associated with two pairs of flagella. At least one flagellum on each karyomastigont has a posterior orientation. The mitotic spindle is intranuclear and surrounded by a permanent nuclear envelope.

The metamonads unite the retortamonads, the diplomonads and the oxymonads. The general organization of their cytoskeleton has certain original characteristics. However, specialists give different values to these characters and some consider this group to be polyphyletic. See Sequenced Genomes.

Retortamonas sp.
20 μm

Giardia lamblia
20 μm

Octomitus neglectus
20 μm

Pyrsonympha sp.
200 μm

Some unique derived features

- The basic unit of the organism is the karyomastigont; fig. 1 shows the karyomastigont ultrastructure of *Hexamita*, a diplomonad with two nuclei (*afl*: anterior flagella, *rfl*: recurrent flagellum, *N*: nucleus).

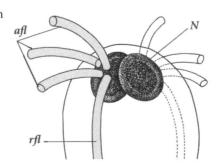

afl

N

rfl

Figure 1

Examples

- RETORTAMONDIDA: *Chilomastix intestinalis*; *Retortamonas* sp.
- DIPLOMONADIDA: *Giardia lamblia*; *Octomitus neglectus*.
- OXYMONADIDA: *Saccinobaculus latus*, *Personympha* sp.

Number of species: 300.
Oldest known fossils: none.
Current distribution: worldwide.

Ecology

The retortamonads are very small, flagellated organisms that live commensally or parasitically in the digestive tube of different animals (insects, leeches, birds, mammals). The diplomonads can be found free-living in fresh water rich in organic matter, but also as parasites in the digestive tube of animals (leeches, vertebrates). The oxymonads are commensal or symbiotic organisms of vertebrates (toads, lizards, guinea pig) and xylophagic insects like termites. *Giardia lamblia*, a human intestinal parasite, is cultivated in the laboratory.

Percolozoa

General Description

The percolozoa are amoeboid-type organisms that lack a Golgi apparatus. They have the ability to change from an amoeboid form to a flagellated form depending on environmental conditions. For each species, it is usually one form or the other that predominates. For example, in *Tetramitus rostratus*, a coprophilic organism that lives in low-oxygen environments, the flagellated form dominates the life cycle. A high concentration of salt or oxygen induces transformation into the amoeboid form that is able to encyst. In contrast, *Naegleria gruberi* lives in the amoeboid form in soils rich in organic matter. When the environment becomes too liquid, it transforms into the flagellated form that neither feeds nor reproduces; this form is used to find a solid milieu where the organism can then transform back into the amoeboid form. See Sequenced Genomes.

Tetramitus rostratus
50 μm

Some unique derived features

■ During a single life cycle, the organism can change from an amoeboid form to a flagellated form and the reverse, as shown in the life cycle of *Naegleria gruberi* (fig. 1: *Am:* amoeboid form, *Fl:* flagellated form, *C:* cyst).

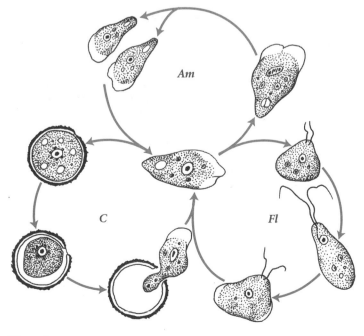

Figure 1

Ecology

Many species live in organically rich environments, sometimes low in oxygen. Some species are parasitic. For example, *Naegleria fowleri*, an amoeba found in polluted swimming pools, penetrates through the nasal passages of swimmers; the amoeboid form migrates along the olfactory nerves to the brain where the parasite proliferates, phagocytizing nerve cells and bringing a rapid death to the host.

Number of species: 20.
Oldest known fossils: none.
Current distribution: worldwide.

Examples

Naegleria fowleri; N. gruberi; Tetramitus rostratus; Vahlkampfia magna.

Unikonta

A few representatives

King bolete
Boletus edulis
15 cm
node 28

African grey parrot
Psittacus erithacus
33 cm
node 32

Amoeba proteus
800 µm
node 34

Echinostelium minutum
2 mm
node 35

Some unique derived features

■ The unicellular organisms and free-living multicellular organisms all have a single flagellum (fig. 1: mammalian spermatozoid; *fl*: flagellum, *ch*: conical head, *N*: nucleus, *acr*: acrosome, *mid*: middle section, *mi*: mitochondria, *cet*: centrioles). As the bikonta are the sister group, we cannot say whether this character is ancestral or derived.

■ Three of the six genes that code for enzymes involved in the biosynthesis of pyrimidine are fused and code for a single protein with multi-enzymatic activities (carbamoyl-phosphate synthase II, dihydroorotase and aspartate carbamoyltransferase).

■ The phosphofructokinase gene is duplicated and the two paralogous genes have fused.

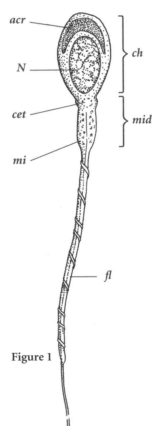

Figure 1

Examples

■ OPISTHOKONTA: baker's yeast: *Saccharomyces cerevisiae;* king bolete: *Boletus edulis;* bath sponge: *Spongia officinalis;* squid: *Loligo vulgaris;* seven-spotted ladybeetle: *Coccinella septempunctata;* African grey parrot: *Psittacus erithacus.*

■ AMOEBOZOA: *Amoeba proteus; Dictyostelium discoideum; Entamoeba histolytica.*

Number of species: 1,313,264.
Oldest known fossils: certain groups (jellyfish, bilaterians) are present in the Ediacara fauna (680 MYA).
Current distribution: worldwide.

Opisthokonta

A few representatives

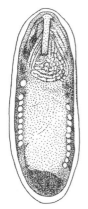
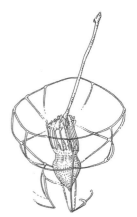

King bolete
Boletus edulis
15 cm
node 28

Nosema lepocreadii
60 μm
node 29

Pleurosiga minima
10 μm
node 31

African grey parrot
Psittacus erithacus
33 cm
node 32

Some unique derived features

- Uniting taxa that include both the eumycetes and the metazoans was never proposed with any real support until the advent of molecular phylogenies. Indeed, several independently analyzed genes produce the opisthokonta clade, and now give it a certain reliability. Other synapomorphies have since become evident.

- Ancestrally, chitin was used as a structural macromolecule. For example, it is found in the cell wall of the eumycetes and in the cuticle of the insects.

- Flagellated cells (such as spermatozoa) always possess a propulsing flagellum that is found in a posterior position. This is where the name of the taxon comes from: *opistho*, behind and *konta*, flagellum. The other known flagellated organisms are all pulled by their flagella, with the flagella therefore being in an anterior position (fig. 1: a. movement of an animal spermatozoid and b. a biflagellated ulvophyte *Chlamydomonas*; the arrows indicate the direction of movement).

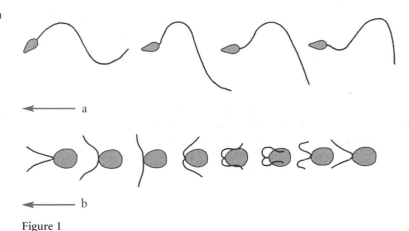

Figure 1

Number of species: 1,312,532.
Oldest known fossils: burrows, probably dug by bilaterian metazoans, were fossilized during the Precambrian (700 MYA). Certain groups (jellyfish, bilaterians) are present in the Ediacara fauna (680 MYA).
Current distribution: worldwide.

Examples

- FUNGI: baker's yeast: *Saccharomyces cerevisiae*; king bolete: *Boletus edulis*; *Penicillium chrysogenum*; *Nosema locustae*; *Diaphanoeca pedicellata*.
- CHOANOZOA: bath sponge: *Spongia officinalis*; squid: *Loligo vulgaris*; earthworm: *Lumbricus terrestris*; seven-spotted ladybeetle: *Coccinella septempunctata*; cat: *Felis catus*.

Fungi

A few representatives

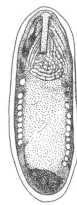

King bolete
Boletus edulis
15 cm
node 28

Morel
Morchella vulgaris
5 cm
node 28

Pin mold
Mucor mucedo
2 mm
node 28

*Nosema
lepocreadii*
60 µm
node 29

Some unique derived features

A molecular phylogeny based on the 18S ribosomal RNA gene had suggested the early emergence of three eukaryotic groups lacking mitochondria: the microsporidia, the parabasalia and the metamonadina. It was thought that these three groups were relic eukaryotes that existed prior to mitochondrial endosymbiosis. It has since been shown that these organisms have lost their mitochondria secondarily and that their position in the 18S rRNA tree was caused by a reconstruction artifact due to their rapid rate of evolution.

Phylogenies based on other genes, like *tubulin* or the protein family *Hsp 70*, bring the microsporidia and the eumycetes closer together. In fact, it is very likely that the microsporidia are eumycetes that underwent cellular simplification due to their parasitic way of life.

Number of species: 100,800.
Oldest known fossils: a type of eumycete fossil, associated with fossilized embryophyte tissue, dates from the Devonian (380 MYA). Glomales were fossilized during the Ordovician (460 MYA).
Current distribution: worldwide.

Examples

- EUMYCETES: *Mucor racemosus;* baker's yeast: *Saccharomyces cerevisiae;* king bolete: *Boletus edulis; Aspergillus niger; Penicillium chrysogenum.*
- MICROSPORIDA: *Glugea stephani; Nosema locustae.*

➡ *see Appendix 1, p. 517*

Eumycetes

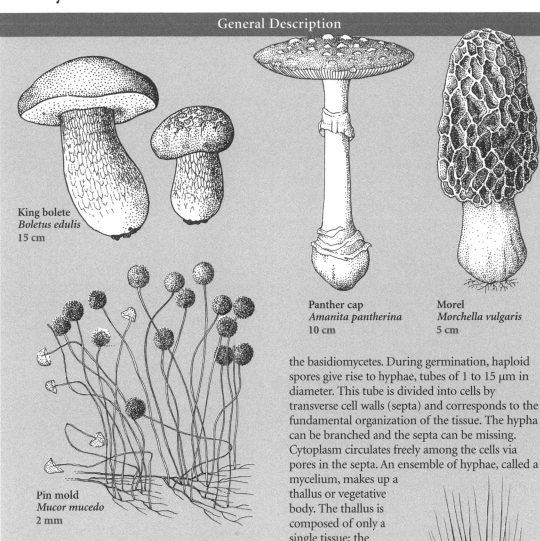

General Description

King bolete
Boletus edulis
15 cm

Pin mold
Mucor mucedo
2 mm

Panther cap
Amanita pantherina
10 cm

Morel
Morchella vulgaris
5 cm

The eumycetes are a diverse group of eukaryotes. They can be unicellular or multicellular. They lack flagella and cilia (except for the chytrids) and produce spores during reproduction. Their cell walls contain chitin, never cellulose. Morphological variation in the group ranges from simple microscopic ellipsoidal cells, like yeasts, to large terrestrial basidiomycete mushrooms whose fruiting body (basidiocarp) can reach several decimeters in size. In fact, certain thalli (*Armillaria*) can cover several thousand hectares. The group contains the ascomycota (yeasts, truffles, certain molds), the basidiomycota (the mushrooms), and the zygomycota (molds, such as pin molds). The names of these groups come from the structure that carries the spores produced by meiosis (ascus, basidium). Eumycete tissues have a filamentous organization. There is no embryonic development or tissue differentiation. At most, we see the beginning of tissue diversification in a few ascomycetes and in the basidiomycetes. During germination, haploid spores give rise to hyphae, tubes of 1 to 15 μm in diameter. This tube is divided into cells by transverse cell walls (septa) and corresponds to the fundamental organization of the tissue. The hypha can be branched and the septa can be missing. Cytoplasm circulates freely among the cells via pores in the septa. An ensemble of hyphae, called a mycelium, makes up a thallus or vegetative body. The thallus is composed of only a single tissue; the eumycetes do not have vascular tissue, even if many species are macroscopic and terrestrial. There is never photosynthesis. In terrestrial forms, the cells are not mobile. Molecular phylogenies place a group of uncertain origin within the eumycetes, the chytridiomycota. These species are mostly algal parasites. See Sequenced Genomes.

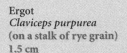

Ergot
Claviceps purpurea
(on a stalk of rye grain)
1.5 cm

Some unique derived features

- The emergence of this group is due principally to molecular phylogenies based on the 18S rRNA gene, along with some protein coding genes (elongation factors, heat-shock proteins).
- A unique metabolic pathway for the biosynthesis of lysine (amino acid) is found in this group.

Ecology

The eumycetes are mostly terrestrial: parasitic or symbiotic. They are largely aerobic and heterotrophic, most often being saprophytic (feeding from decomposing organic matter). They are fundamentally absorbotrophic and typically grow on previously hydrolyzed, organic compounds. Parasitic species feed on the nutrients provided by the host. The eumycetes can secrete lytic enzymes for active extracellular digestion. After exodigestion of the organic compounds, they absorb the soluble nutrients, never solid elements. Approximately 25,000 species (most often ascomycetes) live in symbiosis with green algae or cyanobacteria, forming an association we call lichen. Many lichens are pioneering organisms; by dissolving rock, they are at the origin of soil formation.

Most eumycetes (except the basidiomycetes) have a haploid vegetative body (thallus). They are capable of asexual multiplication by fragmentation or budding. Fertilization occurs by the fusion of hyphal cells of different sexual types; there are never mobile gametes, except in the chytrids. After conjugation, the haploid nuclei remain associated for a time without actually fusing, forming dikaryotic cells. After fusing, the nuclei form an egg (zygote). However, this diploid state is only temporary because meiosis takes place immediately and produces haploid spores. When these spores germinate, they first give rise to hyphae, then to haploid mycelia. This is not true for the basidiomycetes whose dikaryotic phase is the most important in the life cycle. Sexual reproduction has been lost in certain eumycetes.

Many eumycetes require very little to grow. Certain species grow on acidic milieu or in environments lacking nitrogen. Many species cause parasitic diseases, principally in plants. Others are symbiotic, notably in the forest where soil eumycetes intervene in the transport of certain substances to plant roots. In some species, protection against predators has resulted in the synthesis of toxic chemical compounds, notably alkaloids such as those of the *Amanita* that can be lethal for mammals.

The eumycetes are of considerable medical, sanitary and nutritive importance. They can be human parasites (skin mycosis, candidiasis, etc.), but also a source of pharmacological molecules. For example, penicillin is produced by *Penicillium*. Finally, the eumycetes are of a certain economic interest: beer and bread both require the metabolic activity of yeast (*Saccharomyces*) for their production. Many cheeses also require the action of eumycetes. Numerous species are eaten directly by humans (cepes, boletes, morels, etc.). *Neurospora crassa* is famous for its role in the elucidation of gene function.

Number of species: 100,000, but there are certainly more.
Oldest known fossils: a type of eumycete fossil, associated with fossilized plant tissue, dates from the Devonian (380 MYA). Glomales were fossilized during the Ordovician (460 MYA).
Current distribution: worldwide.

Examples

- CHYTRIDIOMYCOTA: *Chytridium lagenaria* (parasite of green algae), *Olpidium viciae* (parasite of vetch).
- ZYGOMYCOTA: *Mucor mucedo, Rhizopus niger, Glomus mossae.*
- ASCOMYCOTA: ergot of rye: *Claviceps purpurea;* morel: *Morchella vulgaris; Neurospora crassa;* baker's yeast: *Saccharomyces cerevisiae;* truffle: *Tuber melanosporum.*
- BASIDIOMYCOTA: panther cap: *Amanita pantherina;* king bolete: *Boletus edulis;* corn smut: *Ustilago maydis;* stem rust: *Puccinia graminis.*

Microsporidia

These unicellular eukaryotes are obligate intracellular parasites that lack mitochondria, chloroplasts, cilia and flagella. They possess one or two nuclei and a well-developed Golgi apparatus. At one point in the life cycle, the cells take the form of resistant spores equipped with chitinous cell walls. These spores contain a protractile filament or "polar tube" that is used to infect host cells by inoculating them with the sporoplasm, the infective agent. This tube, derived from the Golgi apparatus, is often complex and occupies most of the internal space of the spore. See Sequenced Genomes.

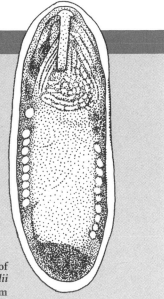

Spore of
Nosema lepocreadii
60 μm

Some unique derived features

- Secondary loss of mitochondria, as shown by the presence of mitochondrial-type genes in the nuclear genome.
- Fusion of the 5.8S and 18S ribosomal RNA genes.
- The polar tube is coiled within the spore (fig.1; *tu*: polar tube).

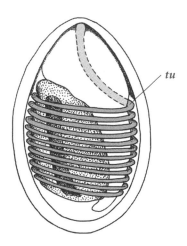

Figure 1

Ecology

The microsporidia are intracellular parasites of a large variety of hosts (apicomplexa, ciliates, cnidaria, nematodes, annelids, mollusks, but principally vertebrates and arthropods). We especially find them in teleost fishes, insects and crustaceans. When a spore germinates, it extrudes its polar tube and inoculates the host cell with sporoplasm. Within parasitized cells, the injected sporoplasm then will undergo mitotic division (the mitotic spindle is contained within the nucleus) to rise to a multi-nucleated plasmal mass (syncytium). This step can then be followed by sporogony around each nucleus of the new "generation." New spores within the host cell are formed in large parasitophorous vacuoles that are surrounded by simple cellular membranes. Sexual reproduction is unknown in the microsporidia, but it is presumed to exist. *Nosema bombycis*, initially recognized as the causal agent of pebrine disease in silkworms by Louis Pasteur, causes significant damage in the natural silk industry. Species of the genus *Glugea* provoke tumors in teleost fishes. In humans, microsporidia are pathogenic only in immunodepressed individuals.

Number of species: 800.
Oldest known fossils: none.
Current distribution: worldwide.

Examples

Glugea stephani; Ichthyosporidium giganteum; Nosema lepocreadii; N. locustae.

Choanozoa

A few representatives

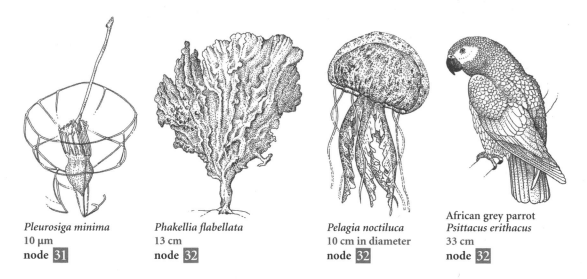

Pleurosiga minima
10 μm
node **31**

Phakellia flabellata
13 cm
node **32**

Pelagia noctiluca
10 cm in diameter
node **32**

African grey parrot
Psittacus erithacus
33 cm
node **32**

Some unique derived features

■ The grouping of the choanoflagellates and the sponges was previously based on similarities in the cellular structure of the choanoflagellates and the choanocytes of the sponges. The hypothesis was centered on the idea that the sponges were derived from colonial choanoflagellates. This simplistic idea was abandoned in the middle of the last century. Molecular phylogenies based on the rRNA 18S gene have since proposed grouping the choanoflagellates with the metazoans, thus rekindling the hypothesis of a structural homology between the two cell types. It remains to be shown whether this hypothesis is truly valid.

Number of species: 1,211,732.
Oldest known fossils: burrows probably dug by bilaterian metazoans were fossilized in the Precambrian (700 MYA). Certain groups (jellyfish, bilaterians) are present in the Ediacara fauna (680 MYA).
Current distribution: worldwide.

Examples

■ CHOANOFLAGELLATA: *Diaphanoeca pedicellata*.
■ METAZOA: bath sponge: *Spongia officinalis;* planaria: *Planaria maculata;* human intestinal roundworm: *Ascaris lumbricoides;* common squid: *Loligo vulgaris;* earthworm: *Lumbricus terrestris;* seven-spotted ladybeetle: *Coccinella septempunctata;* common starfish: *Asterias rubens;* cat: *Felis catus.*

Choanoflagellata

The choanoflagellates are single-celled eukaryotes equipped with a flagellum. They can be solitary or colonial. The cells are often protected by an organic theca or a siliceous lorica. Each cell has a collar and tentacles that aid in feeding.

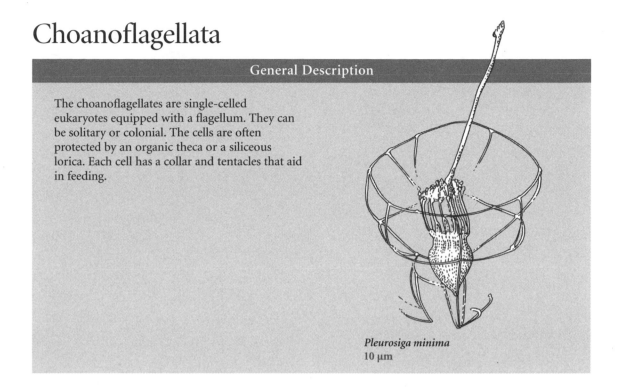

Pleurosiga minima
10 µm

Some unique derived features

■ The cell is equipped with a collar and tentacles (fig. 1; *col:* collar, *te:* tentacles).

■ The cell secretes a theca or lorica (fig.1; *the:* theca).

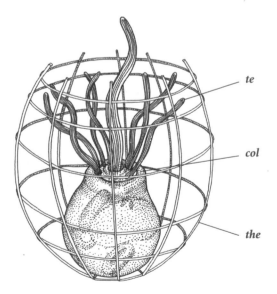

— te

— col

— the

Figure 1

Ecology

The choanoflagellates are principally marine organisms, but some freshwater forms also exist. They can be planktonic or sessile (attached to algae) and typically feed on bacteria. The ecological importance of this group is largely unknown.

Examples

Diaphanoeca pedicellata; Parvicorbicula socialis; Pleurosiga minima.

Number of species: 120.
Oldest known fossils: none.
Current distribution: worldwide, in oceans.

➡ *see chap. 7, p. 196*

Metazoa

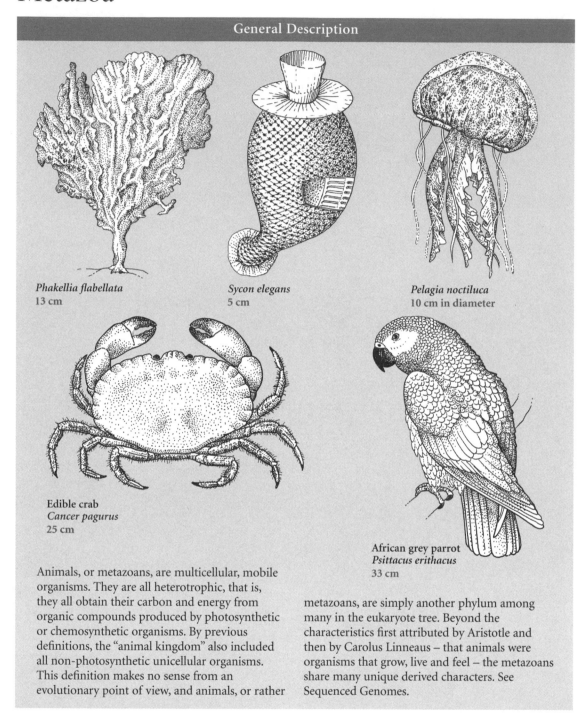

General Description

Phakellia flabellata
13 cm

Sycon elegans
5 cm

Pelagia noctiluca
10 cm in diameter

Edible crab
Cancer pagurus
25 cm

African grey parrot
Psittacus erithacus
33 cm

Animals, or metazoans, are multicellular, mobile organisms. They are all heterotrophic, that is, they all obtain their carbon and energy from organic compounds produced by photosynthetic or chemosynthetic organisms. By previous definitions, the "animal kingdom" also included all non-photosynthetic unicellular organisms. This definition makes no sense from an evolutionary point of view, and animals, or rather metazoans, are simply another phylum among many in the eukaryote tree. Beyond the characteristics first attributed by Aristotle and then by Carolus Linneaus – that animals were organisms that grow, live and feel – the metazoans share many unique derived characters. See Sequenced Genomes.

Some unique derived features

Figure 1

■ Collagen: long, filamentous protein that serves as the extracellular matrix. By linking cells together, it contributes to the construction, architecture and development of multicellular animal organisms.

Fig. 1 shows the structure of the type IV collagen molecule, with two non-collagen regions at the extremities; the triple helix is interrupted by non-helical segments (in color).

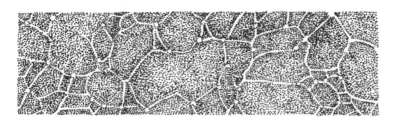

Figure 2

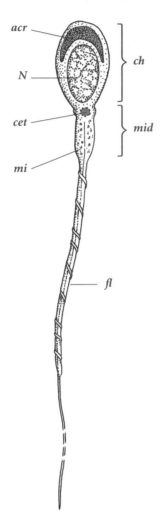

Figure 3

Fig. 2 illustrates an *in vitro* assemblage of type IV collagen molecules (mesh size: 250 nm). Covalent bonds connect the molecules; the resulting network is similar to that of the basal lamina (basement membrane).

■ Spermatozoid structure (fig. 3: mammalian spermatozoid): a conical head (*ch*) contains the nucleus (*N*) and an acrosome (*acr*), and a middle section (*mid*) is made up of mitochondria (*mi*), two centrioles (*cet*) and a flagellum (*fl*).

■ Other molecules, like fibronectin or integrin, are characteristic of metazoans.

■ Centriole: at the base of cilia and flagella, there is an accessory centriole found perpendicular to the basal body.

■ Meiosis directly produces gametes, not spores.

■ During meiosis of the female gamete, only the cell which contains the reserves is functional. The three other cells, or polar bodies, degenerate.

■ Desmosomes: cells can be joined to one another by strong membranous junctions called desmosomes (fig. 4: desmosome of a human keratinocyte: *ifi*: intermediate filament, *pmb*: plasma membrane, *cyp*: cytoplasmic plate).

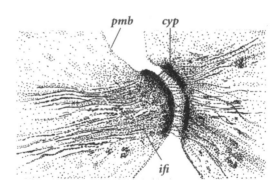

Figure 4

Number of species: 1,211,612.
Oldest known fossils: burrows probably dug by bilaterian metazoans were fossilized in the Precambrian (700 MYA). Certain groups (jellyfish, bilaterians) are present in the Ediacara fauna (680 MYA).
Current distribution: worldwide.

Ecology

The metazoans have colonized all terrestrial, aquatic and marine environments, from the highest mountain to the deepest ocean abyss, from the driest desert to the coldest areas on earth.

Examples

Bath sponge: *Spongia officinalis;* planaria: *Planaria maculata;* human intestinal roundworm: *Ascaris lumbricoides;* common squid: *Loligo vulgaris;* earthworm: *Lumbricus terrestris;* seven-spotted ladybeetle: *Coccinella septempunctata;* common starfish: *Asterias rubens;* cat: *Felis catus.*

Amoebozoa

A few representatives

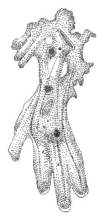

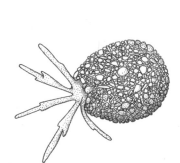

Amoeba proteus
800 μm
node 34

Difflugia oblonga
300 μm
node 34

Echinostelium minutum
2 mm
node 35

Pelomyxa palustris
1 mm
node 35

Some unique derived features

The rhizopods are likely paraphyletic.

■ Ribosomal RNA sequences and those of the actin genes have unique
derived characteristics.

> **Number of species:** 732.
> **Oldest known fossils:** none.
> **Current distribution:** worldwide.

Examples

■ RHIZOPODA: *Amoeba proteus; Difflugia
oblonga.*
■ MYCETOZOA: *Dictyostelium discoideum;*
flowers of tan or dog vomit slime mold:
Fuligo septica; Entamoeba histolytica.

Rhizopoda

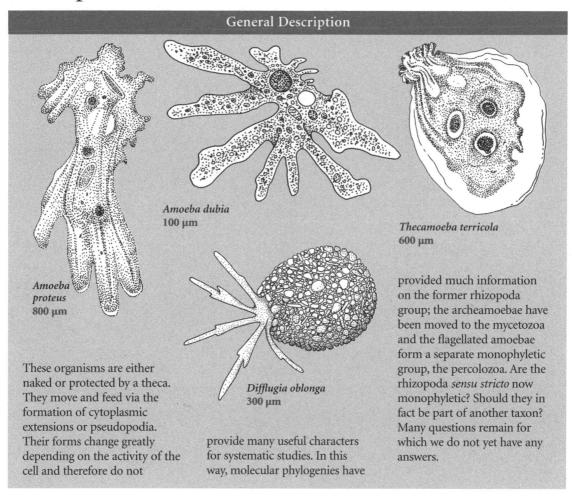

General Description

Amoeba dubia
100 μm

Thecamoeba terricola
600 μm

Amoeba
proteus
800 μm

Diffugia oblonga
300 μm

These organisms are either naked or protected by a theca. They move and feed via the formation of cytoplasmic extensions or pseudopodia. Their forms change greatly depending on the activity of the cell and therefore do not

provide many useful characters for systematic studies. In this way, molecular phylogenies have

provided much information on the former rhizopoda group; the archeamoebae have been moved to the mycetozoa and the flagellated amoebae form a separate monophyletic group, the percolozoa. Are the rhizopoda *sensu stricto* now monophyletic? Should they in fact be part of another taxon? Many questions remain for which we do not yet have any answers.

Some unique derived features

■ Pseudopodia: there are, in reality, no unique derived characters in cytological terms given that recent classifications are based on molecular phylogenies. However, the existence of pseudopodia is potentially a true synapomorphy, with a possible convergence in other groups (percolozoa, mycetozoa, chlorarachniophyta, foraminifera). Fig. 1 shows different types of pseudopodia (a: lobose, b: limax type, with two positions, c: fan-shaped, d: radiating).

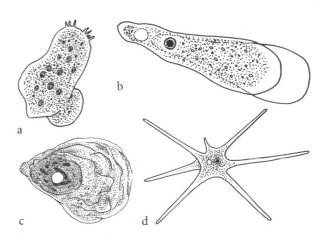

Figure 1

Number of species: 200.
Oldest known fossils: none.
Current distribution: worldwide.

Examples

Amoeba dubia; A. proteus; Thecamoeba terricola; Diffugia oblonga.

Ecology

The rhizopoda are unicellular organisms that we find in solid or liquid environments rich in organic matter. Certain groups have unique ecologies; for example, the thecamoeba use primarily freshwater habitats or live in mosses. Some species are parasitic.

145

Mycetozoa

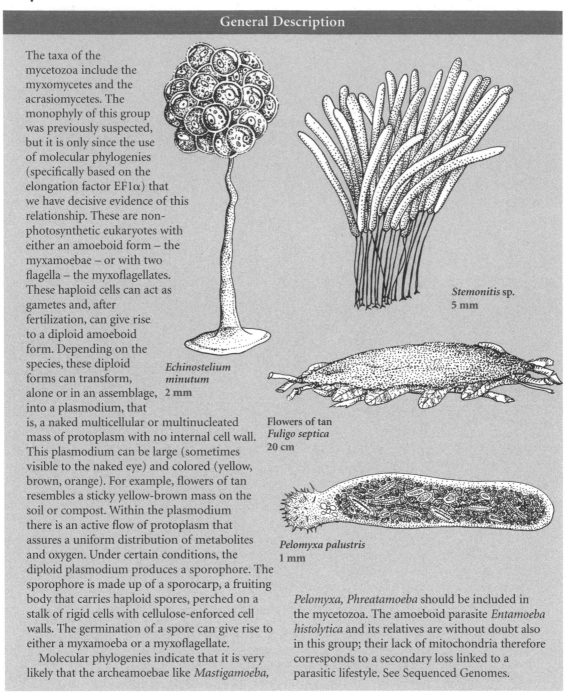

General Description

The taxa of the mycetozoa include the myxomycetes and the acrasiomycetes. The monophyly of this group was previously suspected, but it is only since the use of molecular phylogenies (specifically based on the elongation factor EF1α) that we have decisive evidence of this relationship. These are non-photosynthetic eukaryotes with either an amoeboid form – the myxamoebae – or with two flagella – the myxoflagellates. These haploid cells can act as gametes and, after fertilization, can give rise to a diploid amoeboid form. Depending on the species, these diploid forms can transform, alone or in an assemblage, into a plasmodium, that is, a naked multicellular or multinucleated mass of protoplasm with no internal cell wall. This plasmodium can be large (sometimes visible to the naked eye) and colored (yellow, brown, orange). For example, flowers of tan resembles a sticky yellow-brown mass on the soil or compost. Within the plasmodium there is an active flow of protoplasm that assures a uniform distribution of metabolites and oxygen. Under certain conditions, the diploid plasmodium produces a sporophore. The sporophore is made up of a sporocarp, a fruiting body that carries haploid spores, perched on a stalk of rigid cells with cellulose-enforced cell walls. The germination of a spore can give rise to either a myxamoeba or a myxoflagellate.

Molecular phylogenies indicate that it is very likely that the archeamoebae like *Mastigamoeba,*

Echinostelium minutum 2 mm

Stemonitis sp. 5 mm

Flowers of tan *Fuligo septica* 20 cm

Pelomyxa palustris 1 mm

Pelomyxa, Phreatamoeba should be included in the mycetozoa. The amoeboid parasite *Entamoeba histolytica* and its relatives are without doubt also in this group; their lack of mitochondria therefore corresponds to a secondary loss linked to a parasitic lifestyle. See Sequenced Genomes.

Ecology

The mycetozoans are species that live in humid environments, on old wood and other decaying organic materials. They are heterotrophic and absorb bacteria or decomposing particles by phagocytosis. Asexual reproduction is frequent. Sexual reproduction takes place when the environment becomes unfavorable. In such conditions, the diploid plasmodium produces an erect fruiting body, the sporophore. The aggregation of *Dictyostelium discoideum* myxamoebae into a plasmodium has been heavily studied and modeled; the morphogene responsible for this behavior is cyclic AMP.

Some unique derived features

■ An alternation of unicellular, flagellated or amoeboid, and multicellular phases in the life cycle. The plasmodium (*pl*) produces sporocarps (*spo*). The dispersed spores (*spe*) are haploid and, after germination, give rise to either amoeboid cells (*amc*) or bi-flagellated cells (*flc*), depending on the environmental conditions. These cells then behave like gametes and give rise to a zygote that germinates to produce a plasmodium (fig. 1: *M:* meiosis, *F:* fertilization).

■ The genes of cytochrome oxidase *cox1* and *cox2* are fused together in the mitochondrial genome.

■ The elongation factor gene EF1α has unique characters.

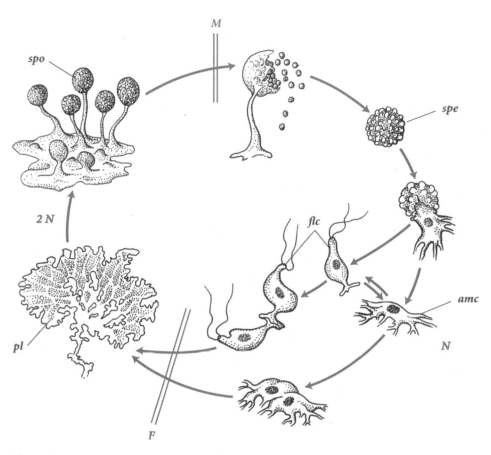

Figure 1

Examples

■ ACRASIOMYCETES: *Dictyostelium discoideum.*
■ MYXOMYCETES: *Didymium iridis;* flowers of tan or dog vomit slime mold: *Fuligo septica; Physarum polycephalum; Stemonitis* sp.
■ ARCHEAMOEBAE: *Mastigamoeba; Pelomyxa palustris; Phreatamoeba; Entamoeba histolytica.*

Chapter 5

CHLOROBIONTA

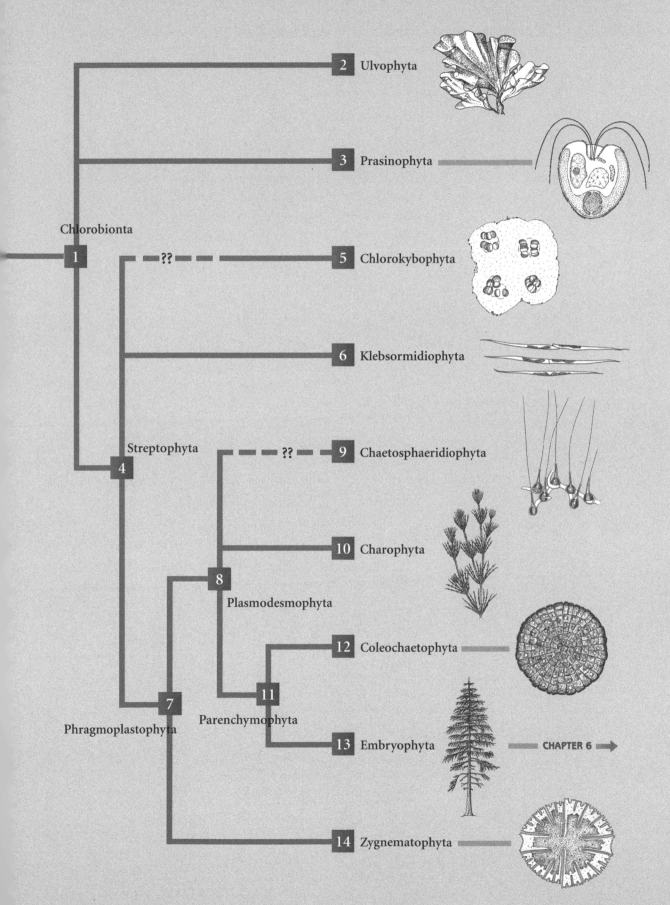

- 2 Ulvophyta
- 3 Prasinophyta
- 1 Chlorobionta
- 5 Chlorokybophyta
- 6 Klebsormidiophyta
- 4 Streptophyta
- 9 Chaetosphaeridiophyta
- 10 Charophyta
- 8 Plasmodesmophyta
- 12 Coleochaetophyta
- 11 Parenchymophyta
- 13 Embryophyta   CHAPTER 6 ⇒
- 7 Phragmoplastophyta
- 14 Zygnematophyta

The chlorobionta constitute a group of considerable importance from both ecological and phylogenetic points of view. Indeed, we find them in the seas and oceans, both as plankton and benthos, but also on all land masses. The terrestrial plants form a monophyletic sub-group of the chlorobionta.

For a long time, the evolutionary origin of the land plants was a mystery. The debate truly started in 1869 when the German botanist W. Hofmeister (1824–1877) discovered some important similarities among the reproductive cycles of some of the large terrestrial plant groups. All embryophytes have digenetic reproductive cycles (fig. 1) with an alternation of multicellular haploid (the gametophyte) and diploid (the sporophyte) phases. A good number of green algae, such as *Ulva lactuca,* also show this type of phase alternation. The majority of well-known botanists (A. H. Church, F. E. Fritsch, N. Pringsheim, etc.) defended what is called the "homologous theory"; the ancestor of the land plants had an isomorphic reproductive cycle with equilibrated haploid and diploid phases.

An alternative hypothesis, called the "antithetic theory," was developed around the same period by F. O. Bower. This theory, based on the work of L. Celakovsky in 1874, proposed that the land plants evolved from freshwater, filamentous green algae with a predominant haploid phase in the life cycle. To make this hypothesis work, certain crucial evolutionary steps are required, such as the retention of the oosphere (female gamete) on the gametophyte and the nourishment of the zygote by this phase. In addition, meiosis of the zygote cannot take place immediately after germination. Rather, the cell would first have to undergo mitosis. Meiosis would therefore be delayed and the sporophyte, the result of zygotic mitosis, would have appeared by the insertion of this phase between gametophytic phases.

Recent work in systematics, including biochemical, ultrastructural and molecular studies, suggests that the coleochaetophyta and the charophyta are the closest groups to the land plants. These groups have haploid life cycles with oogamy, therefore clearly supporting the "antithetic theory" of Celakovsky and Bower. A century was required to come to this solution.

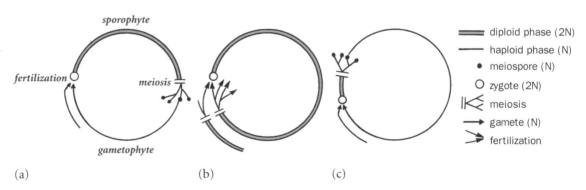

Figure 1: Different possible types of life cycles.

a. Diplohaplontic cycle, with equal parts of the life cycle attributed to the haploid and diploid phases. This is seen, for example, in the ulvophycean *Ulva lactuca;* b. Diplontic cycle, with a predominant diploid phase, as seen in certain brown algae (*Fucus*) and animals; c. Haplontic cycle, with a predominant haploid phase, similar to that of the zygnematophyta, charophyta and coleochaetophyta.

Ulvophyta

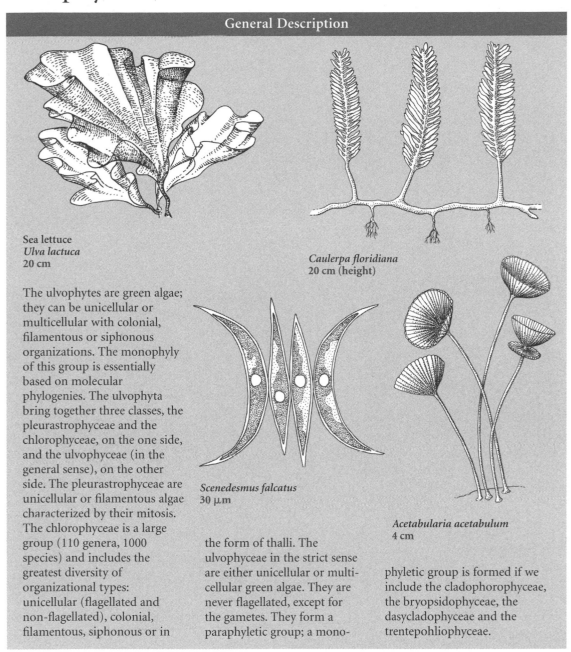

General Description

Sea lettuce
Ulva lactuca
20 cm

Caulerpa floridiana
20 cm (height)

Scenedesmus falcatus
30 μm

Acetabularia acetabulum
4 cm

The ulvophytes are green algae; they can be unicellular or multicellular with colonial, filamentous or siphonous organizations. The monophyly of this group is essentially based on molecular phylogenies. The ulvophyta bring together three classes, the pleurastrophyceae and the chlorophyceae, on the one side, and the ulvophyceae (in the general sense), on the other side. The pleurastrophyceae are unicellular or filamentous algae characterized by their mitosis. The chlorophyceae is a large group (110 genera, 1000 species) and includes the greatest diversity of organizational types: unicellular (flagellated and non-flagellated), colonial, filamentous, siphonous or in the form of thalli. The ulvophyceae in the strict sense are either unicellular or multi-cellular green algae. They are never flagellated, except for the gametes. They form a paraphyletic group; a mono-phyletic group is formed if we include the cladophorophyceae, the bryopsidophyceae, the dasycladophyceae and the trentepohliophyceae.

Some unique derived features

■ Cytoskeleton: The cytoskeleton at the base of the flagella shows one of four arrangements in the chlorobionta, as shown in fig. 1 (a, b, c, or d). The ulvophyta have one of the three arrangements b, c or d, where the bases of the flagella are at 180° to each other and the microtubule bundles (*mit*) are at 90° to the flagella. Fig. 1 shows the four different cytoskeletal arrangements of the bases of the flagella in the green algae; a. streptophyta; b. ulvophyceae (*sensu lato*); c. pleurastrophyceae; d. chlorophyceae (*fl*: flagellum, *prot*: condensed proteins).

Ecology

The ulvophyta are a major component of marine and freshwater phytoplankton. It has been calculated that they fix more than a billon tons of carbon per year by photosynthesis. As they are found at the base of the aquatic ecological food chain, the ulvophyta are an important food source for a multitude of unicellular and multicellular organisms. This group also includes fixed, multicellular forms, certain of which are very commonly found along the coasts. Sexual reproduction varies greatly from one group to another within the ulvophyta. In some species, two

151

mobile gametes fuse (isogamy). In other species, a large immobile gamete is fertilized by a small mobile gamete (oogamy). This biflagellated spermatozoid can sometime greatly resemble the adult (in *Chlamydomonas*, for example). In the chlorophyceae,

some species are haploid for a large part of their life cycle, whereas others have an alternation of haploid and diploid phases. The pleurastrophyceae often live in association with fungi, forming what we call lichen.

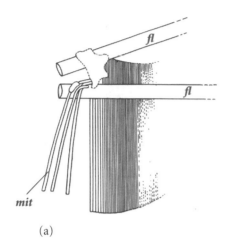

(a)

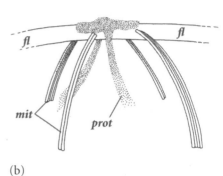

(b)

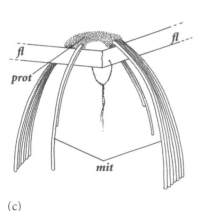

(c)

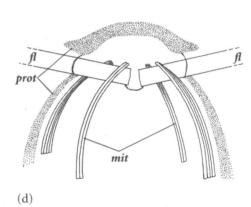

(d)

Figure 1

Number of species: 3,679.
Oldest known fossils: there may be unicellular and filamentous chlorophycean fossils in the rocks of the Bitter Springs formation in central Australia that date from the Precambrian (900 MYA). Fossils that resemble some *Coelastrum* and *Cladophora* species have been found in Spitzbergen (800 MYA).
Current distribution: worldwide.

Examples

- PLEURASTROPHYCEAE: *Trebouxia parmeliae; Friedmannia* sp.; *Myrmecia* sp.
- CHLOROPHYCEAE: *Chlamydomonas reinhardtii; Chlorococcum echinozygotum; Scenedesmus falcatus; Volvox globator.*
- ULVOPHYCEAE (*sensu stricto*): yellow-green hairweed: *Enteromorpha intestinalis;* sea lettuce: *Ulva lactuca.*
- CLADOPHOROPHYCEAE: *Cladophora vagabunda; Chaetomorpha darwinii.*
- BRYOPSIDOPHYCEAE: *Bryopsis tumosa; Caulerpa taxifolia; Codium tomentosum.*
- DASYCLADOPHYCEAE: *Acetabularia acetabulum; Dasycladus vermicularis.*
- TRENTEPOHLIOPHYCEAE: *Trentepohlia aurea.*

Prasinophyta

General Description

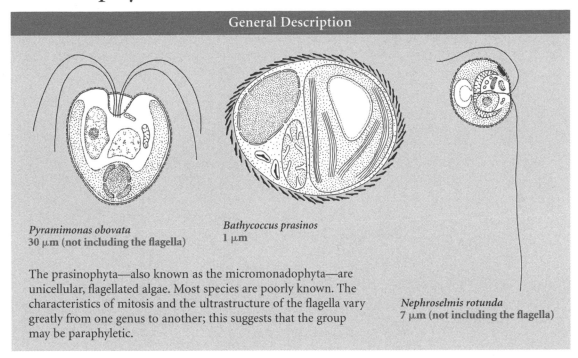

Pyramimonas obovata
30 μm (not including the flagella)

Bathycoccus prasinos
1 μm

The prasinophyta—also known as the micromonadophyta—are unicellular, flagellated algae. Most species are poorly known. The characteristics of mitosis and the ultrastructure of the flagella vary greatly from one genus to another; this suggests that the group may be paraphyletic.

Nephroselmis rotunda
7 μm (not including the flagella)

Some unique derived features

■ Cell and flagellum: these organisms are covered in one or several layers of organic scales synthesized in the Golgi body. These scales have a particular shape and chemical composition. For example, they can be star-shaped (fig. 1a: *Nephroselmis rotunda*), or shield-like (fig. 1b: *Bathycoccus*). The cells carry from one to eight flagella inserted either laterally or apically (fig. 2a: flagellated stage of *Pterosperma*, 5 mm in diameter not including the flagella).

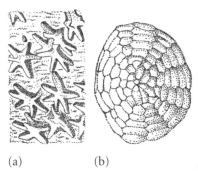

(a)   (b)

Figure 1

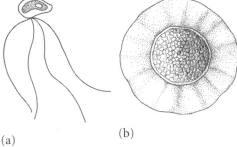

(a)   (b)

Figure 2

■ Cysts or phycomata: these are the resistant forms (fig. 2b: phycoma of *Pterosperma*, 250 μm in diameter)

■ Chloroplasts: they contain chlorophyll c, in addition to chlorophylls a and b.

Number of species: 180.
Oldest known fossils: cysts similar to those of the *Pterosperma* genus are known since the Precambrian (1200 MYA).
Current distribution: worldwide.

Ecology

The prasinophyta are flagellated marine or freshwater algae that make up a large proportion of the marine plankton. In particular, the *Bathycoccus* genus forms the picoplankton (organisms less than 2 μm in diameter). Many prasinophytes are endosymbionts; among the classic examples are *Pedinomonas*, symbiont of the dinophyte *Noctiluca scintillans*, and *Tetraselmis convolutae*, symbiont of the intertidal platyhelminth *Convoluta roscoffensis*. For a long time, only asexual multiplication was documented in this group. Recently, sexual reproduction was observed in a freshwater alga, *Nephroselmis rotunda*. This species undergoes isogamy: two morphologically identical gametes (+ and −) fuse to form a zygote. After a period of stasis, the zygote undergoes meiotic cell division. The life cycle is therefore haplobiontic.

Examples

Bathycoccus prasinos; Nephroselmis rotunda; Pyramimonas lunata; Tetraselmis convolutae.

Streptophyta

A few representatives

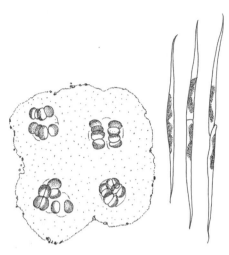

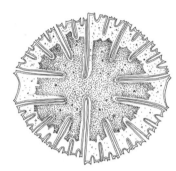

Chlorokybus atmophyticus
200 μm
node 5

Raphidonema longiseta
70 μm (length of one cell)
node 6

Nordmann fir
Abies nordmanniana
50 m
node 13

Micrasterias papillifera
60 μm
node 14

Some unique derived features

■ Sexual reproduction: this occurs by oogamy. In this form of reproduction, only one of the gametes (the male, by definition) is flagellated, while the other (the female) is immobile and filled with reserves. The ancestral state of this character is planogamy (both gametes are mobile).

■ Cytoskeleton: the cytoskeleton at the base of the flagellum has a particular, asymmetrical arrangement that is very homogeneous in the group (fig. 1; *fl*: flagellum, *mit*: microtubules).

■ Flagellated cells (spores or gametes): when seen with an electron microscope, these cells have a multi-layered proteinaceous structure (*MLS*:

multilayered structure) associated with the base of the flagellum (fig. 2: example of *Chaetosphaeridium*).

■ Peroxysomes (*pex*) are typically present and most often found between the nucleus and the plastid (fig. 3a). They contain enzymes like glycolate oxidase and the catalase used to increase the yield from carbon assimilation.

■ Mitosis is open; the nuclear membrane disappears in prophase. Fig. 3 shows four stages of mitosis in *Klebsormidium* (a. prophase, b. metaphase, c. telophase, d. start of interphase; *cet*: centriole, *chl*: chloroplast, *nmb*: nuclear membrane, *mit*: microtubule, *cp*: cell plate). At metaphase (fig. 3b),

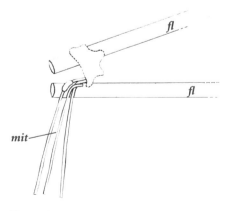

Figure 1

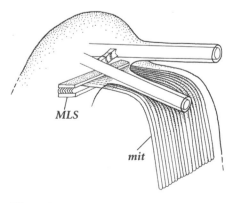

Figure 2

154

the nuclear membrane has disappeared. The division of the cell is guided by the microtubules and occurs by the formation of a groove; the cell plate then forms in this groove. In most chlorobionts, the nuclear membrane remains, even if it is sometimes fenestrated. Fig. 4 shows four stages of mitosis in the non-streptophyte *Chlamydomonas* (a. prophase; b. metaphase; c. telophase; d. start of interphase; *sp:* spindle). At metaphase (fig. 4b), the nuclear membrane is still apparent even though the spindle has formed.

■ Microtubular spindle: the spindle persists at the end of mitosis, when the separation of the nuclei occurs, even though the nuclear membranes have already re-formed (telophase: fig. 3c). In the other chlorobionts (with a few exceptions), the spindle has disappeared at telophase and a system of microtubules that run parallel to the future axis of cell division, called the phycoplast (*phc*, fig. 4c), has appeared.

■ There is the possibility for filamentous growth.

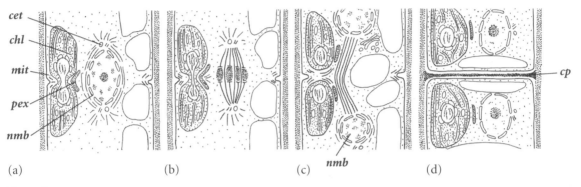

(a)   (b)   (c)   (d)

Figure 3

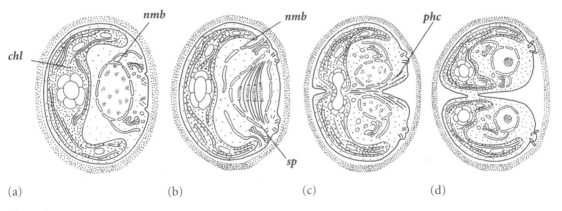

(a)   (b)   (c)   (d)

Figure 4

Number of species: 274,043.
Oldest known fossils: tetrads of embryophyte spores (that is, spores grouped into fours after meiosis) date from the middle Ordovician (460 MYA).
Current distribution: worldwide.

Examples

■ CHLOROKYBOPHYTA: *Chlorokybus atmophyticus.*
■ KLEBSORMIDIOPHYTA: *Klebsormidium flaccidum.*
■ PHRAGMOPLASTOPHYTA: *Desmidium schwartzii; Micrasterias papillifera; Spirogyra fluviatilis; Mougeotia tenuis; Zygnema peliosporum; Chaetosphaeridium minus; Chara vulgaris; Nitella gracilis; Coleochaete scutata;* bracken fern: *Pteridium aquilinum;* corn: *Zea mays;* crab apple: *Malus sylvestris.*

Chlorokybophyta

General Description

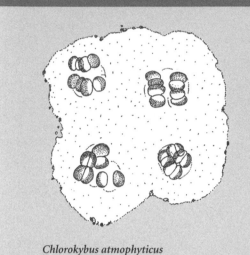

Chlorokybus atmophyticus
200 μm

This group includes only a single monospecific genus, *Chlorokybus*. Previously, this species was placed with the klebsormidiophyta. However, a molecular phylogeny and its simplified organization have led to its separation into a distinct group. Nevertheless, the phylogenetic situation of this group is problematic (*incertae sedis*). The organism is composed of cubical clusters of two to eight spherical or ellipsoid cells surrounded by a gelatinous matrix. This species has only started to be studied recently; for now, sexual reproduction is unknown.

Some unique derived features

- Peroxysome: it is similar to that of the charophyta but is not found between the nucleus and the plastid.

- Chloroplast: two pyrenoids (*py*) can be seen in the cell's chloroplast using optical microscopy (fig. 1).

- After mitosis, the cells remain in clusters (fig. 2; *cw*: cell wall).

Ecology

The only known species grows on rocky, subterrain substrates. Asexual reproduction occurs by the production of biflagellated spores.

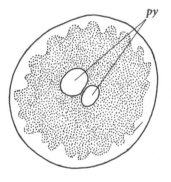

Figure 1

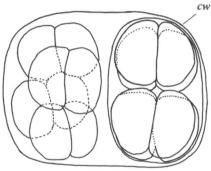

Figure 2

Number of species: 1.
Oldest known fossils: none.
Current distribution: Europe.

Examples

Chlorokybus atmophyticus.

Klebsormidiophyta

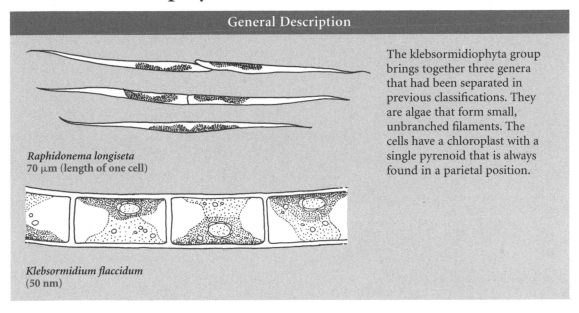

Raphidonema longiseta
70 μm (length of one cell)

Klebsormidium flaccidum
(50 nm)

The klebsormidiophyta group brings together three genera that had been separated in previous classifications. They are algae that form small, unbranched filaments. The cells have a chloroplast with a single pyrenoid that is always found in a parietal position.

Some unique derived features

■ Reproduction: known reproduction occurs by a particular form of oogamy (fig. 1). In *Raphidonema longiseta*, the male gamete (*gam*) is biflagellated (fig. 1a); after fertilization, the zygote (*zy*) has a cyst-like structure (fig. 1b).

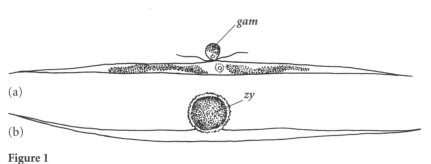

Figure 1

Ecology

The general biology of this group is not well known. For example, sexual reproduction has only been established in a single *Raphidonema* species. These organisms are freshwater species, but certain of them are found in unique habitats. Some *Klebsormidium* and *Stichococcus* have been found in terrestrial habitats (old walls, roofs, rock surfaces). *Raphidonema nivale* grows on the snow of alpine glaciers.

Number of species: 40.
Oldest known fossils: none.
Current distribution: Europe.

Examples

Klebsormidium flaccidum; Stichococcus bacillaris; Raphidonema longiseta; R. nivale.

Phragmoplastophyta

A few representatives

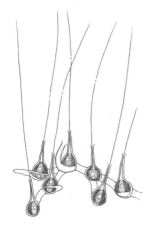

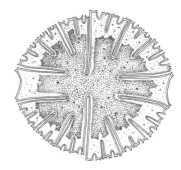

*Chaetosphaeridium
minus*
40 μm without the setae
node 9

Chara hispida
10 cm
node 10

Nordmann fir
Abies nordmanniana
50 m
node 13

Micrasterias papillifera
60 μm
node 14

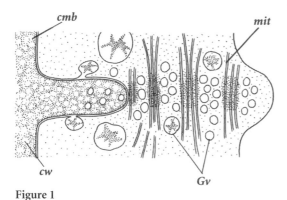

Figure 1

Some unique derived features

■ The separation of the daughter cells at the end of mitosis occurs by formation of a phragmoplast. A new system of microtubules, running parallel to the spindle, forms in the middle of the cell; these microtubules guide the movement of the Golgi vesicles that contain the materials required for the construction of the new cell wall. The details of the phragmoplast of *Spirogyra* are shown in fig. 1 (*cmb:* cell membrane, *mit:* microtubules, *Gv:* Golgi vesicles, *cw:* cell wall).

■ Synthesis of cellulose microfibrils: this is performed by a rosette complex of cellulose

Number of species: 274,002.
Oldest known fossils: tetrads of embryophyte spores (that is, spores grouped into fours after meiosis) date from the middle Ordovician (460 MYA).
Current distribution: worldwide.

synthetases, typically composed of eight sub-units. In *Micrasterias* (fig. 2), the rosette terminals (*ros*) cross the cytoplasmic membrane and take molecules from the cytoplasm (*cyt*). Microfibrils (*mfi*) of 5 nm are then synthesized. The association of these individual microfibrils will produce a cellulose fibril (*cfi*) that will be used in the construction of the cell wall.

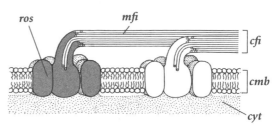

Figure 2

■ Cell wall: contains polyphenolic compounds.

■ The transfer RNA of alanine (tRNA Ala) contains a group II intron; it is not interrupted in the other chlorobionts.

Examples

■ PLASMODESMOPHYTA: *Chaetosphaeridium minus; Chara vulgaris; Nitella gracilis; Coleochaete scutata;* awned haircap moss: *Polytrichum piliferum;* bracken fern: *Pteridium aquilinum;* corn: *Zea mays;* crab apple: *Malus sylvestris.*

■ ZYGNEMATOPHYTA: *Desmidium schwartzii; Micrasterias papillifera; Spirogyra fluviatilis; Mougeotia tenuis; Zygnema peliosporum.*

Plasmodesmophyta

A few representatives

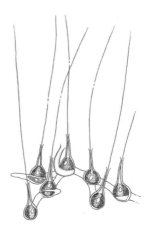

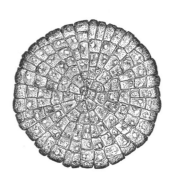

Chaetosphaeridium minus
40 μm without the setae
node 9

Chara hispida
10 cm
node 10

Coleochaete orbicularis
1 mm
node 12

Nordmann fir
Abies nordmanniana
50 m
node 13

Some unique derived features

- During the formation of the phragmoplast, protoplasmic connections between the two daughter cells, the plasmodesmata, are set up. Fig. 1 shows four steps of mitosis in *Coleochaete*. The nuclear membrane disappears at the end of prophase (fig. 1a: early prophase; 1b: metaphase; *cet*: centriole, *nmb*: nuclear membrane). The spindle (*sp*) remains at anaphase (fig. 1c). New microtubules organize the phragmoplast (*phmt*: phragmoplast microtubules). The plasmodesmata

- (*pld*) permit rapid communication between the cells (fig. 1d: start of interphase).

- Apical growth enables branching.

- The oosphere is retained on the organism.

- The zygote grows after fertilization by the transfer of nutrients, mostly lipids and carbohydrates.

- There is a group II intron in the transfer RNA of isoleucine (tRNA Ile), whereas in the other chlorobionts the tRNA is not interrupted.

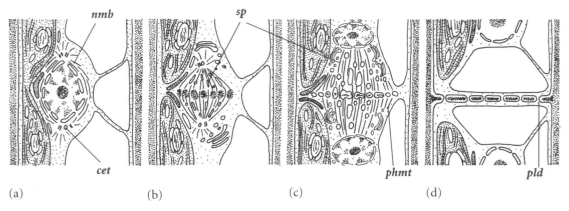

(a) (b) (c) (d)

Figure 1

Number of species: 270,002.
Oldest known fossils: tetrads of embryophyte spores (that is, spores grouped into fours after meiosis) date from the middle Ordovician (460 MYA).
Current distribution: worldwide.

Examples

- CHAETOSPHAERIDIOPHYTA: *Chaetosphaeridium minus.*
- CHAROPHYTA: *Chara vulgaris; Nitella gracilis.*
- PARENCHYMOPHYTA: *Coleochaete scutata;* awned haircap moss: *Polytrichum piliferum;* bracken fern: *Pteridium aquilinum;* corn: *Zea mays;* crab apple: *Malus sylvestris.*

Chaetosphaeridiophyta

General Description

The chaetosphaeridiophyte taxon includes only a single genus, *Chaetosphaeridium*. Although this group was for a long time placed with *Coleochaete* because of their distinctive cytoplasmic extensions (setae), these organisms are nonetheless distinguishable by certain traits, including the structure of several genes. Their taxonomic situation is unclear (*incertae sedis*). They occur in isolated or diffuse groups of cells without any true organization and are partially enveloped in a mucilage.

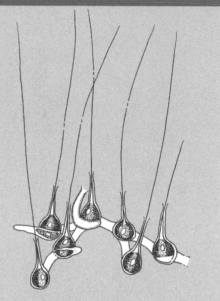

Chaetosphaeridium minus
40 μm without the setae

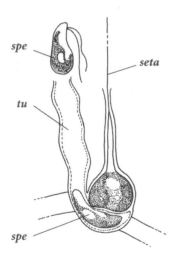

Figure 1

Some unique derived features

It should be pointed out that little is known about this group in general; for example, the ultrastructural details of mitosis have never been studied.

- The spherical cell (fig. 1) has a long cytoplasmic extension, the seta, that is surrounded by a sheath produced by the cell wall. *Chaetosphaeridium* was previously included in the coleochaetophyta on the basis of this characteristic.

- Chloroplast: a single chloroplast lies in a parietal position and has one pyrenoid.

- After mitosis, the biflagellated basal spore (*spe*) can be released via a hyaline tube (*tu*) (fig. 1).

Ecology

The chaetosphaeridiophyta live as epiphytes on large, freshwater algae. Reproduction has been observed only in a single species; it occurs by oogamy with biflagellated, male gametes. Asexual reproduction occurs via biflagellated spores (fig. 1).

Number of species: 4.
Oldest known fossils: none.
Current distribution: Europe and North America.

Examples

Chaetosphaeridium minus.

Charophyta

General Description

All of the members of the charophyta have the same general form, a modular structure with nodes and internodes. As shown in fig. 1, whorls of lateral expansions (falsely called "leaves") give the appearance of a horsetail. In a South American alga, *Nitella cernua*, the overall size ranges from 20 cm to more than 1 m. The internode cells can be very large (up to 25 cm) and may have strongly calcified cell walls.

Chara vulgaris
10 cm

Nitella batrachosperma
10 cm

Chara hispida
10 cm

Some unique derived features

Figure 1

- Cells: each mature cell has a giant vacuole in its center. The cytoplasm has incessant currents that run longitudinally due to the interaction between actin and myosin molecules.

- Chloroplasts: discoid chloroplasts are found in the peripheral layer of the cytoplasm. They lack pyrenoids.

- The antheridia and oogonia are formed at the nodes, at the axils of the lateral extensions, by a complex process (fig. 2: an example of *Chara fragilis*; *ant*: antheridium, *oo*: oogonium). The spermatozoids have a characteristic spiral form and carry two flagella. In the oogonia, the oospheres are surrounded by five spirally arranged cells (*spr*) and crowned by five apical cells (*ace*) that control the opening by which the spermatozoid enters.

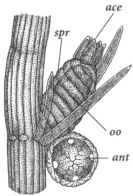

Figure 2

Ecology

The majority of species live in fresh water but a few brackish and saltwater species are known, like *Chara baltica*. These organisms are found in calm waters with soft bottoms that their rhizoids can penetrate. They are haploid; a gametophyte carries the antheridia (producing the biflagellated male gametes) and the oogonia (containing a large oosphere). After fertilization, the zygote is surrounded by a thick wall, falls to the pond floor, and becomes dormant. When germination starts, the first cell division is meiotic.

Examples

Chara vulgaris; Lamprothamnium populosum; Nitella gracilis; Nitellopsis obtusa; Tolypella glomerata.

Number of species: 81, following a recent taxonomic revision (previous authors had proposed 400 species).
Oldest known fossils: the charophyta have left many fossils because certain species have calcified cell walls. The oldest known fossil, *Praesycidium siluricum,* was found in the Ukraine and dates from the upper Silurian (420 MYA). The two major charophyte groups have been separated since the start of the Triassic. The fossils of *Eochara* and *Paleonitella* indicate that the extant forms have existed since the Devonian.
Current distribution: worldwide.

Parenchymophyta

A few representatives

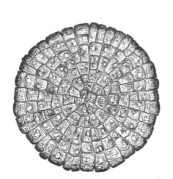

Coleochaete orbicularis
1 mm
node 12

Common polypody
Polypodium vulgare
30 cm
node 13

Nordmann fir
Abies nordmanniana
50 m
node 13

Common wheat
Triticum aestivum
1.2 m
node 13

Some unique derived features

- Zygote cell wall: contains polyphenol compounds that inhibit bacterial attacks, among other things. The synthesis of a large quantity of these compounds is activated by sexual pheromones.

- Zygote: the zygote remains on the organism, at least until sporogenesis occurs. It is surrounded by nutritive cells. We can see the arrangement of the zygotes within a thallus of *Coleochaete scutata* using optical microscopy (fig. 1a; *zy:* zygote). A cross-section of a thallus (fig. 1b) shows how the zygote is protected and nourished. Fig. 2 shows an archegonium of *Marchantia polymorpha* after fertilization; the embryo (*em*) of the sporophyte lives as a parasite on the gametophyte (*gam*).

- The cells can assemble into a parenchyma (or ground tissue), cells of no specialized function. These cells provide metabolic products used by the organism and retain their ability to differentiate (or specialize).

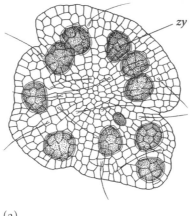

(a)

(b)

Figure 1

Number of species: 269,917.
Oldest known fossils: tetrads of embryophyte spores (that is, spores grouped into fours after meiosis) date from the middle Ordovician (460 MYA).
Current distribution: worldwide.

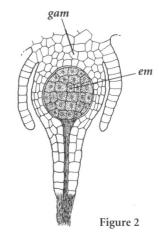

Figure 2

Examples

- COLEOCHAETOPHYTA: *Coleochaete scutata*.
- EMBRYOPHYTA: awned haircap moss: *Polytrichum piliferum*; bracken fern: *Pteridium aquilinum*; stone pine: *Pinus pinea*; corn: *Zea mays*; crab apple: *Malus sylvestris*.

Coleochaetophyta

The coleochaetophytes are tightly bundled organisms that resemble round cushions of a few millimeters in diameter. They are made up of tightly connected cell filaments that are often highly branched (fig. 1 and fig. 2: example of *Coleochaete pulvinata*). Cytoplasmic extensions (setae) are used in defense against herbivores and exude a repulsive substance when they are broken. There is oogamous reproduction. The small antheridia produce biflagellated spermatozoids and the oogonium has a trichogyne that receives the spermatozoid. The zygote is fed by the surrounding cells and then becomes dormant. At germination, it immediately undergoes meiosis producing 16 or 32 biflagellated, haploid spores. These spores will then form new thalli. The life cycle is therefore haplobiontic.

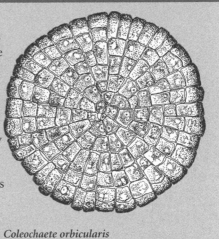

Coleochaete orbicularis
1 mm

Some unique derived features

- Each cell has a single, flattened chloroplast containing a large pyrenoid.

- Certain cells have a long cytoplasmic extension (a seta), similar to that of the chaetosphaeridiophytes.

- The oogonium (*oo*) has a large trichogyne (*trg*) (fig. 1; *ant*: antheridium).

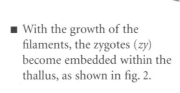

- With the growth of the filaments, the zygotes (*zy*) become embedded within the thallus, as shown in fig. 2.

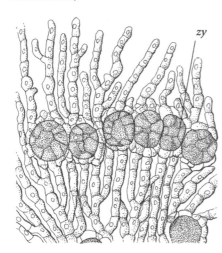

Figure 2

Figure 1

Ecology

The coleochaetophytes live as epiphytes in the shallow littoral zone of freshwater lakes. To see them, however, you must microscopically examine large algae, the leaves of aquatic plants such as pondweed, or inorganic substrates like bottles or tin cans. As they are sensitive to eutrophication, these organisms are good pollution indicators.

Number of species: 15.
Oldest known fossils: *Parka decipiens*, a Scottish fossil that dates from the upper Silurian (415 MYA), seems to be morphologically and ecologically identical to the extant coleochaetes. However, this species could reach up to 7 cm in diameter.
Current distribution: Europe and North America. Certain poorly described species may have been found in Australia and in tropical regions.

Examples

Coleochaete divergens; C. irregularis; C. nitellarum; C. orbicularis; C. pulvinata; C. scutata; C. soluta.

⇒ *see chap. 6, p. 168*

Embryophyta

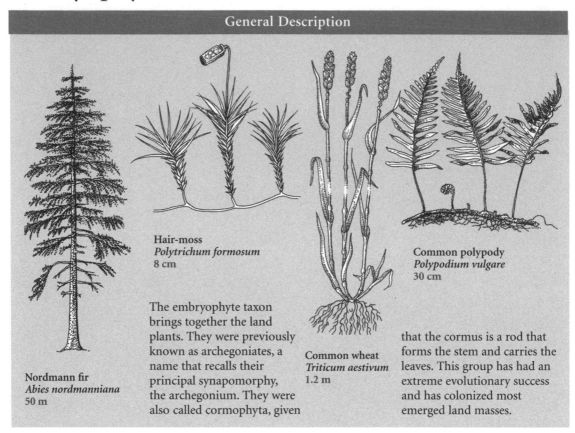

General Description

Hair-moss
Polytrichum formosum
8 cm

Common polypody
Polypodium vulgare
30 cm

The embryophyte taxon brings together the land plants. They were previously known as archegoniates, a name that recalls their principal synapomorphy, the archegonium. They were also called cormophyta, given

Common wheat
Triticum aestivum
1.2 m

Nordmann fir
Abies nordmanniana
50 m

that the cormus is a rod that forms the stem and carries the leaves. This group has had an extreme evolutionary success and has colonized most emerged land masses.

Some unique derived features

■ The female sexual organs are called archegonia. Fig. 1 shows an archegonium of *Marchantia polymorpha* (fig. 1a) and that of *Ophioglossum vulgatum* (fig. 1b). We note the venter (*v*) that protects the oosphere (*oo*) and the canal (*ca*) of the neck (*ne*) that the spermatozoids move through to reach the oosphere. The archegonium can be more or less embedded within the gametophyte (*gam*).

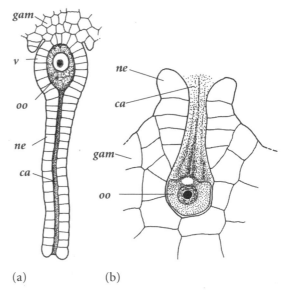

(a)   (b)

Figure 1

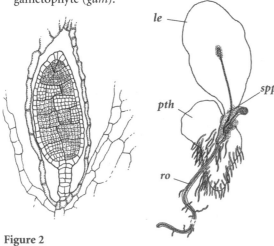

Figure 2

Figure 3

■ The male sexual organs are called antheridia. Fig. 2 illustrates an antheridium of *Marchantia polymorpha*. The grid-like structure within the antheridium is a result of cell division of the mother cells.

■ The life cycle has a diploid, multicellular phase. The diploid sporophyte, the result of fertilization, relies on the haploid gametophyte (the result of spore development), at least at the beginning of its existence. On fig. 3, we note the diploid sporophyte (*spp*) of a fern with its first root (*ro*) and leaf (*le*) developing on the haploid gametophyte (*pth:*

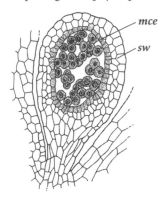

Figure 4

prothallus); this gametophyte will quickly degenerate.

■ Sporangia: the spores are contained within a sporangium with multicellular walls (*sw*) (fig. 4: example of *Lycopodium selago* close to maturity; *mce:* spore mother cells).

■ The spore cell walls contain a particular molecule, sporopollenin.

■ The epidermis is covered by a cuticle.

■ Spermatozoids: they have certain characteristic ultrastructural details, like the presence of a centrosome with a double centriole.

■ At the beginning of mitosis, the cortical microtubules come together to form a particular structure, the preprophase band. Four stages of mitosis are shown in fig. 5 (a. interphase;

b. start of prophase; c. start of telophase, *G:* Golgi body; d. end of telophase). At interphase, the microtubules (*mit*) are at the surface of the cell, lying below the cytoplasmic membrane. At the beginning of prophase, they come together at the center of the cell to form the preprophase band (*ppb*) that surrounds the nucleus (*N*). This band then disappears during the synthesis of the spindle (*sp*). In telophase, we can see the phragmoplast (*phr*) that constructs the new cell wall (*cw*) and the plasmodesmata (*pld*).

■ The *tufA* gene that codes for a chloroplastic elongation factor is found in the nuclear genome of the embryophytes. In the other streptophytes, this gene is found in the chloroplastic genome.

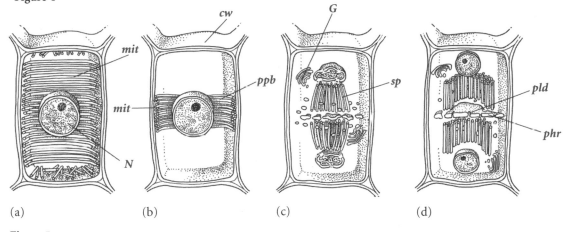

(a) (b) (c) (d)

Figure 5

Ecology

The embryophytes have colonized all possible terrestrial environments. Some species (eelgrass, pondweeds) have secondarily acquired a completely aquatic life style. The spores formed at meiosis are initially enclosed within the sporangia of the diploid sporophyte. After germination, each spore gives rise to a haploid organism, the gametophyte, that produces the gametes. The reunion of the gametes produces a zygote that gives rise to the sporophyte. The initial development of the sporophyte depends on nourishment from the gametophyte.

Number of species: 269,902.
Oldest known fossils: tetrads of embryophyte spores (that is, spores grouped into fours after meiosis) date from the middle Ordovician (460 MYA). The oldest known embryophyte macrofossil is *Cooksonia pertoni* from the end of the Silurian (410 MYA) from Scotland. We find traces of this genus starting in the middle Silurian (425 MYA).
Current distribution: worldwide.

Examples

Awned haircap moss: *Polytrichum piliferum; Marchantia polymorpha;* lake quillwort: *Isoetes lacustris;* field horsetail: *Equisetum arvense;* bracken fern: *Pteridium aquilinum;* stone pine: *Pinus pinea;* sago palm: *Cycas revoluta;* horsetail ephedra: *Ephedra equisetiformis;* corn: *Zea mays;* crab apple: *Malus sylvestris.*

Zygnematophyta

General Description

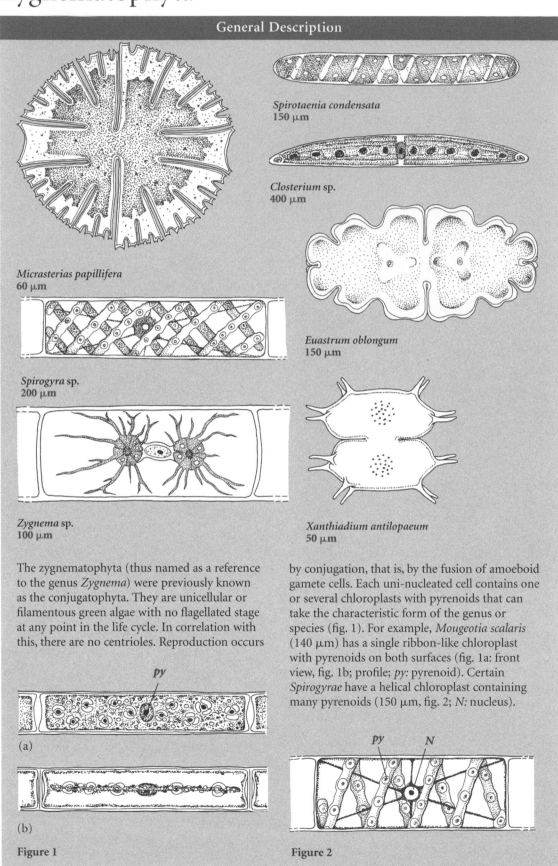

Micrasterias papillifera
60 μm

Spirotaenia condensata
150 μm

Closterium sp.
400 μm

Euastrum oblongum
150 μm

Spirogyra sp.
200 μm

Zygnema sp.
100 μm

Xanthiadium antilopaeum
50 μm

The zygnematophyta (thus named as a reference to the genus *Zygnema*) were previously known as the conjugatophyta. They are unicellular or filamentous green algae with no flagellated stage at any point in the life cycle. In correlation with this, there are no centrioles. Reproduction occurs by conjugation, that is, by the fusion of amoeboid gamete cells. Each uni-nucleated cell contains one or several chloroplasts with pyrenoids that can take the characteristic form of the genus or species (fig. 1). For example, *Mougeotia scalaris* (140 μm) has a single ribbon-like chloroplast with pyrenoids on both surfaces (fig. 1a: front view, fig. 1b; profile; *py:* pyrenoid). Certain *Spirogyrae* have a helical chloroplast containing many pyrenoids (150 μm, fig. 2; *N:* nucleus).

py

(a)

(b)

Figure 1

py   *N*

Figure 2

Some unique derived features

- There is no centriole.
- Sexual reproduction: reproduction is performed by conjugation. These algae are always haploid. Fig. 3 outlines three stages in the conjugation of *Zygogonium ericetorum*. Initially, the cells of two filaments send out cell extensions (fig. 3a; *ga*: gamete). The cell walls then fuse, forming a type of bridge that enables cell passage from one filament to the other (fig. 3b; *F*: fertilization). The gamete-cells fuse and form a zygote (*zy*, fig. 3c).
- There are only two types of organization; the species are either unicellular or filamentous, never branched.
- The small ribosomal subunit RNA gene (RNA 18S) contains a group I intron.

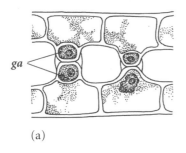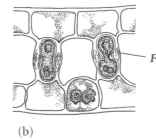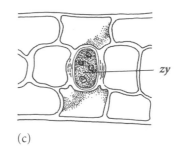

(a)   (b)   (c)

Figure 3

Ecology

The majority of species are freshwater algae. Certain species can tolerate brackish waters and a few others are terrestrial. At conjugation, the zygote, formed by the fusion of two cells, becomes surrounded in a thick outer wall and enters into a dormant period. Its germination starts with meiosis.

Number of species: 4,000.
Oldest known fossils: one fossil, *Paleoclosterium leptum*, dates from the middle Devonian (380 MYA). Fossilized zygotes have been found in Carboniferous rocks.
Current distribution: worldwide.

Examples

Closterium lunula; Cylindrocystis brebissonii; Cosmarium botrytis; Desmidium schwartzii; Micrasterias papillifera; Spirogyra fluviatilis; Mougeotia tenuis; Zygnema peliosporum.

Chapter 6

EMBRYOPHYTA

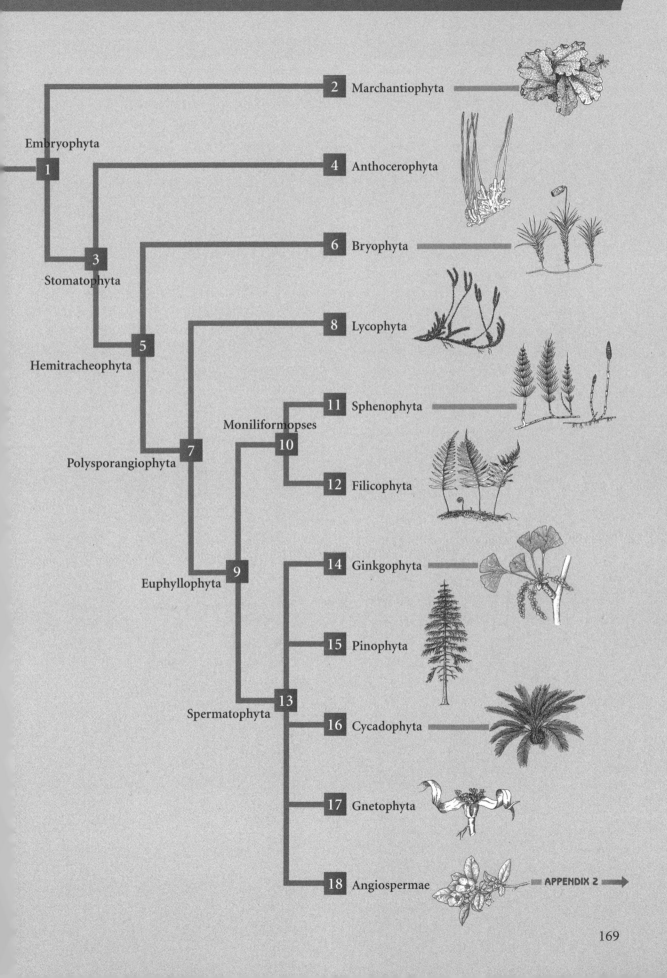

Embryophyta

1

2 Marchantiophyta

4 Anthocerophyta

Stomatophyta

3

6 Bryophyta

Hemitracheophyta

5

8 Lycophyta

Moniliformopses

11 Sphenophyta

10

12 Filicophyta

Polysporangiophyta

7

Euphyllophyta

9

14 Ginkgophyta

15 Pinophyta

Spermatophyta

13

16 Cycadophyta

17 Gnetophyta

18 Angiospermae

APPENDIX 2 →

For a long time the phylogeny of the embryophytes was interpreted according to two general theories. The first considers that the group is polyphyletic, whereas in the second it is monophyletic.

If we accept the "homologous theory"—i.e., the embryophytes evolved from a green alga with an isomorphic life cycle (equal development of both haploid and diploid phases)—we are obliged to recognize two distinct evolutionary lineages. In one of these lineages, the gametophyte became dominant, giving rise to the "mosses" *sensu lato* (or bryophytes, as they were formerly known). The other lineage developed a dominant sporophyte phase; this lineage would lead to the polysporangiophyta, a group that includes the ferns and spermatophytes.

The validation of the "antithetic theory"—i.e., the formation of the diploid phase *de novo* by a delay in the meiotic division of the zygote—requires a different interpretation of the "mosses." Indeed, their poorly developed sporophyte phase (fig. 1) suggests that they may be an intermediate stage between the hypothetical ancestor that first gave rise to the sporophyte and the ferns with their reduced gametophyte phase. However, certain authors continue to suggest a polyphyletic origin for the embryophytes based on different ancestors related to the charophytes and the coleochaetophytes. In this case, characters like the acquisition of the archegonium and antheridium would be homoplastic.

Morphological and molecular phylogenetic analyses tend to support the monophyly of the embryophytes, with the "mosses" as a paraphyletic group. This provides some evidence of the progressive adaptation of the sporophyte to terrestrial life.

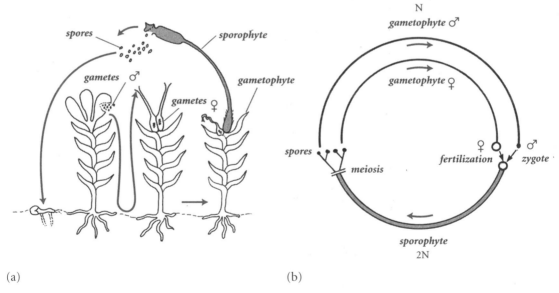

(a)　　　　　　　　　　　　　　　　(b)

Figure 1

Bryophyte life cycles: a. the sporophyte is never independent and always develops on the gametophyte; b. Both the haploid and diploid phases of the life cycle are important, but the haploid phase is dominant.

Marchantiophyta

General Description

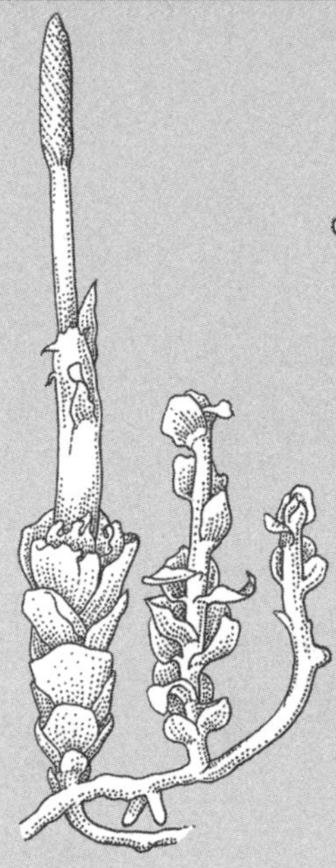

Marchantia polymorpha
5 cm

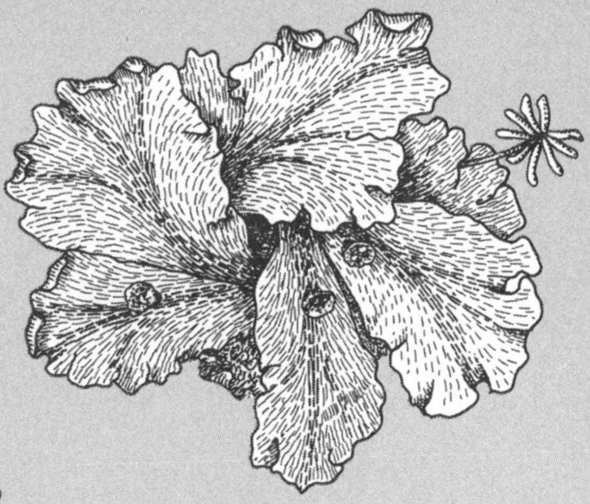

Haplomitrium sp.
1.5 cm

The marchantiophytes (from the genus name *Marchantia*), or liverworts, are small plants with no roots or vascular system. The haploid phase of the life cycle is dominant. The structure of the gametophyte is variable; some have a leafy structure, whereas others have a thalloid structure, sometimes with liver-like lobes (hence the name "liverworts"). The sporophyte capsule is simple, without a columella. Cells that will become elaters develop between the spores; the elaters aid in dispersal. The gametophyte has unicellular rhizoids. There are no true stomata, although openings that allow gas exchange do exist. The cross-section of the gametophyte of *Marchantia polymorpha* (fig. 1) shows a pore in the epidermis that leads to an air chamber containing chlorophyll filaments; the unicellular rhizoids are visible on the lower surface.

Some unique derived features

- Gametophytes: the gametophyte has oil bodies (*ob*) in its cells (fig. 1; *ap*: air pore, *rz*: rhizoid).

- Lunularic acid is synthesized.

- Elaters: before the formation of the spore mother cells, numerous cells within the capsule differentiate into elaters, elongated cells with spiral bands within the cell wall (fig. 2: end of an elater). A homology with the pseudo-elaters of the anthocerophyta is sometimes suggested.

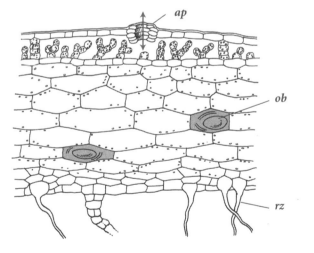

Figure 1

Figure 2

Number of species: 9,100.
Oldest known fossils: microspores thought to belong to *Dyadospora* (Sphaerocarpales) date from the lower Silurian (435 MYA). The oldest indubitable macrofossil is that of *Pallavicinites* from the upper Devonian (375 MYA).
Current distribution: worldwide (but more abundant in the tropics).

Ecology

The liverworts grow in humid environments, on soil, rocks or tree trunks. They can also live as epiphytes on the leaves of other plants.

Examples

Lunularia cruciata; Marchantia polymorpha; Riccia glauca.

Stomatophyta

A few representatives

Anthoceros fusiformis
3 cm
node 4

Hair moss
Polytrichum formosum
8 cm
node 6

Common polypody
Polypodium vulgare
30 cm
node 12

Quince
Cydonia vulgaris
4 m (tree)
node 18

Some unique derived features

■ Stomata: stomata are present (fig. 1) and characterized by two, bean-shaped guard cells (*gc*) containing chlorophyll. Because of their shape, these cells are able to open and close an opening or ostiole (*ost*) in the leaf surface that allows gas exchange between the atmosphere and a substomatal chamber (*sch*). The stomata therefore enable the organism to regulate gas exchange.

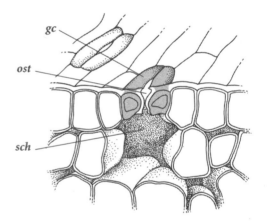

Figure 1

■ Sporangia: the sporangia have columellae. A longitudinal section of a sporangium of *Funaria hygrometrica* (fig. 2) shows the columella (*com*) surrounded by spore mother cells (*smc*).

■ Three new introns appear in the mitochondrial genome: two in the *cox2* gene and one in the *nad1* gene.

com

smc

Figure 2

Number of species: 260,802.
Oldest known fossils: *Stomatophytes* spores date from the upper Ordovician (455 MYA). Fossilized cuticles of stomatophyte plants have also been found in the upper Silurian (415 MYA).
Current distribution: worldwide.

Examples

■ ANTHOCEROPHYTA: *Anthoceros fusiformis*.

■ HEMITRACHEOPHYTA: cord moss: *Funaria hygrometrica;* juniper hair-cap moss: *Polytrichum juniperinum;* sphagnum moss: *Sphagnum subsecundum;* stag's horn clubmoss: *Lycopodium clavatum;* lake quillwort: *Isoetes lacustris;* bracken fern: *Pteridium aquilinum;* field horsetail: *Equisetum arvense;* ginkgo: *Ginkgo biloba;* Scots pine: *Pinus sylvestris;* elder: *Sambucus nigra;* corn: *Zea mays.*

Anthocerophyta

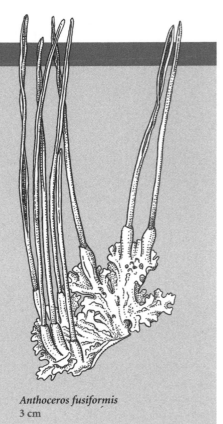

General Description

The name *anthocerophyta* comes from the genus name *Anthoceros*. The monophyly of this small group is well supported. For long time, they were placed with the marchantiophytes (6.2) because of their overall resemblance to the members of this group. The gametophyte looks like a lobed thallus with dichotomous branches. The sporophyte is unusual in that it displays very strong developmental polarity: the basal area is still in the juvenile stage when the apical end has already produced mature spores. The upright sporophyte is sometimes considered to be the result of an independent acquisition (convergence with the hemitracheophyta).

Anthoceros fusiformis
3 cm

Some unique derived features

- Biochemical character: the anthocerophytes are not able to synthesize flavonoids.

- Cyanobacteria of the *Nostoc* genus live as symbionts in depressions on the ventral surface of the thallus.

- Development of the antheridia: an antheridium develops from a sub-epidermal cell and not from a superficial cell as in the bryophytes and the marchantiophytes. Three stages in the development of an antheridium of *Anthoceros pulcherrimus* are shown in fig. 1. We see on fig. 1a that the origin of this structure is sub-epidermal (*epd:* epidermis). The antheridium (*atr*) then pierces the epidermis (fig. 1b) such that it appears to lie within a chamber that opens to the exterior (fig. 1c, at a smaller magnification).

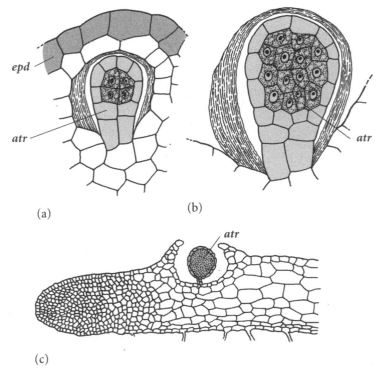

(a)

(b)

(c)

Figure 1

■ Archegonia: they lie within the thallus and never at the surface. Two developmental stages of an archegonium of *Anthoceros fusiformis* (fig. 2) show its placement within the tissues of the gametophyte. The basal cell is the oosphere (oos).

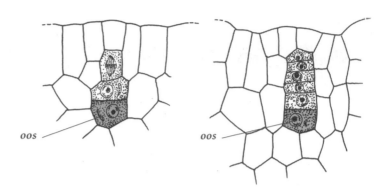

Figure 2

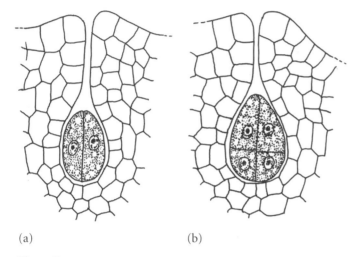

(a)   (b)

Figure 3

■ Zygote: the first cell division of the zygote is typically vertical, whereas it is transverse in other embryophytes. On fig. 3, a young embryo of *Anthoceros fusiformis* undergoes a first longitudinal cell division (fig. 3a) and a second, transverse division (fig. 3b).

■ Sporangia: pseudo-elaters develop in the sporangium. Their homology to the elaters of the marchantiophyta is debated.

Number of species: 300.
Oldest known fossils: anthocerophyte spores date from the upper Silurian (415 MYA). The oldest anthocerophyte macrofossil is *Notothylacites* from the lower Cretaceous (120 MYA).
Current distribution: temperate and tropical regions.

Ecology

The anthocerophytes (or hornworts) grow directly on the ground. Certain tropical species are epiphytes on the leaves, branches and trunks of trees.

Examples

Anthoceros fusiformis; Dendroceros endivaefolius; Notothylas flabellatus.

Hemitracheophyta

A few representatives

Hair moss
Polytrichum formosum
8 cm
node 6

Stag's horn clubmoss
Lycopodium clavatum
12 cm (height)
node 8

Field horsetail
Equisetum arvense
25 cm
node 11

Quince
Cydonia vulgaris
4 m (tree)
node 18

Some unique derived features

The hemitracheophytes form a monophyletic group that unites the bryophytes and the tracheophytes (or vascular plants). The bryophytes *sensu stricto* were previously placed with the marchantiophytes and the anthocerophytes in a paraphyletic group (bryophytes *sensu lato*) on the basis of shared ancestral characters. It now appears that the bryophytes (*sensu stricto*) are more closely related to tracheophytes than to anthocerophytes. They share several derived characters with tracheophytes:

■ The sporophyte has a straight, vertical stem. Fig. 1a shows the upright sporophyte of a hair moss with its stalk (*st*); fig. 1b shows a reconstituted drawing of *Horneophyton lignieri*, a Devonian fossil with a straight stalk and terminal, bifurcated sporangia (*spo*).

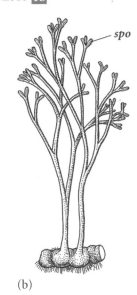

(a)   (b)

Figure 1

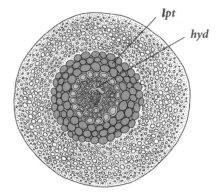

Figure 2

■ The stem of the gametophyte and/or the sporophyte contains transporting elements. In the bryophytes (fig. 2: example of *Polytrichum juniperinum*), the hydroid (*hyd*) and the leptoid (lpt) tissues are considered to be homologous to the xylem (conducting tissue of the xylem sap) and phloem (conducting tissue of the phloem sap), respectively.

■ Spores: the spore wall is surrounded by a well-differentiated external layer called the perine.

■ The gametangia (antheridia and archegonia) are found in an apical position on the gametophyte. This is a weak character, however, with some exceptions, such as the hypnobryales.

Number of species: 260,502.
Oldest known fossils: *Cooksonia* sp. dates from the middle Silurian (430 MYA).
Current distribution: worldwide.

Examples

■ BRYOPHYTA: juniper hair-cap moss: *Polytrichum juniperinum*.
■ POLYSPORANGIOPHYTA: lake quillwort: *Isoetes lacustris;* bracken fern: *Pteridium aquilinum;* field horsetail: *Equisetum arvense;* ginkgo: *Ginkgo biloba;* stone pine: *Pinus pinea;* elder: *Sambucus nigra;* corn: *Zea mays.*

Bryophyta

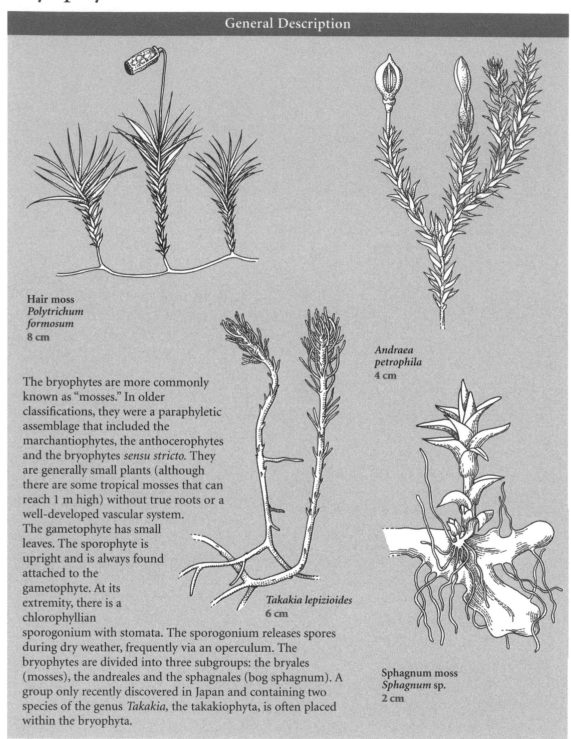

General Description

Hair moss
Polytrichum
formosum
8 cm

Andraea
petrophila
4 cm

The bryophytes are more commonly known as "mosses." In older classifications, they were a paraphyletic assemblage that included the marchantiophytes, the anthocerophytes and the bryophytes *sensu stricto*. They are generally small plants (although there are some tropical mosses that can reach 1 m high) without true roots or a well-developed vascular system. The gametophyte has small leaves. The sporophyte is upright and is always found attached to the gametophyte. At its extremity, there is a chlorophyllian sporogonium with stomata. The sporogonium releases spores during dry weather, frequently via an operculum. The bryophytes are divided into three subgroups: the bryales (mosses), the andreales and the sphagnales (bog sphagnum). A group only recently discovered in Japan and containing two species of the genus *Takakia*, the takakiophyta, is often placed within the bryophyta.

Takakia lepizioides
6 cm

Sphagnum moss
Sphagnum sp.
2 cm

Ecology

The mosses typically live in moist, temperate or tropical environments. They play an important role in the water regulation of other plants. Sphagnum mosses form the bogs of cold, temperate countries.

The mosses are among the first pioneering plants that form the humus layer. Many species are very sensitive to environmental conditions and are thus good bio-indicators of atmospheric and freshwater pollution.

Some unique derived features

- The gametophyte has small, pointed "leaves." Fig. 1 shows the leafy base of *Funaria hygrometrica* with a developing sporophyte (*spp*). We note the blade-shaped leaves (*bla*) and the rhizoids (*rz*).

- The rhizoids are multicellular. The young *Sphagnum* gametophyte of fig. 2 shows the extending spore filament (*spe*), the future leafy axis (*la*) as it grows, and a multicellular rhizoid (*rz*).

- Sporangial capsule: its dehiscence is accomplished by the opening of an operculum (*o*) (fig. 3: capsule of a hair moss).

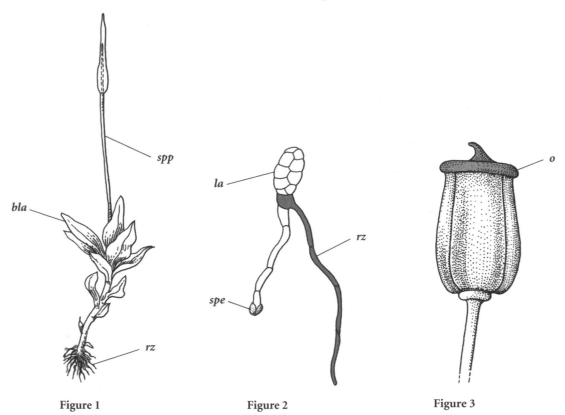

Figure 1   Figure 2   Figure 3

Number of species: 15,000.
Oldest known fossils: *Sporogonites* is a macrofossil from the lower Devonian (395 MYA).
Current distribution: worldwide.

Examples

Andraea petrophila; cord moss: *Funaria hygrometrica*; juniper hair-cap moss: *Polytrichum juniperinum*; sphagnum moss: *Sphagnum subsecundum*; *Takakia lepidozioides*.

Polysporangiophyta

A few representatives

Stag's horn clubmoss
Lycopodium clavatum
12 cm (height)
node 8

Field horsetail
Equisetum arvense
25 cm
node 11

Nordmann fir
Abies nordmanniana
50 m
node 15

Quince
Cydonia vulgaris
4 m (tree)
node 18

Some unique derived features

If we do not take the fossils into consideration, the polysporangiophyte clade merges with that of the tracheophytes and eutracheophytes because the groups that emerged between the polysporangiophyte and euphyllophyte nodes are now extinct. The first polysporangiophyte group to emerge is represented by the *Cooksonia* genus, followed by the *Horneophyton* genus and *Aglaophyton major* (previously called *Rhynia major*). The most recent deposit is that of the Rhynie Chert in northern Scotland that dates from the middle Devonian (although certain species may be found in slightly older deposits). This clade seems to provide key elements for our understanding of how plants conquered the terrestrial environment because within it we see the rapid appearance of the main characters found in the modern eutracheophytes: dominance of the diploid phase, branching of the sporophyte, formation of the tracheids, storage of lignin, etc. These features enabled the plants to remain upright and conduct water to their higher parts.

The eutracheophytes were previously divided into the vascular cryptogams (or pteridophytes) and the spermatophytes. The spermatophytes form a monophyletic group. By contrast, the vascular cryptogams seem to be paraphyletic.

Because of their evolutionary interest, the unique characters of each of these two embedded clades are described below with the fossil information taken into account.

Polysporangiophyte clade

- The diploid phase of the life cycle is dominant and the sporophyte becomes independent from the gametophyte at a certain point in its existence.

- The archegonia lie within the gametophyte. It is thought that this character was derived independently in the

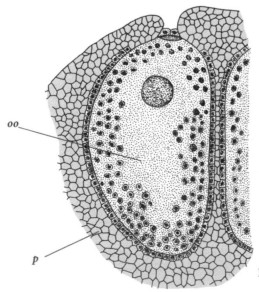

oo

p

Figure 1

anthocerophytes. Fig. 1 shows an archegonium of *Pinus lambertiana* deeply embedded in the parenchyma (*p*) of the female gametophyte (*oo:* oosphere).

■ The sporophyte is branched, meaning there are multiple sporangia (*spo*), as shown in the reconstituted sketch of *Aglaophyton* (*Rhynia*) *major* (fig. 2).

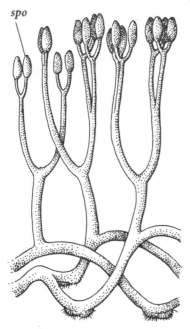

spo

Figure 2

■ The protoxylem (first to be differentiated) is diffuse or centrarch, that is, it is central in relation to the metaxylem (secondarily differentiated).

Number of species: 245,502.
Oldest known fossils: fossils of *Cooksonia hemisphaerica* and *Caia langii* have been found starting at the end of the Silurian (425 MYA).
Current distribution: worldwide.

Tracheophyte clade

(included with the polysporangiophytes, if we do not take fossil species into account)

■ The cells that transport the xylem sap—or tracheids—have cell walls with ring-shaped (fig. 3a) or helical bands (fig. 3b: example of *Foeniculum*).

■ The tracheids have completely lignified cell walls.

■ The sporangium no longer has a columella.

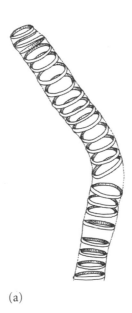

(a)   (b)

Figure 3

Eutracheophyte clade

(included with the polysporangiophytes, if we do not take fossil species into account)

■ The dehiscence of the sporangium is by a simple vent.

■ The stem has a sterome, a peripheral zone of several cell layers with thick, resistant cell walls.

Examples

■ Lycophyta: Stag's horn clubmoss: *Lycopodium clavatum*; lake quillwort: *Isoetes lacustris*.
■ Euphyllophyta: bracken fern: *Pteridium aquilinum*; field horsetail: *Equisetum arvense*; ginkgo: *Ginkgo biloba*; Scots pine: *Pinus sylvestris*; elder: *Sambucus nigra*; corn: *Zea mays*.

Lycophyta

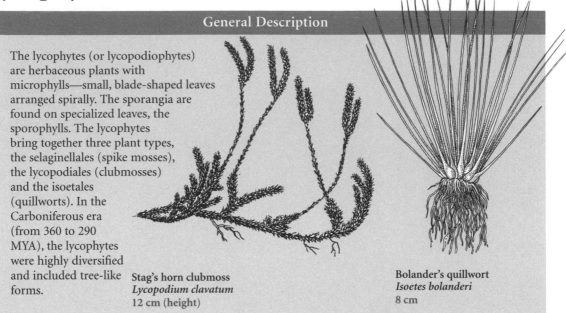

General Description

The lycophytes (or lycopodiophytes) are herbaceous plants with microphylls—small, blade-shaped leaves arranged spirally. The sporangia are found on specialized leaves, the sporophylls. The lycophytes bring together three plant types, the selaginellales (spike mosses), the lycopodiales (clubmosses) and the isoetales (quillworts). In the Carboniferous era (from 360 to 290 MYA), the lycophytes were highly diversified and included tree-like forms.

Stag's horn clubmoss
Lycopodium clavatum
12 cm (height)

Bolander's quillwort
Isoetes bolanderi
8 cm

Some unique derived features

- Leaves: they are microphylls, small and blade-shaped (fig. 1: a branch of *Selaginella selaginoides* with leaves (*le*) and strobili (spore-bearing tips; *str*). The leaves of certain fossil species were very large.

- Sporangia: found on the upper surface or axils of the sporophylls. The spore-bearing tip of *Selaginella willdenovi* (fig. 2a) has macrosporophylls (*ma*) at the base and micro-sporophylls (*mi*) at the tip. The macrosporophylls (fig. 2b) are leaves that carry a macrosporangium (*mas*) containing four female macro-spores on their upper surfaces. The microsporophylls (fig. 2c) carry a microsporangium (*mis*) containing many male microspores.

Figure 1

Figure 2

Number of species: 1,275.
Oldest known fossils:
Baragwanathia dates from the upper Silurian (420 MYA).
Current distribution: worldwide.

- The cells that lie along the dehiscence line (distal or transverse) of the sporangium have thick cell walls.

- Primary xylem: it has an exarch arrangement (fig. 3), that is, the protoxylem (*px*) lies external to the metaxylem (*mx*). On a cross-section of a stem of *Lycopodium lucidulum*, the small-celled protoxylem lies in a peripheral position, whereas the large-celled metaxylem is in a central position (*endo:* endoderm, *pec:* pericycle, *phl:* phloem).

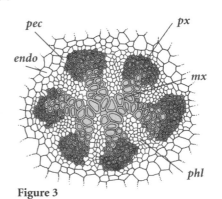

Figure 3

Ecology

The lycophytes can be terrestrial herbs, epiphytes or aquatic plants (like the quillworts).

Examples

Stag's horn clubmoss: *Lycopodium clavatum; Selaginella denticulata;* lake quillwort: *Isoetes lacustris.*

Euphyllophyta

A few representatives

Field horsetail
Equisetum arvense
25 cm
node 11

Nordmann fir
Abies nordmanniana
50 m
node 15

Sago palm
Cycas revoluta
1.5 m
node 16

Quince
Cydonia vulgaris
4 m (tree)
node 18

Some unique derived features

- Leaves, or megaphylls, are formed from lateral leaf branches.

- The branches show phyllotaxy, meaning that the leaves or branches are arranged spirally around the stem. A transverse section of a ginkgo bud is shown in fig. 1 outlining the spiral arrangement of the leaf primordia.

- The sporangia are arranged in pairs.

- The metaxylem tracheids have pitted walls (fig. 2: an example of a summer tracheid of a sequoia, with a cross-section of a pit (*pi*). Around each such pit, the lignified secondary cell wall (*scw*) detaches from the primary cellulose cell wall (*cw*). The primary cell wall then develops a central thickening in this open region, called the torus (*to*). These circular openings of the cell wall enable the movement of sap from cell to cell.

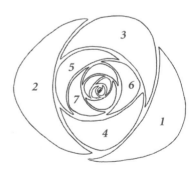

Figure 1

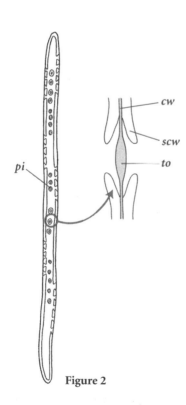

Figure 2

Number of species: 244,227.
Oldest known fossils: *Pertica* and *Psilophyton* date from the lower Devonian (390 MYA).
Current distribution: worldwide.

Examples

- MONILIFORMOPSES: field horsetail: *Equisetum arvense*; bracken fern: *Pteridium aquilinum*.
- SPERMATOPHYTA: stone pine: *Pinus pinea*; ginkgo: *Ginkgo biloba*; sago palm: *Cycas revoluta*; Chinese ephedra: *Ephedra sinica*; corn: *Zea mays*; elder: *Sambucus nigra*.

Moniliformopses

A few representatives

Field horsetail
Equisetum arvense
25 cm
node **11**

Common polypody
Polypodium vulgare
30 cm
node **12**

Polystichum lonchitis
60 cm
node **12**

Hart's-tongue fern
Phyllitis scolopendrium
60 cm
node **12**

Some unique derived features

■ This clade is debatable and is mainly supported from a molecular point of view. Morphologically, there is really only one character that defines the group. This character is linked to the histology of the conductive tissues, and more precisely to the relative positions of the proto- and meta-xylems. In the moniliformopses, the protoxylem (*px*) is the first tissue to differentiate and is called "mesarch" (fig. 1d). This means that it is surrounded by the metaxylem (*mx*), but without being perfectly central—or "centrarch" (1a; 1b: lobed centrarch)—or completely peripheral—"exarch" (1c). The primitive state is either a diffuse and axial protoxylem or a centrarch protoxylem.

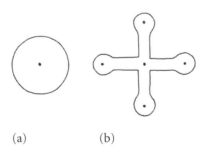

(a)   (b)   (c)   (d)

Figure 1

Number of species : 9,520.
Oldest known fossils: *Ibyka* is a macrofossil that dates from the middle Devonian (380 MYA).
Current distribution : worldwide.

Examples

■ SPHENOPHYTA: field horsetail: *Equisetum arvense*.
■ FILICOPHYTA: royal fern: *Osmunda regalis;* bracken fern: *Pteridium aquilinum;* common polypody: *Polypodium vulgare; Psilotum triquetrum.*

Sphenophyta

General Description

The sphenophyta (or equisetophyta) form a very homogeneous group made up of two genera. The horsetails, herbaceous plants with jointed upright stems and whorled leaves, have fertile non-branching stems that lack chlorophyll. These plants greatly diversified during the Carboniferous (from 360 to 290 MYA) with tree-like forms such as the calamites.

Field horsetail
Equisetum arvense
25 cm

Some unique derived features

- Branches: on sterile stems, the branches are arranged in whorls that alternate regularly from one node to another.

- Sporangiophores: they are grouped together and form a strobilus or cone with a particular anatomical structure. Fig. 1a shows a strobilus (*str*) of *Equisetum arvense*. The internal structure of a sporangiophore (*spg*) of *Equisetum telmateia* with open sporangia is shown in fig. 1b.

- Leaves: very small microphylls are whorled and attached to the stem by a basal sheath.

- Spores: they have a perispore, meaning that the nutritive cells in the sporangia (the tapetum) remain attached to the spore wall.

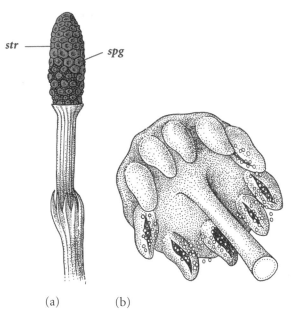

str   *spg*

(a)   (b)

Figure 1

Number of species: 20.
Oldest known fossils: *Ibyka* is a macrofossil that dates from the middle Devonian (380 MYA).
Current distribution: worldwide.

Ecology

The horsetails are found in humid habitats, along rivers or in degraded environments (moist ditches and embankments, along roads, etc.).

Examples

Field horsetail: *Equisetum arvense; Hippochaete hyemale.*

Filicophyta

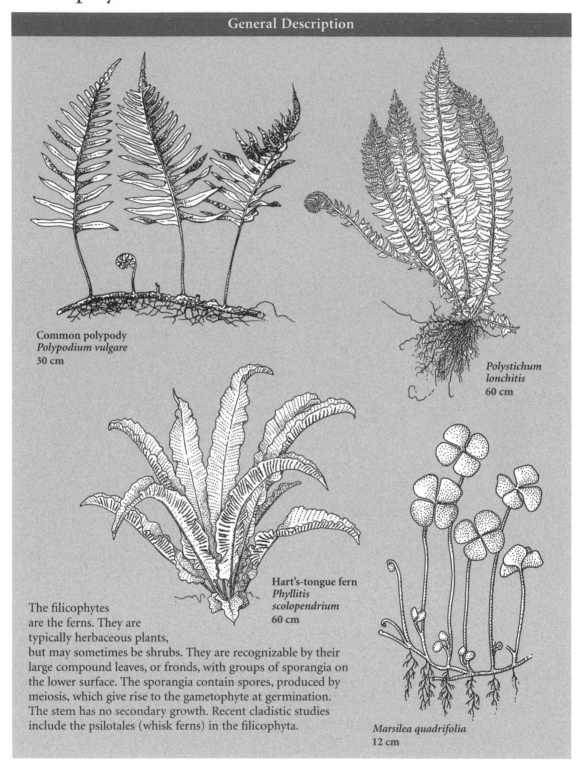

General Description

Common polypody
Polypodium vulgare
30 cm

*Polystichum
lonchitis*
60 cm

Hart's-tongue fern
*Phyllitis
scolopendrium*
60 cm

The filicophytes
are the ferns. They are
typically herbaceous plants,
but may sometimes be shrubs. They are recognizable by their
large compound leaves, or fronds, with groups of sporangia on
the lower surface. The sporangia contain spores, produced by
meiosis, which give rise to the gametophyte at germination.
The stem has no secondary growth. Recent cladistic studies
include the psilotales (whisk ferns) in the filicophyta.

Marsilea quadrifolia
12 cm

Ecology

We find ferns in all humid
environments, both tropical and
temperate.

Examples

■ *Marsilea quadrifolia;* royal fern: *Osmunda regalis;* bracken fern:
Pteridium aquilinum; common polypody: *Polypodium vulgare;*
Psilotum nudum.

Some unique derived features

■ The compound leaf, or frond, has a characteristic form; the pinnae and pinnules are arranged on either side of a central axis, or rachis. The lower surface of part of a frond of *Dryopteris filix-mas* is shown in fig. 1 (*pin:* pinnule, *so:* sori, clusters of sporangia).

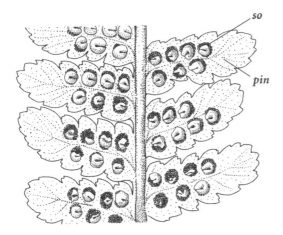

Figure 1

■ The frond unfurls in a spiral during growth.

■ Sporangia: they are normally found on the lower surface of the frond and may sometimes have a dehiscence ring (*dr*). This ring is clearly seen on the sporangium of a polypodiaceae (fig. 2). We can see the spores within the sporangia using microscopy.

■ The rhizoids of the gametophyte are multicellular.

■ The antheridium is formed from an outer surface cell. It has an apical cell that acts as an operculum. Fig. 3 summarizes the development of an

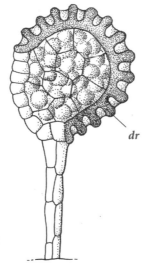

Figure 2

antheridium of *Woodsia*. The spermatogenic cells (*spc*) are found within the antheridium whose apical cell (*apc*) is already obvious (fig. 3a). The spermatozoids (sp) are mature and the cells of the antheridium wall are compressed (fig. 3b). The antheridium is mature and the cuticle (*cui*) of the apical cell opens (fig. 3c). Finally, the antheridium opens via the apical cell, which acts as an operculum. The spermatozoids are released and the cells of the antheridium wall expand (fig. 3d).

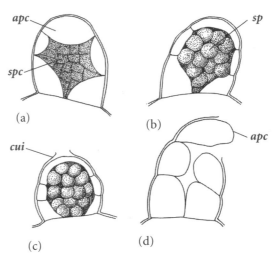

Figure 3

■ The conducting system (xylem and phloem) includes parts of the stele (vascular tissue). These elements are fundamentally arranged in a siphono-dictyostele fashion, intermediate between the simple siphonostele (fig. 4a; *lt:* leaf trace) and the dictyostele (fig. 4b; *lg:* leaf gap).

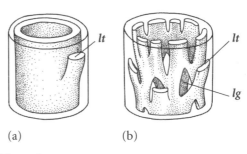

Figure 4

Number of species: 9,500.
Oldest known fossils: *Rhacophyton* dates from the middle Devonian (375 MYA). The sporangia of *Senftenbergia* have been found in rocks of the lower Carboniferous (350 MYA).
Current distribution: worldwide.

Spermatophyta

A few representatives

Nordmann fir
Abies nordmanniana
50 m
node 15

Sago palm
Cycas revoluta
1.5 m
node 16

*Welwitschia
mirabilis*
5 m (leaf length)
node 17

Quince
Cydonia vulgaris
4 m (tree)
node 18

Some unique derived features

If we ignore the fossil species, the spermatophytes, or seed plants (*sperma:* semen), merge with the lignophytes, plants with secondary phloem produced by the functioning of the bifacial vascular cambium (secondarily lost in the monocotyledons). For certain authors, the gnetophytes (formerly included with the gymnosperms) and the angiosperms form the anthophyte clade. However, molecular data, particularly that from the developmental MADSbox genes, are contradictory and tend to support the monophyly of the gymnosperms. Overall, the phylogeny of this group is not stable at the present time.

The seed is the sexual unit of dispersal that includes an embryonic sporophyte, nutritive reserves and one (or several) integument layers. We often superficially distinguish between mature and immature seeds, depending on when nutrient reserves are laid down (before or after fertilization) and whether there is a true dormancy that permits seed dissemination. In order to have a seed, the female gametophyte must be reduced and entirely retained on the sporophyte within the ovule. The male gametophyte is also reduced to the size of a pollen grain; this pollen grain will lead the male gamete to the ovule.

■ The stem has secondary growth, meaning that the bifacial cambium produces wood and phloem. Fig. 1 shows part of a cross-section of a young pine stem; the bifacial cambium (*bca*) gives rise to wood (*wo*) in an inward direction and to phloem

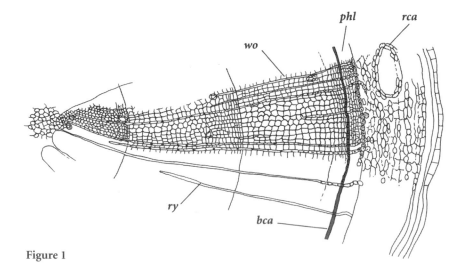

Figure 1

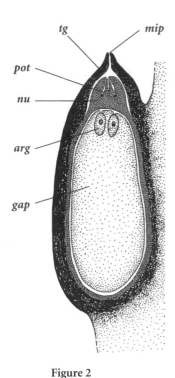

Figure 2

- After fertilization, the ovule is transformed into a seed.

- The male gametophyte is greatly reduced, the prothallus being made up of only a few cells. The pollen grain of *Ephedra trifurca* (fig. 3) has two prothallial nuclei (*pN*); the other two nuclei will be used in reproduction; the reproductive nucleus (*rN*) gives

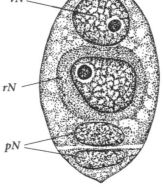

Figure 3

rise to spermatozoids and the vegetative nucleus (*vN*) to the pollen tube.

- Pollen: a pollen grain germinates to give rise to a pollen tube that enters into the nucellus. However, fertilization is by internal zoidogamy; the ciliated gametes are released and move freely within the ovule. A cross section of an ovule of *Dioon edule* (cycadophyta) at fertilization is shown in fig. 4 (*tg*: integument, *gap*: gametophyte). The pollen grains entered before the micropyle closed (*mip*). The pollen tube (*pot*) has implanted in the nucellus (*nu*) and the pollen grain on the right has freed its two ciliated spermatozoids (*sp*) and the liquid that they swim in. These spermatozoids will then find and fertilize the oosphere (*oos*).

(*phl*) in an outward direction (*rca*: resin canal; *ry*: xylem ray).

- Branches are produced by buds at the leaf axils.

- The ovule is a unit made up of the nucellus surrounded by one or two integument layers and containing the female gametophyte. The ovule of *Pinus lambertiana*, shown in fig. 2, contains a gametophyte (*gap*) with two archegonia (*arg*) that lie within the nucellus (*nu*). The nucellus is surrounded by an integument (*tg*). This integument contains a micropyle (*mip*), an opening that enables the entrance of the pollen tubes (*pot*).

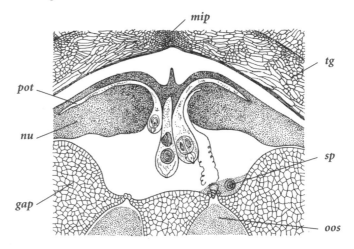

Figure 4

Number of species: 234,707.
Oldest known fossils: *Crossia* dates from the middle Devonian (380 MYA), and *Elkinsia* from the upper Devonian (370 MYA).
Current distribution: worldwide.

Examples

- GINKGOPHYTA: ginkgo: *Ginkgo biloba*.
- PINOPHYTA: stone pine: *Pinus pinea*.
- CYCADOPHYTA: sago palm: *Cycas revoluta*.
- GNETOPHYTA: Chinese ephedra: *Ephedra sinica*.
- ANGIOSPERMAE: common wheat: *Triticum aestivum*; meadow oatgrass: *Avena pratensis*; olive: *Olea europaea*; elder: *Sambucus nigra*.

Ginkgophyta

General Description

The ginkgophytes, although greatly diversified and abundant in the Jurassic, are currently represented by a single species, *Ginkgo biloba*. It is an elegant tree that can reach up to 40 m high. The deciduous leaves are bilobed and fan-shaped with a soft green color in the spring and golden yellow color in the fall. The leaf venation is openly dichotomous, forking from the base of the leaf. This tree is dioecious, that is, there are separate male and female trees. The ginkgo is a botanical curiosity; although it superficially resembles an angiosperm, there are a good number of plesiomorphies linked to reproduction: zoidogamy, acquisition of reserves in the ovule before fertilization, fertilization after the ovule is separated from the sporophyte, etc.

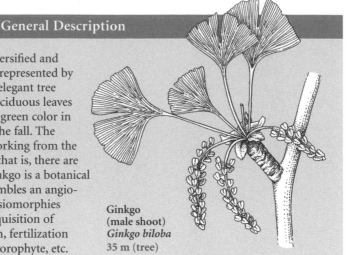

Ginkgo
(male shoot)
Ginkgo biloba
35 m (tree)

Some unique derived features

- Leaf: the venation pattern is dichotomous (fig. 2).

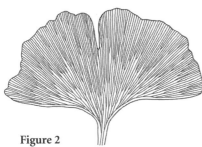

Figure 2

- Branches: there are two branch types on the tree. One grows indefinitely and has internodes that separate the leaves (fig. 3a); the other has limited growth with the leaves originating at the same place on the branch, with no internodes (fig. 3b; *ou:* ovule).

- Dioecious species: the male tree (fig. 1a) produces catkins carried on short shoots. The female tree (fig. 1b) produces ovules on stalks. This may be a derived character of the spermatophytes because it is also found in the cycadophytes.

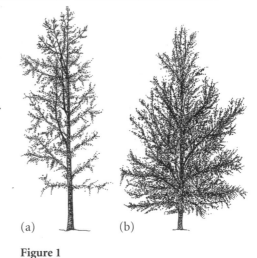

(a)   (b)

Figure 1

Ecology

The ginkgo is a tree of temperate countries. The leaves are green, but become golden yellow in the fall before falling to the ground. As it is relatively tolerant of atmospheric pollution, it is frequently used as an ornamental tree in cities. It is the living organism that best resisted the nuclear explosion of Hiroshima (1945).

Examples

Gingko: *Ginkgo biloba.*

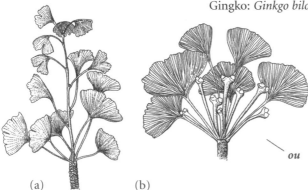

(a)   (b)

Figure 3

Number of species: 1.
Oldest known fossils: the Jurassic species *Ginkgoites lunzensis* is the oldest trace of bilobed leaves. *Sphenobaiera* from the Permian (270 MYA) is sometimes considered part of the ginkgoales.
Current distribution: *Ginkgo biloba* is a relic species that originated in the mountains of Southeast China. Wild populations may no longer exist. The species has been protected by Buddhist monks since the twelfth century and was introduced to Europe in the eighteenth century. It is currently cultivated in all temperate regions.

Pinophyta

General Description

The pinophytes are the conifers, common trees such as pines, firs, spruces, larches, junipers, yews, etc. These trees typically have needles and can be extremely tall, like the sequoia (100 m) or the Douglas pine (110 m). In contrast to those in the Northern hemisphere, the pinophytes of the Southern hemisphere are sprawling trees or bushes that can have very large leaves. The seeds are true seeds from a botanical perspective, meaning the food reserves are acquired after fertilization and the embryo has a dormancy period. The seeds are protected within cones; these cones are typically lignified (fig. 2).

Nordmann fir
Abies nordmanniana
50 m

Maritime pine
Pinus pinaster
35 m

Yew
Taxus baccata
25 m

Cedar of Lebanon
Cedrus libani
30 m

Ecology

Pinophytes are frequently encountered in temperate and humid climates. We also find them in cold regions, where they are often the dominant tree type. Some of these trees can have surprisingly long lifetimes: 5000 years for *Pinus longaeva*, 4000 years for *Wellingtonia gigantea* from California, 2000 years for *Taxodium mucronatum* from Mexico.

189

Some unique derived features

- Pollen: they are siphonogamous—the male gametes are led to the oosphere by a pollen tube. The development of the pollen tube is shown in fig. 1 for *Pinus nigra* subsp. *laricio*. The pollen grain (1a) has a prothallial cell (*pth*), a generative cell (*gce*) and a vegetative nucleus (*vN*). After germination in the ovule (1b), the pollen tube extends, carrying the vegetative nucleus with it. The generative cell divides to give rise to the spermatogenous cell (*spc*) that will then divide to form two sperm. It is likely that siphonogamy in the pinophytes is not homologous to that of the gnetophytes and angiosperms.

- Ovules (then seeds): they are carried on the upper surface of woody scales. The ensemble of these scales forms a female cone, or pine cone

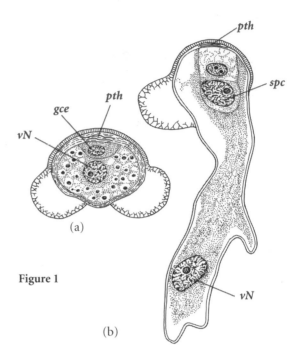

Figure 1

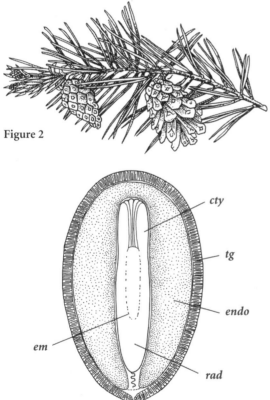

Figure 2

(fig. 2). Fig. 3 shows a cross-section of a *Pinus edulis* seed. The embryo (*em*) is surrounded by the endosperm (*endo*) that is produced by the development of the gametophyte. A hard integument (*tg*) protects the seed. Each seed possesses numerous cotyledons (*cty*) and a radicle (*rad*).

- The development of the proembryo is characteristic. On fig. 4 (example of *Pinus banksiana*), we observe the primary suspensor cells (*psu*) and the secondary suspensor cells (*ssu*) that push the developing embryo into the nutritive tissues. Only the terminal cell (*tec*) of the proembryo will give rise to the embryo proper.

- The secondary xylem is homoxylous, made up of only tracheids.

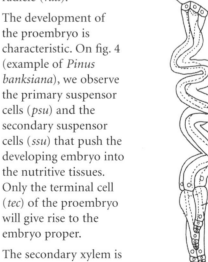

Figure 4

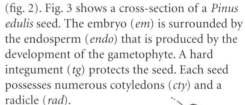

Figure 3

Number of species: 600.
Oldest known fossils: *Walchia* dates from the upper Carboniferous (310 MYA).
Current distribution: worldwide.

Examples

Scots pine: *Pinus sylvestris;* stone pine: *Pinus pinea;* silver fir: *Abies alba;* Atlas cedar: *Cedrus atlantica;* European larch: Larix decidua; Norway spruce: *Picea abies;* common juniper: *Juniperus communis;* cypress: *Cupressus fastigiata.*

Cycadophyta

General Description

The cycadophytes, or cycads, are plants with pinnately compound leaves that resemble a palm tree or tree fern in form; they have thick trunks and are typically unbranching. The plants are dioecious. Like the ginkgo, the cycads are botanical curiosities that show many plesiomorphies from a reproductive perspective: zoidogamy, acquisition of reserves before fertilization, etc.

Sago palm
Cycas revoluta
1.5 m

Cycas media
5 m

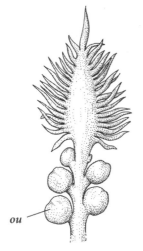

ou

Figure 1

Some unique derived features

- Dioecious species: the plant has distinct male and female trees (interpreted as a convergence with the ginkgoes). This feature could also be a derived character of the spermatophytes because it is present in the ginkgophytes.

- Ovules: they are found in two rows on seed-bearing leaves that form a female cone (fig. 1: example of *Cycas revoluta; ou:* ovule).

- There are unique biochemical characters. In particular, these plants synthesize toxic glycosides called cycasins that protect them against bacteria and fungi.

Ecology

The cycads have a tropical and sub-tropical distribution in the Americas, Asia and Australia. They have a particular root system with coralloid roots, thus named because they resemble coral. These roots house symbiotic cyanobacteria that fix atmospheric nitrogen.

Number of species: 305.
Oldest known fossils: *Taeniopteris taiyuanensis* dates from the Permian from China (270 MYA).
Current distribution: worldwide.

Examples

Sago palm: *Cycas revoluta; Cycas media; Dioon edule.*

Gnetophyta

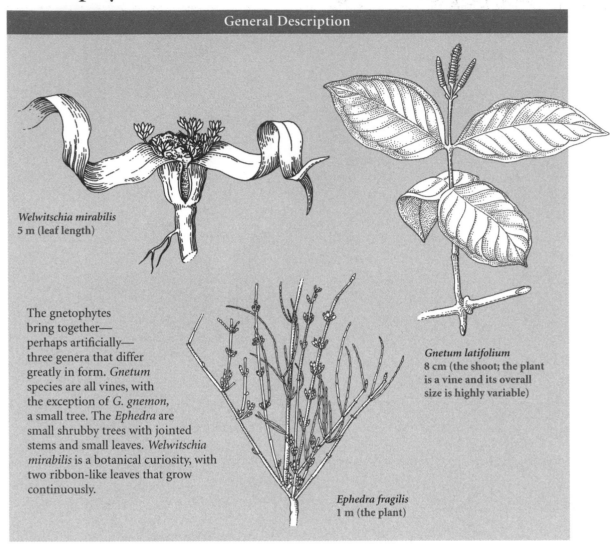

General Description

Welwitschia mirabilis
5 m (leaf length)

The gnetophytes bring together—perhaps artificially—three genera that differ greatly in form. *Gnetum* species are all vines, with the exception of *G. gnemon*, a small tree. The *Ephedra* are small shrubby trees with jointed stems and small leaves. *Welwitschia mirabilis* is a botanical curiosity, with two ribbon-like leaves that grow continuously.

Gnetum latifolium
8 cm (the shoot; the plant is a vine and its overall size is highly variable)

Ephedra fragilis
1 m (the plant)

Some unique derived features

- Staminate flower: all the parts of the staminate flower of *Ephedra viridis* (fig. 1) are not likely homologous to those of an angiosperm flower (*sta:* stamen).

- Pollen grain: the exine has obvious longitudinal grooves. These grooves are lost in *Gnetum*.

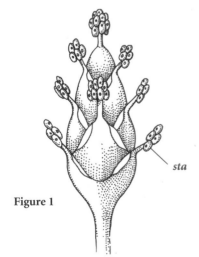

sta

Figure 1

■ There is siphonogamy, that is, the sperm are led directly to the oosphere by the pollen tube without the release of free-swimming flagellated male gametes. Fig. 2 shows a germinating gametophyte of *Ephedra trifurca* with its two reproductive nuclei (*rN*) moving along the pollen tube (*pot*). They are accompanied by two vegetative nuclei (*vN*). It is likely that the siphonogamy of the gnethophytes is not homologous to that of the pinophytes and angiosperms.

■ Ovule: the ovule is surrounded by an envelope. However, this envelope is not continuous and has a greatly elongated micropyle (*mip*). The micropyle along with two integument layers (*tg*) are clearly visible in a longitudinal cross-section of an ovule of *Ephedra trifurca* (fig. 3).

■ There is double fertilization. Both of the spermatozoids produced by the male gametophyte move through the pollen tube and fuse with nuclei of the female gametophyte; the second fusion gives rise to an extra zygote. This double fertilization is not likely homologous to that of the angiosperms.

■ The lignin gives a positive Mäule reaction, resulting in an intense red color after successive reactions with potassium permanganate and ammonia. This character is not likely homologous to that of the angiosperms.

■ The xylem has true vessel elements. This character is not likely homologous to that of the angiosperms.

■ The shoot apical meristem has a separation between the tunica and corpus regions. This character is not likely homologous to that of the angiosperms.

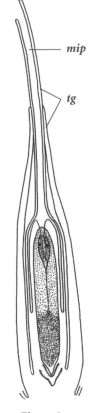

mip

tg

Figure 3

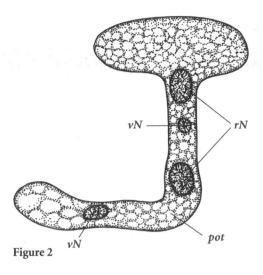

vN — *rN*

vN

pot

Figure 2

Ecology

The genus *Gnetum* is found in tropical rain forests. The genus *Ephedra* unites the xerophytes, plants that can tolerate dry conditions (desert or semi-arid zones, beaches or rocky areas). Ephedrine, a vasoconstrictor that mimics adrenaline, is extracted from *Ephedra equisitina*. *Welwitschia mirabilis* is a desert plant.

Examples

Chinese ephedra: *Ephedra sinica*; *Gnetum gnemon*; *Welwitschia mirabilis*.

Number of species: 91.
Oldest known fossils: pollen grains of *Equisetosporites* from China date from the Permian (270 MYA).
Current distribution: *Gnetum* species live in the tropics. *Ephedra* are found in Mediterranean and warm, temperate climates of the Northern hemisphere. *Welwitschia mirabilis* is endemic to southwest Angola and Namibia, close to the Kalahari Desert.

⇒ *See appendix 2, p. 518*

Angiospermae

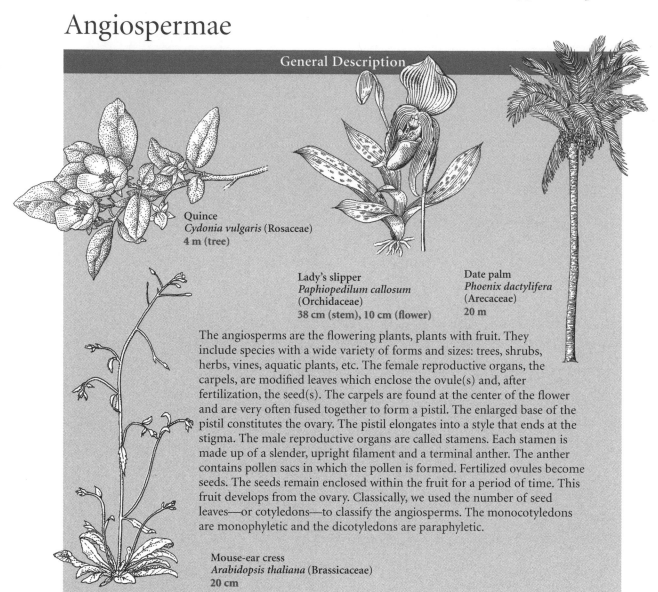

General Description

Quince
Cydonia vulgaris (Rosaceae)
4 m (tree)

Lady's slipper
Paphiopedilum callosum
(Orchidaceae)
38 cm (stem), 10 cm (flower)

Date palm
Phoenix dactylifera
(Arecaceae)
20 m

The angiosperms are the flowering plants, plants with fruit. They include species with a wide variety of forms and sizes: trees, shrubs, herbs, vines, aquatic plants, etc. The female reproductive organs, the carpels, are modified leaves which enclose the ovule(s) and, after fertilization, the seed(s). The carpels are found at the center of the flower and are very often fused together to form a pistil. The enlarged base of the pistil constitutes the ovary. The pistil elongates into a style that ends at the stigma. The male reproductive organs are called stamens. Each stamen is made up of a slender, upright filament and a terminal anther. The anther contains pollen sacs in which the pollen is formed. Fertilized ovules become seeds. The seeds remain enclosed within the fruit for a period of time. This fruit develops from the ovary. Classically, we used the number of seed leaves—or cotyledons—to classify the angiosperms. The monocotyledons are monophyletic and the dicotyledons are paraphyletic.

Mouse-ear cress
Arabidopsis thaliana (Brassicaceae)
20 cm

Some unique derived features

- Flower: it is complex and classically composed of four basic structures. The non-reproductive parts are external: the sepals make up the calyx, and the often colorful petals form the corolla. The fertile parts are internal: the stamens collectively form the androecium, and the carpels make up the gynoecium. A cross-section of a flower of *Helianthemum corymbosum* (fig. 1) shows the sepals (*sep*), petals (*pet*), stamens (*sta*) and pistil (*pis*).

- Angiospermy (*angeion:* vessel, *sperma:* seed): the ovule is completely protected within the carpel, which will give rise to the fruit that will protect the seed.

- Female gametophyte: it is extremely reduced and forms an embryonic sac with four nuclei (an

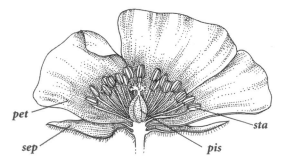

Figure 1

embryonic sac with eight nuclei is a synapomorphy of the euangiosperms).

- The type of double fertilization is very unusual. Whereas the first fertilization gives rise to the

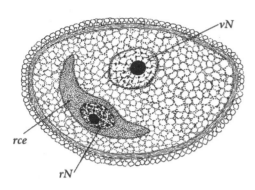

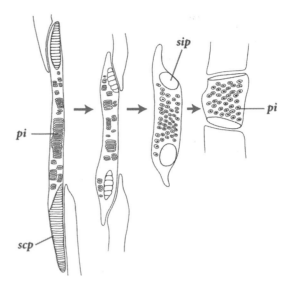

Figure 2

Figure 3

zygote that will transform into the embryo, the second leads to the formation of an extra diploid tissue, the albumen (the triploid albumen is a synapomorphy of the euangiosperms).

- The pollen grain, or male gametophyte, is reduced and contains only three nuclei, one vegetative and two sperm nuclei. The vegetative nucleus (*vN*) and reproductive cell (*rce*) with its reproductive nucleus (*rN*) are indicated in a cross-section of a pollen grain of *Lilium auratum* (fig. 2). The reproductive nucleus will divide to give rise to the two spermatozoids.

- The lignin gives a positive Mäule reaction, resulting in an intense red color after successive reactions with potassium permanganate and ammonia. This character is not likely homologous to that of the gnetophytes.

- Xylem: it contains true vessel elements that result from the differentiation of a column of cells linked together by their end walls. The tracheids, on the other hand, are the result of the differentiation of single cells. Fig. 3 shows the different states of specialization of the vessel elements in the angiosperms: the primitive state is represented by a long, straight element with beveled, scalariform perforation plates (*scp*) containing numerous bars. In the derived state, the vessel element is short and wide, with two horizontal terminal partitions containing a large number of simple perforation plates (*sip*) and evenly distributed pits (*pi*). This character is not likely homologous to that of the gnetophytes.

- Wood: when it exists, it is complex with vessel elements and fibers. The fibers are cells with thick cell walls that are used for support.

- Phloem: the sieve tubes possess companion cells that help with the circulation of the phloem sap.

- The shoot apical meristem, that is, the embryonic tissue responsible for branch growth, has a tunica (fig. 4). This tunica is made up of two cell layers that differ by the plane of cell division. T_1 and T_2 give rise to the epidermis and sub-epidermal tissues (*lpr*: leaf primordium) respectively. The corpus (*cor*) gives rise to more underlying tissues. We can distinguish another meristem zone linked to the degree of mitotic activity, that outlined by the dashed line in fig. 4; the axial zone (*az*) is less active than the peripheral zone (*pz*). This character is not likely homologous to that of the gnetophytes.

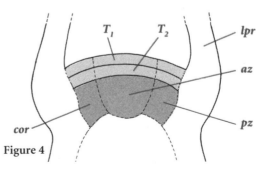

Figure 4

Ecology

Angiosperms have colonized all terrestrial habitats, and some have returned to an aquatic environment: fresh and salt water. This group is of vital economic importance because almost all cultivated plants are angiosperms.

Number of species: 233,885. **Oldest known fossils:** pollen grains from *Clavatipollenites* and *Ficophyllum* date from the lower Cretaceous (135 MYA). **Current distribution:** worldwide.

Examples

Corn: *Zea mays;* common wheat: *Triticum aestivum;* elder: *Sambucus nigra;* olive: *Olea europaea;* crab apple: *Malus sylvestris;* cork oak: *Quercus suber.*

195

Chapter 7

metazoa

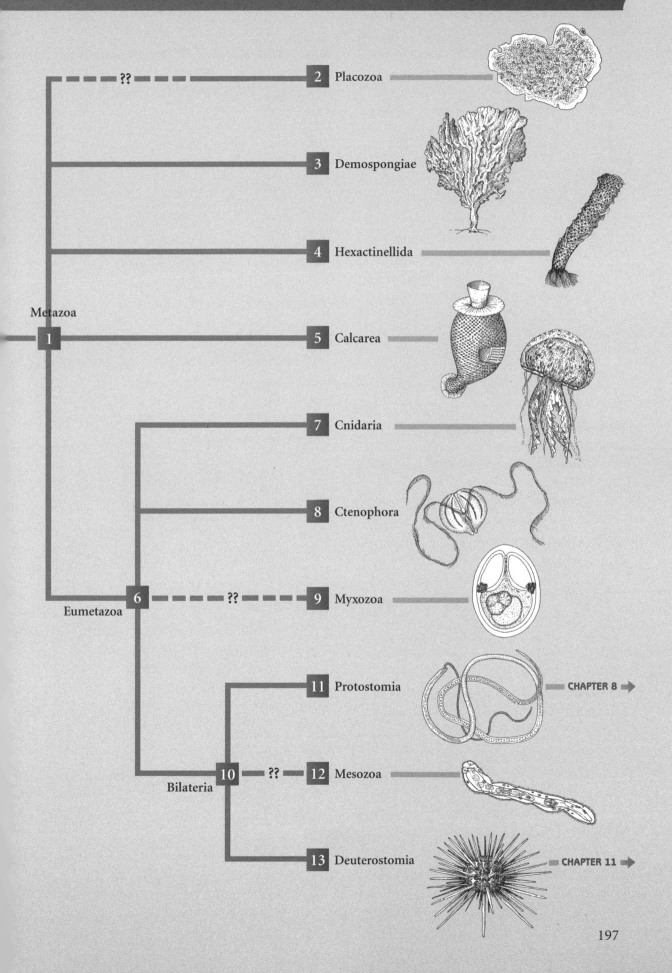

1 Metazoa

2 Placozoa

3 Demospongiae

4 Hexactinellida

5 Calcarea

6 Eumetazoa

7 Cnidaria

8 Ctenophora

9 Myxozoa

10 Bilateria

11 Protostomia CHAPTER 8

12 Mesozoa

13 Deuterostomia CHAPTER 11

??

??

??

The "classical" classification of the metazoans, founded on the principle of increasing complexity, has been in disfavor for the last decade. This is due as much to the birth of molecular phylogenies as to the logical use of new anatomical and embryological characters in addition to major inconsistencies in the definitions of complexity.

Pre-cladistic classification was dichotomous, based on the presence or absence of a character. One of the problems with this method was that the absence of a character was always considered to be the ancestral state. For example, acoelomates were seen as descendants of a triploblastic ancestor that existed before the appearance of the coelom. However, from an evolutionary perspective, secondary losses frequently occur. Now, all possible hypotheses are envisioned before phylogenetic reconstruction; in this way, we consider without bias that an organism without a coelom can either be the descendant of an acoelomate ancestor or the descendant of a coelomate ancestor that has lost its coelom secondarily.

Let us review the major points that have led to the collapse of the former classification:

- the sponges seem to be paraphyletic.
- the eumetazoa are characterized by the presence of muscle and nerve cells and, without doubt, true embryonic tissue layers.
- all triploblastic organisms seem to be coelomates and the concepts of acoelomate and pseudocoelomate are no longer used; the platyhelminthes, nemertea, nematoda, nematomorpha, entoprocta, rotifera, gastrotricha, kinorhyncha are all classified as protostomians.

- the protostomia are largely divided into two main groups, if we ignore the gastrotricha: the lophotrochozoa and the ecdysozoa. In this context, the annelids and arthropods become quite distant, invalidating the hypothesis that the arthropods descended from the annelids.

Despite the progress made in our understanding of this group, many important questions remain unanswered. For example:

- Did the coelom appear a single time, or several times? Here, we consider that the coeloms of the protostomians and the deuterostomians are homologous; this is currently being discussed.
- Are the protostomians truly monophyletic? There is currently some doubt about this.
- Are the segments of the protostomians and deuterostomians homologous or has metamerism appeared independently several times? This question goes beyond the objectives of this book because it entails a long discussion on the origin of the implicated developmental genes.

Finally, certain groups are difficult to classify; in the tree, question marks highlight these difficulties. This is particularly the case for the placozoans, the myxozoans, the mesozoans and, in the protostomian tree, the chaetognatha.

We should emphasize that in certain cases there is disagreement among zoologists; we have therefore taken one side indicating this in the description of the taxon. For example, this is the case for the ctenophora, the gastrotricha and the entoprocta in the protostomian tree.

Placozoa

General Description

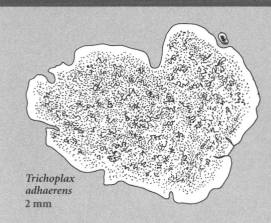

Trichoplax
adhaerens
2 mm

The placozoan group contains only a single species, *Trichoplax adhaerens,* discovered in 1883 in a saltwater aquarium in Austria. It is a soft-bodied, microscopic, disc-shaped organism, made up of a few thousand cells. Like the sponges, there is no basal membrane. The body is composed of dorsal and ventral layers of ciliated cells. These two layers are different enough that we distinguish dorsal-ventral sides. These layers enclose a fluid-filled cavity and a loose mesenchyme layer. There are no organs or differentiated tissues. Certain authors have suggested that the upper and lower layers of the animal are homologous to the ectoderm and endoderm respectively. The primitive characters exhibited by *T. adhaerens* are difficult to interpret. Nonetheless, some of them, like the absence of a digestive cavity, may correspond to a secondary loss. Certain authors suggest that this animal is quite ancient and represents a living relic. However, for others, it is a simplified sponge or cnidarian. Unfortunately, the placement of the animal in molecular phylogenies using the 18S rRNA gene is uncertain. However, this gene does contain some characters that suggest that *T. adhaerans* is a eumetazoan.

Some unique derived features

- Two cell layers delimit a space containing nucleated mesenchyme cells in an extracellular matrix (fig. 1: cross-section of *T. adhaerens; cil:* cilium, *dl:* dorsal layer, *vl:* ventral layer, *fc:* fiber cell, *gc:* glandular cell, *mic:* microvillus).

- External digestion (exodigestion) takes place by the temporary formation of a digestive chamber on the ventral surface. This has led some to consider the ventral layer as homologous to the endoderm (fig. 2: *ds:* digestive sac).

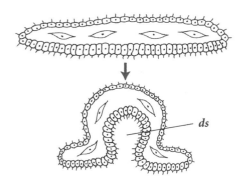

Figure 2

Number of species: 1.
Oldest known fossils: none.
Current distribution: likely worldwide.

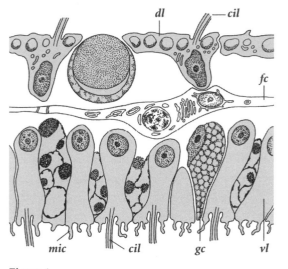

Figure 1

Ecology

This is a marine organism. It moves by its cilia and by deforming its body. It feeds on organic debris and unicellular algae, absorbing the products of the external digestion. Digestive enzymes are released from the glandular cells into the area between the ventral surface and the substrate. Reproduction can occur by fission, but sexual reproduction has also been observed.

Examples

Trichoplax adhaerans.

Demospongiae

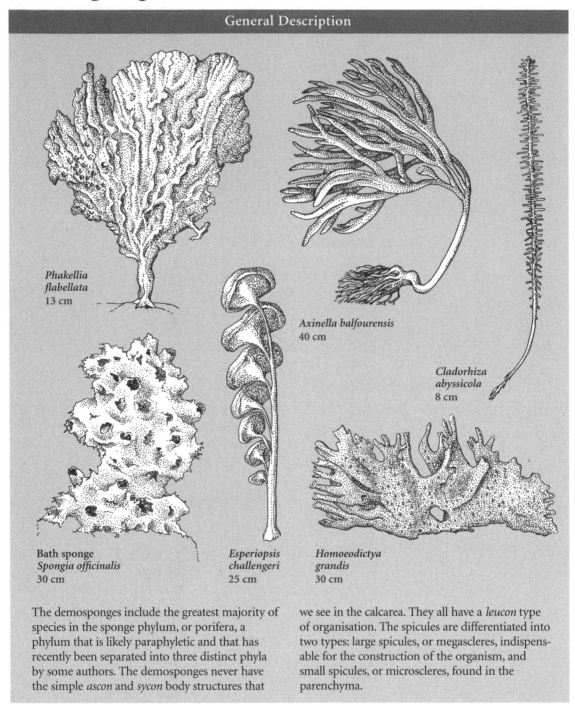

General Description

Phakellia
flabellata
13 cm

Axinella balfourensis
40 cm

Cladorhiza
abyssicola
8 cm

Bath sponge
Spongia officinalis
30 cm

Esperiopsis
challengeri
25 cm

Homoeodictya
grandis
30 cm

The demosponges include the greatest majority of species in the sponge phylum, or porifera, a phylum that is likely paraphyletic and that has recently been separated into three distinct phyla by some authors. The demosponges never have the simple *ascon* and *sycon* body structures that we see in the calcarea. They all have a *leucon* type of organisation. The spicules are differentiated into two types: large spicules, or megascleres, indispensable for the construction of the organism, and small spicules, or microscleres, found in the parenchyma.

Ecology

The demosponges are present in all aquatic environments: fresh and salt water, polar and tropical seas, intertidal zones and ocean abysses (up to 8600 m!). Like the other sponges, they feed and take in oxygen by creating a current of water in their internal chambers. This current is created by the beating of the flagella of the choanocytes. Their food consists of plankton (in particular, dino-flagellates and bacteria) and particles of organic detritus. These food particles are picked up by the choanocytes. Digestion is intracellular and waste products are evacuated by exhalant pores. Certain carnivorous sponges, like *Cladorhiza*, have lost this system of acquiring food. Oxygen diffuses through the tissues.

The sponges are often hermaphroditic. Sperm is released to the external environment by the exhalant pore. The spermatozoids swim in the water and some of them will enter another sponge with the flow of water via the inhalant pore. Fertilization takes place in the recipient sponge. Some demosponges are oviparous, whereas others incubate the embryos. The larvae are planktonic before settling on a substrate.

The demosponges contain a large number of symbionts and commensal organisms. For example, in Florida, a single individual of *Spheciospongia vesparia* was found to harbor 16,000 shrimp. In the Gulf of California, *Geodia mesotriaena* was found to host 100 different species of algae and animals, not including the bacteria. Pharmacologists are becoming increasingly interested in the molecules emitted by demosponges, diverse types of toxins and growth factors.

Some unique derived features

- The skeleton is formed by two components: siliceous spicules including from one (monoactine) to four (tetractine) rays (fig. 1: spicules with one and four rays) and spongin, a particular type of collagen, that is either dispersed throughout the parenchyma or structured into large fibers.

- There is never a simple *ascon* or *sycon* type of body structure; we only find a *leucon*-type organization (fig. 2a: *ascon* type; b: *sycon* type; c: *leucon* type; *at:* atrium, *cch:* choanocyte chamber).

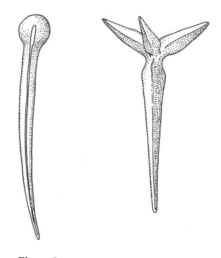

Figure 1

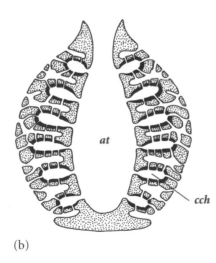
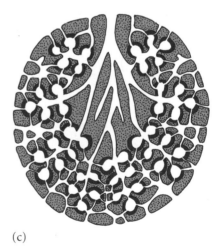

(a)   (b)   (c)

Figure 2

Number of species: 8,000.
Oldest known fossils: *Hazelia* dates from the middle Cambrian (520 MYA).
Current distribution: worldwide.

Examples

Axinella balfourensis; Cladorhiza abyssicola; Esperiopsis challengeri; Homoeodictya grandis; Phakellia flabellata; bath sponge: *Spongia officinalis;* freshwater sponge: *Spongilla fluviatilis.*

Hexactinellida

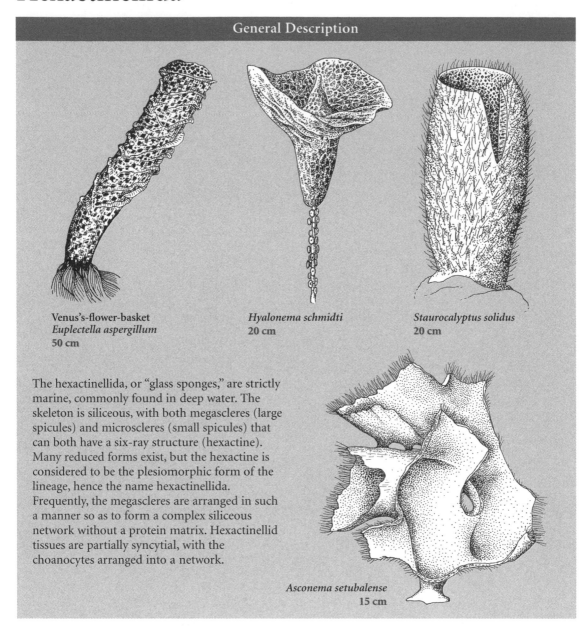

General Description

Venus's-flower-basket
Euplectella aspergillum
50 cm

Hyalonema schmidti
20 cm

Staurocalyptus solidus
20 cm

The hexactinellida, or "glass sponges," are strictly marine, commonly found in deep water. The skeleton is siliceous, with both megascleres (large spicules) and microscleres (small spicules) that can both have a six-ray structure (hexactine). Many reduced forms exist, but the hexactine is considered to be the plesiomorphic form of the lineage, hence the name hexactinellida. Frequently, the megascleres are arranged in such a manner so as to form a complex siliceous network without a protein matrix. Hexactinellid tissues are partially syncytial, with the choanocytes arranged into a network.

Asconema setubalense
15 cm

Ecology

The solubility of silica decreases with temperature and, as a consequence, with ocean depth. This explains why we find hexactinellids in abundance in deep water, 200 m and greater, and in cold polar seas. Detailed studies have been carried out in the Antarctic, where the hexactinellids are found in abundance at depths of between 30 and 60 m. Some of these sponges are several hundred years old. Hexactinellids have been found by researchers of the Endoume marine station in an underwater cave close to Marseilles, France. They have shown that the embryon of *Oopsacas minuta* undergoes gastrulation by primary delamination, the only known case in the sponges. *Euplectella* is a famous example of commensalism. A pair of shrimp enters into the sponge when they are small. After growing, they find themselves imprisoned and must therefore remain there for the rest of their lives. This sponge, with its two guests, is a traditional wedding gift in Japan.

Some unique derived features

- Spicules: the hexactinellids possess spicules with three principal axes, or six rays, the hexactines (fig. 1: hexactine of *Calycosoma validum*).

- The organization of the hexactines creates a complex and harmonious construction (fig. 2: cross-section of *Euplectella* sp.; *cch:* choanocyte chamber, *he:* hexactine).

- The hexactinellids do not have a pinacoderm, an external layer formed by pinacocytes that we find in other sponges (fig. 2: *emb:* external membrane).

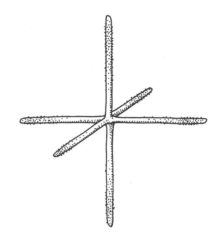

Figure 1

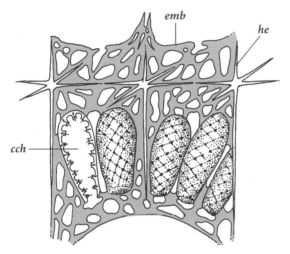

Figure 2

- There is no choanoderm, a cellular layer made up of choanocytes. Instead, we have a choano-syncytium, a continuous layer that lodges structures resembling anucleated choanocytes (fig. 3: choanosyncytium of *Aphrocallistes vastus; col:* collar, *fl:* flagellum, *N:* nucleus).

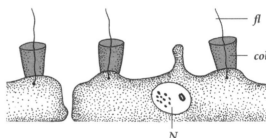

Figure 3

Number of species: 1,000.
Oldest known fossils: *Protospongia* is the first perfectly recognizable hexactinellid and was found in the Burgess shale fauna (start of the Cambrian, 540 MYA).
Current distribution: worldwide.

Examples

Aphrocallistes vastus; Asconema setubalense; Venus's-flower-basket: *Euplectella aspergillum; Hyalonema schmidti; Staurocalyptus solidus.*

Calcarea

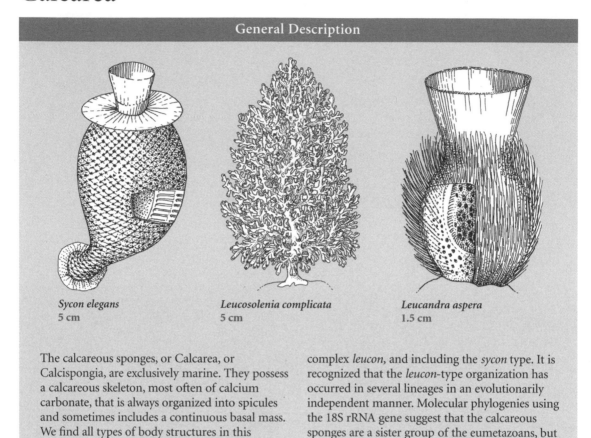

General Description

Sycon elegans
5 cm

Leucosolenia complicata
5 cm

Leucandra aspera
1.5 cm

The calcareous sponges, or Calcarea, or Calcispongia, are exclusively marine. They possess a calcareous skeleton, most often of calcium carbonate, that is always organized into spicules and sometimes includes a continuous basal mass. We find all types of body structures in this lineage, from the most simple *ascon* to the most complex *leucon*, and including the *sycon* type. It is recognized that the *leucon*-type organization has occurred in several lineages in an evolutionarily independent manner. Molecular phylogenies using the 18S rRNA gene suggest that the calcareous sponges are a sister group of the eumetazoans, but this notion has no solid support for now.

Some unique derived features

- The skeleton is made up of calcareous spicules.
- The spicules are not differentiated into megascleres and microscleres.

Number of species: 1,000.
Oldest known fossils: *Chancellaria* sp. is described from the start of the Cambrian (540 MYA).
Current distribution: worldwide.

Examples

Grantia compressa; Leucandra aspera; Leucosolenia complicata; Sycon elegans.

Ecology

As the solubility of calcium carbonate increases with depth, the secretion of a calcareous skeleton therefore becomes more difficult when one leaves the surface waters. This is likely the principal reason that calcareous sponges show a preference for shallow waters (less than 100 m in depth), even though some species do occur in abyssal waters. In contrast to the demosponges and the hexactinellids, the calcareous sponges require a hard substrate on which to anchor and develop. The gastrulation process has been observed in *Sycon*. The larva attaches itself by the anterior end and a small sponge of asconoid structure forms. This will later change into a *sycon*-type sponge.

Eumetazoa

A few representatives

Pelagia noctiluca
10 cm (diameter)
node **7**

Sea gooseberry
Pleurobrachia pileus
2 cm
node **8**

Edible crab
Cancer pagurus
25 cm
node **11**

African grey parrot
Psittacus erithacus
33 cm
node **13**

Some unique derived features

■ The extracellular matrix contains type IV collagen up to the level of the epithelium and is organized into a basal membrane, assuring the existence of true tissues (fig. 1: cross section of a pig's skin; *bcp:* blood capillary, *kc:* keratinized cells, *col:* collagen, *epd:* epidermis, *bm:* basal membrane, *ctis:* connective tissue).

■ The lacunar junctions (gap junctions), formed in part by connexin, allow small molecules to pass from one cell to another.

■ Digestion: the organism has a differentiated digestive cavity. The products of digestion are distributed among all the cells. The mouth is the entrance to this digestive apparatus. In all other groups, digestion is either extracellular (placozoans) or intracellular (sponges).

■ Gastrulation gives rise to the embryonic germ layers, the ectoderm and the endoderm, that have predefined embryonic fates.

■ The endoderm contains secretory cells that produce active digestive enzymes in the extracellular environment (exoenzymes).

■ Cellullar differentiation is elaborate with the formation of muscle cells, nerve cells and sensory cells.

■ Nervous system: chemical synapses enable the construction of a nervous system (fig. 2: nervous network of *Hydra* sp., *neu:* neuron).

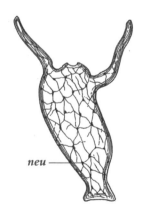

Figure 2

■ There are no choanocytes.

kc
epd
bm
ctis
col *bcp*

Figure 1

Number of species: 1,201,611.
Oldest known fossils: burrows dug by eumetazoans were fossilized in the Precambrian (700 MYA). Some eumetazoans were definitely present in the Ediacara fauna (680 MYA).
Current distribution: worldwide.

Examples

■ CNIDARIA: green hydra: *Hydra viridis.*
■ CTENOPHORA: beroe: *Beroe cucumis.*
■ BILATERIA: planaria: *Planaria maculata;* human intestinal roundworm: *Ascaris lumbricoides,* European brown garden snail: *Helix aspersa;* common squid: *Loligo vulgaris;* earthworm: *Lumbricus terrestris;* house spider: *Tegenaria domestica;* tench: *Tinca tinca.*

Cnidaria

General Description

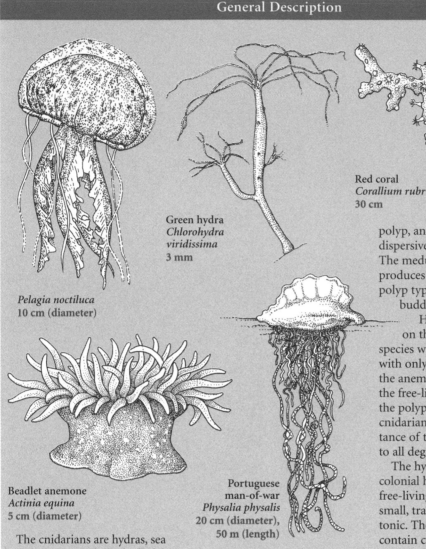

Pelagia noctiluca
10 cm (diameter)

Green hydra
*Chlorohydra
viridissima*
3 mm

Red coral
Corallium rubrum
30 cm

Beadlet anemone
Actinia equina
5 cm (diameter)

Portuguese
man-of-war
Physalia physalis
20 cm (diameter),
50 m (length)

The cnidarians are hydras, sea anemones, corals, and jellyfish. They are all characterized by very specialized stinging cells, the cnidocytes. These radially symmetrical animals are fundamentally composed of two embryonic germ layers; we say that they are diploblastic. Between the two germ layers, there is a jelly-like layer called the mesoglea. The well-differentiated digestive cavity has a single orifice and opens to the exterior by a mouth surrounded by tentacles. There is no circulatory system or excretory apparatus. Nerve cells form a network without a central nervous system. This branch has two principal morphologies, the polyp and the

medusa. A polyp is a hollow tube, closed off at the bottom and open at the top (mouth). From exterior to interior, the cell wall is made up of the ectoderm, the mesoglea and the endoderm. A polyp is fixed to a substrate by its base. The free-swimming medusa has more of a disc shape, a result of the flattening and enlargement of the polyp tube. In this form, the mesoglea layer is greatly thickened and the body organization is reversed: the mouth is now on the ventral side of the body, but still surrounded by tentacles.

During a typical cnidarian life cycle, there is an alternation between the fixed stage, the

polyp, and the free-living, dispersive stage, the medusa. The medusa has gonads and produces the gametes. The polyp typically reproduces by budding, that is, asexually. However, depending on the group, we can see species with mixed life cycles, with only the sessile phase (like the anemones), or with only the free-living stage (i.e., without the polyp form). Within the cnidarians, the relative importance of these two stages varies to all degrees.

The hydrozoans are typically colonial hydras, with a reduced free-living stage; the medusa is small, transparent and plank-tonic. The mesoglea does not contain cells and only the epidermis (ectoderm) carries cnidocytes. The gonads are derived from the epidermis. The scyphozoans are pelagic; the medusoid stage is dominant or uniquely present. These medusae are large and have mesoglea containing cells. Both the endoderm and the ectoderm carry cnidocytes. The gonads are formed by the endodermis. The anthozoans include the sea anemones and the corals. They are typically colonial and lack the free-living stage. The gastrovascular cavity has a pharynx and is partially divided into multiple chambers by septa. Corals are colonial polyps that secrete an external calcareous exoskeleton.

Some unique derived features

- Cnidocyte: this is a specialized cell used for food capture and defense. The cnidocyte (*cny*) encloses a vesicle, the nematocyst (*nem*), that contains a small, spring-coiled tubule carrying a paralyzing toxin (fig. 1: nematocyst before and after discharge; *cni*: cnidocil, *sp*: spine, *N*: nucleus, *o*: operculum, *sty*: stylet, *tu*: tubule).

- The musculature originates from both the ectoderm and the endoderm, with myoepithelial cells.

- Reproductive cycle: with the exception of the anthozoans, there is an alternation between polyp and medusoid stages.

However, numerous variations on this life cycle exist.

- There is typically a planula larva with a ciliated epidermis (fig. 2: longitudinal section of a 300 µm planula larva of *Gonothyraea* sp.; *cil*: cilia, *ecto*: ectoderm, *endo*: endoderm).

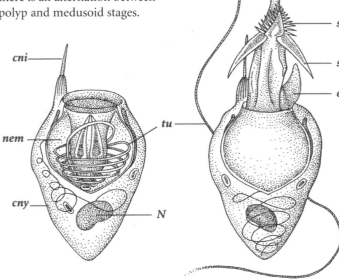

Figure 1

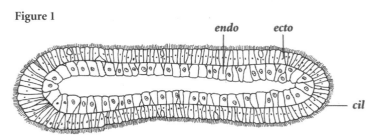

Figure 2

Number of species: 9,000.
Oldest known fossils:
Siphonophora medusae are
part of the Ediacara fauna
(680 MYA).
Current distribution:
worldwide.

Ecology

Sessile forms are found along coastlines and make up the coral reefs of tropical seas. Few species live in fresh water. The cnidarians are predators of planktonic organisms, small invertebrates and sometimes small fishes. The tentacles of the polyps and medusae are covered in cnidocytes, cells containing a stinging harpoon-like tubule that paralyzes prey with a toxin. The tentacles then bring the prey to the central mouth.

The general reproductive cycle goes through two phases, one sessile and one free-living. The polyp produces medusae by budding or, in the cubozoans, by complete metamorphosis. In many species, a polyp can produce new polyps by fission or asexual planula larvae that give rise to other polyps. The medusa, whether it is produced by a polyp or a planula larva, swims and carries gonads. Once the gametes are mature, they are evacuated into the external environment where fertilization takes place. The free-swimming and ciliated planula larva is initially planktonic. It then anchors itself to a substrate and transforms into a very small polyp. Depending on the species, the polyp and medusoid stages can be reduced in size or in lifespan, or they can disappear completely. Certain species have separate sexes, while others are hermaphroditic. Colonial living has evolved independently several times in this group. Certain colonies even have polyps with specialized functions that work for the ensemble, acting in an organ-like manner.

Examples

- HYDROZOA: green hydra: *Hydra viridis; Hydractinia equinata; Obelia geniculata;* Portuguese man-of-war: *Physalia physalis.*
- SCYPHOZOA: *Aurelia aurita; Pelagia noctiluca.*
- ANTHOZOA: *Adamsia palliata; Aneumonia viridis;* black coral: *Antipathes subpinnata;* beadlet anemone: *Actinia equina;* scarlet-and-gold-star coral: *Balanophyllia regia;* red coral: *Corallium rubrum; Urticina felina.*
- CUBOZOA: sea wasp: *Chiroplasmus quadrigatus; Chironex fleckeri.*

Ctenophora

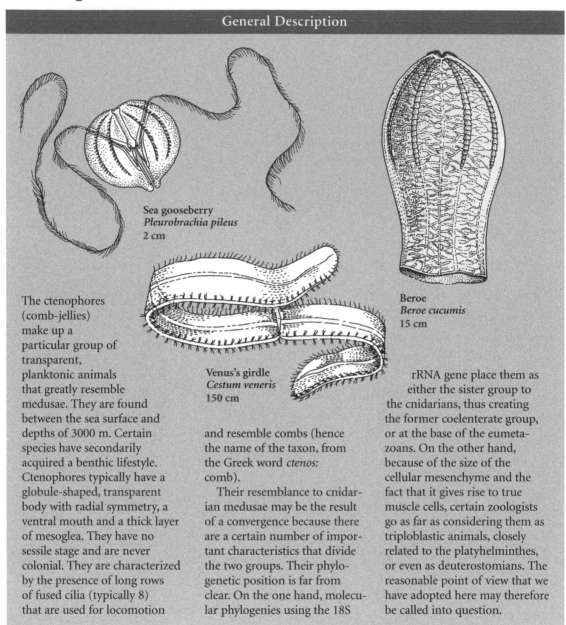

General Description

Sea gooseberry
Pleurobrachia pileus
2 cm

Venus's girdle
Cestum veneris
150 cm

Beroe
Beroe cucumis
15 cm

The ctenophores (comb-jellies) make up a particular group of transparent, planktonic animals that greatly resemble medusae. They are found between the sea surface and depths of 3000 m. Certain species have secondarily acquired a benthic lifestyle. Ctenophores typically have a globule-shaped, transparent body with radial symmetry, a ventral mouth and a thick layer of mesoglea. They have no sessile stage and are never colonial. They are characterized by the presence of long rows of fused cilia (typically 8) that are used for locomotion and resemble combs (hence the name of the taxon, from the Greek word *ctenos:* comb).

Their resemblance to cnidarian medusae may be the result of a convergence because there are a certain number of important characteristics that divide the two groups. Their phylogenetic position is far from clear. On the one hand, molecular phylogenies using the 18S rRNA gene place them as either the sister group to the cnidarians, thus creating the former coelenterate group, or at the base of the eumetazoans. On the other hand, because of the size of the cellular mesenchyme and the fact that it gives rise to true muscle cells, certain zoologists go as far as considering them as triploblastic animals, closely related to the platyhelminthes, or even as deuterostomians. The reasonable point of view that we have adopted here may therefore be called into question.

Some unique derived features

- The symmetry is biradial, if we take into account both the combs and the tentacles (fig. 1: the two symmetrical planes of the cydippid larva; 1: sagittal plane, 2: transverse plane, *cna:* canal, *com:* comb, *dt:* digestive tube, *te:* tentacle).

- Colloblasts: the animals have specialized adhesive cells called colloblasts, often carried on the tentacles.

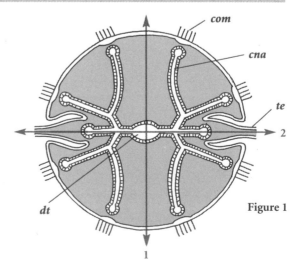

Figure 1

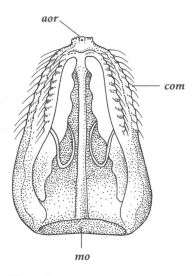

Figure 2

- Combs: a ctenophore normally has eight comb rows that are controlled by a unique apical sensory organ.
- A cydippid larva is characteristic of the taxon (fig. 2: *mo:* mouth, *aor:* apical sensory organ).

Ecology

The greatest majority of ctenophores are planktonic animals that are difficult to capture because of their fragility. They are rarely found in good condition after trawling. Yet, in certain areas of the world, they make up the majority of plankton biomass. They are carnivores that principally feed on zooplankton. After having fed on jellyfish, *Haeckelia* sp. retains the prey's cnidocytes and incorporates them into its own epidermis.

Number of species: 100.
Oldest known fossils: one fossil that resembles a cydippid larva has been recorded from the Devonian (380 MYA).
Current distribution: worldwide.

Examples

Beroe: *Beroe cucumis;* Venus's girdle: *Cestum veneris; Haeckelia rubra;* Sea-gooseberry: *Pleurobrachia pileus.*

Myxozoa

General Description

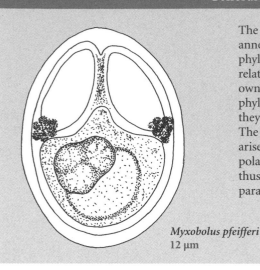

The myxozoans are simple organisms that parasitize annelids and poikilothermic vertebrates. Their phylogenetic situation remains quite uncertain; until relatively recently, these organisms were placed in their own phylum, distinct from the eukaryotes. Molecular phylogenies based on the 18S rRNA gene now show that they are, in fact, metazoans, and very likely eumetazoans. The simplicity of these organisms must therefore have arisen secondarily as a result of a parasitic way of life. The polar capsules resemble the cnidocytes of the cnidarians, thus supporting the hypothesis that the myxozoans are parasitic cnidarians.

Myxobolus pfeifferi
12 μm

Some unique derived features

- The spore has cells containing polar filaments that resemble the cnidocytes of the cnidarians (fig. 1: schematic cross-section of a spore with two terminal cells containing polar filaments; *pof:* polar filament; *sw:* spore wall; *N:* nucleus).

- The 18S rRNA gene contains unique derived characters.

- The absence of embryonic germ layers may be the result of a secondary loss due to a parasitic lifestyle.

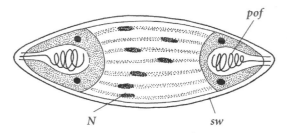

Figure 1

Number of species: 1,200.
Oldest known fossils: none.
Current distribution: worldwide.

Examples

Myxobolus pfeifferi (parasite of freshwater fishes); *Triactinomyxon ignotum* (parasite of the freshwater oligochaete *Tubifex*).

Ecology

The myxozoans are parasites whose spores are ingested by the hosts. Once ingested, the polar filaments are expelled and used to slow down the transit through the digestive tube. The spore wall opens and liberates the parasite that crosses the intestinal endothelium using amoeboid movements and migrates to its final destination. Certain myxozoans that parasitize vertebrates have a complex life cycle involving a protostomian intermediate host.

Bilateria

A few representatives

Edible crab
Cancer pagurus
25 cm
node **11**

Paper nautilus
Argonauta argo
10 cm
node **11**

Dicyema truncatum
400 µm
node **12**

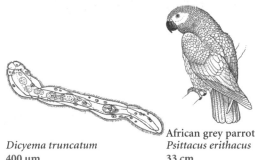

African grey parrot
Psittacus erithacus
33 cm
node **13**

Some unique derived features

■ Bilateral symmetry: the animal has two main polar axes (anterior-posterior and dorsal-ventral) linked to a bilateral symmetry.

■ The anterior-posterior axis is parallel to the sense of locomotion and to the movement of food in the digestive tract, entering at the anterior end by the mouth and exiting at the posterior end at the anus (except in cases of secondary simplifications).

■ A cephalization process concentrates the sensory and prehension organs around the mouth. This is the head (except in cases of secondary losses).

■ The blastopore of the gastrula gives rise to at least one of the orifices of the digestive tube.

■ A third germ layer, the mesoderm, occurs between the endoderm and the ectoderm. It contains a cavity called the coelom (except in cases of secondary losses).

■ The protonephridium, a system of excretory cells, is made up of three differentiated cells in its primitive state: a terminal cell with a flagellum, a canal cell, and a nephridiopore cell (fig. 1: protonephridium without the nephridiopore cell;

cac: canal cell, *tc:* terminal flame cell, *fl:* flagellum). This basic plan is typically more complex.

■ All synapses are unidirectional with a characteristic acetylcholine / acetylcholinesterase system.

■ Central nervous system: a central nervous system is organized around a cephalic ganglion and nerve cord.

■ Developmental genes of the *Hox* family are grouped into a complex. These genes obey the colinearity rule: the order of the genes in the complex corresponds to their site of action on the animal, from anterior to posterior (fig. 2: the *Hox* complex of *Drosophila melanogaster; ANT-C:* Antennapedia complex, *BX-C:* Bithorax complex). The ancestral role of this complex was most likely to organize the interactions between the central nervous system and the rest of the body along the anterior-posterior axis.

Figure 1

tc
fl
cac

Number of species: 1,191,311.
Oldest known fossils: burrows probably dug by a coelomate or pseudocoelomate animal were fossilized in the Precambrian (700 MYA); to be preserved, this organism must have had an internal cavity. Indisputable bilaterians were present in the Ediacara fauna (680 MYA). *Vernanimalcula,* from Yunnan, China, dates from the Precambrian (600 MYA).
Current distribution: worldwide.

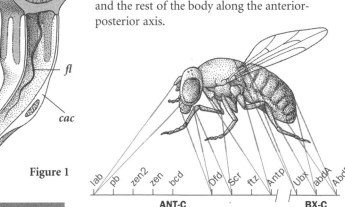

lab pb zen2 zen bcd   Dfd Scr ftz Antp   Ubx abdA AbdB

ANT-C   **BX-C**

Figure 2

Examples

■ PROTOSTOMIA: human intestinal roundworm: *Ascaris lumbricoides,* common cuttlefish: *Sepia officinalis,* earthworm: *Lumbricus terrestris;* seven-spotted ladybeetle: *Coccinella septempunctata.*

■ DEUTEROSTOMIA: common starfish: *Asterias rubens;* perch: *Perca fluviatilis.*

211

➡ *See chap. 8, p. 216*

Protostomia

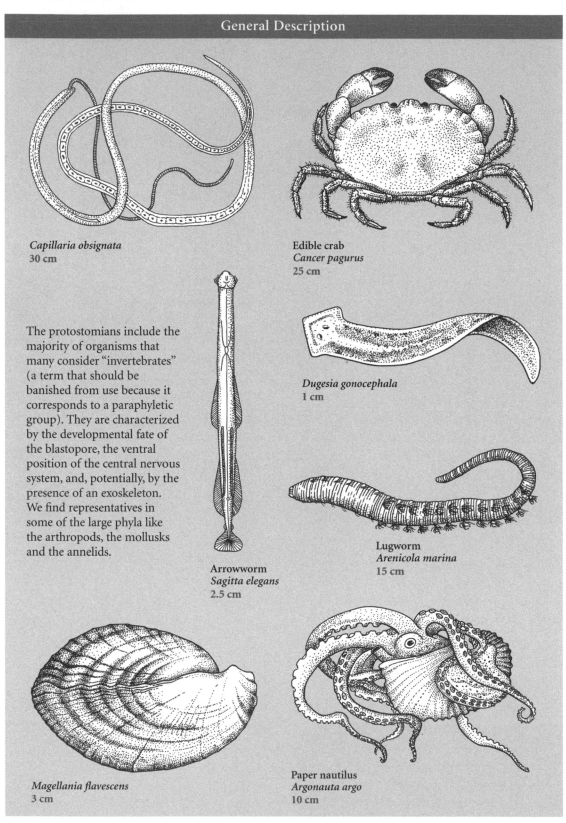

General Description

Capillaria obsignata
30 cm

Edible crab
Cancer pagurus
25 cm

The protostomians include the majority of organisms that many consider "invertebrates" (a term that should be banished from use because it corresponds to a paraphyletic group). They are characterized by the developmental fate of the blastopore, the ventral position of the central nervous system, and, potentially, by the presence of an exoskeleton. We find representatives in some of the large phyla like the arthropods, the mollusks and the annelids.

Dugesia gonocephala
1 cm

Arrowworm
Sagitta elegans
2.5 cm

Lugworm
Arenicola marina
15 cm

Magellania flavescens
3 cm

Paper nautilus
Argonauta argo
10 cm

Some unique derived features

- The blastopore becomes the future mouth and anus; the mouth is thus formed in a primitive manner, hence the name of the taxon (fig. 1: the blastopore (*blp*) of the onychophoran *Peripatopsis capensis* gives rise to the mouth (*mo*) and the anus (*a*), *POS*: posterior, *ANT*: anterior).

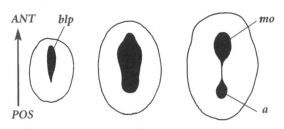

Figure 1

- The mesoderm develops from a mesentoblast.
- The coelom is formed by schizocoely, that is, masses of mesentoblastic cells separate to form a cavity (fig. 2: frontal view of an annelid trochophore larva showing the initial teloblasts

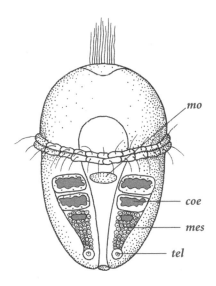

Figure 2

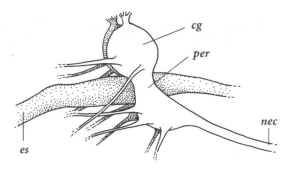

Figure 3

that give rise to the mesoderm. The mesoderm then enlarges to form the coelom; *coe*: coelom, *mes*: mesoderm, *tel*: teloblast).

- The nervous system is in a ventral position with the exception of the cephalic ganglion. The nerve cords lie below the digestive tube (the hyponeurian condition) and there is a periesophageal ring (fig. 3: anterior part of the nervous system of the pycnogonid *Nymphon* sp.; *nec*: nerve cord, *per*: periesophageal ring, *cg*: cerebral ganglion, *es*: esophagus).

- Exoskeleton: when a skeleton is present, it is external.

Ecology

The protostomians have conquered all environments, from mountain summits to deep ocean abysses, and all latitudes. Insects, the taxon with the greatest number of representatives, are protostomians.

Examples

- Lophotrochozoa: *Philodina roseola; Seison annulatus; Moniliformis dubius* (infests mice, rats, dogs and cats*); Symbion pandora; Loxosomella elegans; Pedicellina cernua;* human tapeworm: *Taeniarhynchus Saginatus; Dugesia gonocephala;* liver fluke: *Fasciola hepatica; Lineus longissimus; Chiton tuberculatus;* European brown garden snail: *Helix aspersa;* razor clam: *Solen marginatus;* paper nautilus: *Argonauta argo; Sipunculus robustus;* lugworm: *Arenicola marina; Membranipora membranacea; Lingula anatina; Magellania flavescens; Phoronis architecta.*
- Cuticulata: velvet worm: *Peripatus jamaicensis;* cross orb weaver: *Araneus diadematus;* edible crab: *Cancer pagurus;* seven-spotted ladybeetle: *Coccinella septempunctata; Capillaria obsignata;* human intestinal roundworm: *Ascaris lumbricoides; Trichinella spiralis* (causal agent of trichinosis*); Caenorhabditis elegans; Gordius robustus; Nectonema agile; Tubiluchus corallicola; Priapulus caudatus; Cateria styx; Kinorhynchus* sp.; *Nanaloricus mysticus; Lepidoderma squamatum.*

Number of species: 1,132,930.
Oldest known fossils: a joint that may have belonged to an arthropod has been found in the Ediacaran fauna (680 MYA). Several trilobite species (arthropoda) are known from before the Cambrian (580 MYA). In addition, an echiurian, *Protechiurus edmonsi*, is also from the Precambrian era (upper Vendian from Namibia, 560 MYA).
Current distribution: worldwide.

Mesozoa

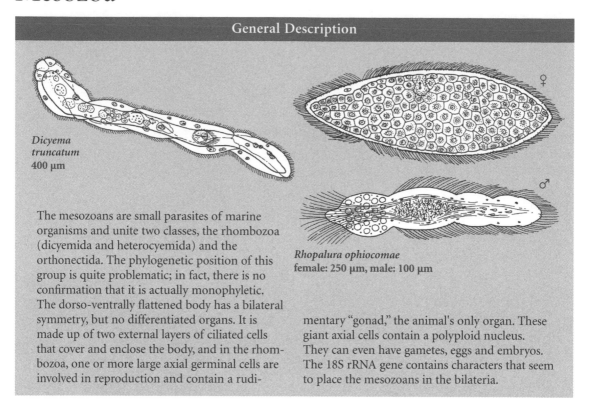

Dicyema truncatum
400 μm

Rhopalura ophiocomae
female: 250 μm, male: 100 μm

The mesozoans are small parasites of marine organisms and unite two classes, the rhombozoa (dicyemida and heterocyemida) and the orthonectida. The phylogenetic position of this group is quite problematic; in fact, there is no confirmation that it is actually monophyletic. The dorso-ventrally flattened body has a bilateral symmetry, but no differentiated organs. It is made up of two external layers of ciliated cells that cover and enclose the body, and in the rhombozoa, one or more large axial germinal cells are involved in reproduction and contain a rudi- mentary "gonad," the animal's only organ. These giant axial cells contain a polyploid nucleus. They can even have gametes, eggs and embryos. The 18S rRNA gene contains characters that seem to place the mesozoans in the bilateria.

Some unique derived features

- There is a loss of tissues (mesoderm) and organs (digestive tube, nervous system) in relation to the parasitic lifestyle.

In the rhombozoa:

- One (or several) large cell(s) containing a polyploid nucleus is found in an axial position (fig. 1: cross-section of *Conocyema*; the axoblasts (*axb*) give rise to the embryos (*em*)).

- Stem cells are nested within the axial cell. In certain cases, the growth and differentiation of these cells gives rise to a larva. The development of the embryos is thus intracellular.

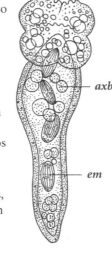

— *axb*

— *em*

Figure 1

Number of species: 50, divided into two classes.
Oldest known fossils: none.
Current distribution: worldwide, because mesozoans parasitize widely distributed and diverse marine organisms.

Ecology

The class rhombozoa brings together parasites of the excretory system of cephalopod mollusks. The orthonectida are parasites of marine platy- helminthes, nemerteans, polychaete annelids, bivalve mollusks, echinoderms. The mesozoans absorb the nutrients from diverse host body fluids by diffusion, explaining the absence of digestive organs. The complex life cycle involves an alternation of sexual and asexual generations, both infective for the hosts. Certain specialists have interpreted the particular anatomical features of the mesozoa as those of degenerated platyhelminthes. Others believe that they may represent an earlier group, independent of the platyhelminthes. The name mesozoa was given to them because it was thought that they represented intermediates between unicellular and multicellular organisms. Their phylogenetic position, even though still uncertain, is definitely not at the base of the metazoan tree, as we once believed, but rather with the bilaterians.

Examples

Conocyema polymorphum; Dicyema truncatum; Microcyema gracile; Rhopalura ophiocomae.

➡ *see chap. 11, p. 324*

Deuterostomia

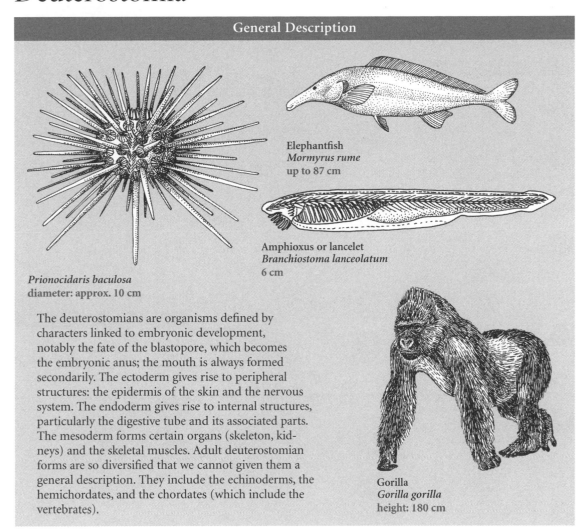

General Description

Elephantfish
Mormyrus rume
up to 87 cm

Amphioxus or lancelet
Branchiostoma lanceolatum
6 cm

Prionocidaris baculosa
diameter: approx. 10 cm

The deuterostomians are organisms defined by characters linked to embryonic development, notably the fate of the blastopore, which becomes the embryonic anus; the mouth is always formed secondarily. The ectoderm gives rise to peripheral structures: the epidermis of the skin and the nervous system. The endoderm gives rise to internal structures, particularly the digestive tube and its associated parts. The mesoderm forms certain organs (skeleton, kidneys) and the skeletal muscles. Adult deuterostomian forms are so diversified that we cannot given them a general description. They include the echinoderms, the hemichordates, and the chordates (which include the vertebrates).

Gorilla
Gorilla gorilla
height: 180 cm

Some unique derived features

- Mouth: it is formed secondarily, hence the name of the group (*deutero* = "secondary" and *stoma* = "mouth"). The first embryonic orifice becomes the anus in the larva and, most often, in the adult.

- The coelom is formed by enterocoely and the mesoderm from the archenteron (fig. 1: invagination of the archenteron (*arc*) wall gives rise to a pair of coelomic pouches (*cpo*); *blp*: blastopore).

- The skeleton is internal.

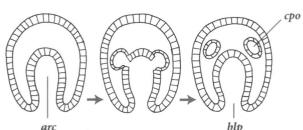

arc   *blp*

Figure 1

Ecology

The deuterostomians contain a vast diversity of forms. They are present in all environments (aerial, terrestrial, marine and aquatic), at all latitudes and at all altitudes (from 11,000 m below to 6,000 m above sea level).

Number of species: 58,331.
Oldest known fossils: *Helicoplacus* sp. from Nevada dates from the start of the Cambrian (540 MYA).
Current distribution: worldwide.

Examples

- ECHINODERMATA: common starfish: *Asterias rubens;* purple sea urchin: *Paracentrotus lividus.*
- HEMICHORDATA: acorn worm: *Balanoglossus clavigerus.*
- CHORDATA: *Ciona intestinalis;* amphioxus or lancelet: *Branchiostoma lanceolatum;* crested newt: *Triturus cristatus;* human: *Homo sapiens.*

Chapter 8

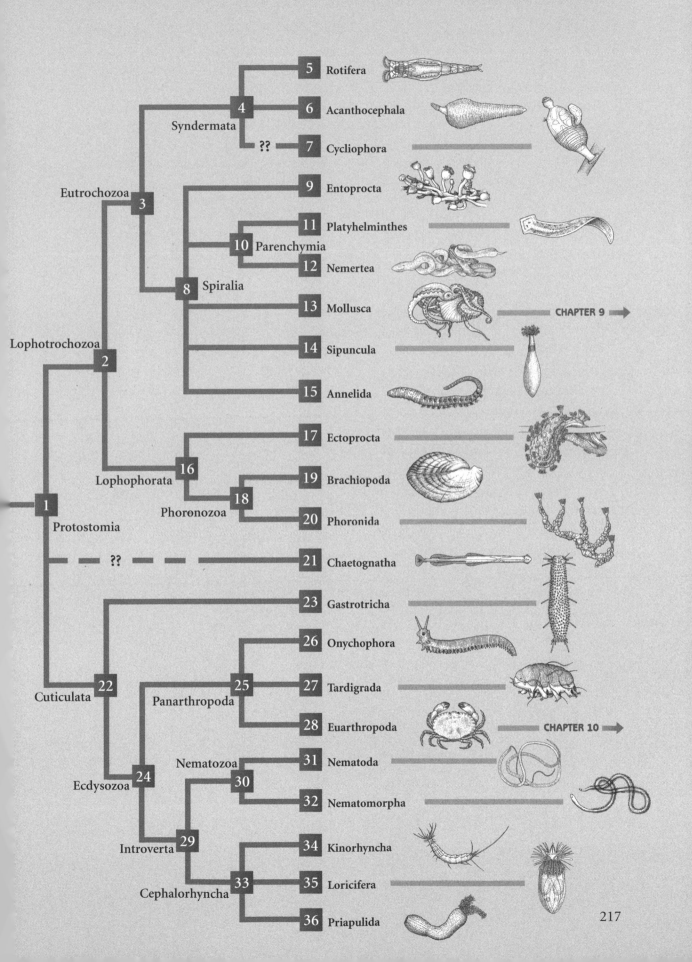

			5	Rotifera
		4	**6**	Acanthocephala
	Syndermata		**??**	**7**	Cycliophora
Eutrochozoa	**3**		**9**	Entoprocta
			11	Platyhelminthes
		10 Parenchymia	**12**	Nemertea
		8 Spiralia	**13**	Mollusca
Lophotrochozoa			**14**	Sipuncula
2			**15**	Annelida
			17	Ectoprocta
	Lophophorata **16**	**19**	Brachiopoda
	Phoronozoa	**18**	**20**	Phoronida
1				
Protostomia	??		**21**	Chaetognatha
			23	Gastrotricha
			26	Onychophora
Cuticulata **22**	Panarthropoda **25**	**27**	Tardigrada
			28	Euarthropoda
Ecdysozoa **24**	Nematozoa	**31**	Nematoda
	30	**32**	Nematomorpha
Introverta **29**		**34**	Kinorhyncha
	Cephalorhyncha **33**	**35**	Loricifera
		36	Priapulida

CHAPTER 9 →

CHAPTER 10 →

217

Protostomia

From a historical perspective, the protostomians represent a unique taxon. Although first delineated in the nineteenth century on the basis of the absence of a character (non-deuterostomian), it is one of the rare taxonomic groups defined in this way that seems, at least for the moment, to be monophyletic. The classification of the protostomians has been completely revisited over the last decade, primarily from the perspective of molecular phylogenies based on the 18S rRNA gene. Nonetheless, this work is still far from complete and has yet to resolve certain important points.

In comparison with the classic vision of this group, the phylogeny that we propose here integrates much new data; among the changes, we can highlight:

- the elimination of the "acoelomate" and "pseudocoelomate" groups, and their integration into the protostomia;
- the separation of the protostomians into two main taxa (not taking the gastrotricha into account), the lophotrochozoa and the ecdysozoa;
- the proximity of the nematodes and the arthropods;
- the proximity of the mollusks and the annelids.

Although the phylogeny presented here replaces the former gradist vision of increasing complexity, it is still not able to answer all questions. Indeed, this phylogeny raises numerous evolutionary questions, such as the origin of the coelom, of metamery, and, more generally, of the principal body plans. It is beyond the scope of this book to deal with these issues as they require a considerable amount of background information and a long discussion.

Nevertheless, in this book, we have considered that the hollowed mesoderm of the coelom is a bilaterian synapomorphy; this means that protostomian and deuterostomian coeloms are homologous. This view is not shared by everyone. If we take only the formation process into consideration (schizocoely or enterocoely), we can assume that the coelom was acquired twice independently. This hypothesis also means that all protostomian organisms that lack a coelom have lost it secondarily—another idea debated by scientists. This hypothesis nonetheless obtains substantial support from the fact that the monophyly of the protostomians is rarely challenged. However, it should be noted that the morphological and embryological characters justifying the protostomian clade are weak and could be interpreted as plesiomorphies. We wager that in the years to come the final blow to the traditional classification of the metazoans will be the disappearance of the protostomians once and for all.

Lophotrochozoa

A few representatives

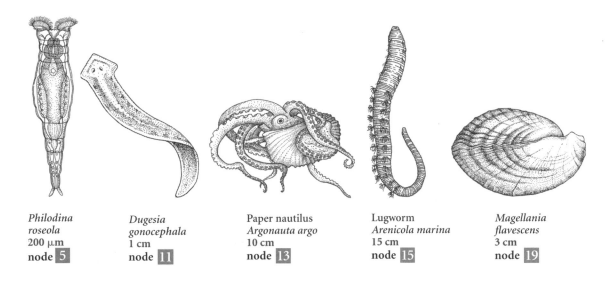

Philodina
roseola
200 μm
node 5

Dugesia
gonocephala
1 cm
node 11

Paper nautilus
Argonauta argo
10 cm
node 13

Lugworm
Arenicola marina
15 cm
node 15

Magellania
flavescens
3 cm
node 19

Some unique derived features

■ 18S and *Hox* genes: Molecular phylogenies, based largely on the 18S ribosomal RNA gene, and the characteristics of certain developmental genes (in particular, the posterior *Hox* genes) unite those animals with a trochophore-type larva and possessing a lophophore (ring of tentacles surrounding the mouth). The name of the group comes from these two characteristics.

Number of species: 154,806.
Oldest known fossils: the most ancient fossil is an echiurian (annelid), *Protechiurus edmonsi*, dating from the Precambrian era (upper Vendian from Namibia, 560 MYA). The platyhelminthes *Platypholinia pholiata* and *Vladimissa missarzhevskii* have been found in the region of the White Sea and date from the upper Precambrian (550 MYA). Several classes of mollusks are known from the lower Cambrian (540 MYA): bivalves like *Heraultipegma* sp. from Siberia and *Projetaia runnegari* from South Australia and China, and monoplacophorans like *Anabarella plana* from Siberia and *Yochelcionella* sp. from South Australia.
Current distribution: worldwide.

Examples

■ Eutrochozoa: *Philodina roseola; Seison annulatus; Stephanoceros* sp.; *Corynosoma* sp. (infests aquatic birds and seals); *Moniliformis dubius* (infests mice, rats, dogs and cats); *Symbion pandora; Loxosomella elegans; Pedicellina cernua;* human tapeworm: *Taeniarhynchus saginatus* (hosts: human, cattle); *Dugesia gonocephala;* planarian: *Planaria maculata;* liver fluke: *Fasciola hepatica; Lineus longissimus; Chiton tuberculatus;* European brown garden snail: *Helix aspersa;* razor clam: *Solen marginatus;* squid: *Loligo vulgaris; Sipunculus robustus;* lugworm: *Arenicola marina;* sandworm: *Nereis diversicolor;* earthworm: *Lumbricus terrestris;* medicinal leech: *Hirudo officinalis.*
■ Lophophorata: *Cristatella mucedo; Membranipora membranacea; Lingula anatina; Magellania flavescens; Terebratulina retusa; Phoronis architecta; P. harmeri.*

Eutrochozoa

A few representatives

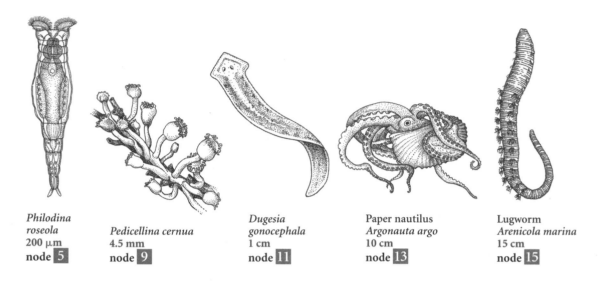

*Philodina
roseola*
200 μm
node 5

Pedicellina cernua
4.5 mm
node 9

*Dugesia
gonocephala*
1 cm
node 11

Paper nautilus
Argonauta argo
10 cm
node 13

Lugworm
Arenicola marina
15 cm
node 15

Some unique derived features

- Trochophore-type larvae: These larvae are top-shaped and characterized by the presence of a band of locomotory cilia anterior to the mouth (the prototroch) (fig. 1: *mo*: mouth, *i*: intestine, *mes*: mesoderm, *ptt*: prototroch, *apt*: apical tuft). Depending on the group, this larva can either undergo segmentation (metamerization), as in the annelida, or transform without segmentation, as in the mollusca. This larval type may be a synapomorphy of the lophotrochozoans because the ectoprocta, the brachiopoda and the phoronida all have larvae with similar anatomies.

Number of species: 149,956.
Oldest known fossils: the oldest known fossil is an echiurian (annelid), *Protechiurus edmonsi*, dating from the Precambrian era (upper Vendian from Namibia, 560 MYA). The platyhelminthes *Platypholinia pholiata* and *Vladimissa missarzhevskii* have been found in the region of the White Sea and date from the Upper Precambrian (550 MYA). Several classes of mollusks are known from the lower Cambrian (540 MYA): bivalves like *Heraultipegma* sp. from Siberia and *Projetaia runnegari* from South Australia and China, and monoplacophorans like *Anabarella plana* from Siberia and *Yochelcionella* sp. from South Australia.
Current distribution: worldwide.

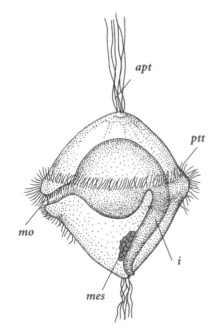

Figure 1

Examples

- SYNDERMATA: *Philodina roseola; Seison annulatus; Stephanoceros* sp.; *Corynosoma* sp. (infests aquatic birds and seals); *Moniliformis dubius* (infests mice, rats, dogs and cats); *Symbion pandora; Loxosomella elegans.*
- SPIRALIA: *Pedicellina cernua;* human tapeworm: *Taeniarhynhus saginatus* (hosts: human, cattle); *Dugesia gonocephala;* planarian: *Planaria maculata;* liver fluke: *Fasciola hepatica; Lineus longissimus; Chiton tuberculatus;* European brown garden snail: *Helix aspersa;* razor clam: *Solen marginatus;* paper nautilus: *Argonauta argo;* squid: *Loligo vulgaris; Sipunculus robustus;* lugworm: *Arenicola marina;* sandworm: *Nereis diversicolor;* earthworm: *Lumbricus terrestris;* medicinal leech: *Hirudo officinalis.*

Syndermata

A few representatives

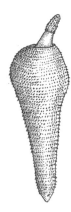
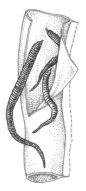
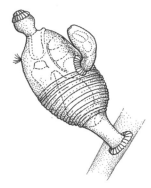

Philodina roseola
200 μm
node 5

Rhinops fertöensis
300 μm
node 5

Corynosoma sp.
3 cm
node 6

Macracanthorhynchus hirudinaceus
1 m in pigs
node 6

Symbion pandora
350 μm
node 7

Some unique derived features

■ The keratinous cuticule, produced by the syncytial epidermal cells, is intracellular. Fig. 1 shows an electron microscope view of the structure of the syncytial epidermis of the rotifer *Asplanchna sieboldi* (*icu*: intracellular cuticle, *cmb*: cytoplasmic membrane, *po*: pore). This character is found in the rotifers and the acanthocephalans but is not shared by the cyclophorans.

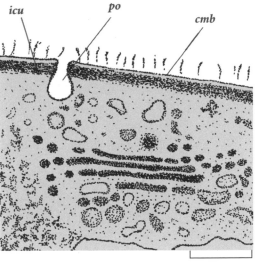

Figure 1   0.5 μm

Number of species: 2,951.
Oldest known fossils: a rotifer, *Keratella* sp., has been found in South Australia and dates from the middle Eocene (45 MYA).
Current distribution: worldwide.

Examples

■ ROTIFERA: *Bryceelia tenella; Lecane luna; Philodina roseola; Seison annulatus; Stephanoceros* sp.
■ ACANTHOCEPHALA: *Corynosoma* sp. (infests aquatic birds and seals); *Macracanthorhynchus hirudinaceus* (infests swine); *Moniliformis dubius* (infests mice, rats, dogs and cats).
■ CYCLIOPHORA: *Symbion pandora.*

Rotifera

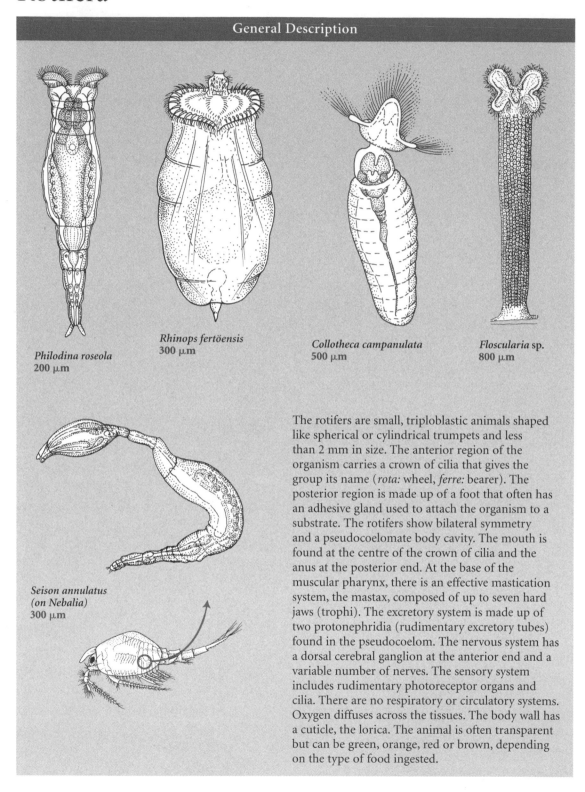

Philodina roseola
200 μm

Rhinops fertöensis
300 μm

Collotheca campanulata
500 μm

Floscularia sp.
800 μm

Seison annulatus
(on Nebalia)
300 μm

The rotifers are small, triploblastic animals shaped like spherical or cylindrical trumpets and less than 2 mm in size. The anterior region of the organism carries a crown of cilia that gives the group its name (*rota:* wheel, *ferre:* bearer). The posterior region is made up of a foot that often has an adhesive gland used to attach the organism to a substrate. The rotifers show bilateral symmetry and a pseudocoelomate body cavity. The mouth is found at the centre of the crown of cilia and the anus at the posterior end. At the base of the muscular pharynx, there is an effective mastication system, the mastax, composed of up to seven hard jaws (trophi). The excretory system is made up of two protonephridia (rudimentary excretory tubes) found in the pseudocoelom. The nervous system has a dorsal cerebral ganglion at the anterior end and a variable number of nerves. The sensory system includes rudimentary photoreceptor organs and cilia. There are no respiratory or circulatory systems. Oxygen diffuses across the tissues. The body wall has a cuticle, the lorica. The animal is often transparent but can be green, orange, red or brown, depending on the type of food ingested.

Some unique derived features

- The anterior region carries a ciliated crown of characteristic shape used for locomotion and food capture (fig. 1: *Stephanoceros* sp.; *clc:* ciliated crown). Fig. 2 shows two characteristic crown types.

- The mastax is composed of 7 mobile cuticular pieces that are characteristic of the group. On fig. 3, these pieces are colored (*ph:* pharynx, *es:* esophagus).

- A retrocerebral organ is found in association with the cerebral ganglion; the function of this glandular structure is unknown.

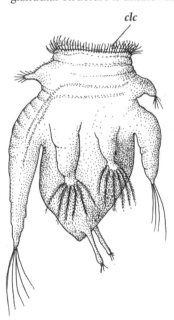

Figure 1

Ecology

Most rotifers are freshwater animals. A few, like *Seison annulatus,* are marine, whereas others live in moist terrestrial habitats. Certain species are parasites of crustaceans, mollusks and annelids. Rotifers are a major component of freshwater zooplankton and are an extremely important source of nutrients in freshwater ecosystems. In terrestrial ecosystems, they participate in soil decomposition. We find them among the plankton of seas, lakes and rivers, in the sand of beaches, in peat and even in mosses, lichens and frozen mud. These organisms are able to resist desiccation. They are either free-swimming or sessile and fixed to a substrate by their foot. Certain benthic species are free-swimming but can temporarily anchor themselves to a substrate. Planktonic species have a reduced foot. Rotifers feed on diverse unicellular organisms, other rotifers and suspended organic materials. The sexes are separate. Monogonont males are smaller than the females. Fertilization is internal. The spermatozoid reaches the ovule by passing through the body wall of the female. The egg has radial cleavage and development is direct. In many monogonont species, there are two types of eggs: resting eggs and rapidly hatching eggs. The bdelloids have populations that contain only females and that reproduce by parthenogenesis. The two egg types, as well as parthenogenesis, are seen as adaptations allowing rapid colonization of temporarily favorable environments. Resting eggs can survive adverse conditions and hatch as soon as favorable conditions return.

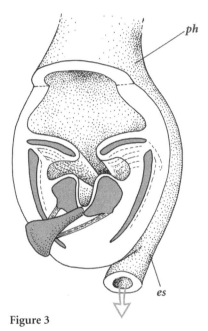

Figure 3

Number of species: 1,800 separated among 3 classes.
Oldest known fossils: *Keratella* sp. has been found in South Australia and dates from the middle Eocene (45 MYA).
Current distribution: worldwide.

Examples

Asplanchna priodonta; Bryceelia tenella; Collotheca campanulata; Floscularia sp.; *Lecane luna; Philodina roseola; Pleurotrocha petromyzon; Rhinops fertöensis; Seison annulatus; Stephanoceros* sp.; *Trochosphaera solstitialis.*

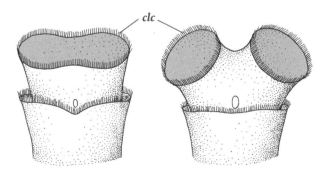

Figure 2

Acanthocephala

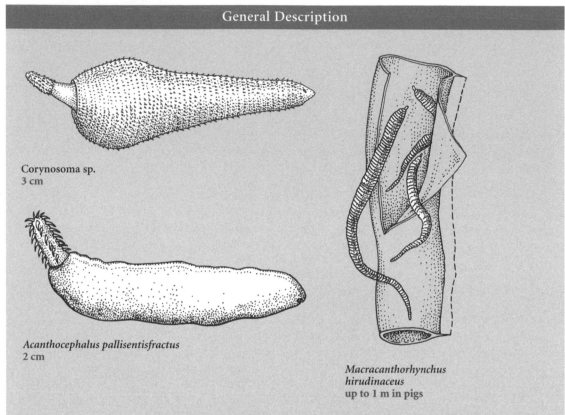

Corynosoma sp.
3 cm

Acanthocephalus pallisentisfractus
2 cm

Macracanthorhynchus
hirudinaceus
up to 1 m in pigs

The acanthocephalans are small obligate parasites of vertebrates. They are vermiform in shape with an anterior, retractile horn, or proboscis, armed with curved spines. This hooked proboscis can be retracted into a muscular cavity. Overall, the body is cylindrical and shows bilateral symmetry. Although annulations are sometimes superficially visible, there is no segmentation. The trunk of the body can carry papillae or spines. Most acanthocephalans are a few millimeters in size; however, one species, a parasite of swine—*Macracanthorhynchus hirudinaceus*—can reach close to one meter in length. The body cavity, the pseudocoelom, is large. There is no mouth, digestive tract, circulatory or respiratory system. The muscular, nervous and excretory systems are reduced. The external color is variable: off-white, yellow, orange or red. Many zoologists feel that the acanthocephalans are really just rotifers that have adopted a parasitic way of life.

Some unique derived features

- The horn, or proboscis, is armed with hooks that enable the animal to attach to the intestinal wall of its host; it can be retracted into a muscular cavity (fig. 1: *Acanthocephalus pallisentisfractus; mcv:* muscular cavity, *hk:* hook). Fig. 2 shows the horn retracted and extended.

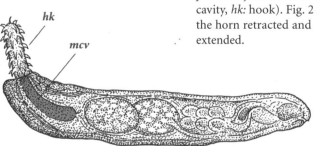

hk

mcv

Figure 1

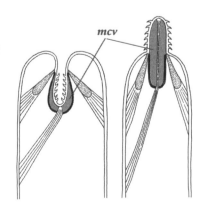

mcv

Figure 2

Ecology

The acanthocephalans are all arthropod parasites in the juvenile stage and vertebrate intestinal parasites in the adult stage. Their life cycle always requires two hosts. There is no free-living stage. Nutrients from the host's body fluids are absorbed directly into the body cavity of the parasite, passing through the body wall. The horn is used for moving within the host in the larval stage, whereas in the adult stage it is used to attach to the host's intestinal wall. The vertebrate hosts are often teleost fishes. The sexes are separate. The male has an eversible penis and fertilization is internal. The eggs develop in the female's pseudocoelom. During this period of the life cycle, the adult worms are in the vertebrate digestive tract. The eggs are released to the external environment with the feces. If an insect or crustacean ingests an egg, the larva will hatch and live in the arthropod's body cavity. This acanthor larva is armed with a hooked rostrum that enables it to penetrate the host's tissues and go from the digestive tract into the body cavity. Here, the larva will then undergo two transformations to give rise to the successive stages, the acanthella and the cystacanth. If a vertebrate consumes an infected arthropod (rat eating an insect, bird or fish eating an aquatic crustacean), its digestive tract will destroy the shell of the cystacanth and the juvenile acanthocephalan will attach to the intestinal wall of its new host. Acanthocephalans cause major damage to the intestinal walls of their hosts when found in large numbers. For example, a thousand individual parasites have been counted in a duck's intestine, and 1154 in a seal's intestine.

Number of species: 1,150 divided among three orders.
Oldest known fossils: none.
Current distribution: worldwide.

Examples

- *Acanthocephalus pallisentisfractus; Corynosoma* sp. (infests aquatic birds and seals); *Leptorhynchoides thecatus* (infests the teleost fish *Micropterus salmoides*); *Macracanthorhynchus hirudinaceus* (infests pigs); *Moniliformis dubius* (infests mice, rats, dogs and cats); *Polymorphus minutus* (parasite of waterfowl and domestic ducks).

Cycliophora

General Description

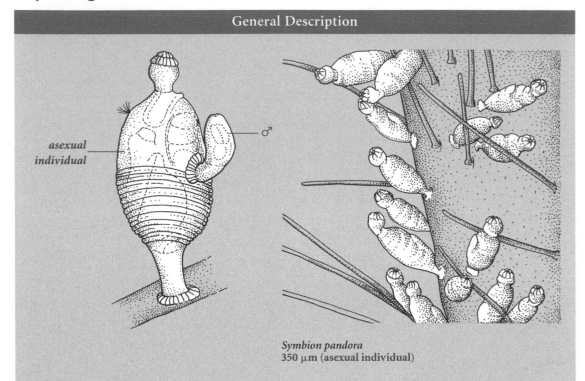

asexual individual

♂

Symbion pandora
350 μm (asexual individual)

This is the most recently described animal phylum (December 1995). The only described species lives on the bristles surrounding the mouth of the Norway lobster *Nephrops norvegicus*. These animals are acoelomates with bilateral symmetry. They are sessile in the feeding stages and take in food by filtration using a ring of cilia that surrounds the mouth. The males are dwarfed and live attached to the females. The phylogenetic position of this group has been uncertain since its description. It has been described as a phylum "showing affinities with the entoprocta and the ectoprocta"; this is an absurd description. Molecular phylogenies actually place them close to (or within) the snydermata, whereas larval morphology suggests that they are a sister group of the entoprocta.

Some unique derived features

■ The buccal apparatus is characteristic.

■ Males are dwarfed.

■ There are two larval types, the pandora larva that results from asexual multiplication and the chordoid larva that results from fertilization.

Number of species: 1.
Oldest known fossils: none.
Current distribution: North Atlantic, for the moment.

Ecology

The only described species of this group lives attached to the Norway lobster. However, we have observed cycliophorans morphologically identical to *Symbion pandora* on the mandibular bristles of the European lobster *Homarus gammarus* in Roscoff, France. Reproduction is complex; there is a cycle of asexual reproduction with the liberation of *pandora* larvae. Sexual reproduction takes place between a sessile female and a dwarf male.

Examples

Symbion pandora.

Spiralia

A few representatives

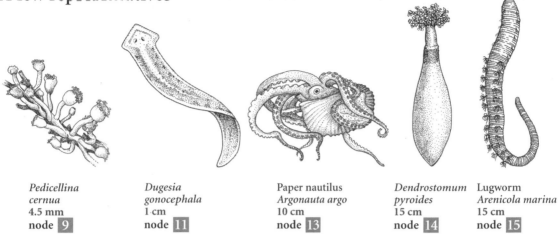

Pedicellina
cernua
4.5 mm
node **9**

Dugesia
gonocephala
1 cm
node **11**

Paper nautilus
Argonauta argo
10 cm
node **13**

Dendrostomum
pyroides
15 cm
node **14**

Lugworm
Arenicola marina
15 cm
node **15**

Some unique derived features

■ The egg undergoes spiral cleavage. Cell division is oblique to the animal-vegetal polar axis of the egg. The initial cell divisions give rise to small cells (the micromeres, *d*) near the animal pole and large cells (the macromeres) near

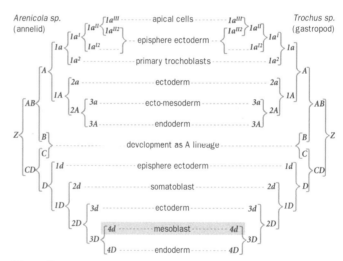

Figure 2

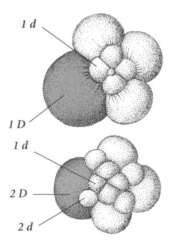

Figure 1

the vegetal pole. Fig. 1 outlines two steps in the spiral cleavage of the gastropod *Ilyanassa* sp.; Fig. 2 shows that the cell lineages of the annelid *Arenicola* sp. and the gastropod *Trochus* sp. are identical. The micromere 4d, obtained at the sixth division, will give rise to the mesoderm.

Number of species: 147,005.
Oldest known fossils: *Protechiurus edmonsi*, an echiurian (annelid), dating from the Precambrian era (upper Vendian from Namibia, 560 MYA). The platyhelminthes *Platypholinia pholiata* and *Vladimissa missarzhevskii* have been found in the region of the White Sea (Upper Precambrian, 550 MYA). Several classes of mollusks are known from the lower Cambrian (540 MYA): bivalves like *Heraultipegma* sp. from Siberia and *Projetaia runnegari* from South Australia and China, and monoplacophorans like *Anabarella plana* from Siberia and *Yochelcionella* sp. from South Australia.
Current distribution: worldwide.

Examples

■ ENTOPROCTA: *Loxosomella elegans; Pedicellina cernua.*
■ PARENCHYMIA: human tapeworm: *Taeniarhynchus saginatus* (hosts: humans, cattle); *Dugesia gonocephala;* planarian: *Planaria maculata;* liver fluke: *Fasciola hepatica; Lineus longissimus.*
■ MOLLUSCA: *Chiton tuberculatus;* European brown garden snail: *Helix aspersa;* razor clam: *Solen marginatus;* paper nautilus: *Argonauta argo:* squid: *Loligo vulgaris.*
■ SIPUNCULA: *Dendrostomum pyroides; Sipunculus robustus.*
■ ANNELIDA: lugworm: *Arenicola marina;* sandworm: *Nereis diversicolor;* earthworm: *Lumbricus terrestris;* medicinal leech: *Hirudo officinalis.*

227

Entoprocta

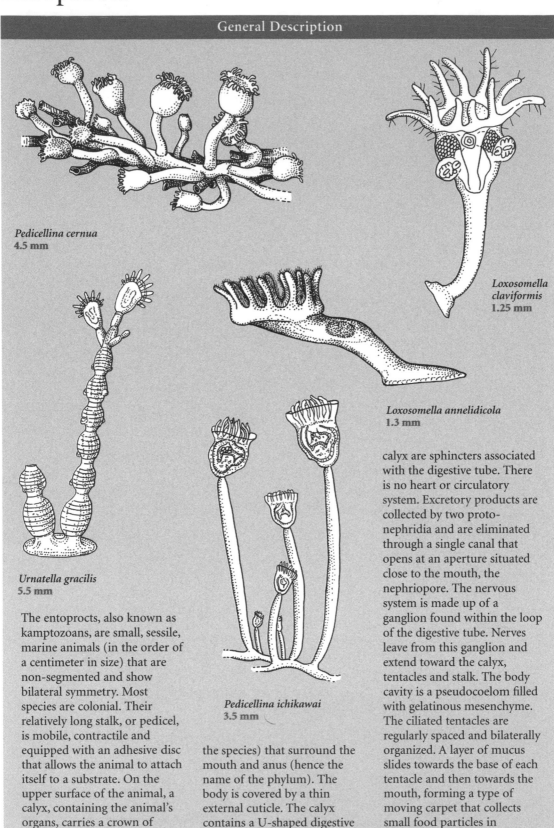

Pedicellina cernua
4.5 mm

Loxosomella claviformis
1.25 mm

Urnatella gracilis
5.5 mm

Loxosomella annelidicola
1.3 mm

Pedicellina ichikawai
3.5 mm

The entoprocts, also known as kamptozoans, are small, sessile, marine animals (in the order of a centimeter in size) that are non-segmented and show bilateral symmetry. Most species are colonial. Their relatively long stalk, or pedicel, is mobile, contractile and equipped with an adhesive disc that allows the animal to attach itself to a substrate. On the upper surface of the animal, a calyx, containing the animal's organs, carries a crown of tentacles (6 to 36 depending on the species) that surround the mouth and anus (hence the name of the phylum). The body is covered by a thin external cuticle. The calyx contains a U-shaped digestive tube. The only muscles of the calyx are sphincters associated with the digestive tube. There is no heart or circulatory system. Excretory products are collected by two proto-nephridia and are eliminated through a single canal that opens at an aperture situated close to the mouth, the nephriopore. The nervous system is made up of a ganglion found within the loop of the digestive tube. Nerves leave from this ganglion and extend toward the calyx, tentacles and stalk. The body cavity is a pseudocoelom filled with gelatinous mesenchyme. The ciliated tentacles are regularly spaced and bilaterally organized. A layer of mucus slides towards the base of each tentacle and then towards the mouth, forming a type of moving carpet that collects small food particles in suspension.

Some unique derived features

- The anus and mouth are found within the ring of tentacles (fig. 1: *a:* anus, *mo:* mouth, *gg:* ganglion, *L:* larva, *te:* tentacle). From an anatomical perspective, this is the most striking difference between the entoprocts and the ectoprocts (or bryozoans).

- There is a characteristic larva.

Ecology

These are solitary or colonial sedentary animals that live in shallow waters. Many species are commensal. Only a single genus is found living in fresh water (*Urnatella*) and only one species has a free-swimming adult form (*Loxosomella davenporti*). The entoprocts feed on suspended marine particles that they capture using the cilia and mucus covering the tentacles. The mucus traps the particles and slides them toward the base of the tentacles and then toward the mouth. The tentacles are not retractile. Instead, they roll in on themselves and are covered by the inter-tentacular membrane. Asexual multiplication occurs by budding, often from the stolon, the common axis linking the animals within a colony. Certain segments of the stolon generate the new calyces of the colony. Buds can also sometimes form on the stalk. In solitary species, a bud forms on the calyx, detaches and settles somewhere else. The entoprocts are hermaphroditic. The calyx has two ovaries and two testes. Gametes are released through a single gonopore that opens close to the nephriopore. Fertilization is internal, but without

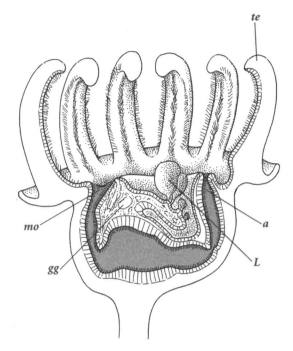

Figure 1

copulation. Eggs are incubated in a pouch that lies between the gonopore and the anus. The embryos are fed by cells on the body surface of the parent. There is a ciliated and free-swimming trochophore larva, similar to the larva we find in the mollusks and annelids. After a free-swimming period, this larva will anchor itself to a substrate and undergo a metamorphosis into the adult form.

Number of species: 150.
Oldest known fossils: *Barentsia* sp. was found in England and dates from the upper Jurassic (145 MYA).
Current distribution: coastal waters of the Americas, Africa, Europe, Asia and the Arctic.

Examples

Barentsia laxa; Loxosomella annelidicola; L. claviformis; Pedicellina cernua; P. ichikawai; Urnatella gracilis.

Parenchymia

A few representatives

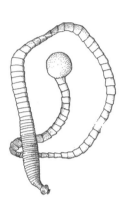

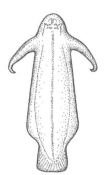

Dugesia gonocephala
1 cm
node **11**

Taenia taeniaeformis
50 cm
node **11**

Cerebratulus californiensis
20 cm
node **12**

Nectonemertes mirabilis
10 cm
node **12**

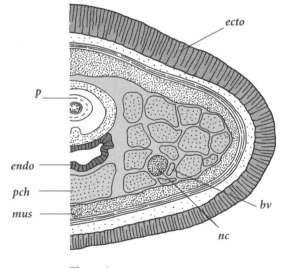

Figure 1

Some unique derived features

■ The mesoderm consists of a loose tissue, the parenchyma, that fills the space between the ectoderm and the endoderm. Fig. 1 shows a cross-section near the proboscis of a nemertean (*nc*: nerve cord, *ecto*: ectoderm, *endo*: endoderm, *mus*: muscles, *p*: proboscis, *pch*: parenchyma, *bv*: blood vessel). It is likely that the general body organization is that of a triploblastic animal in which the coelom was filled secondarily.

Number of species: 14,680.
Oldest known fossils: the platyhelminthes *Platypholinia pholiata* and *Vladimissa missarzhevskii*, dating from the upper Precambrian (550 MYA), have been found in the region of the White Sea.
Current distribution: worldwide.

Examples

■ PLATYHELMINTHES: *Dugesia gonocephala;* planarian: *Planaria maculata* (freshwater); human tapeworm: *Taeniarhynchus saginatus* (hosts: humans, cattle); *T. taeniaeformis;* schistosome: *Schistosoma mansoni* (host: humans); liver fluke: *Fasciola hepatica.*
■ NEMERTEA: *Cerebratulus californiensis; Lineus longissimus.*

Platyhelminthes

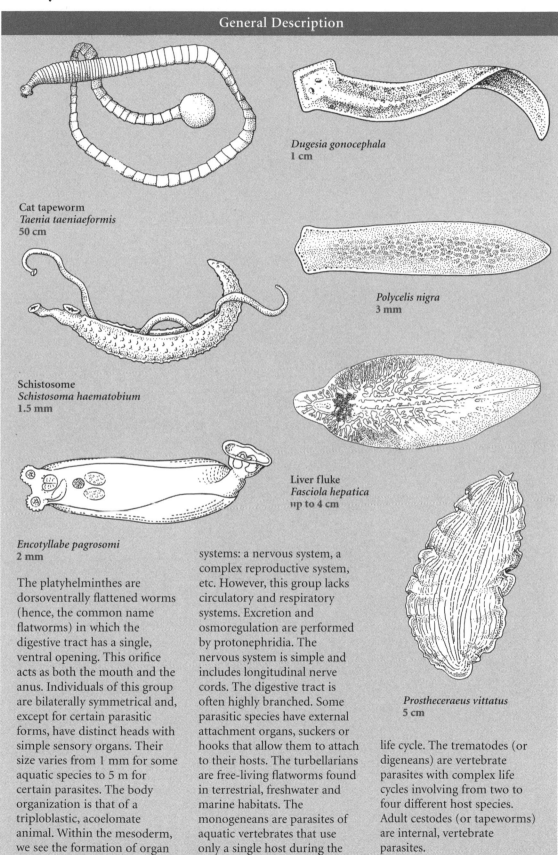

General Description

Cat tapeworm
Taenia taeniaeformis
50 cm

Schistosome
Schistosoma haematobium
1.5 mm

Encotyllabe pagrosomi
2 mm

Dugesia gonocephala
1 cm

Polycelis nigra
3 mm

Liver fluke
Fasciola hepatica
up to 4 cm

Prostheceraeus vittatus
5 cm

The platyhelminthes are dorsoventrally flattened worms (hence, the common name flatworms) in which the digestive tract has a single, ventral opening. This orifice acts as both the mouth and the anus. Individuals of this group are bilaterally symmetrical and, except for certain parasitic forms, have distinct heads with simple sensory organs. Their size varies from 1 mm for some aquatic species to 5 m for certain parasites. The body organization is that of a triploblastic, acoelomate animal. Within the mesoderm, we see the formation of organ systems: a nervous system, a complex reproductive system, etc. However, this group lacks circulatory and respiratory systems. Excretion and osmoregulation are performed by protonephridia. The nervous system is simple and includes longitudinal nerve cords. The digestive tract is often highly branched. Some parasitic species have external attachment organs, suckers or hooks that allow them to attach to their hosts. The turbellarians are free-living flatworms found in terrestrial, freshwater and marine habitats. The monogeneans are parasites of aquatic vertebrates that use only a single host during the life cycle. The trematodes (or digeneans) are vertebrate parasites with complex life cycles involving from two to four different host species. Adult cestodes (or tapeworms) are internal, vertebrate parasites.

Some unique derived features

- The animals of this group are triploblastic, often with a complex digestive system, and have only a single orifice. On fig. 1, we see the pharynx (*ph*) of *Monocelis galapagoensis*.

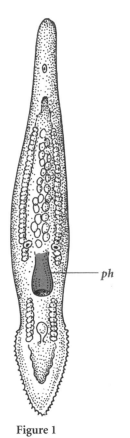

Figure 1

Ecology

The platyhelminthes contain both free-living and parasitic species. Most species are aquatic, although some are terrestrial. As a group, they can tolerate temperatures that range from -50 to 47 °C and can be found in a large diversity of climates. They can be either carnivores or detritivores. Most species are hermaphroditic. Sexual organs are often complex and enable internal fertilization. In certain species, the spermatozoids are bi-flagellated. The life cycle of parasitic species is often quite complex, involving several host organisms, numerous larval stages and phases of asexual multiplication. Most flatworms are capable of regenerating lost or damaged tissue. Within the group, there are a number of important vertebrate parasites, certain of which cause major human health problems at a global scale. For example, *Schistosoma mansoni* is responsible for bilharziasis in tropical countries.

Number of species: 13,780. **Oldest known fossils:** this group has representatives starting in the upper Precambrian (550 MYA) in the White Sea region, *Platypholinia pholiata* and *Vladimissa missarzhevskii*.
Current distribution: worldwide. Freshwater flatworms tend to live in cooler, temperate zones, whereas marine species can be found in all seas and oceans of the world. The vast terrestrial distribution of this group is largely due to parasitic forms that are widely dispersed by their hosts. The number of truly terrestrial species is limited; most of these species live in humid soils in tropical zones.

Examples

- CESTODA: human tapeworm: *Taeniarhynchus saginatus* (hosts: humans, cattle); large tapeworm: *T. solium* (hosts: humans, pigs); *T. taeniaeformis*.
- TURBELLARIA: *Dugesia gonocephala*; planarian: *Planaria maculata* (freshwater); *Polycelis nigra*; *Thysanozoon brocchii* (marine).
- TREMATODA: liver fluke: *Fasciola hepatica* (infests the liver of sheep); *Polystomum integerrimum* (host: frogs); schistosome: *Schistosoma mansoni* (host: humans); *S. haematobium*.
- MONOGENEA: *Benedenia melleni* (host: freshwater "fishes"); *Encotyllabe pagrosomi*.

Nemertea

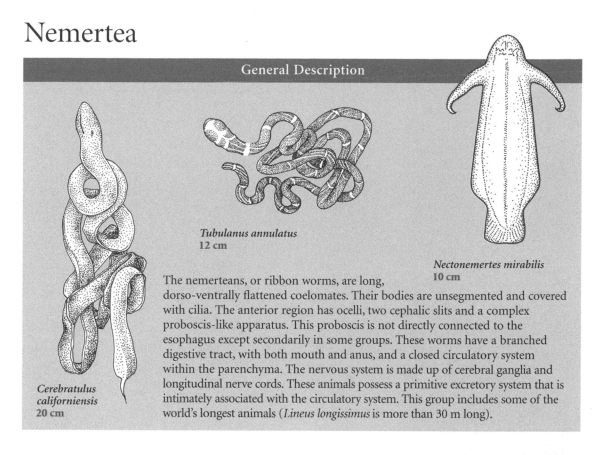

General Description

Tubulanus annulatus
12 cm

Nectonemertes mirabilis
10 cm

*Cerebratulus
californiensis*
20 cm

The nemerteans, or ribbon worms, are long, dorso-ventrally flattened coelomates. Their bodies are unsegmented and covered with cilia. The anterior region has ocelli, two cephalic slits and a complex proboscis-like apparatus. This proboscis is not directly connected to the esophagus except secondarily in some groups. These worms have a branched digestive tract, with both mouth and anus, and a closed circulatory system within the parenchyma. The nervous system is made up of cerebral ganglia and longitudinal nerve cords. These animals possess a primitive excretory system that is intimately associated with the circulatory system. This group includes some of the world's longest animals (*Lineus longissimus* is more than 30 m long).

Some unique derived features

- The eversible, trunk-like proboscis is not connected to the digestive tube (fig. 1; *mo*: mouth, *pr*: proboscis).

- The rhynchocoel (fig. 1: *rhy*) is a tubular, hydrostatic coelomic cavity that surrounds the proboscis. An increase in pressure within the rhynchocoel, caused by muscular contraction, triggers the extension of the proboscis (a, b, c).

rhy

(a)

pr

(b)

(c)

mo

Figure 1

Number of species: 900 divided into two classes.
Oldest known fossils: the oldest fossil which is without doubt a nermertean is *Archisymplectes rhothon*, found in Illinois and dating from the Carboniferous (310 MYA). However, *Amiskwia* sp. from the Cambrian period (from 540 to 500 MYA) may also be a nemertean.
Current distribution: worldwide.

Examples

Cerebratulus californiensis; Lineus ruber; L. longissimus; Nectonemertes mirabilis; Paranemertes peregrina; Tubulanus annulatus.

Ecology

The nemerteans are mostly benthic, marine worms, except for a few pelagic species like *Nectonemertes mirabilis* and a few terrestrial and freshwater species. Most are found in crevices or under rocks, but certain species dig burrows in soft marine substrates. They move via external cilia or body undulations. They are active predators that use their proboscis, abruptly everted by the hydrostatic pressure of the rhynchocoel, for hunting. Prey is captured, or injured, by a venomous stylet found on the tip. Terrestrial species can use the proboscis for locomotion. The nemerteans are oviparous with separate sexes. Fertilization is external. Eggs are laid in the seawater, sometimes protected by a gelatinous mass. Development can be direct or indirect, with a pilidium larva that undergoes an unusual type of metamorphosis.

⇒ *see chap. 9, p. 268*

Mollusca

General Description

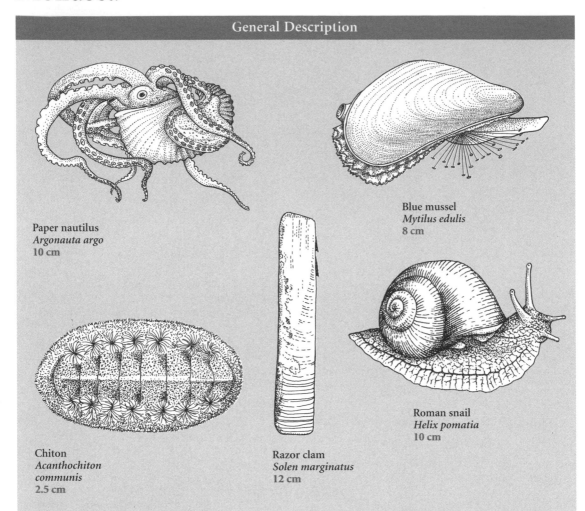

Paper nautilus
Argonauta argo
10 cm

Blue mussel
Mytilus edulis
8 cm

Chiton
Acanthochiton communis
2.5 cm

Razor clam
Solen marginatus
12 cm

Roman snail
Helix pomatia
10 cm

The mollusks are triploblastic, coelomate metazoans with a greatly reduced coelom in the adult stage. They include organisms of diverse shapes, but with a homogeneous body plan. The often muscular foot has taken on many different functions. The dorsal and lateral regions are modified into a mantle that secretes the shell and/or calcareous spicules. In certain species, the shell can be secondarily absent or become internal and incorporated into the mantle. A visceral mass, located dorsal to the foot, is associated with a cavity formed by a fold in the mantle, the pallial (or mantle) cavity. This cavity opens to the exterior. It is the end-point for the products of the excretory and reproductive systems and, on the dorsal side, the anus. It also contains chemoreceptors (the osphradia) and a pair of ciliated gills. The body is unsegmented and the initial bilateral symmetry can be secondarily altered by torsion. The digestive tube is simple. The mouth contains two chitinous jaws and a toothed ribbon, the radula. The circulatory system is open and contains vessels and sinuses. The dorsal heart controls the circulation of the blood through these structures. The nervous system includes a nerve ring surrounding the esophagus, several pairs of ganglia, a pair of nerve cords that innervate the foot (pedal cords) and visceral cords that innervate the mantle and visceral mass. Modifications to the general structure provide the specific characteristics that define each of the 8 classes of this branch.

Some unique derived features

- A specialized integument, the mantle, secretes calcareous (calcium carbonate) formations. These formations can be spicules, when the secretory glands are distributed across the mantle, or plates, when the glands are grouped together into secretion zones.

- A chitinous buccal structure, the radula, is hard and toothed and is used for scraping up food.

234

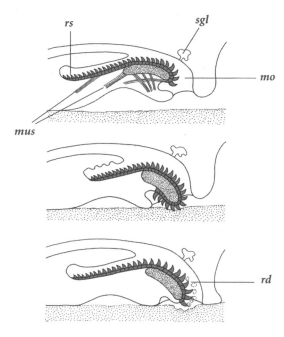

Figure 1

Fig. 1 shows the "conveyor belt" type movement of the radula (*mo:* mouth, *sgl:* salivary gland, *mus:* muscle, *rd:* radula, *rs:* radula sac). The shape of this toothed, chitinous band can change depending on the group.

■ The gills, called ctenidia, are comb-shaped.

Ecology

The mollusks live in aquatic or moist environments: sediments, silt, high seas, ocean floor, forest leaf litter, diverse soils, rivers, lakes. They can burrow into the sediments, like the bivalves, and capture microorganisms and suspended food particles using their gills. They can be phytophagous (gastropods, polyplacophorans), grazing on algae or terrestrial plants. In these instances, the retracting motion of the radula shreds the vegetative material and brings it towards the esophagus. They can also be carnivorous, feeding on sessile animals such as colonial ectoprocts (or bryozoans). Some mollusks are active hunters with rapid movements and sophisticated vision. This is the case for the cephalopods, predators of fish, crustacea or other cephalopods. Some gastropod species are parasitic on other marine animals. The mollusks typically have separate sexes; some gastropods and bivalves are hermaphrodites, but almost all undergo cross-fertilization. In the bivalves, there are sometimes periodic sex changes. Fertilization is typically external, but can sometimes be internal (gastropods). In the cephalopods, a nuptial parade is followed by the male transferring his spermatophores (sacs of spermatozoids) into the mantle cavity of the female using a specialized arm. Females are most often oviparous, but there are a few examples of viviparity. Eggs are often deposited within a protective jelly, sac, capsule, etc. The female octopus incubates, cleans, oxygenates and defends her eggs. The initial larvae are free-swimming, ciliated trochophore larvae; the trochophore larvae transform into veliger larvae, and then into the adult form. The length of the different larval stages can vary and some stages may be absent. In certain groups, there is direct development; this is the case for terrestrial gastropods and cephalopods. Although the smallest mollusk is the size of a grain of sand, this group also includes some very large organisms: with its tentacles, the cephalopod *Architeuthis* can measure up to 20 m. Humans consume cephalopods, gastropods and lamellibranchs. Snails and oysters are animals of intense agricultural production. By their species richness, diversity and abundance, the mollusks constitute an important group in many ecosystems. They can be intermediate hosts for numerous parasites (schistosome is an example that presents a danger to humans). They are prey to many cetaceans, fishes, seals and walrus. The terrestrial gastropods are also prey to many tetrapods.

Number of species: 117,495.
Oldest known fossils: several molluskan classes are known from the lower Cambrian (540 MYA): bivalves like *Heraultipegma* sp. from Siberia or *Projetaia runnegari* from South Australia and China, monoplacophorans like *Anabarella plana* from Siberia or *Yochelcionella* sp. from South Australia.
Current distribution: the mollusks have colonized most marine, freshwater and terrestrial environments. They are absent only from the driest and coldest deserts.

Examples

■ Solenogastres: *Proneomenia aglaopheniae.*
■ Caudofoveata: *Chaetoderma nitidulum.*
■ Polyplacophora: chiton: *Acanthochiton communis; Chiton tuberculatus.*
■ Monoplacophora: *Neopilina galathea.*
■ Gastropoda: European brown garden snail: *Helix aspersa;* flat periwinkle: *Littorina littoralis.*
■ Cephalopoda: paper nautilus: *Argonauta argo;* squid: *Loligo vulgaris;* nautilus: *Nautilus pompilius.*
■ Bivalvia (or lamellibranchia): razor clam: *Solen marginatus;* blue mussel: *Mytilus edulis.*
■ Scaphopoda: tusk shell: *Dentalium entalis.*

Sipuncula

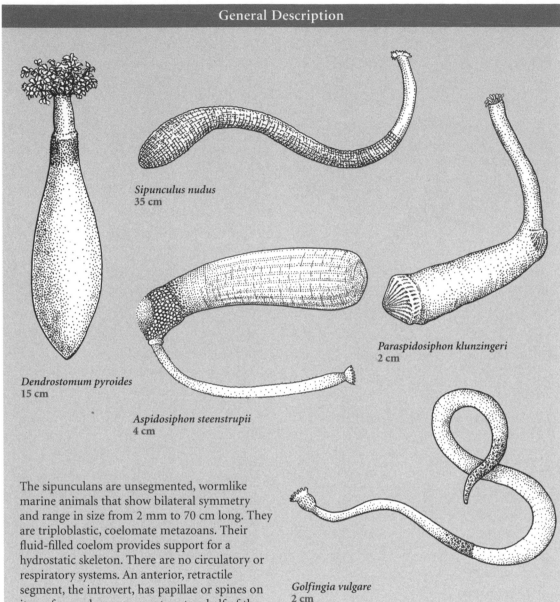

Sipunculus nudus
35 cm

Dendrostomum pyroides
15 cm

Aspidosiphon steenstrupii
4 cm

Paraspidosiphon klunzingeri
2 cm

Golfingia vulgare
2 cm

The sipunculans are unsegmented, wormlike marine animals that show bilateral symmetry and range in size from 2 mm to 70 cm long. They are triploblastic, coelomate metazoans. Their fluid-filled coelom provides support for a hydrostatic skeleton. There are no circulatory or respiratory systems. An anterior, retractile segment, the introvert, has papillae or spines on its surface and can represent up to a half of the total body length of the animal. The tip of the introvert terminates in a mouth surrounded by ciliated tentacles or lobes that collect suspended particles. The introvert can also be used for burrowing in the sediment. The animal uses the pressure of the coelomic liquid to extend the introvert into the sediment. The tip is then inflated and anchors the animal. Then, one to four retractor muscles contract the introvert, which causes the posterior part of the body to advance. The tentacles have a separate system of hydraulic canals. The digestive tube is U-shaped and the anus is situated at the base of the introvert on the dorsal side of the animal. The excretory system is made up of one or two metanephridia that remove nitrogen wastes via one or two nephridiopores that empty ventrally at the base of the introvert. The nervous system includes a bilobed brain located above the esophagus and connected to a single, ventral nerve cord with no ganglia. The sipunculans have ciliated, photoreceptive cells. The body wall includes a cuticle, an epidermis, along with longitudinal and circular transverse muscles. The external color is variable but is typically dull.

Some unique derived features

- The introvert carries a ring of tentacles that surround the mouth and forms the characteristic anterior part of the animal (fig. 1: dissection of *Golfingia vulgare; nc:* nerve cord, *int:* introvert, *es:* esophagus, *rec:* rectum, *te:* tentacles, *dt:* digestive tube).

- The epidermal canals play a role in respiration.

Ecology

The sipunculans live on the ocean floor and dig tunnels in the sand or silt that they line with mucus. They can also live in rock crevices, coral reefs, excavations, abandoned mollusk shells, under rocks or among mangrove roots. Certain species are annelid ectoparasites. This group feeds on diatoms, diverse unicellular organisms, larvae, organic detritus from the sediment, and even on microscopic films of algae adhering to rocks. Food is trapped by the mucus that covers the tuft of tentacles. The tentacular cilia bring the food to the central mouth. An abundant internal ciliature moves food through the digestive system. The sexes are separate, except in a single species that is hermaphroditic. The animals have a single gonad and gametes are released from the coelom to the external environment via the nephridiopore. Fertilization is external. In certain species, eggs give rise to a free-swimming, ciliated trochophore larva with a relatively short planktonic life. It rapidly falls to the sea bottom and undergoes metamorphosis. In other species, the larva has a relatively longer pelagic lifetime (pelagosphera larva). Again other species have direct development. The sipunculans can sometimes undergo asexual multiplication by transversally dividing the body.

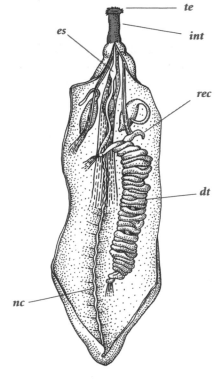

Figure 1

Examples

Aspidosiphon steenstrupii; Dendrostomum pyroides; Golfingia elongata; G. vulgaris; Paraspidosiphon klunzingeri; Phascolion strombi; Sipunculus nudus; S. robustus.

Number of species: 320.
Oldest known fossils: *Trypanites* sp. dates from the upper Silurian (415 MYA).
Current distribution: most sipunculans live in shallow, warm seas. However, a few species are found in polar waters. They are typically found at the intertidal zone, but some species may also be found at depths of up to 7000 m.

Annelida

General Description

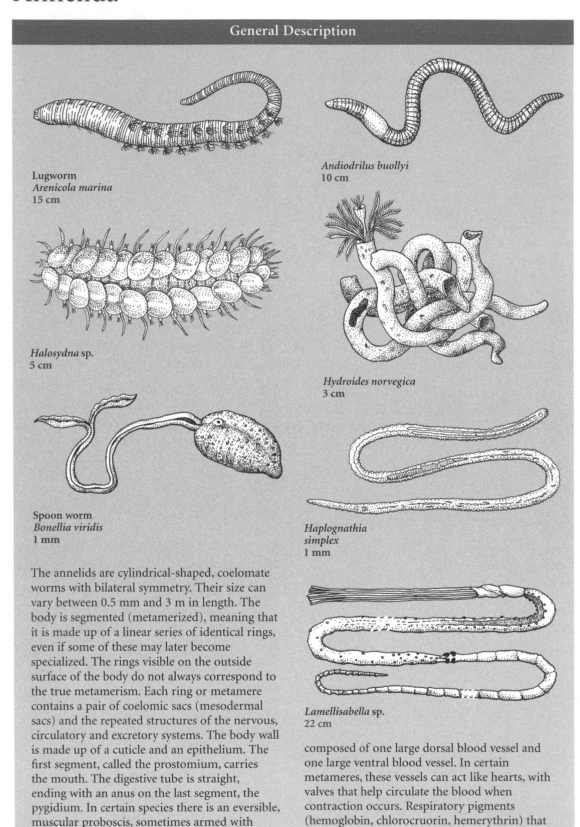

Lugworm
Arenicola marina
15 cm

Andiodrilus buollyi
10 cm

Halosydna sp.
5 cm

Hydroides norvegica
3 cm

Spoon worm
Bonellia viridis
1 mm

Haplognathia simplex
1 mm

Lamellisabella sp.
22 cm

The annelids are cylindrical-shaped, coelomate worms with bilateral symmetry. Their size can vary between 0.5 mm and 3 m in length. The body is segmented (metamerized), meaning that it is made up of a linear series of identical rings, even if some of these may later become specialized. The rings visible on the outside surface of the body do not always correspond to the true metamerism. Each ring or metamere contains a pair of coelomic sacs (mesodermal sacs) and the repeated structures of the nervous, circulatory and excretory systems. The body wall is made up of a cuticle and an epithelium. The first segment, called the prostomium, carries the mouth. The digestive tube is straight, ending with an anus on the last segment, the pygidium. In certain species there is an eversible, muscular proboscis, sometimes armed with teeth or chitinous mandibles. The coelomic fluid provides support for a hydrostatic skeleton. The circulatory system is closed. It is composed of one large dorsal blood vessel and one large ventral blood vessel. In certain metameres, these vessels can act like hearts, with valves that help circulate the blood when contraction occurs. Respiratory pigments (hemoglobin, chlorocruorin, hemerythrin) that transport oxygen are often present. Gas exchange occurs across the moist body wall, by the gills, or, in the polychaetes, by the parapodia (lateral

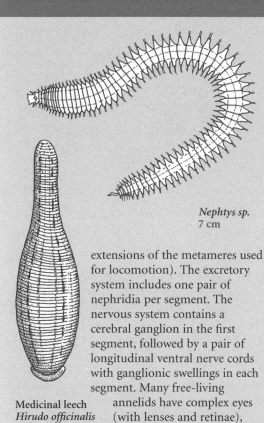

Nephtys sp.
7 cm

Medicinal leech
Hirudo officinalis
8 cm

extensions of the metameres used for locomotion). The excretory system includes one pair of nephridia per segment. The nervous system contains a cerebral ganglion in the first segment, followed by a pair of longitudinal ventral nerve cords with ganglionic swellings in each segment. Many free-living annelids have complex eyes (with lenses and retinae), anterior tentacles and balance organs found close to the brain. They often have chitinous bristles (or setae) on the body surface. Annelids of the class polychaeta are marine worms in which each segment has locomotory extensions called parapodia, each with a bundle of hard, chitinous setae. The head is made up of several modified segments that have fused together and has antennae, eyes, palps, mandibles, and tentacular cirri. This group includes both sedentary and wandering worms (nereids, arenicolids, sabellids). The myzostomids, crinoid parasites, are likely small, derived polychaetes. The oligochaetes have a prostomium without any prominent sensory structures and lack parapodia. They are covered with only a few simple chitinous setae. The reproductive organs are confined to a few segments. These worms possess a glandular region on the skin, the clitellum, that secretes a cocoon to protect the embryos during development. Most are freshwater or marine organisms, but there are some terrestrial species like the earthworms (nightcrawlers) and the manure worms. The hirudinea, or leeches, sometimes have an anterior sucker in addition to their characteristic posterior sucker. They are ventrally flattened worms that have neither setae nor parapodia. Sanguivorous species are included in this group.

Modern phylogenies show that the polychaetes are a paraphyletic group. The oligochaetes and leeches form the clitellata clade.

Finally, certain organisms that were previously described as part of separate phylums seem to be derived polychaetes. This is the case for the pogonophorans and the vestimentiferans, organisms that live at great depths. This is equally the case for the echiurians (spoon worms), animals that have fascinated zoologists for centuries. This may also be true for the gnathostomulids, animals of the meiofauna (living among the grains of marine sediment), but this question is still under debate.

Some unique derived features

Despite the homogeneity in the body structure, there are no major unique derived characters for this group. We can suggest two:

- The animals are fundamentally segmented. Fig. 1 shows a stage of the metamorphosis of a trochophore larva; the colored area is segmented (*mo:* mouth, *nc:* nerve cord, *met:* metamere, *dt:* digestive tube). The modification of metamerism is always secondary; *Branchiura* has nearly complete metamerism, the segments being almost identical except that gills (*g*, fig. 2) are present at the posterior end.

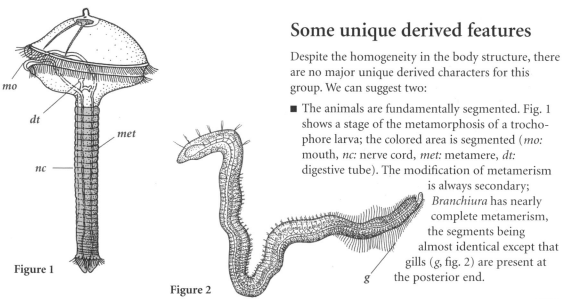

mo

dt

met

nc

Figure 1

Figure 2

g

239

■ They are the only protostomians (along with the cephalopod mollusks) with a closed circulatory system (fig. 3: *capi*: capillaries, *pad*: parapodium, *dbv*: dorsal blood vessel, *dt*: digestive tube, *vbv*: ventral blood vessel).

Ecology

The annelids include marine, freshwater and terrestrial species. The polychaetes can be free-swimming, tube-inhabiting or benthic burrowing animals. They are predators that capture diverse animals and eggs. However, they can also feed on the microorganisms found in sand and silt. Tube-inhabiting filter-feeders live within calcareous tubes or tubes formed of sand grains held together by mucus. These worms filter seawater using their tentacles and respire by extending their gills out beyond the end of the tube. The tentacles have a ciliated canal that brings food particles filtered from the seawater to the mouth.

The oligochaetes are marine, freshwater or terrestrial. They are microphagous. Terrestrial oligochaetes are burrowing animals that like moist environments. They feed on decomposing organic matter found in humus that they cover in saliva before absorption. They are extremely important for soil aeration and the recycling of minerals and organic matter.

The hirudinea are both aquatic and terrestrial. They are ectoparasites of aquatic animals and terrestrial vertebrates. Blood-sucking leeches attach to their hosts by one or two suckers and absorb the blood. Certain terrestrial species are common in tropical rain forests.

The annelids can reproduce by both sexual and asexual means. The hirudinea are exceptional in that they can undergo a division and regenerate the missing piece. In polychaetes, the sexes are separate and fertilization is external. The egg gives rise to a ciliated free-swimming larva, a trochophore larva.

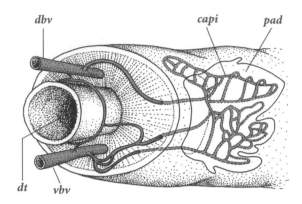

Figure 3

A few polychaete species incubate their eggs. In certain cases, the males protect and provide oxygen to the brood. In many species, individuals undergo a transformation just before reproduction. A modified individual is produced, the epitokal form, that is specialized for gamete production. The endocrine-controlled morphological changes involved in this process are so extreme that specialists initially believed that the immature and reproductive forms were different species. Both the oligochaetes and the leeches are hermaphroditic but practice cross-fertilization. Eggs are incubated within mucus cocoons and give rise to juveniles that resemble the adult forms; there is no larval stage. This is linked to a general adaptation to terrestrial life.

The pogonophorans and the vestimentiferans are deep-sea animals with a unique ecology that often live in hydrothermal vent areas.

Among the echiurans, the spoonworm *Bonellia* shows great sexual dimorphism; the male is dwarfed and parasitic on the female.

The simplification of the body in the gnathostomulids (in particular, the absence of the coelom) is linked to their specialized life style as organisms of the meiofauna.

Number of species: 14,360 (polychaetes: 10,000, oligochaetes: 3,500, leeches: 500, echiurians: 135, pogonophorans and vestimentiferans: 145, gnathostomulids: 80).
Oldest known fossils: the oldest recorded fossil is an echiurian: *Protechiurus edmonsi* from the end of the Precambrian (from Namibia during the upper Vendian, 560 MYA). We find annelids at the beginning of the Cambrian: *Palaeoscoleca sinensis* (530 MYA) from Yunnan, China, *Scolelepsis* sp. (540 MYA) from Australia. Some Precambrian fossils of the Ediacara fauna may also be annelids.
Current distribution: marine annelids are present in all oceans. Aquatic and terrestrial annelids have colonized almost all continents, but are not found in Madagascar or Antarctica.

Examples

■ POLYCHAETA: sea mouse: *Aphrodite aculeata*; lugworm: *Arenicola marina*; *Halosydna* sp.; *Hydroides norvegica*; sandworm: *Nereis diversicolor*; *Owenia fusiformis*; *Protula tubularia*; peacock worm: *Sabella pavonina*.
■ OLIGOCHAETA: *Andiodrilus buollyi*; red wiggler or compost worm: *Eisenia foetida*; earthworm: *Lumbricus terrestris*; *Megascolides australis*; *Tubifex tubifex*.
■ HIRUDINEA: *Haementeria officinalis*; *Haemopis sanguisuga*; medicinal leech: *Hirudo officinalis*.
■ ECHIURIA: green spoon worm: *Bonella viridis*; *Thalassema hartmani*.
■ POGONOPHORA: *Lamellisabella* sp.
■ VESTIMENTIFERA: *Riftia pachyptila*.

Lophophorata

A few representatives

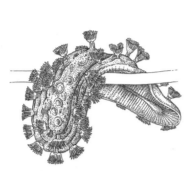

Cristatella mucedo
20 cm (entire colony)
node **17**

Lingula
Lingula anatina
30 cm
node **19**

Magellania flavescens
3 cm
node **19**

Phoronis psammophila
5 cm
node **20**

Some unique derived features

- The body is fundamentally divided into three parts, the reduced prosome called the epistome, the mesosome and the metasome. Each part is organized around a separate coelomic cavity.

- The digestive tube is U-shaped.

- The lophophore surrounds the mouth only and contains coelomic extensions (fig. 1: apical view of a phoronidian lophophore; *a:* anus, *mo:* mouth, *epis:* epistome, *nrp:* nephriopore, *te:* cut tentacles, also see fig. 1, p. 244).

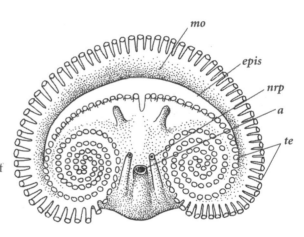

Figure 1

Number of species: 4,850.
Oldest known fossils: The brachiopods are found from the beginning of the Cambrian (540 MYA). They left behind numerous fossils (12,000 described fossil species) during all the periods that followed, but particularly so during the Ordovician (from 500 to 420 MYA).
Current distribution: worldwide.

Examples

- Ectoprocta: *Cristatella mucedo; Membranipora membranacea.*
- Phoronozoa: *Lingula anatina; Lacazeiia mediterranea; Magellania flavescens; Terebratulina retusa; Phoronis architecta; Phoronis psammophila; Phoronopsis harmeri.*

Ectoprocta

General Description

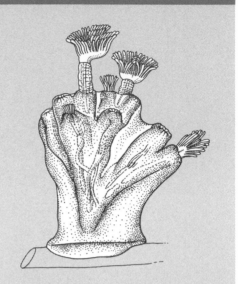

Cristatella mucedo
1.5 cm, up to 20 cm for an entire colony

Lophopus sp.
up to 3 cm for a colony

The ectoprocts, also known as bryozoans, are small, colonial animals that live anchored to substrates in aquatic environments. A colony contains a physical continuum of individuals, called zooids, generated by budding starting from a single individual. Each zooid is housed in a chitinous, gelatinous or chitino-calcareous exoskeleton that is continuous among all members of the colony. The size of a colony can range from one to several centimeters, whereas a single zooid is approximately 0.5 mm in length. Depending on the species, these colonies can either be flat or erect and branching with a diversity of shapes. To the naked eye, they look similar to aquatic mosses, hence the name bryozoa (moss animals). A zooid has a non-segmented, bilaterally symmetrical body that is connected to the neighboring zooids. Each has a calyx with a lophophore, gonads and a U-shaped digestive tube. The mouth is found within the ring of tentacles, whereas the anus, while still anterior, empties outside this ring. The epidermis secretes the external skeleton. The lophophore enters and exits this skeleton by a tubular opening. A pair of retractor muscles attaches the calyx to the skeleton and enables the tentacles to be retracted. There are no circulatory, respiratory, or excretory apparatuses. Gas exchange occurs across the tissues. The nervous system is simple. It includes a ring encircling the esophagus from which nerves arise. Sometimes the zooids of a colony specialize for certain functions (food, defense, egg incubation), to the benefit of the entire colony.

Some unique derived features

- Within the lophophorata, this group is mostly characterized by secondary losses: the epistome, the anterior end that contains the first coelomic cavity, is absent. In addition all parts linked to the circulatory, respiratory and excretory systems have disappeared.

- The lophophore is retractile (fig. 1: *Bowerbankia; a:* anus, *mo:* mouth, *lop:* lophophore).

Ecology

The ectoprocts are marine animals for the most part. Nevertheless, a few dozen species live in fresh water. The colonies form gelatinous or hard masses on the surface of rocks or certain algal species. The ectoprocts feed principally on phytoplankton. The tentacles are extended beyond the skeleton by increasing hydrostatic pressure in the coelomic cavity. This change in pressure is created by the compression of certain areas of the body. Plankton is captured by the cilia on the tentacles. These cilia also create a current that brings food towards the mouth. The tentacle ring can be retracted rapidly by the action of a pair of specialized muscles attached to the skeleton. The freshwater ectoprocts have hermaphroditic zooids with alternating production of sperm and ovules. In all ectoprocts, the gametes are released into the coelom and then exit the body via diverse modes; there is no specialized genital duct. Fertilization is internal, but without copulation. Most ectoprocts are oviparous and release small eggs into the aquatic environment. Free-swimming larvae hatch from these eggs. After a pelagic phase, these larvae settle and metamorphose into the adult form. Certain species incubate their eggs and have specialized reproductive zooids that have lost their tentacles and digestive tubes. Fertilization either occurs in the coelomic cavity or in a specialized chamber. The eggs are released through a coelomic pore or through a pore associated with a specialized organ that exits within the tentacles. In species that live in temperate freshwater habitats, the colonies die in the fall leaving behind statoblasts, packages of cells protected by a resistant outer shell that are able to survive the winter period. Zooids arise from these cells in the spring. Asexual multiplication of a single individual by budding can give rise to an entire colony.

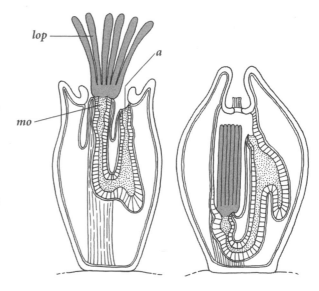

Figure 1

Number of species: 4,500 divided among 4 classes and 9 orders.
Oldest known fossils: different fossils have been found from the lower Ordovician (480 MYA): *Wolinella baltica* from around the Baltic Sea, *Profistulipora arctica* from Russia, *Xenotrypa primaeva* and *Revalotrypa eugeniae* from Estonia.
Current distribution: worldwide.

Examples

Bugula neritina; Cellaria sinuosa; Cristatella mucedo; Fiustrella hispida; Lophopus sp.; *Membranipora membranacea; Plumatella casmiana.*

Phoronozoa

A few representatives

Magellania flavescens
3 cm
node 19

Lingula
Lingula anatina
3 cm
node 19

Phoronis psammophila
5 cm
node 20

Phoronis architecta
5 cm
node 20

Some unique derived features

■ The prosome is reduced and transformed into a fleshy lobe, the epistome, that overhangs the mouth (fig. 1: organization of the apical region of a phoronidian; *a:* anus, *mo:* mouth, *epis:* epistome, *int:* intestine, *mec:* mesocoelom, *nr:* nephridium, *es:* esophagus, *protc:* protocoelom, *te:* cut tentacles of the lophophore, *bv:* blood vessels).

Number of species: 350.
Oldest known fossils: at least 12 families of brachiopods are known from the start of the Cambrian (540 MYA). *Skolithos* is a tube that was probably produced by a phoronidian; it has been found in Germany and northern Australia and dates from the lower Cambrian (540 MYA).
Current distribution: worldwide.

Examples

■ Brachiopoda: *Lingula anatina; Lacazella mediterranea; Magellania flavescens; Terebratulina retusa.*
■ Phoronida: *Phoronis architecta; P. psammophila; Phoronopsis harmeri.*

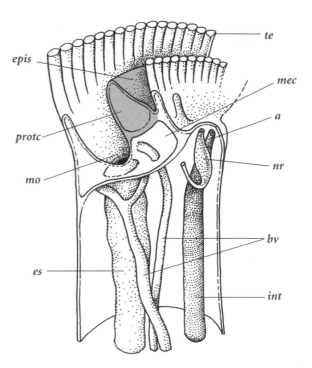

Figure 1

Brachiopoda

General Description

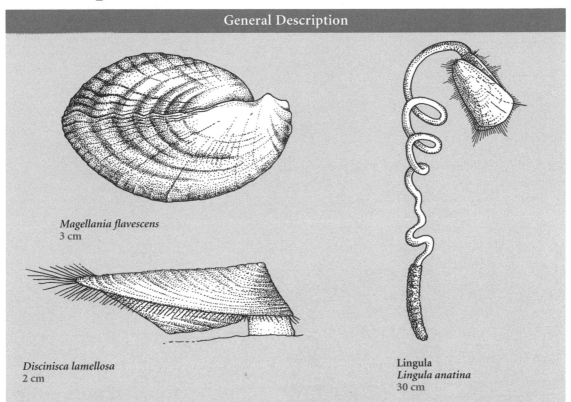

Magellania flavescens
3 cm

Discinisca lamellosa
2 cm

Lingula
Lingula anatina
30 cm

The brachiopods are non-segmented, bilaterally symmetrical animals whose body is covered by a dorso-ventrally oriented shell composed of two unequally sized valves. In contrast to the right and left valves of the bivalves, the brachiopods have dorsal and ventral valves. These animals have flexible stalks (pedicle) that enable them to anchor to a substrate and orient the ventral surface upwards. The combined pedicle and shell measure from 1 to 30 cm in length. The shell is composed of chitin together with either calcium carbonate or calcium phosphate. It is secreted by an extension of the body wall, the mantle (not homologous to the mantle of the mollusks). Chitinous setae are found around the mantle margin.

Inside the shell, most of the space is taken up by the lophophore, which is supported by only a thin skeletal part. Depending on the species, the digestive tube can either be blind or have an anus that exits into the mantle cavity. The different organs occupy a smaller space in the shell than does the lophophore. The coelom is large and extends into the mantle cavity and tentacles. The circulatory system is open and includes a small, contractile heart. Dissolved oxygen is transported in the blood. The excretory system is made up of one or two pair of metanephridia found on either side of the intestine. These animals possess three muscle pairs; one opens and closes the valves and the other two are associated with the pedicle. The nervous system includes a periesophageal ring from which the nerves arise. There are no individual gonads but, rather, dorsal and ventral pairs of aggregated germinal cells. The brachiopods are divided into two classes. The inarticulata have no hinge linking the two valves, but they have an anus. The articulata have a hinge and no anus.

The phylogenetic situation of the brachiopods was quite problematic. Their unusual development led some to believe that they were deuterosomians.

Some unique derived features

- The shell is secreted by the mantle.
- The two valves of the shell are oriented perpendicular to the plane of symmetry, such that one valve is dorsal and the other is ventral (fig. 1: sagittal section of *Magellania* sp.; *coe:* coelom, *st:* stomach, *int:* intestine, *lop:* lophophore, *man:* mantle, *dvl:* dorsal valve, *vvl:* ventral valve, *ped:* pedicle).

- The pedicle extends from the ventral valve and anchors the animal to a substrate. Fig. 2 shows *Lingula* (frontal and side views) in its natural habitat, attached by its pedicle (*ped*) to the bottom of its burrow.

Ecology

The brachiopods are exclusively marine animals that typically live at great depths. Most are sessile. A few species, like *Lingula*, can dig long vertical galleries in the sediment. The members of this group prefer cold seas and can be found from the intertidal zone to the deepest ocean depths. They are filter-feeders, feeding on suspended organic particles. The sexes are separate and fertilization is external. There is radial cleavage in the egg that gives rise to a free-swimming, ciliated larva. This larva then metamorphoses into the adult form. The only enemies of the brachiopods seem to be starfish.

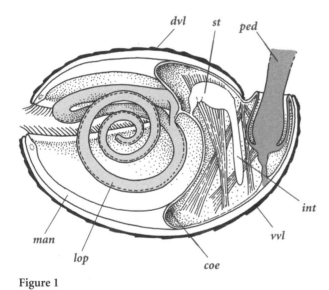

Figure 1

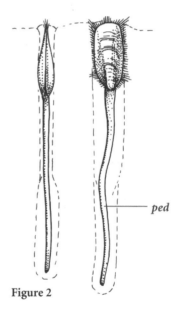

Figure 2

Number of species: 335, divided among 5 orders and 2 classes.
Oldest known fossils: at least 12 families of brachiopods are described from the start of the Cambrian (540 MYA). This group has left numerous fossils (12,000 described fossil species) during all the periods that have followed, but particularly during the Ordovician (from 500 to 435 MYA). They were greatly affected by the mass extinctions at the end of the Permian (245 MYA); the diversity of species and their geographic distribution have remained low ever since this event. The extant genus *Lingula* has been around since the Ordovician.
Current distribution: these organisms have a worldwide, but punctuated, distribution and are typically found in cold waters: Antarctic and sub-Antarctic seas, coastal waters of South America, Australia and New Zealand. However, they are also present in the Mediterranean Sea and the English Channel.

Examples

- INARTICULATA or ECARDINA: *Lingula anatina; Discinisca lamellosa.*
- ARTICULATA or TESTICARDINA: *Lacazella mediterranea; Magellania flavescens; Gryphus vitreus; Terebratulina retusa.*

Phoronida

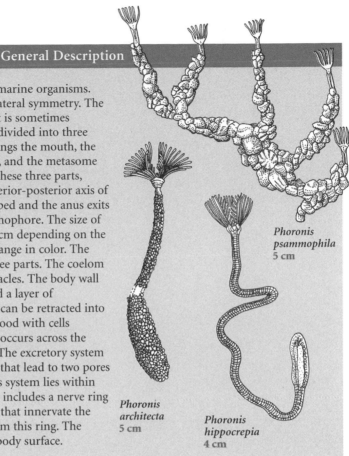

General Description

The phoronids are solitary worm-like marine organisms. They are non-segmented and show bilateral symmetry. The body is housed in a chitinous tube that is sometimes covered with sand grains. The body is divided into three general parts: the epistome that overhangs the mouth, the mesosome that carries the lophophore, and the metasome that makes up the rest of the animal. These three parts, however, do not correspond to the anterior-posterior axis of the animal; the digestive tube is U-shaped and the anus exits dorsally, close (but exterior) to the lophophore. The size of these animals varies from 1 mm to 50 cm depending on the species. They can be yellow, pink or orange in color. The coelomic cavity is also divided into three parts. The coelom of the mesosome extends into the tentacles. The body wall contains a layer of circular muscles and a layer of longitudinal muscles. The lophophore can be retracted into the tube. The circulatory system has blood with cells containing hemoglobin. Gas exchange occurs across the tissues, principally at the lophophore. The excretory system is made up of a pair of metanephridia that lead to two pores opening close to the anus. The nervous system lies within the body wall, under the epithelium. It includes a nerve ring that surrounds the mouth. The nerves that innervate the tentacles and other structures leave from this ring. The phoronids have sensory cells on their body surface.

Phoronis psammophila
5 cm

Phoronis architecta
5 cm

Phoronis hippocrepia
4 cm

Some unique derived features

- An actinotroch larva (400 μm) is characteristic of the group; the metasomal sac will greatly develop before reaching the adult form (fig. 1: *mo:* mouth, *int:* intestine, *ms:* metasomal sac).

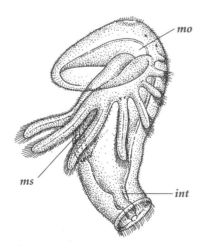

mo

ms

int

Figure 1

Ecology

The phoronids live in shallow marine waters, just below the low-tide line. The adult is anchored to a substrate and feeds on plankton and suspended organic particles. This food is brought towards the mouth by the cilia on the tentacles. These tentacles also have the capacity to select and remove unwanted particles. Most species are hermaphroditic. The gonads of the two sexes are well separated and lie in the posterior area of the coelomic cavity. The gametes are released into the environment via the nephridia. The ovules are led towards the area delimited by the tentacles. Here, external fertilization takes place with free-swimming sperm from another individual. Some species incubate their eggs among their tentacles. There is a free-swimming, ciliated larva. This larva leads a pelagic existence before fixing to a substrate and undergoing metamorphosis. These organisms are also capable of asexual multiplication by fission and regeneration.

Number of species: 15.
Oldest known fossils: *Skolithos*, a tube that was probably produced by a phoronid, has been found in Germany and northern Australia (lower Cambrian, 540 MYA). Phoronids that have been identified with certainty are recorded from the start of the Devonian (420 MYA).
Current distribution: in all the seas of the world.

Examples

Phoronis architecta; P. hippocrepia; P. psammophila; Phoronopsis harmeri.

Chaetognatha

General Description

The chaetognaths (or arrow worms) are small, planktonic organisms of 0.5 to 12 cm in length showing bilateral symmetry. Some species of the *Spadella* genus are benthic. The transparent body is arrow-shaped and divided into three body parts. The head (armed with eyes, a mouth and powerful, chitinous hooks) can be entirely covered by a hood created by a fold in the integument of the neck. This hood makes the animals more streamlined for moving through the water. The trunk of the body has two lateral fins. The tail region, behind the anus, carries a single horizontal fin. The chaetognaths have no circulatory or excretory apparatuses. Their phylogenetic position is uncertain. The details of their embryonic development lead one to think that they could belong to the deuterostomia. Molecular phylogenies invalidate this hypothesis. They are protostomians with an unusual development. However, for now, nothing is suggested about their situation within the protostomian clade.

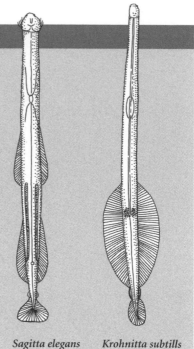

Sagitta elegans
2.5 cm

Krohnitta subtills
0.9 cm

Some unique derived features

- The three body areas are very homogeneous with characteristic fins (fig. 1: ventral view of *Spadella*; *f*: fin, *he*: head).

- The mouth is surrounded by chitinous hooks (teeth and spines) of particular shape and arrangment (fig. 2: head of *Sagitta*; *ho*: hood, *sp*: spine).

- A hood can cover the head and protect the hooks (fig. 1).

Ecology

These organisms are found from the water's surface, among the plankton, to depths of up to 900 meters. They are predators that feed principally on small crustaceans (copepods), which are detected by the characteristic vibrations that they create while swimming. The chaetognaths can also hunt fish. The hooks, uncovered by protective hood, are used to capture, injure and inject prey with venom. These animals are hermaphrodites. Individuals recognize each other through a signaling behavior. Fertilization is internal and occurs by passing a spermatophore from one individual to another. There is direct development in the egg.

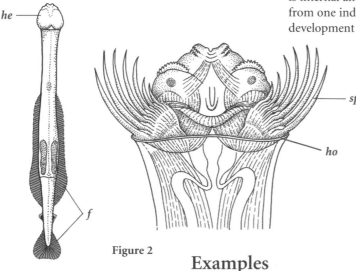

he

f

Figure 2

Figure 1

Examples

Krohnitta subtilis; Sagitta elegans; Spadella cephaloptera.

sp

ho

Number of species: 100.
Oldest known fossils: if we reject the diagnosis that *Amiskwia* is a chaetognath (Burgess Shale deposit, Cambrian, 520 MYA), the oldest known fossil is that of *Paucijaculum* (Carboniferous, 340 MYA). However, un-named species from British Columbia, Canada date from the Cambrian.
Current distribution: all the marine waters of the world, but most species are found in tropical waters.

Cuticulata

A few representatives

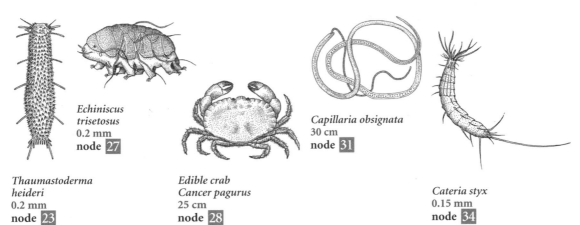

Echiniscus trisetosus
0.2 mm
node 27

Capillaria obsignata
30 cm
node 31

Thaumastoderma heideri
0.2 mm
node 23

Edible crab Cancer pagurus
25 cm
node 28

Cateria styx
0.15 mm
node 34

Some unique derived features

- All cuticulates possess a two-layered cuticle with an epicuticle and a procuticle. The epicuticle has three lamellar layers, like those we find in all insects: (from outside to inside) a layer of cement, a layer of wax and a complex layer of cuticulin (fig. 1: section of an insect epicuticle; *can:* wax secreting canal, *w:* wax, *cem:* cement, *cui:* cuticulin, *pr:* procuticle).

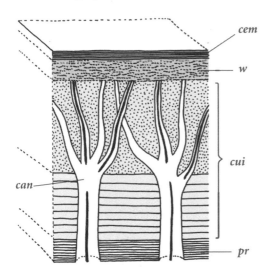

Figure 1

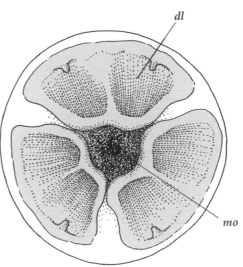

Figure 2

- The opening of the mouth is terminal.

- The pharynx has three radial muscles whose presence often correlates with that of a three-lipped mouth (fig. 2: anterior view of the buccal extremity of the nematode *Ascaris* sp. showing the three lips; *mo:* mouth, *dl:* dorsal lip).

Number of species: 978,024.
Oldest known fossils: an articulation that may have belonged to an arthropod has been found in the Ediacara fauna of the Precambrian (680 MYA). Several trilobite-like arthropod species are described from before the Cambrian (580 MYA).
Current distribution: worldwide.

Examples

- GASTROTRICHA: *Lepidoderma squamatum; Thaumastoderma heideri.*
- ECDYSOZOA: *Heteroperipatus engelhardi;* velvet worm: *Peripatus jamaicensis;* cross orb weaver spider: *Araneus diadematus;* scorpion: *Scorpio maurus;* edible crab: *Cancer pagurus;* seven-spotted ladybeetle: *Coccinella septempunctata; Echiniscus trisetosus; Macrobiotus hufelandi;* human intestinal roundworm: *Ascaris lumbricoides; Trichinella spiralis* (causes trichinosis); *Caenorhabditis elegans; Gordius robustus; Nectonema agile; Tubiluchus corallicola; Priapulus bicaudatus; Cateria styx; Kinorhynchus* sp.; *Nanaloricus mysticus.*

Gastrotricha

General Description

The gastrotrichs are bilaterally symmetrical animals with a lobed head and an elongated body that can be either straight or bottle-shaped. They range in size from 50 μm to 4 mm long. The ventral surface is flattened and covered with cilia, whose arrangement is used for classification. The body is covered with an ornamented, non-chitinous cuticle. The dorsal surface of the animal has a special kind of cilia; each cell is monociliated. The back and sides are covered in spines or scales. The posterior end of the animal is forked or carries adhesive tubes (up to 250). These tubes secrete a substance used to temporarily anchor the animal in the sand or to a substrate. The body is transparent. The gastrotrichs have a complete digestive tube and a muscular pharynx. They have no skeleton or respiratory or circulatory systems. They regulate their osmolarity with the help of protonephridia, rudimentary excretory tubes. They possess both circular and longitudinal muscles. The nervous system is well-developed. It encircles the pharynx at the anterior end and continues posteriorly with a pair of longitudinal nerve cords. These animals have sensory spines and setae. In addition, certain species also have red photoreceptors.

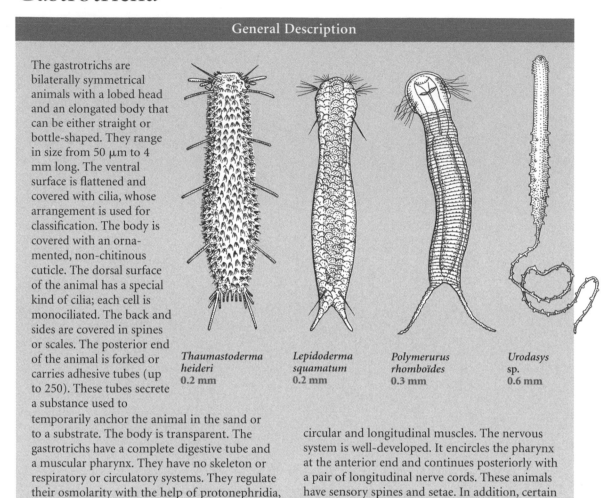

Thaumastoderma heideri
0.2 mm

Lepidoderma squamatum
0.2 mm

Polymerurus rhomboïdes
0.3 mm

Urodasys sp.
0.6 mm

Some unique derived features

- The epicuticle is made up of numerous layers.
- The cuticle covers the entire body, including the cell cilia.

Ecology

The gastrotrichs are principally marine animals found in the intertidal zone, but some freshwater species also exist. They live in the interstitial spaces of sediments (sand, silt), on the surface of superficial detritus or submerged plants and animals, in mosses, soil water or calm freshwater bodies. They feed on organic debris, forams, and diatoms that they bring to their mouth using their ventral cilia. They themselves are prey to amoeba, hydra, turbellarians, insects, crustaceans and annelids.

Marine gastrotrichs are hermaphroditic; they alternate between the production of ovules and spermatozoids using separate gonads. The individual acting as the male transfers a spermatophore (sac of spermatozoids) to the individual acting as female. A "female" will produce only between 1 and 5 large eggs in its lifetime. Development is direct. Freshwater gastrotrichs often reproduce parthenogenetically. They produce two types of eggs: thin-walled eggs that develop as soon as they are laid, and thick-walled, resistant eggs that must undergo exposure to extreme temperatures or desiccation before development occurs. The resistant eggs are used in the colonization of temporarily hostile environments.

Number of species: 430, separated into two orders.
Oldest known fossils: none.
Current distribution: worldwide.

Examples

Lepidoderma squamatum; Polymerurus rhomboides; Thaumastoderma heideri; Tubanella ocellata; Urodasys sp.

Ecdysozoa

A few representatives

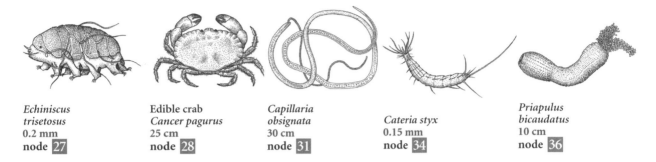

Echiniscus
trisetosus
0.2 mm
node 27

Edible crab
Cancer pagurus
25 cm
node 28

Capillaria
obsignata
30 cm
node 31

Cateria styx
0.15 mm
node 34

Priapulus
bicaudatus
10 cm
node 36

Some unique derived features

Molecular phylogenies based on 18S rRNA and
Hox genes separated the protostomians into the
lophotrochozoans and ecdysozoans without taking
the gastrotrichs into account. The gastrotrichs may
be the sister group of the ecdysozoans. Regardless,
there are many important synapomorphies that
justify the ecdysozoan clade:

■ The locomotory cilia of the epidermal cells have
been lost.

■ Molt: as these organisms are covered in a rigid
cuticle, they grow by molting (*ecdysis:* molting);
the molts are controlled by ecdysteroid hormones.

■ The cuticle contains α chitin and is made up of
three layers, the epicuticle, the exocuticle and the
endocuticle (fig. 1: *endoc:* endocuticle, *epic:*
epicuticle, *epd:* epidermis, *exoc:* exocuticle).

■ The epicuticle is secreted by the epidermal
microvilli.

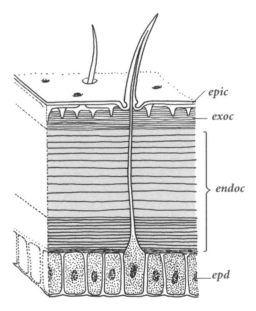

epic

exoc

endoc

epd

Figure 1

Examples

■ PANARTHROPODA: *Heteroperipatus engelhardi;* velvet worm:
Peripatus jamaicensis; cross orb weaver spider: *Araneus
diadematus;* scorpion: *Scorpio maurus;* edible crab: *Cancer
pagurus;* seven-spotted ladybeetle: *Coccinella septempunctata;*
Echiniscus trisetosus; Macrobiotus hufelandi.

■ INTROVERTA: human intestinal roundworm: *Ascaris lumbricoides;*
Caenorhabditis elegans, Capillaria obsignata; Trichinella spiralis
(causes trichinosis); *Gordius robustus; Nectonema agile; Tubiluchus
corallicola; Priapulus bicaudatus; Cateria styx; Kinorhynchus* sp.;
Nanaloricus mysticus.

Panarthropoda

A few representatives

*Heteroperipatus
engelhardi*
8 cm
node 26

*Echiniscus
trisetosus*
0.2 mm
node 27

Edible crab
Cancer pagurus
25 cm
node 28

Cross orb weaver
Araneus diadematus
1.5 cm
node 28

Some unique derived features

■ The panarthropoda are segmented metazoans, often with a rigid external skeleton. This group includes the euarthropods (node 8.28) and two closely related groups that up to now were difficult to place, the onychophorans (8.26) and the tardigrades (8.27). It is because of this difficulty that the tardigrades were sometimes considered close relatives of the nematodes and the onychophorans close relatives of the annelids. The panarthropoda clade seems solid and refutes the hypothesis of an annelid origin for the arthropods.

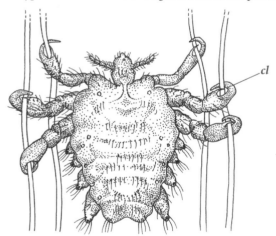

Figure 1

■ These animals have paired, detached appendages armed with claws at the extremities (fig. 1: ventral view of a crab louse *Phthirus pubis*, 1 mm: *cl*: claw).

■ Food is obtained with the help of modified anterior locomotory appendages.

■ Heart: blood is driven through the body by a dorsal heart with lateral ostia.

■ The main body cavity is the hemocoel, a result of the fusion of the coelom and blastocoel.

■ The complex brain includes a protocerebrum that innervates the lateral eyes and one or two ganglia that innervate the antennae (fig. 2: illustration of a crustacean brain; *vnc*: ventral nerve cord, *1an*: nerve to the first pair of antennae, *2an*: nerve to the second pair of antennae, *on*: optic nerve, *es*: esophagus, *protc*: protocerebrum). The third ganglion, the tritocerebrum, encircles the esophagus of the animal.

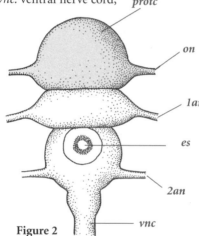

Figure 2

Number of species: 957,094.
Oldest known fossils: an articulation that may have belonged to an arthropod has been found in the Ediacara fauna of the Precambrian (680 MYA). Several trilobite-like arthropod species are described from before the Cambrian (580 MYA).
Current distribution: worldwide.

Examples

■ ONYCHOPHORA: *Heteroperipatus engelhardi; Peripatus jamaicensis.*

■ EUARTHROPODA: *Pygnogonum littorale;* cross orb weaver: *Araneus diadematus; Scorpio maurus; Graphidostreptus* sp.; edible crab: *Cancer pagurus;* seven-spotted ladybeetle: *Coccinella septempunctata;* migratory locust: *Locusta migratoria.*

■ TARDIGRADA: *Echiniscus trisetosus; Macrobiotus hufelandi.*

Onychophora

General Description

The onychophorans, or velvet worms, are cylindrical, worm-like animals ranging from 1 to 30 cm long. They are segmented. Each segment carries a pair of short, unjointed appendages, the lobopoda, held rigid by internal hydrostatic pressure. The end of each lobopodium is armed with a small claw (the name of the group comes from the Greek word *onyx*: nail). The body is covered in a chitinous cuticle that is thin, flexible and permeable. This cuticle has papillae and scales giving the animal a velvety, iridescent appearance and can be of variable color (blue-black, grey, orange, green, whitish). There are also a number of tubercles, each ending in a sensory bristle. The head has two segmented antennae, eyes and a ventral mouth armed with a pair of mandibles and mucus glands, called "glue glands." The trunk is made up of 14 to 43 segments. The onychophorans initially possess true segmentation, but this disappears in the adult stage. The only remains we see are the paired appendages and certain metamerized internal organs, like the excretory and nervous systems. The coelom is reduced in the adult form, giving rise to only the nephridia and the genital cavities and ducts. The

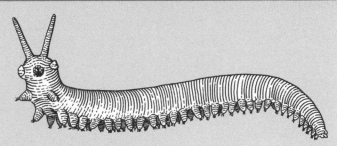

Heteroperipatus engelhardi
8 cm

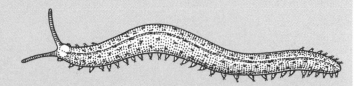

Macroperipatus geayi
10 cm

blood is colorless and circulates in the general cavity or hemocoel. It moves throughout the body by the movements of the body and through the action of the two dorsal tubular hearts. Subcutaneous hemal channels are linked to the hemocoel but do not cross the muscle layer internally. Each segment has a pair of nephridia that empty at the base of each lobopodium by two pores. Oxygen travels across the body wall and enters into tracheae. The nervous system includes an anterior bilobed brain lying above the pharynx and longitudinal ventral nerve cords with transverse commissures. The long digestive tube terminates at the anus on the ventral side of the last segment.

Some unique derived features

- Glue: glands that secrete an adhesive substance used for defense and prey capture are found on the oral papillae. This substance, the glue, is projected, hardens in the air and transforms into an elastic, sticky gum (fig. 1: ventral view of the head of *Peripatus* sp.; *aa:* antenna, *mo:* mouth, *scn:* salivary canal, *cl:* claw, *opa:* oral papillae).

- Tracheae (not homologous to those of insects and spiders) are used for gas exchange.

- There is a continuous layer of longitudinal smooth muscle.

- The body has a hemocoel and subcutaneous circulatory system, the hemal channels. These channels lie between the circular muscle layer and the cuticle, as we can see in cross-section, and continue as longitudinal canals (fig. 2: cross section of an onychophoran; *hcn:* hemal channel, *nc:* nerve cord, *hem:* hemocoel, *mus:* muscle, *nr:* nephridium, *dt:* digestive tube, *dbv:* dorsal blood vessel).

Ecology

The onychophorans are exclusively terrestrial, even though the Paleozoic ancestors of the group were marine organisms. They only live in humid areas and, in particular, in tropical rain forests. They are slow-moving animals and are found among humus or rotten wood, under leaves or rocks. They are predators of small insects like termites or crustaceans that they capture by entangling them in a stream of special saliva, the glue. They can project this sticky glue up to 15 cm. They can also feed on decomposing vegetation.

During unfavorable periods, onychophorans are inactive. The sexes are separate. Each individual has two metamerized gonads. Fertilization can be internal or external depending on the species. Oviparous species have ovipositors. Females deposit large, chitinous-shelled eggs rich in vitellus in humid habitats. In certain species, females incubate their eggs in a type of uterus. The eggs develop within the female, but depend on the vitelline reserves for energy. In viviparous species, the eggs contain little vitellus and the embryos feed on secretions produced by the uterine wall.

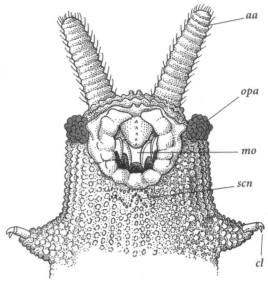

Figure 1

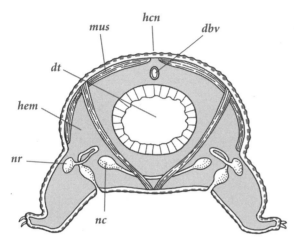

Figure 2

Examples

Heteroperipatus engelhardi; Peripatus jamaicensis; Peripatopsis capensis.

Number of species: 80.
Oldest known fossils: *Aysheaia pedunculata* and *Hallucigenia sparsa* were found in the Burgess Shale deposit (middle Cambrian, 520 MYA).
Current distribution: the geographic distribution is fragmented and confined to equatorial and tropical zones. Almost no species live north of the Tropic of Cancer. We find these species in Central America, South America, Caribbean, South Africa, Madagascar, Gabon, Congo, Himalayas, tropical India, South-East Asia, Australia and New Zealand. This fragmentation is linked not only to climatic factors but also to the ancient continent of Gondwana, of the Paleozoic era, that broke up approximately 200 MYA. It is because of this that South African species are more closely related to species in Chile than to those of equatorial Africa.

Tardigrada

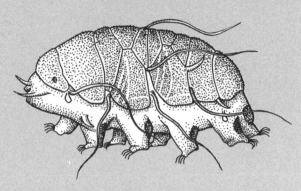

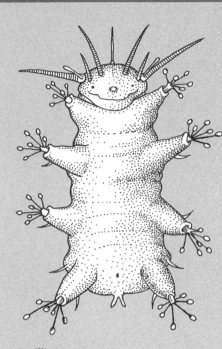

Echiniscus trisetosus
0.2 mm

Batillipes noerrevangi
0.16 mm (ventral view)

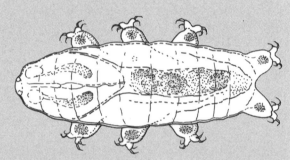

Macrobiotus hufelandi
0.1 mm (dorsal view)

The tardigrades (from the Latin words *tardus:* slow and *gradi:* walker) are small animals that vary in size from 50 μm to 1.2 mm. Their bodies are broad and cylindrical with four pairs of short, stubby legs, each armed with 4 to 8 mobile claws. They have no neck; the head is directly attached to the trunk. The general appearance of this animal has given it the common English name "water bear." Its body is covered in a complex cuticle made up of chitin, proteins and mucopolysaccharides. This cuticle can be arranged into dorsal plates that may be armed with spines or cirri. The animal has a single pair of photoreceptors and a mouth with two perforating stylets. Adult forms have a reduced coelomic cavity, confined to the gonadal cavity. The principal body cavity is a pseudocoelom filled with a body fluid that provides support for a hydrostatic skeleton. This cavity is a vestige of the first embryonic cavity, the blastocoel. The body arrangement of the tardigrades made it difficult to recognize them as true coelomates. Tardigrades have no circulatory or respiratory organs. Gas exchange occurs across the moist body surface and diffuses throughout the tissues. There are only smooth longitudinal muscles, hence the resulting slow movement. The digestive system includes a mouth with its perforating stylets, a muscular pharynx that functions like a pump, an esophagus, a large stomach, a short rectum and a terminal anus. Sac-like excretory glands on the dorsal surface open into the rectum. The nervous system is made up of a bi-lobed dorsal brain connected to a ring that encircles the pharynx. These are then linked to a pair of longitudinal ventral nerve cords that carry a chain of four ganglia (one for each pair of legs). The metamerization of the nervous system may lead one to hypothesize that ancestrally the tardigrades had a head followed by five segments; the first segment fused with the head and the tarsal claws have become the buccal stylets.

Some unique derived features

- A nervous connection unites the lobes of the protocerebrum to the first ventral ganglion (fig. 1: profile sketch of *Batillipes*; *nc:* nerve cord, *con:* connection, *cgg:* cephalic ganglia).

- The photoreceptors have a particular structure.

- The buccal stylets may be transformed claws.

Ecology

The tardigrades include aquatic and terrestrial species found from Arctic regions to hot springs. Marine species live in the interstitial spaces of the sediment from the intertidal zone to deep waters. Freshwater species live among algae and fine mud and on the surface of lake mosses. More than 100 species live in moist terrestrial environments, interstitially in soils and in the water films surrounding plant leaves, particularly those of mosses and liverworts. The tardigrades feed on the contents of the plant cells that they pierce using their stylets. Some attack the nematodes and rotifers that share their habitat. They pierce their prey and suck up the body fluids. Other tardigrades feed on detritus and organic particles of the humus or sediment.

The tardigrades are extremely resistant to variations in temperature and humidity. If a drought occurs, the animal loses its water, contracts, and goes into suspended animation. It can remain in this inert, but living, state for several years and can withstand extreme temperatures (from −273°C to 151°C). If food resources are low, they can also

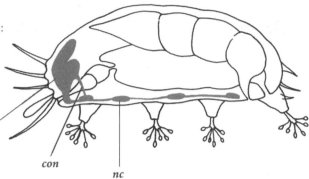

Figure 1

encyst. In this case, the animal loses part of its internal organs and retracts into its old cuticle. A new cuticle then forms the cyst wall. These animals are also surprisingly resistant to X-rays. The lethal dose for a tardigrade is one thousand times superior to that of humans. Russian zoologists have reported that these animals survived exposure to outer space.

The tardigrades have a single gonad, found dorsally. The sexes are separate. Fertilization can be internal or external, according to the species. Most species are oviparous. They lay a few dozen eggs. These eggs can be of two different types. In unfavorable conditions, the eggs are thick-walled and resistant. During the favorable season, they are thin-walled with a sticky outer surface. These thin-walled eggs immediately undergo direct development. Certain species are parthenogenetic.

Number of species: 600, divided into three orders.
Oldest known fossils: we find tardigrades encased in amber. For example, *Beorn leggi* was found in Lake Manitoba, Canada (Cenomanian, 95 MYA).
Current distribution: worldwide.

Examples

Batillipes noerrevangi; Echiniscus trisetosus; Macrobiotus hufelandi; Wingstrandarctus corallinus.

see chap. 10, p. 300

Euarthropoda

General Description

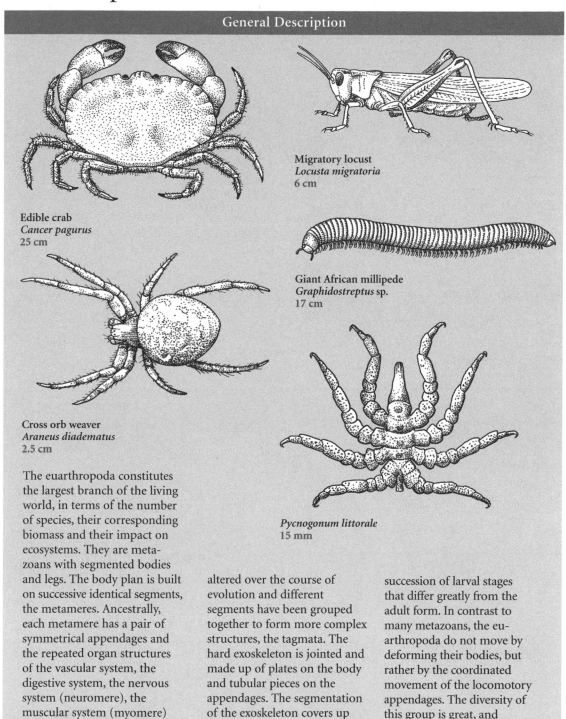

Edible crab
Cancer pagurus
25 cm

Migratory locust
Locusta migratoria
6 cm

Giant African millipede
Graphidostreptus sp.
17 cm

Cross orb weaver
Araneus diadematus
2.5 cm

Pycnogonum littorale
15 mm

The euarthropoda constitutes the largest branch of the living world, in terms of the number of species, their corresponding biomass and their impact on ecosystems. They are metazoans with segmented bodies and legs. The body plan is built on successive identical segments, the metameres. Ancestrally, each metamere has a pair of symmetrical appendages and the repeated organ structures of the vascular system, the digestive system, the nervous system (neuromere), the muscular system (myomere) and the excretory system (nephromere). The metamerization of this group has been altered over the course of evolution and different segments have been grouped together to form more complex structures, the tagmata. The hard exoskeleton is jointed and made up of plates on the body and tubular pieces on the appendages. The segmentation of the exoskeleton covers up the internal segmentation. The development of the euarthropoda typically involves a succession of larval stages that differ greatly from the adult form. In contrast to many metazoans, the euarthropoda do not move by deforming their bodies, but rather by the coordinated movement of the locomotory appendages. The diversity of this group is great, and individuals can vary in size from 0.1 mm to more than 1 m.

Some unique derived features

■ Exoskeleton: the metamerized animal is covered in an exoskeleton subdivided into jointed pieces, the sclerites (tergites, pleurites and sternites) (fig. 1: cross-section of an arthropod segment; *nc*: nerve cord, *he*: heart, *mus*: muscles, *pleu*: pleurite, *ster*: sternite, *ter*: tergite, *dt*: digestive tube).

- Jointed appendages: fundamentally, each segment carries a pair of jointed appendages. Ancestrally, these are used for locomotion (fig. 1).

- The euarthropoda have at least one pair of lateral compound eyes formed by numerous independent photoreceptive units, the ommatidia (fig. 2: *cryc:* crystalline cone, *cor:* cornea, *fac:* facet, *nef:* nerve fiber, *oma:* ommatidium, *rha:* rhabdomere).

- The cells lack cilia and flagella, with the exception of spermatozoids in certain groups.

Ecology

The euarthropoda have colonized almost all terrestrial and aquatic environments, regardless of climate, altitude or depth, or latitude; only the polar deserts lack euarthropods. Many species live in close association with humans (mites, cockroaches, fleas, flies, etc.). The euarthropods are of vital importance in all terrestrial and aquatic ecosystems and represent a considerable biomass. They are often primary consumers of plants and constitute an important food source for many metazoans. For example, larval crustaceans make up a large proportion of the zooplankton. Many flowering plants depend on flying insects for sexual reproduction (pollination). The euarthropods are also parasitic; for example, ticks are blood-sucking ectoparasites of vertebrates.

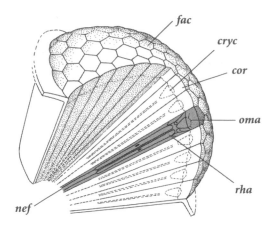

Figure 2

The euarthropods are of considerable economic interest. They are a food source (crustaceans), but also plant pests and vectors of plant pathogens (eubacteria and fungi). In addition, they are vectors of human pathogens: for example, fleas transmit the plague. Likewise, *Plasmodium vivax* (malaria), *Leishmania* (leishmaniasis), and *Trypanosoma* (sleeping sickness) are all human pathogens transmitted by biting insects. The impact of these diseases on the human populations of tropical regions is considerable. Finally, many euarthropods are commensal with humans. Hundreds of thousand of microscopic mites, responsible for many allergies, live in the dust, beds, etc., of our homes and feed on our dead body cells; cockroaches, flies, and ants feed on our garbage and food reserves.

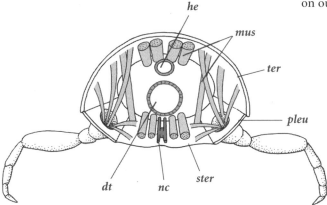

Figure 1

Examples

- CHELICERIFORMES: *Nymphon gracile, Pycnogonum littorale;* horseshoe crab: *Limulus polyphemus;* scorpion: *Androctonus australis; Buthus occitanus;* cross orb weaver spider: *Araneus diadematus.*
- MANDIBULATA: *Polydesmus* sp.; centipede: *Geophilus electricus;* goose barnacle: *Lepas anatifera; Daphnia pulex;* common prawn: *Palaemon serratus;* edible crab: *Cancer pagurus; Ctenosta bastardi;* swallowtail: *Papilio machaon.*

> **Number of species:** 956,414.
> **Oldest known fossils:** a joint that may have belonged to a euarthropod was found in the Precambrian Ediacara fauna (680 MYA). *Parvancorina,* from the Vendian seas (approx. 600 MYA), is sometimes considered to be an arthropod. It is certain that six ostracod families were already present at the start of the Cambrian (540 MYA). For example, we can cite *Bradoria scrutator* and *Hipponicharion eos* from Canada, *Cambria sibirica* from east Siberia, *Kunmingella maxima* from Yunnan, China, *Uskarella prisca* from Kazakhstan, and *Hesslandona* sp. and *Strenuella* sp. from England. The phyllocarids date from the same period, like *Hymenocaris* sp. from the start of the Cambrian from Europe, North America, Australia, and New Zealand (540 MYA), or *Perspicaris* sp. from the late Cambrian from Yunnan, China (540 MYA). In addition, forty-nine trilobite families were already present by the Cambrian period (540 MYA).
> **Current distribution:** worldwide.

Introverta

A few representatives

Capillaria obsignata
30 cm
node 31

Nectonema agile
2 cm
node 32

Cateria styx
0.15 cm
node 34

Nanaloricus mysticus
250 μm
node 35

Some unique derived features

- Introvert: these animals possess an introvert, that is, the animal's anterior end can be retracted into the body (introvert: fig. 1a of the nematode *Kinonchulus*, 1b of the nematomorph *Gordius*, 1c of the priapulid *Priapulus*).

- The pharynx has radial symmetry.

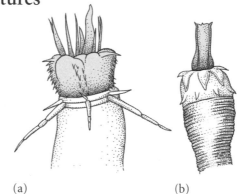

(a)   (b)   (c)

Figure 1

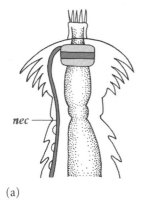

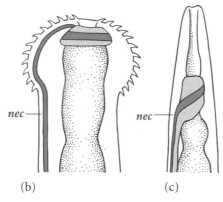

(a)   (b)   (c)

Figure 2

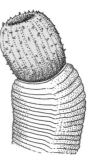

- The brain encircles the pharynx like a collar and is divided into three regions (brain: fig. 2a of a kinorhynch, 2b of a priapulid, 2c of a nematode; *nec*: nerve cord). It is for this reason that this clade is sometimes called Cycloneuralia.

Number of species: 20,500.
Oldest known fossils: *Maotianshania cylindrica* is a priapulid from the lower Cambrian from Yunnan, China (540 MYA).
Current distribution: worldwide.

Examples

- NEMATOZOA: human intestinal roundworm: *Ascaris lumbricoides; Caenorhabditis elegans; Capillaria obsignata; Chondronema passali; Trichinella spiralis* (causes trichinosis); *Gordius robustus; Nectonema agile.*
- CEPHALORHYNCHA: *Tubiluchus corallicola; Priapulus bicaudatus; Cateria styx; Kinorhynchus* sp.; *Nanaloricus mysticus.*

259

Nematozoa

A few representatives

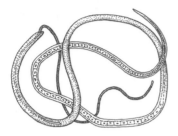

Capillaria obsignata
30 cm
node 31

Chondronema passali
0.2 mm (larva)
node 31

Nectonema agile
2 cm
node 32

Some unique derived features

- Molecular phylogenies, especially those based on the 18S ribosomal RNA gene, almost always group the nematodes and the nematomorphs together. The general resemblance of these organisms is so great that certain scientists think that the nematomorphs could be included in the nematodes.

- There is almost a complete loss of the chitinous cuticle, which has been replaced by a collagen-based cuticle.

- There are only longitudinal muscles.

Number of species: 20,325.
Oldest known fossils: *Heydonius antiquus* is a nematode from the Eocene (50 MYA) from Germany.
Current distribution: worldwide.

Examples

- NEMATODA: human intestinal roundworm: *Ascaris lumbricoides; Caenorhabditis elegans; Capillaria obsignata; Chondronema passali; Draconema* sp.; *Greeffiella* sp.; *Wuchereria bancrofti* (causes elephantiasis); *Trichinella spiralis* (causes trichinosis); *Ankylostoma duodenale.*
- NEMATOMORPHA: *Gordius robustus; Nectonema agile.*

Nematoda

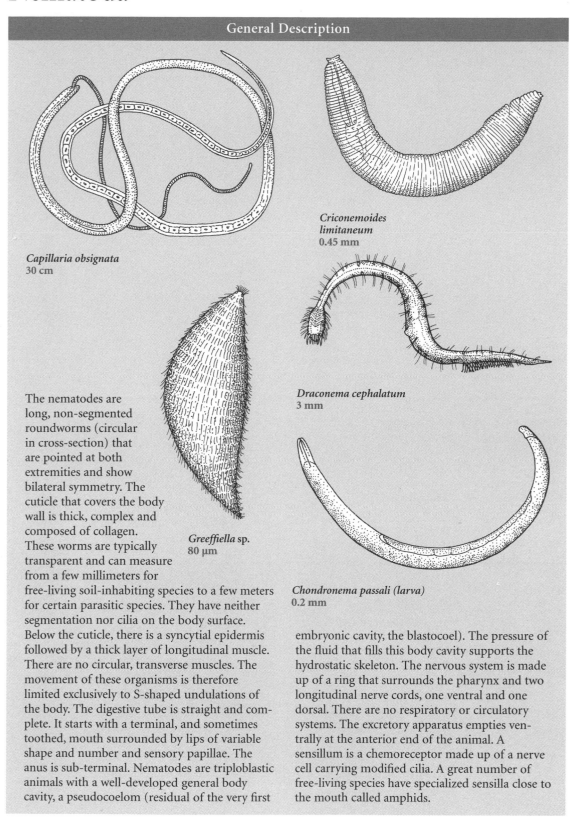

General Description

Capillaria obsignata
30 cm

Criconemoides limitaneum
0.45 mm

Draconema cephalatum
3 mm

Greeffiella sp.
80 µm

Chondronema passali (larva)
0.2 mm

The nematodes are long, non-segmented roundworms (circular in cross-section) that are pointed at both extremities and show bilateral symmetry. The cuticle that covers the body wall is thick, complex and composed of collagen. These worms are typically transparent and can measure from a few millimeters for free-living soil-inhabiting species to a few meters for certain parasitic species. They have neither segmentation nor cilia on the body surface. Below the cuticle, there is a syncytial epidermis followed by a thick layer of longitudinal muscle. There are no circular, transverse muscles. The movement of these organisms is therefore limited exclusively to S-shaped undulations of the body. The digestive tube is straight and complete. It starts with a terminal, and sometimes toothed, mouth surrounded by lips of variable shape and number and sensory papillae. The anus is sub-terminal. Nematodes are triploblastic animals with a well-developed general body cavity, a pseudocoelom (residual of the very first embryonic cavity, the blastocoel). The pressure of the fluid that fills this body cavity supports the hydrostatic skeleton. The nervous system is made up of a ring that surrounds the pharynx and two longitudinal nerve cords, one ventral and one dorsal. There are no respiratory or circulatory systems. The excretory apparatus empties ventrally at the anterior end of the animal. A sensillum is a chemoreceptor made up of a nerve cell carrying modified cilia. A great number of free-living species have specialized sensilla close to the mouth called amphids.

Some unique derived features

■ Sensillum rings: the nematodes have three rings of sensilla surrounding the mouth. The two anterior rings carry six sensilla each and the posterior ring, only four (fig. 1: anterior view of a nematode; *amh*: amphid, *mo:* mouth, *li:* lip, *sen:* sensillum).

■ Amphids: we see amphids close to the mouth. These chemoreceptors are composed of 4 to 13 sensory cells (fig. 1).

■ The mouth is surrounded by three to six lips (fig. 1).

■ The excretory system includes renette cells that empty to the external environment via an excretory pore.

■ The number of cells and cell lineages are (most likely) constant for each species. For example, hermaphroditic forms of *Caenorhabditis elegans*, a nematode commonly used in laboratory studies, have 959 somatic cells, whereas males have 1,031.

Ecology

Free-living species can be found in seas (pelagic or burrowing in sediment), beach sand, lakes, hot springs, moist terrestrial habitats and soils rich in humus. Nematodes make up the greatest proportion of the soil biomass. The diversity is large in terms of the number of species. These organisms play an essential role in soil aeration and in the cycling of organic and mineral matter. A large number of species are parasites of plants, insects and vertebrates. They can infest the leaf tissue, roots or fruits of plants. In animals, nematodes parasitize many different organs. In vertebrates, they are often found in the digestive tube, where the adult worms live, and in the striated red muscle, where they form calcified cysts (resistant forms). Ascarid worms copulate and lay their eggs in the digestive tube of a carnivorous animal. The eggs exit with the feces and are then ingested by a herbivore. The eggs hatch in the herbivore's digestive tract and the juveniles are transported via the circulatory system to the striated muscles, where they encyst. A new carnivore is infected when it feeds on the parasitized muscle tissue of the herbivore. In the digestive tube of the carnivore, the cysts quickly give rise to adult forms that are capable of reproducing. Host-parasite relationships and the details of the life cycles are extremely diverse in nematodes. For example, nematodes may be endoparasites or ectoparasites of plants, saprophagic animal parasites (the juveniles feed on the host tissue after the host dies), juvenile animal parasites (the adult form is free-living), juvenile animal parasites with adults infesting plants and the reverse, animal parasites with a single host, with two successive hosts, etc. Many nematode parasites pose considerable human health problems (particularly in tropical zones): blindness, filariasis, trichinosis and elephantiasis. *Ascaris* is the most common human parasite.

Nematodes typically have separate sexes. The male is generally smaller than the female and carries copulatory spicules. Fertilization is internal and development is direct. A few species are parthen-ogenetic. The female lays her eggs via a gonopore. In many species, the reproductive potential is enormous; nematode females have been recorded to contain up to 27 million eggs with a laying capacity of 200,000 eggs per day. Many species have extremely resistant eggs capable of withstanding more than ten years of unfavorable conditions.

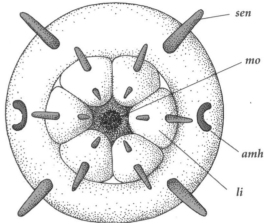

Figure 1

Number of species: 20,000, divided into 2 classes and 20 orders. Some zoologists estimate that several million species likely exist.
Oldest known fossils: *Heydonius antiquus* from the Eocene (50 MYA) has been found in Germany.
Current distribution: the nematodes inhabit all regions of the world, all climates (arid deserts to polar zones), all marine, terrestrial and freshwater environments, at all depths, altitudes and latitudes.

Examples

Human intestinal roundworm: *Ascaris lumbricoides; Caenorhabditis elegans; Capillaria obsignata; Chondronema passali; Criconemoides limitaneum; Draconema* sp.; *Eubostrichius dianae; Greeffiella* sp.; eye worm: *Loa loa* and *Onchocera volvulus* (parasites of the human eye); *O. gibsoni; Pirofilaria immitis; Ankylostoma duodenale; Trichinella spiralis* (causes trichinosis); *Wuchereria bancrofti* (causes elephantiasis).

Nematomorpha

General Description

The nematomorphs are extremely long, thin worms. They are also known as horsehair worms because it was believed they come from horses, or Gordian worms because they resemble a complex knot. These worms are non-segmented, black, gray or yellowish in color with a lighter-colored head that is the same width as the body. They measure 0.5 to 2.5 mm in diameter and 10 to 70 cm in length. The females are longer than the males. The body wall is covered with a cuticle of collagen. Round or polygon-shaped thickenings ornament the surface of the cuticle. Below the cuticle there is an epidermis followed by a layer of longitudinal muscle. There is no circular transverse muscle. The body cavity is a pseudocoelom. This cavity tends to be reduced in size because it is filled with mesenchyme. There are no excretory, respiratory or circulatory systems. The digestive tube is atrophied. Only the cloaca remains intact and serves for reproduction. Adult nematomorphs absorb soluble nutrients across their body wall. These nutrients either come from the external aquatic

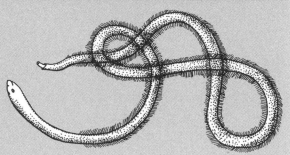

Nectonema agile
2 cm

environment or, for parasitic worms, from the body fluids of the host. The nervous system includes a ring around the pharynx and a single ventral nerve cord. Some species have photoreceptors. Swimming is facilitated by the action of the longitudinal muscles and the presence of a hydrostatic skeleton supported by the pseudocoelomic fluid. Some zoologists believe that the nematomorphs may be highly specialized nematodes.

Some unique derived features

- The adult either lacks a digestive tube or has a reduced and non-functional digestive tube.
- The larvae are always arthropod parasites (fig. 1: a nematomorph parasite in the process of leaving its insect host).

Number of species: 325, divided into two orders.
Oldest known fossils: *Gordius tenuifibrosus* has been described from the Eocene in Germany (Lutetian, 45 MYA).
Current distribution: we find nematomorphs in all aquatic environments: oceans, lakes, and rivers of temperate and tropical regions, high-altitude rivers, moist soils, etc. They are absent from the arid deserts.

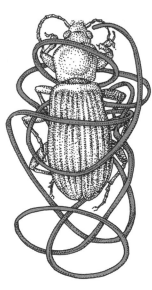

Figure 1

Examples

Gordius robustus;
Nectonema agile.

Ecology

The adults are free-living in fresh water or in moist soils. A few species live in marine environments. The juveniles parasitize different arthropod species. Adults absorb dissolved nutrients through their body wall. The sexes are separate. The spermatozoids are released into the environment via the rectum. There are no copulatory spicules. The ovules are released through the cloaca. The male is more active than the female. During reproduction, the male winds himself around the female and releases his sperm in the proximity of the female's cloaca. He dies soon after. The female can release a string of millions of eggs surrounded by a protective jelly. This brood is then attached to an aquatic plant. Fifteen to eighty days later, the larvae will hatch, armed with an eversible proboscis covered in curved hooks and stylets. A larva enters the body cavity of an arthropod (aquatic crustacean, millipede, cricket, etc.) and undergoes metamorphosis within the host. The worm-like adult nematomorph then exits this host. When the host is terrestrial, like a cricket, the parasite manipulates the host by some unknown means to approach a water source. Depending on the species, the life cycle can also include two hosts.

Cephalorhyncha

A few representatives

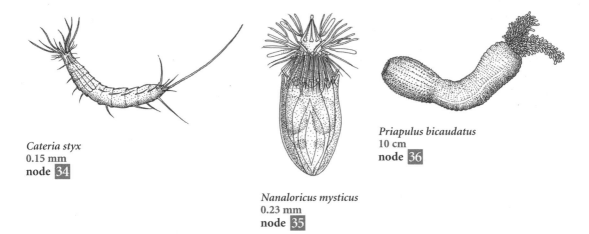

Cateria styx
0.15 mm
node 34

Nanaloricus mysticus
0.23 mm
node 35

Priapulus bicaudatus
10 cm
node 36

Some unique derived features

■ The introvert carries rings of small spines or scalids (fig. 1, *scal*: scalid).

■ Two retractor muscle rings link the buccal region to the area where the collar-like brain is; one ring passes inside the nerve ring and the other outside (fig. 1: section of the anterior end of the priapulid *Tubiluchus; mo:* mouth, *br:* brain, *rmus:* retractor muscles).

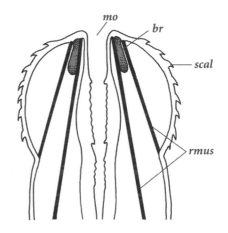

Figure 1

Number of species: 175.
Oldest known fossils: *Maotianshania cylindrica* is a priapulid from the lower Cambrian from Yunnan, China (540 MYA).
Current distribution: worldwide.

Examples

■ KINORHYNCHA: *Cateria styx; Kinorhynchus* sp.; *Echinoderes aquilonius.*
■ LORICIFERA: *Nanaloricus mysticus.*
■ PRIAPULIDA: *Tubiluchus corallicola; Priapulus bicaudatus.*

Kinorhyncha

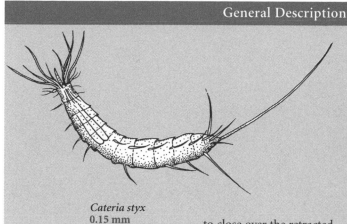

Cateria styx
0.15 mm

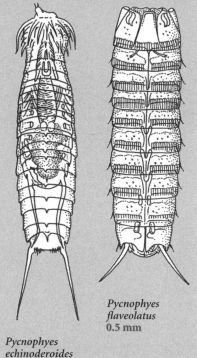

*Pycnophyes
flaveolatus*
0.5 mm

*Pycnophyes
echinoderoides*
0.5 mm

The kinorhynchs are small, free-living animals that are yellow-brown in color, less than 1 mm in length with an elongated body. The segmented cuticle is ornamented with many spines oriented towards the posterior end. The body is bilaterally symmetrical and triangular in cross-section. The anterior end includes a conical head with a mouth, an initial tuft of long spines, and then a ring of large spines that surround the head and start on the neck. The entire head can be withdrawn and protected by the body trunk; the cuticular plates of the trunk are adapted to close over the retracted head. The trunk carries 13 superficial segments, the zonites, that correspond to the segmentation of the nervous and muscular systems. The rigid cuticular plates found on each zonite have neither cells nor cilia. They are hinged by the very flexible skin. The dorsal plates carry large mobile spines. On the most posterior plate are two large terminal spines. The digestive tube is complete. There are no circulatory or respiratory systems. The body cavity is a pseudocoelom, with fluid that functions in respiration, circulation and skeletal support. Fluid pressure is also used for burrowing. Excretion via two protonephridia empties to the external environment at the eleventh segment. A nerve ring surrounds the pharynx. There are also ganglia and a ventral nerve cord. Certain species have red photoreceptors.

Some unique derived features

■ The body is covered in a chitinous cuticle divided into 13 zonites; the head and neck each form one zonite (fig. 1: *Echinoderes*, with its head exposed).

Number of species: 150, divided into 2 orders.
Oldest known fossils: none.
Current distribution: worldwide.

Examples

Cateria styx; Kinorhynchus sp.; *Echinoderes aquilonius; Pycnophyes echinoderoides; P. flaveolatus.*

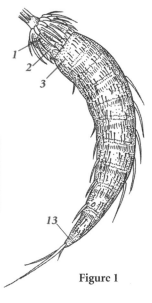

Figure 1

Ecology

The kinorhynchs are free-living marine animals. We find them from the intertidal zone to depths of over 1000 m. They are burrowing animals that live in the interstitial spaces of marine sands and silt. The kinorhynchs feed principally on diatoms and organic particles found in the sediment. These animals do not swim, but dig in the sediment by varying the hydrostatic pressure in their spiny heads.

The sexes are separate. Fertilization is internal; the males have a spiny penis. The eggs are carried externally. The larvae do not have a spiny head. Their digestive tube is incomplete and they must undergo five molts before reaching the adult stage.

Loricifera

General Description

The loriciferan branch was only recently discovered (1983) and is based on an organism collected off the coast of Roscoff, France. These are short, microscopic animals (250 μm) made up of a conical anterior region with eight large stylets surrounding the buccal cone and a posterior region called the abdomen or trunk. This posterior region is embedded in a type of cuticular carapace, the lorica, composed of four plates. The cone-shaped anterior end bears many small, recurved spines, or scalids, on its lateral surface. These spines are identical to those that cover the neck. The digestive tube is complete. A relatively large nerve ganglion is found in the anterior conical region. The general organization of the body corresponds to that of a pseudocoelomate. Some authors have proposed that the loriciferans may be neotenic priapulids.

Nanaloricus mysticus
250 μm

Pliciloricus enigmaticus
250 μm

Some unique derived features

■ The chitinous spines, the scalids, have an intrinsic musculature.

■ The cuticular carapace, the lorica, is made up of four plates: one ventral, one dorsal and two lateral. Loricifera means "lorica bearer."

Number of species: 9 (but there are certainly at least a hundred).
Oldest known fossils: none.
Current distribution: worldwide.

Ecology

The loriciferans live in the interstitial spaces of marine gravel, in warm, temperate and cold seas. They adhere very tightly to sand grains, making it difficult to isolate them. The sexes are separate, each with a pair of gonads. Mating has not yet been described. The larval form resembles the adult form, except that it lacks spines on the thorax and stylets on the buccal cone and it has two posterior feet used for swimming.

Examples

Nanaloricus mysticus; Pliciloricus enigmaticus.

Priapulida

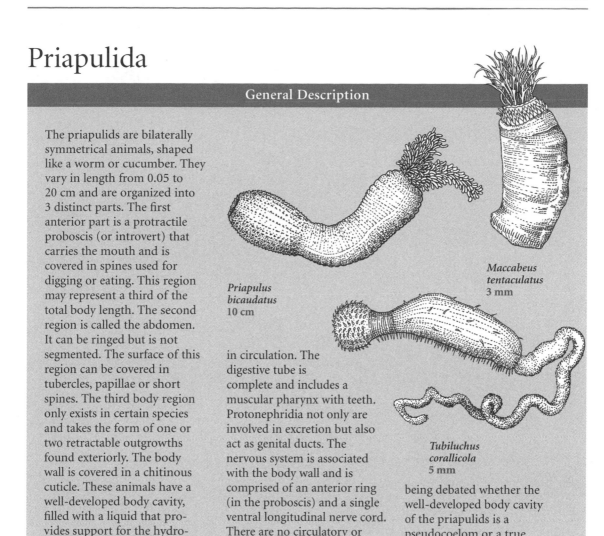

General Description

The priapulids are bilaterally symmetrical animals, shaped like a worm or cucumber. They vary in length from 0.05 to 20 cm and are organized into 3 distinct parts. The first anterior part is a protractile proboscis (or introvert) that carries the mouth and is covered in spines used for digging or eating. This region may represent a third of the total body length. The second region is called the abdomen. It can be ringed but is not segmented. The surface of this region can be covered in tubercles, papillae or short spines. The third body region only exists in certain species and takes the form of one or two retractable outgrowths found exteriorly. The body wall is covered in a chitinous cuticle. These animals have a well-developed body cavity, filled with a liquid that provides support for the hydrostatic skeleton and plays a role in circulation. The digestive tube is complete and includes a muscular pharynx with teeth. Protonephridia not only are involved in excretion but also act as genital ducts. The nervous system is associated with the body wall and is comprised of an anterior ring (in the proboscis) and a single ventral longitudinal nerve cord. There are no circulatory or respiratory systems. It is still being debated whether the well-developed body cavity of the priapulids is a pseudocoelom or a true coelom.

Priapulus bicaudatus
10 cm

Maccabeus tentaculatus
3 mm

Tubiluchus corallicola
5 mm

Some unique derived features

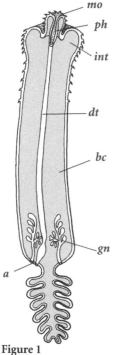

Figure 1

■ The large body cavity (*bc*), possibly a coelom, contains a fluid rich in amoebocytes and erythrocytes (fig. 1: longitudinal section; *a*: anus, *mo*: mouth, *int*: introvert, *gn*: gonad, *ph*: pharynx, *dt*: digestive tube).

Number of species: 16, divided into 2 classes.
Oldest known fossils: *Maotianshania cylindrica* dates from the lower Cambrian from Yunnan, China (540 MYA). *Ancalagon minor* and *Fieldia lanceolata* are priapulids from the Burgess Shale deposit (520 MYA).
Current distribution: worldwide.

Ecology

The priapulids are marine animals that we find in estuaries and at depths of up to 500 m. They burrow in the sediment or silt by contracting their bodies and by extending and retracting the introvert. The priapulids are predators of other burrowing animals, such as polychaete annelids, and sometimes other priapulids. The prey is grabbed using the curved hooks that surround the mouth and swallowed whole. During digestion, the mouth and introvert are continuously extended and retracted, helping the prey pass through the muscular, toothed pharynx. The food passes through the straight intestine and wastes are evacuated by the anus. The sexes are separate. Fertilization is external. The egg undergoes radial cleavage. The priapulid larva is distinct from the adult. It has a carapace composed of eight circular plates and strongly resembles a loriciferan. This larva lives in the silt and undergoes numerous molts before reaching the adult stage.

Examples

Maccabeus tentaculatus; Tubiluchus corallicola; Priapulus bicaudatus.

Chapter 9

MOLLUSCA

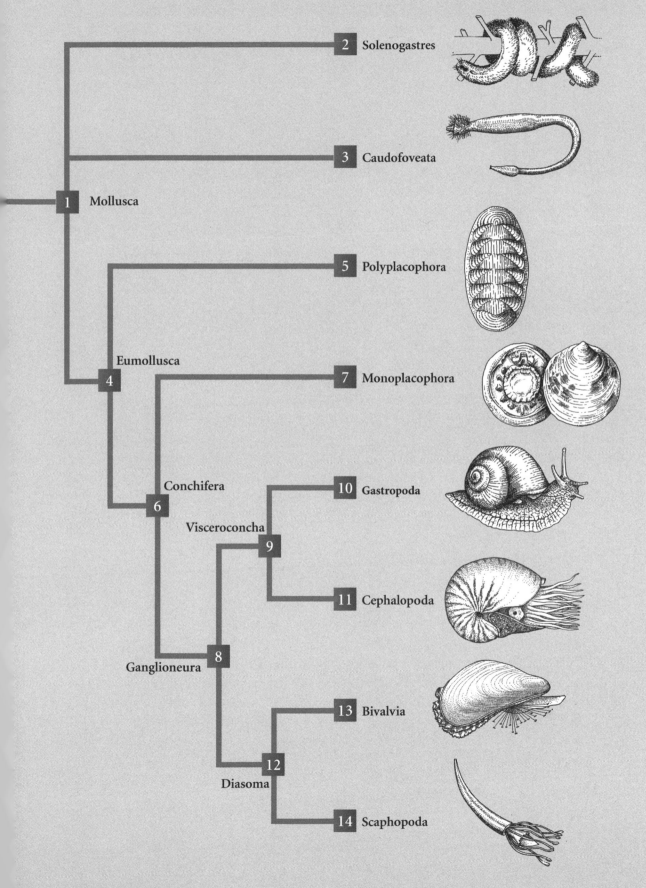

- 2 Solenogastres
- 3 Caudofoveata
- 1 Mollusca
- 5 Polyplacophora
- Eumollusca
- 4
- 7 Monoplacophora
- Conchifera
- 6
- 10 Gastropoda
- Visceroconcha
- 9
- 11 Cephalopoda
- 8
- Ganglioneura
- 13 Bivalvia
- 12
- Diasoma
- 14 Scaphopoda

This clade is made up of some of the strangest and most familiar metazoans. Everyone has a general idea about what a snail, an octopus or a mussel is. However, the mollusks also include the tusk shells, monoplacophorans, chitons, along with strange worm-like animals covered in calcareous spicules—the solenogastres and the caudofoveata. In addition, we also underestimate the diversity of forms contained within the familiar groups, particularly within the gastropoda. The divergence of the different molluscan groups takes us back to the start of the Cambrian. The greatest difficulties with their phylogeny are to determine the position of the root of the tree and to imagine how we might interpret the anatomy of the solenogasters and caudofoveats. These two groups were previously referred to as "aplacophora," terminology aimed to underline the absence of the shell. Indeed, these animals may be shell-less mollusks whose origin dates from before the appearance of the shell. In this case, they should root at the base of the molluscan tree. The other possibility is that the two groups are quite derived, having lost the shell secondarily. The easiest way to determine which of these two scenarios is correct would be to examine the development of the mantle in the solenogasters and the caudofoveats; if no shell is seen or if there is not even a clustering of glandular cells, it would suggest a basal position for these groups. If, on the other hand, initial shell formation occurs, the hypothesis of a secondary loss is possible; the "aplacophora" condition would therefore be derived. The only developmental observation of a solenogaster, dating from the nineteenth century, notes the formation of eight glandular clusters, precursors to the shell plates of the mantle and similar to the eight plates of the polyplacophorans. Unfortunately, this initial observation has never been successfully repeated.

If we consider that this second scenario is the correct one, we still need to determine the relationship of the solenogasters and caudofoveats to the extant molluscan groups. A close relationship to the polyplacophorans is a strong possibility for two reasons. The first is the organization of the eight plates seen in the embryological observation described above. The second is the anatomy of the vermiform polyplacophoran *Cryptoplax*, whose eight small rudimentary plates are embedded in a dorsal fold in the mantle. Instead of having a creeping sole, this species has an elongated foot that lies within a longitudinal, ventral groove formed by the mantle. This convergence with the solenogasters, such as we see it, suggests how we might move from a chiton-like structure to that of a solenogaster. If the solenogasters are derived polyplacophorans, the synapomorphies of the eumollusca (the large, flat creeping foot, the pallial fold, the secretory glands of the mantle that produce calcareous spicules and, when grouped into discrete glandular clusters, can synthesize the shell plates, and the presence, at least primitively, of 16 pairs of dorso-ventral muscles) would become those of the mollusks, with a secondary loss of all or part of them in the "aplacophora."

We still have yet to interpret the presence of eight dorsal shell plates: are they a synapomorphy unique to the polyplacophora, solenogastres and caudofoveata, attesting to the relatedness of these three groups, or are they a primitive condition of shell formation for all mollusks (i.e., is it a molluscan synapomorphy)? As there is no non-molluscan group sufficiently close for comparing and interpreting these possibilities, the rooting of the tree will most certainly rely on molecular phylogenies. This molecular information will then enable us to interpret how this character evolved and to clarify the precise phylogenetic position of the solenogasters and the caudofoveats.

Solenogastres

General Description

The solenogasters are bilaterally symmetrical, worm-shaped mollusks. They are more or less circular in cross-section and, depending on the species, can vary in length from 0.15 to 30 cm. The head is indistinct, lacks sensory organs (no eyes or statocysts) and has a ventral mouth. The radula has been lost secondarily in some species, particularly in those that feed by sucking up digested food material. The intestine has numerous successive lateral evaginations that are continuous from front to back of the animal. The mantle is well-developed and secretes a thick, shiny cuticle containing calcareous spicules that project outward, giving a velvety appearance to the surface of the animal. The color, light or dark, is quite variable according to the species and seems to depend in part on the color of the prey that they feed upon: grey, light or dark brown, yellowish, red. The poorly developed foot is a simple ciliated ridge that lies within a ventral pedal groove formed by the lateral margins of the mantle. The dorsal pallial cavity is well-developed into a cloacal cavity (or "cloaca") that contains two rudimentary gills. Both the anus and the two uro-genital pores empty into this cavity. The gills are not comb gills (ctenidia), but rather simple folds with a respiratory function.

There are no independent gonoducts. The reproductive system empties into a single reno-pericardial cavity. It is therefore the nephridia (or coelomoducts) that assure the evacuation of the gametes through the uro-genital pores; these pores empty into the cloacal cavity. There is a copulatory organ and, in certain species, copulatory stylets (or cloacal stylets), stimulatory organs that act in concert with the copulatory

Rhopalomenia aglaopheniae
1 to 3.5 cm

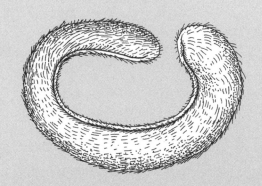

Proneomenia aglaopheniae
1 to 15 cm

organs. These organs are not found in any other molluscan group. The embryo undergoes spiral cleavage and develops into a trochophore larva. This larva has eight dorsal plates, similar to the larva of the polyplacophorans; these plates will ultimately disappear. However, it should be noted that development in this group has only been described a single time.

Some unique derived features

- Ventral pedal groove: a pedal groove, formed by the lateral margins of the mantle, runs longitudinally on the ventral surface of the animal. Pedal glands produce mucus that empties into this groove. The groove starts at a ciliated fossette behind the mouth and ends at the cloaca. It encloses a ciliated ridge-shaped foot. Locomotion is carried out by the movements of the cilia covering the ridge and by the flow of mucus secreted by glands lying behind the mouth. Fig. 1 shows *Proneomenia aglaopheniae* (1a) and *Rhopalomenia aglaopheniae* (1b): *mo*: mouth, *mar*: lateral margin of the mantle, *clo*: cloaca, *cif*: ciliated fossette, *cr*: cephalic region, *pdg*: pedal groove, *spi*: spicules.

Ecology

The solenogasters are free-living marine animals that move around close to the sea floor, not usually in the sediment, but rather across sessile metazoans whose tissues they graze upon. They can sometimes live in the sediment (*Biserranenia sammobionta*) and, in this case, also be burrowing animals (*Neomenia carinata)*. Most species are predators of ectoprocts and cnidarians. For example, *Rhopalomenia aglaopheniae,* a common species found around the European coasts at depths of 50 to 100 m, always attacks the cnidarian *Lytcarpia myriophyllum*. In contrast, *Nematomenia banyulensis,* an animal of the same size (1 to 3.5 cm) and found in the same places, attacks other hydrozoans. Both of these species have lost their radula. Nutrients are therefore digested by the action of enzymes—rich glandular secretions. In other species, the radula can take on a variety of forms according to the feeding mode. For example, *Genitoconia rosea* (0.3 cm) and *Eleutheromenia sierra* (1 cm) both possess bipartite, prehensile radula. A few species are omnivorous, such as *Proneomenia sluiteri* (1 to 15 cm). In many of the species that predate cnidarians, the stinging cells (the typical nematocysts of the cnidarians) pass through the digestive tube without discharging. We don't yet understand how the solenogasters are able to protect themselves against the noxious substances of the nematocysts. It has been suggested that glandular secretions block the evagination of the nematocysts, but this remains to be tested. The cuticle, covered in spicules, is also a protective shield against the effects of the cnidarian stinging cells. Careful observation of a 0.2 cm long solenogaster from the Red Sea, *Forcepimenia protecta,* showed that the mantle of this animal was completely covered in the nematocysts of its prey.

The solenogasters are hermaphroditic with internal fertilization. The larvae are free swimming. *Epimenia verrucosa, Pruvotina providens* and *Halomenia gravida* protect their young by keeping them within pockets of their pallial cavity.

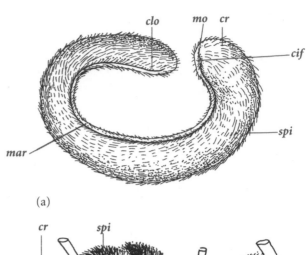

(a)

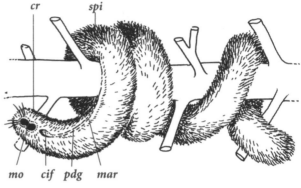

(b)

Figure 1

Number of species: 350.
Oldest known fossils: there is a single, unnamed case from Illinois (USA) that dates from the upper Carboniferous (310 MYA).
Current distribution: we find solenogasters in all the seas of the world, including Arctic and Antarctic, and at all depths. *Pachymenia abyssorum* was collected from a depth of 4000 m off the coast of California.

Examples

Nematomenia banyulensis; Genitoconia rosea; Proneomenia aglaopheniae; P. sluiteri; Rhopalomenia aglaopheniae; Pruvotina providens; Strophomenia lacazei; Eleutheromenia sierra; Forcepimenia protecta; Epimenia verrucosa; Halomenia gravida; Biserramenia psammobionta; Neomenia carinata; Pachymenia abyssorum.

Caudofoveata

General Description

The caudofoveata are bilaterally symmetrical, vermiform mollusks that are almost circular in cross-section. They range in length from 0.3 to 14 cm depending on the species and lack both a foot and a ventral pedal groove. In fact, the mantle covers the entire body. They secrete a cuticle, generally of a grey-brown color, that contains calcareous scales. The posteriorly situated pallial cavity is well-developed into a cloacal cavity (cloaca) that can be completely sealed off by a muscular ring. This cavity contains two true ctenidia (comb gills). The anus and the two urogenital pores empty into this cavity. There are no independent gonoducts. The single gonad empties into the reno-pericardial cavity. It is therefore the nephridia (or coelomoducts) that ensure the release of the gametes via the cloacal cavity.
The mantle is bell-shaped and is protected by long spines. At the anterior end, the mouth is surrounded by a characteristic pedal shield. The radula can take on a variety of different forms according to the species. For example, it can be made up of one or two pair of pincers, denticles of variable size or plate-like formations. It can also have secondarily disappeared, most notably in the genus *Chaetoderma*. The digestive tube includes a stomach that is either laterally or ventrally associated with a single digestive gland and a straight rectum. There are no copulatory organs.

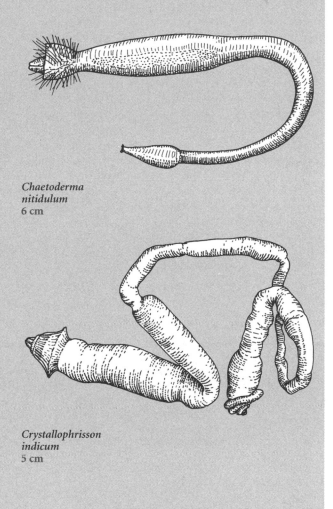

Chaetoderma nitidulum
6 cm

Crystallophrisson indicum
5 cm

Some unique derived features

■ Pedal shield: around the mouth and/or behind it, there is a pedal shield made up of a plate that has both sensory and locomotory functions. This plate is sometimes divided into two pieces and is thought to be the anterior remains of the foot. It is used for digging and sorting prey items. On fig. 1 the lateral (1a) and anterior-lateral (1b) views of the cephalic region of *Chevroderma turnerae* are shown. The oral shield *(osh)* is divided into two pieces and surrounds the mouth *(mo)*. In contrast, *Scutopus megaradulatus* has a shield made up of a single piece.

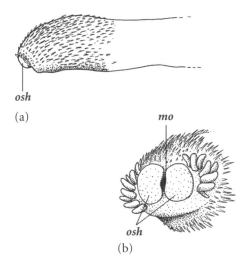

Figure 1

Ecology

The caudofoveata are marine microphagous animals that live buried vertically in the sediment with only their posterior end in the water. Respiration occurs across this end of the animal with the help of the ctenidium held out into the external waters. Fig. 2 shows the position of a caudofoveat in the sediment (*cr:* cloacal ring or muscular ring that closes off the pallial cavity, *mo:* mouth, *osh:* oral shield, *ctn:* ctenidium). Their galleries can also include horizontal branches, such as those of *Limifossor talpoideus* for example. They feed on unicellular organisms, algae and small metazoans living in the sediment. *Prochaetoderma raduliferum* (0.3 to 0.4 cm)

advances while holding its radula out like pincers with its mouth wide open. It is possible that in this case, the radula acts as a prehensile organ to pick up prey; the sensory shield will then sort them. Burrowing species dig themselves into the sediment using alternating contractions of the anterior end of their body, but they cannot go "backwards." It is likely that the posteriorly oriented scales and spines stop the animal from slipping backwards while digging, thus increasing the gain from the digging effort. The caudofoveata have separate sexes with external fertilization. Their development is largely unknown. The larvae are free-swimming.

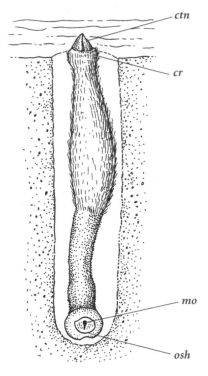

Figure 2

Number of species: 100.
Oldest known fossils: none.
Current distribution: we find the caudofoveats in all the seas of the world

Examples

Limifossor talpoideus; Scutopus megaradulatus; Falcidens crossotus; Prochaetoderma raduliferum; Chaetoderma nitidulum, Chevroderma turnerae; Crystallophrisson indicum.

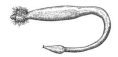

Eumollusca

A few representatives

Acanthochiton
communis
3 cm
node 5

Neopilina
galatheae
3.5 cm
node 7

Roman snail
Helix pomatia
10 cm
node 10

Common octopus
Octopus vulgaris
up to 300 cm
node 11

Some unique derived features

- Pedal foot: The eumolluscan foot is made up of a large, well-developed, muscular sole. This sole is shown in fig. 1 (sol) on a ventral view of *Chiton tuberculatus*.

- The mantle forms a specialized fold, the pallial fold, that extends over the entire perimeter of the body (fig 1, *paf*). This fold delineates the dorsal limit between the mantle and its products (spicules, shell plates or shell) and ventrally creates a spacc between the fold and the foot, called the pallial groove *(pag)* or pallial cavity (= mantle cavity). In the ancestral state, this groove houses the ctenidia *(ctn)* used for respiration, the excretory and genital orifices and the anus *(a)*. On fig 2, we see schematic lateral views (left side) of two eumollusks, a polyplacophoran (2a) and a monoplacophoran (2b). On these diagrams, we note the pedal sole, the pallial fold, and the pallial groove (*mo:* mouth, *shp:* shell plate, *nsy:* nervous system).

- The mantle glands that secrete the calcareous spicules go from a scattered state to form discrete glandular clusters capable of synthesizing the shell plates (fig. 2a). The number of plates is ancestrally 7 (found in certain fossils) or 8 (the more typical state) and has been secondarily transformed into a single plate in the conchiferans (fig. 2b).

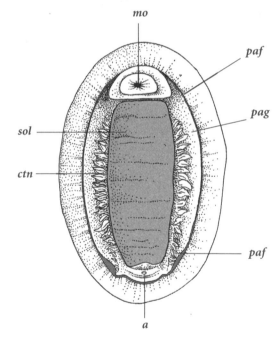

Figure 1

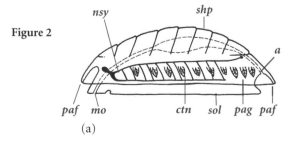

Figure 2

(a)

(b)

■ In the primitive condition, there are 16 dorso-ventral muscles. However, this innovation is weak; indeed, these muscles seem to have been lost in the solenogasters and caudofovoeats. This characteristic, if confirmed, would be a trait that appeared at the origin of all mollusks.

Number of species: 117,045.
Oldest known fossils: mollusks with strange shells, such as the circothecid *Circotheca longiconica* from Yunnan, China and the globorilid *Wyattia reedensis* from California, are found among the Ediacara fauna (680 MYA). Attributing these ancient fossils to one of the extant molluscan classes is difficult. Many of the large molluscan groups exist since the lower Cambrian (530 MYA). These include the helcionelloid monoplacophorans *Latouchella korobkovi, Latouchella memorabilis, Anabarella plana* and *A. indecora* from Siberia, and the praenuculid bivalve *Projetaia runnegari* from South Australia and China.
Current distribution: worldwide.

Examples

- POLYPLACOPHORA: *Chiton tuberculatus; Acanthochiton communis.*
- CONCHIFERA: *Neopilina galatheae; Laevipilina rolani;* common limpet: *Patella vulgata;* common periwinkle: *Littorina littorea;* panther cowry: *Cypraea pantherina;* leopard slug: *Limax maximus;* European brown garden snail: *Helix aspersa;* sea hare: *Aplysia fasciata; Vermetus adansoni;* nautilus: *Nautilus pompilius;* squid: *Loligo vulgaris;* common cuttlefish: *Sepia officinalis;* common octopus: *Octopus vulgaris;* paper nautilus: *Argonauta argo;* nut shell: *Nucula nucleus;* blue mussel: *Mytilus edulis;* blacklip pearl oyster: *Pinctada margaritifera;* great scallop: *Pecten maximus;* zebra mussel: *Dreissena polymorpha;* razor clam: *Solen marginatus; Cuspidaria typus;* common tusk shell: *Dentalium vulgare; D. stenoschizum; Entalina quinquangularis; Siphonodentalium galatheae; Fissidentalium plurifissuratum.*

Polyplacophora

General Description

The polyplacophora, or chitons, are bilaterally symmetrical, dorso-ventrally flattened mollusks, with a distinct head, foot and visceral mass. Their size varies from 0.3 to 33 cm long. The foot is a flat, oval-shaped, creeping sole. At the edge of the mantle cavity, a canal-like pallial cavity almost completely surrounds the foot and contains several pairs of secondary serial gills (6 to 88 pairs depending on the species). Mucus glands empty into this cavity. At the posterior end of the cavity, urinary pores and separate genital orifices open on either side of the anus. The dorsal region of the mantle secretes a shell made up of eight articulating, calcareous plates. The mantle also secretes calcareous spicules around the edges of these dorsal plates. In the acanthochitons, we also find a variable number of pairs of tactile bristle tufts. The mouth opens ventrally at the center of a buccal disc. A protractile organ, called the subradular organ, enables the animal to taste food before scraping it from the substrate. The radula typically has a rasping form composed of several rows of 17 teeth. The stomach is flanked by two hepatopancreases (or digestive

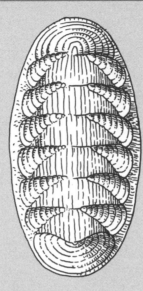

Lepidopleurus cajetanus 2.5 cm

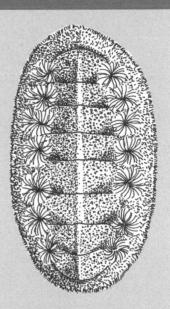

Common chiton *Acanthochiton communis* 3 cm

glands). The ciliated intestine has two loops. The anus is carried by a papilla. The coelomic cavity is reduced to a genital cavity and a reno-pericardial cavity with independent coelomoducts, gonoducts and nephroducts. The excretory apparatus is made up of two nephridia associated with glandular caecums. The nervous system has an anterior concentration of ganglia. The four posterior nerve cords (two pallial and two pedal) are connected by numerous, regularly spaced transverse commissures.

The polyplacophorans do not have eyes, strictly speaking. However, the dorsal plates are covered by numerous epidermal papillae that have a tactile or olfactive function (microaesthetes), or by several crystalline photoreceptors, the macroaesthetes. The chitons include animals of a wide variety of colors and are sometimes ornamented with bands and spots.

Some unique derived features

■ Gills: the polyplacophorans have multiple pairs of secondary gills lodged in the pallial groove. These gills are not homologous to the ctenidia of the other mollusks. They are shown in fig. 1 on a ventral view of *Chiton tuberculatus* (*a:* anus, *mo:* mouth, *g:* gills, *bd:* buccal disc, *paf:* pallial fold, *pag:* pallial groove, *sol:* pedal sole).

■ The shell is subdivided into seven or eight articulating plates. Each plate corresponds to a glandular cluster in the mantle. This condition could be primitive in mollusks, assuming that the solenogasters and caudofoveats are highly derived polyplacophorans. These plates are shown in fig. 2 on the dorsal view of *Acanthochiton communis* (*gr:* marginal area of the mantle called the "girdle," which secretes the calcareous spicules, *P1:* anterior plate, *P8:* posterior plate, *st:* tufts of sensory bristles).

■ Sensory photoreceptors, the aesthetes, are embedded in the dorsal plates. These structures are found in large numbers and provide information to the animal on the quantity of light (chitons do not have cephalic eyes). They are mantle cells that move up through vertical canals and develop in the pores (megalopores or micropores). They open to the outside in the last layers of the shell. The details of how they function are unknown. Fig. 3 shows the longitudinal section of an aesthete of *Acanthopleura spiniger;* this is one of the most sophisticated photoreceptive organs that we find in the polyplacophora (*cor:* cornea, *phc:* photosensitive cells, *pl:* pigment layer, *len:* lens, *meg:* megalopore, *n:* "nerve," *st:* sensory terminal, *fz:* fibrillar zone).

- The radula contains 17 teeth per row.

Ecology

The polyplacophorans are algal grazers and prefer to live on hard substrates. The majority of species are found in the inter-tidal zone or just below this zone. Certain species can be found in coral reefs and some others at depths of between 1500 and 3000 m (lepidopleurids). Their articulated plates enable them to roll up into a ball like pill bugs. When they are open, they are so firmly attached to rocks that they are extremely difficult to dislodge. This ability to adhere to rocks protects them from drying out during low tide. Even though the overwhelming majority of polyplacophorans graze on micro-organisms, certain species like *Mopalia hindsi* or *Placiphorella velata* are exceptions. These species, despite their slow speed, are predators of worms and small crustaceans. The mantle edge of *Placiphorella velata* is enlarged at the anterior end. This part of the mantle is kept raised and immobile until a prey

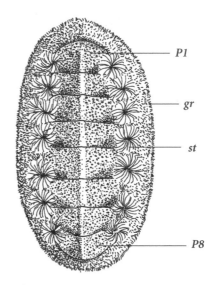

Figure 2

passes below it. The chiton then lowers its trap, thus capturing a prey more rapid then itself. The prey will then be slowly digested. In certain species, like *Cryptochiton stelleri,* the mantle covers over the dorsal plates. In other species, the plates are reduced. For example, *Cryptoplax larvaeformis* has secondarily adopted a worm-like appearance by the reduction of the foot into the pallial groove. This is a surprising convergence with the solenogasters.

The sexes are separate in the chitons and fertilization is external. Spiral cleavage gives rise to a typical trochophore larva that quickly shows traces of the eight skeletal plates and a ventral outcropping that will form the foot. The veliger stage does not exist. The larva quickly loses its cilia and settles to the bottom. Certain species incubate their larvae in the pallial cavity. In the case of *Callistochiton viviparous,* the young leave the maternal pallial cavity already in the adult form. *Chiton tuberculatus* can live over 12 years.

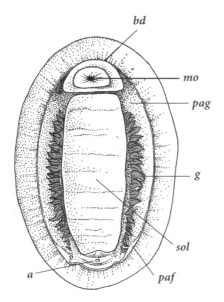

Figure 1

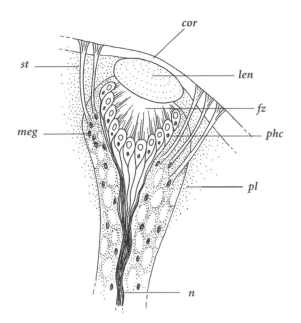

Figure 3

Number of species: 900.
Oldest known fossils: the mattheviids *Matthevia variabilis* and *M. walcotti* and the preacanthochitonids *Preacanthochiton cooperi* and *P. productus*, from Wisconsin and Missouri, respectively, are from the upper Cambrian (United States, 510 MYA). The polyplacophorans have left a moderately rich fossil record that includes 24 families.
Current distribution: worldwide, marine, found at depths of between 0 and 3000 m.

Examples

Lepidopleurus luridus; Hemiarthrum setulosum; lschnochiton imitator; Mopalia hindsi; Placiphorella velata; Callistochiton viviparous; Chiton tuberculatus; Cryptochiton stelleri; Cryptoplax larvaeformis; Acanthochiton communis; Acanthopleura spiniger.

Conchifera

A few representatives

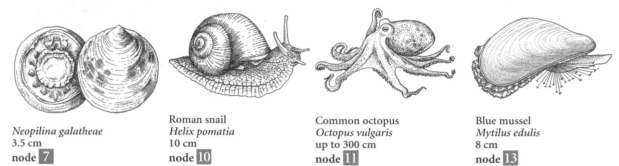

Neopilina galatheae
3.5 cm
node 7

Roman snail
Helix pomatia
10 cm
node 10

Common octopus
Octopus vulgaris
up to 300 cm
node 11

Blue mussel
Mytilus edulis
8 cm
node 13

Some unique derived features

■ The shell is made up of a single piece, at least in the larval state and, in the majority of cases, also in the adult state. This shell is formed from a single glandular region of the mantle. On fig. 1 (*a:* anus, *mo:* mouth, *sh:* shell, *ft:* foot, *vl:* velum) we see the lateral view of two veliger larvae (size: 150 μm), one of the gastropod *Patella* (1a) and the other of the bivalve *Teredo* (1b). We note that the bivalve condition has been acquired secondarily from a univalve condition.

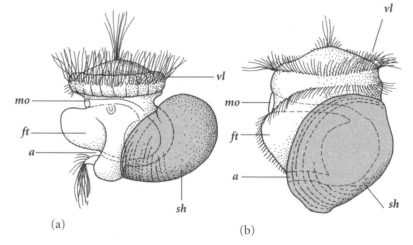

(a)   (b)

Figure 1

■ The margin of the mantle is composed of three parallel folds one above the other. Each fold has a specific function. The outer lobe (*oul,* fig. 2) secretes the two most external layers of the shell, the periostracum *(pem)* and the prismatic layer *(prl);* the middle lobe *(mil)* has a sensory function, and the inner lobe *(inl)* is muscular. The section shown in fig. 2

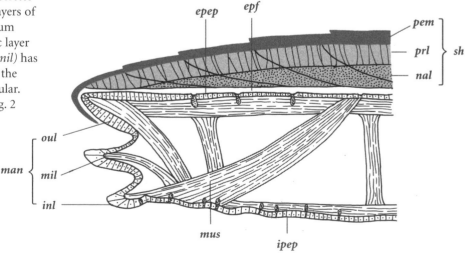

Figure 2

outlines the relationship between the mantle and the shell (*sh:* shell; *nal:* nacreous layer, *epep:* external pallial epithelium, *ipep:* internal pallial epithelium, *epf:* extra-pallial fluid, *man:* mantle, *mus:* muscles).

■ The shell is made up of three layers (fig. 2): the periostracum, the most exterior, the prismatic layer and the nacreous layer, the most interior. The nacreous layer is synthesized by the external pallial epithelium whereas the two other layers are synthesized by the outer lobe of the mantle.

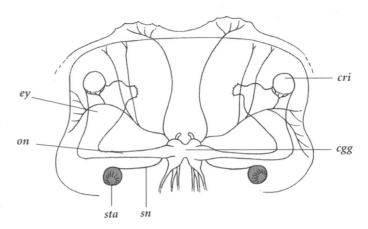

Figure 3

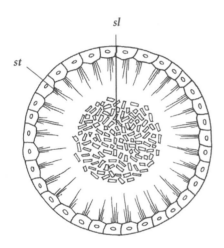

Figure 4

■ In the conchifera, dorso-ventral muscle bundles fuse to form eight pairs of foot retractor muscles, at least in the primitive state.

■ Statocysts: the conchiferans have statocysts, sensory organs used for balance. The position of these spherical organs and their connection to the nervous system are indicated in fig. 3 on the pelagic gastropod *Pterotrachea mutica* (*cri:* crystalline style, *cgg:* cerebral ganglion, *on:* optic nerve, *ey:* eye, *sn:* statocystic nerve, *sta:* statocyst). A statocyst can be thought of as a sphere whose surface is covered by nerve filaments linked to a statocystic nerve. This sphere contains statoliths (crystals) at its center and tufts of sensory bristles that start from its internal side, oriented in a concentric manner towards the statoliths. Fig. 4 shows a transverse section of a statocyst (*sl:* statoliths, *st:* sensory bristle tufts). All movements move the statolith crystals, which will then make contact with the sensory bristles and stimulate the sensory cells lining the cavity. A nerve influx is then directed toward the statocystic nerve.

Number of species: 116,145.
Oldest known fossils: mollusks with strange shells, such as the circothecid *Circotheca longiconica* from Yunnan, China and the globorilid *Wyattia reedensis* from California, are part of the Ediacara fauna (680 MYA). These organisms were without doubt conchiferans. Nonetheless, attributing these ancient fossils to one of the extant conchiferan classes is difficult. Many of the large conchiferan groups are known since the start of the Cambrian (530 MYA). These include the helcionelloid monoplacophorans *Latouchella korobkovi, L. memorabilis, Anabarella plana* and *A. indecora* from Siberia, and the praenuculid bivalve *Projetaia runnegari* from South Australia and China.
Current distribution: worldwide.

Examples

■ MONOPLACOPHORA: *Neopilina galatheae; Laevipilina rolan.*

■ GANGLIONEURA: *Patella vulgata;* common periwinkle: *Littorina littorea;* panther cowry: *Cypraea pantherina;* leopard slug: *Limax maximus;* European brown garden snail: *Helix aspersa;* sea hare: *Aplysia fasciata; Vermetus adansoni;* nautilus: *Nautilus pompilius;* squid: *Loligo vulgaris;* common cuttlefish: *Sepia officinalis;* common octopus: *Octopus vulgaris;* paper nautilus: *Argonauta argo;* nut shell: *Nucula nucleus;* blue mussel: *Mytilus edulis;* blacklip pearl oyster: *Pinctada margaritifera;* great scallop: *Pecten maximus;* zebra mussel: *Dreissena polymorpha;* razor clam: *Solen marginatus; Cuspidaria typus;* common tusk shell: *Dentalium vulgare; D. stenoschizum; Entalina quinquangularis; Siphonodentalium galatheae; Fissidentalium plurifissuratum.*

Monoplacophora

General Description

The name monoplacophora was created in 1940 to describe certain Cambrian fossils, the tryblidiida. These animals were bilaterally symmetrical with a patelloid shell. The internal surface of their shells displayed eight pairs of muscle scars, suggesting an internal segmentation or "metamerism." In May 1952, the Danish *Galathea* expedition discovered a living representative of the tryblidiida, *Neopilina galatheae*, off the Pacific coast of Costa Rica at a depth of 3590 m. This species possesses a thin shell, approximately 3.5 cm wide and 3.7 cm long, marked with growth striations and with an anteriorly directed shell apex. Other species, found later, increased the size range of these organisms to between 0.15 and 3.7 cm long. The large creeping foot has a distinct edge and is completely surrounded by the pallial cavity followed by the mantle margin. In the pallial cavity, there are 5 or 6 symmetrical pairs of gills; there is a nephridiopore at the base of each gill. The anus opens at the posterior end of the body. The anterior mouth is surrounded by upper and lower lips covered in a thick cuticle and carrying large ciliated lateral lobes (the velum). The lower lip carries fan-shaped tentacles. The radula has 11 rows of teeth. In addition to salivary glands and

Neopilina galatheae
3.5 cm

two large anterior stomach pouches, the digestive system includes a short posterior caecum containing a crystalline style that produces digestive enzymes. This organ is also found in the gastropods and bivalves. There are eight pairs of pedal retractor muscles. The nervous system has no ganglionic concentrations. The coelom is more highly developed than that of the polyplacophora: it includes two pericardial cavities, two genital cavities and a dorsal visceral cavity arranged into two flattened sacs. The segmented excretory system is made up of 6 pairs of nephridia. The first four pairs are directly linked to the genital cavities and the last two to the pericardial cavities. Two pairs of gonads release their

gametes to the external environment via two pairs of gonoducts connected to the third and fourth pair of nephridia.

As the mollusks lack all traces of internal segmentation, the successive organ series of the monoplacophorans was seen as traces of a metamerized ancestor that would link the mollusks to the annelids or arthropods. This idea has been largely abandoned for several reasons. First, the repeated organs of the monoplacophorans do not depend on one another and not all coelomic cavities show this "segmentation." Second, the eight pairs of retractor muscles could simply reflect the eight ancestral plates that have fused into a single plate in the conchifera.

Some unique derived features

■ The internal anatomy shows a characteristic repetition with 8 pairs of pedal retractor muscles, 6 pairs of nephridia, 5 pairs of gills and 2 pairs of gonads. The third and fourth pairs of nephridia act as genital canals. Fig. 1 shows ventral *(left)* and dorsal *(right)* views of *Neopilina galatheae* (*a:* anus, *shap:* anterior shell apex, *mo:* mouth, *sh:* shell, *ctn:* ctenidium, *ftm:* foot margin, *paf:* pallial fold, *pag:* pallial groove, *sol:* sole, *pot:* pre-oral tentacle, *tt:*

post-oral tentacular tuft, *vl:* velum). Fig. 2 shows an exposed ventral view of *Neopilina* (*ao:* aorta, *ar:* auricle, *gn:* gonad, *prm:* pedal retractor muscle, *nr:* nephridium (kidney), *fnr:* "fertile" nephridium, *ve:* heart ventricle). The monoplacophorans are the only mollusks with this body organization. However, the eight retractor muscles of the foot are not a characteristic unique to the monoplacophorans, but rather to all conchiferans (just as the presence of an aorta is characteristic of the eumollusca). This characteristic has been secondarily modified in the ganglioneurans.

Ecology

The monoplacophorans typically live at great ocean depths, never above 200 m and most often between 2500 m and 4000 m; a few species are found at depths of 6500 m. They feed on organic detritus, especially radiolarians and diatoms. The spicules of sponges and spines from echinoderms were found in the stomach of *Neopilina ewingi*. The sexes are separate and fertilization is most likely external. The monoplacophorans have a free-swimming veliger larva.

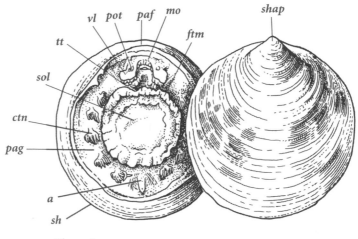

Figure 1

Number of species: 15.
Oldest known fossils: helcionelloid monoplacophorans such as *Latouchella korobkovi*, *L. memorabilis*, *Anabarella plana* and *A. indecora* are known from the start of the Cambrian (530 MYA) from Siberia. The oldest known fossil from the extant family Tryblidiidae is *Helcionopsis* sp. from the lower Cambrian in Australia (520 MYA). The monoplacophorans have left a moderately rich fossil record with 8 families.
Current distribution: marine depths of the eastern Pacific coast, Gulf of Aden in the Arabian Sea, central Atlantic Ocean, Mediterranean Sea.

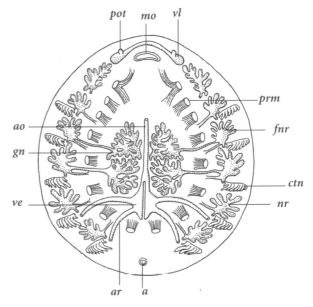

Figure 2

Examples

Neopilina galatheae; N. bruuni; N. ewingi; N. veleronis; N. bacescui; Laevipilina rolani.

Ganglioneura

A few representatives

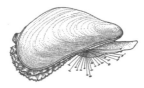

Roman snail
Helix pomatia
10 cm
node **10**

Common octopus
Octopus vulgaris
up to 300 cm
node **11**

Blue mussel
Mytilus edulis
8 cm
node **13**

*Pulsellum
lofotense*
5 cm
node **14**

Some unique derived features

■ Retractor muscles: the number of pedal
retractor muscles is reduced from eight pairs to
one or two pairs. On fig. 1 we see a ventral view
of the non-ganglioneuran layout of *Neopilina*
(monoplacophora) with its eight pairs of
retractor muscles. In fig. 2 we see the layout of
the pedal retractor muscle *(prm)* in a pulmonate
gastropod (single) and in fig. 3 that for a
scaphopod (paired).

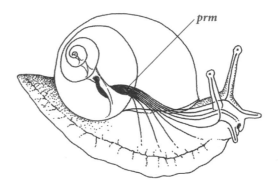

Figure 2

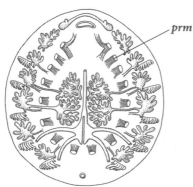

Figure 1

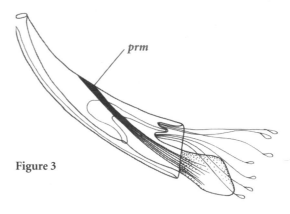

Figure 3

Number of species: 116,130.
Oldest known fossils: the ganglioneurans are known
from the early Cambrian (530 MYA) from the
praenuculid bivalve *Projetaia runnegari* from South
Australia and China, and the ribeiriid rostroconchs
(a fossil group, closely related to bivalves)
Heraultipegma varensalense from the Herault
département in France and *Watsonella crosbyi* from
Massachusetts, United States.
Current distribution: worldwide.

Examples

■ VISCEROCONCHA: common limpet: *Patella vulgata*;
common periwinkle: *Littorina littorea*; panther cowry:

Cypraea pantherina; leopard slug: *Limax maximus*;
European brown garden snail: *Helix aspersa*; sea hare:
Aplysia fasciata; *Vermetus adansoni*; nautilus: *Nautilus
pompilius*; squid: *Loligo vulgaris*; common cuttlefish:
Sepia officinalis; common octopus: *Octopus vulgaris*;
paper nautilus: *Argonauta argo*.

■ DIASOMA: nut shell: *Nucula nucleus*; blue mussel:
Mytilus edulis; blacklip pearl oyster: *Pinctada
margaritifera*; great scallop: *Pecten maximus*; zebra
mussel: *Dreissena polymorpha*; razor clam: *Solen
marginatus*; *Cuspidaria typus*; common tusk shell:
Dentalium vulgare; *D. stenoschizum*; *Entalina
quinquangularis*; *Siphonodentalium galatheae*;
Fissidentalium plurifissuratum.

Visceroconcha

A few representatives

Roman snail
Helix pomatia
10 cm
node **10**

Limacia clavigera
3 cm
node **10**

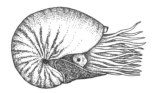

Chambered nautilus
Nautilus macromphalus
up to 20 cm
node **11**

Common octopus
Octopus vulgaris
up to 300 cm
node **11**

Some unique derived features

■ The head is distinct and well-developed with a concentration of the nervous system in the anterior region due to the fusion of the nerve ganglia. Fig. 1 shows schematic drawings of the left lateral view of a polyplacophoran (1a), a monoplacophoran (1b), a gastropod (1c), a cephalopod (1d), a bivalve (1e), and a scaphopod (1f) (*a*: anus,

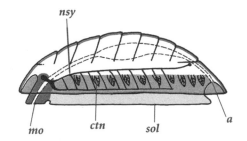

(a)

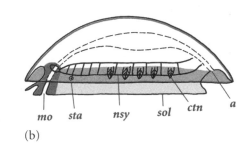

(b)

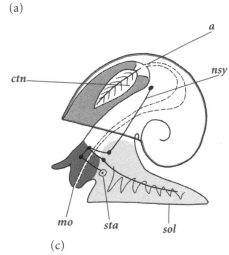

(c)

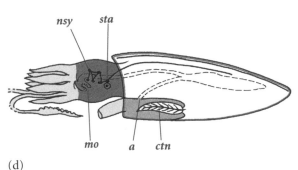

(d)

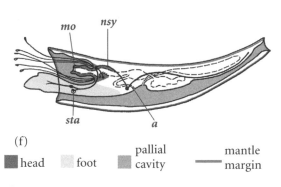

(f)

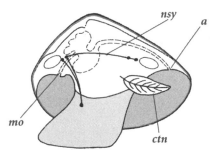

(e)

■ head　　■ foot　　pallial cavity　　— mantle margin

Figure 1

mo: mouth, *ctn:* ctenidium, *nsy:* nervous system, *sol:* pedal sole, *sta:* statocyst). In the non-ganglioneurans (1a, 1b), the head is almost indistinguishable and the nervous system is distributed throughout the body. The head is greatly reduced or absent in the diasomans (1e, 1f), whereas, in the visceroconchs (1c, 1d), it is prominent and the nerve ganglia are concentrated anteriorly in the cephalic region.

- In the visceroconchs, the mantle covers only the visceral part of the body. In the primitive condition, the shell therefore protects only the visceral mass. This is the case for the gastropods and the nautiloid, ammonitoid (fossils) and belemnitoid (fossils) cephalopods. However, the shell can be secondarily reduced or covered by the mantle, as seen in the nudibranch gastropods, the terrestrial slugs, or in extant, non-nautiloid cephalopods. On fig. 1, we note that the mantle protects the entire body except for the pedal foot in the non-ganglioneurans (1a, 1b) and it covers the entire animal in the diasoma (1e, 1f). In the visceroconchs, that is, the gastropods (1c) and cephalopods (1d), the mantle can be fundamentally compared to a sac that encloses the visceral mass.

- An endogastric bend gives rise to a U-shaped intestine.

Number of species: 103,730.
Oldest known fossils: visceroconch fossils are found starting in the upper Cambrian (510 MYA). These fossils include the hyperstrophic onychochilid gastropod *Kobayashiella circe* from Shantung, China, the maclurite gastropod *Macluritella walcotti* from Nevada, the sinuitid bellerophont *Sinuella minuta* from Texas, along with *Sinuopea sweeti* from Wisconsin, *Schizopea typica* from Missouri and *Prohelicotoma uniangularia* from New York.
Current distribution: worldwide.

Examples

- GASTROPODA: common limpet: *Patella vulgata;* common periwinkle: *Littorina littorea;* panther cowry: *Cypraea pantherina;* leopard slug: *Limax maximus;* European brown garden snail: *Helix aspersa;* sea hare: *Aplysia fasciata; Vermetus adansoni.*
- CEPHALOPODA: nautilus: *Nautilus pompilius;* squid: *Loligo vulgaris;* common cuttlefish: *Sepia officinalis;* common octopus: *Octopus vulgaris;* paper nautilus: *Argonauta argo.*

Gastropoda

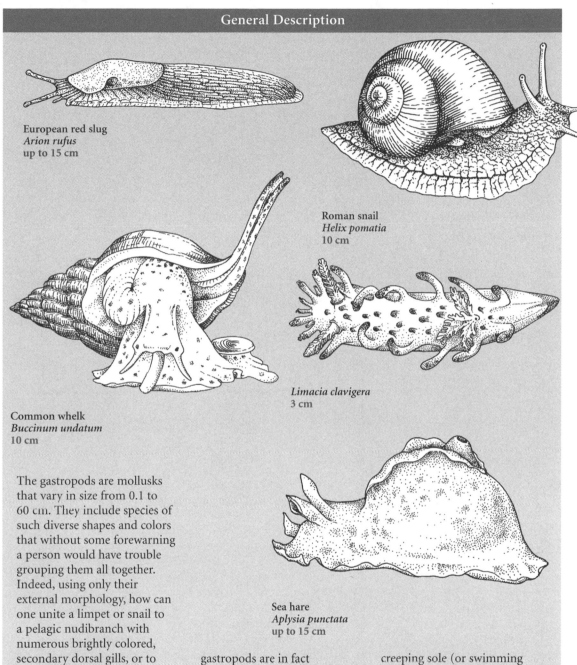

European red slug
Arion rufus
up to 15 cm

Roman snail
Helix pomatia
10 cm

Limacia clavigera
3 cm

Common whelk
Buccinum undatum
10 cm

Sea hare
Aplysia punctata
up to 15 cm

The gastropods are mollusks that vary in size from 0.1 to 60 cm. They include species of such diverse shapes and colors that without some forewarning a person would have trouble grouping them all together. Indeed, using only their external morphology, how can one unite a limpet or snail to a pelagic nudibranch with numerous brightly colored, secondary dorsal gills, or to a giant worm shell with its atypical, corkscrew-shaped, tubular shell? What about the strange sea butterflies, like *Cavolinia* and *Clio*, with their loosely spiraled, globular shells and large pallial lobes, or the obligate parasites *Gasterosiphon* and *Entocolax* that have been greatly modified by the parasitic life style? Despite their spectacular diversity, the gastropods are in fact recognizable by their anatomy. They are mollusks with distinct heads, most often with a dorsal shell composed of a single coiled piece. The head carries one or two pairs of dorsal tentacles and two eyes found either at the base of the peduncle or on the tip of the tentacles. The foot makes up most of the fleshy mass that we see exteriorly. It forms a creeping sole (or swimming flaps) and is stretched posteriorly to form a type of tail. The foot is usually rich in ciliated mucus cells that can be distributed evenly across the sole or grouped together. In terrestrial species, the foot undergoes waves of muscular undulations. In marine species, the only foot margins undulate and there can be pedal appendages, parapodia that enable the animal to swim using flying strokes. The visceral mass and the shell are rolled up into an asymmetric spiral.

Primitively, during development, the visceral mass undergoes torsion: a 180° twist that moves the pallial cavity from a posterior position to an anterior position above or to the side of the head. Organs that were initially on the left side are now found toward the right side and are reduced or have disappeared (auricle, gill, kidney, osphradium—a chemoreceptor). The organs that were initially on the right side are now found toward the left side and develop normally. The nervous system shows the traces of this torsion and is found more centrally than in the non-conchiferan mollusks. There is a single radula that presents a large diversity of different forms. The anterior part of the digestive tube is modified into a sucking

proboscis (or snout). The middle intestine often has mastication plates. The stomach is usually associated with a crystalline style that synthesizes and releases amylase. The hepatopancreas forms two lobes with independent ducts. The two nephridia form sacs that empty into the pallial cavity. The nephridium that was embryonically situated on the right side is found on the left (a result of the torsion) and is generally the only functional nephridium.

The single genital apparatus, found on the right side, shows a wide variety of different organizations and levels of complexity. The remaining gonoduct does not empty directly in the pallial cavity, but rather into the renal coelomoduct. Males typically have a copulatory organ situated anteriorly on the right side and linked to the uro-genital orifice by a groove. The female

genital tracts of certain species are divided into separate parts, one used for copulation and one reserved for the eggs. In terrestrial gastropods (pulmonates), the pallial cavity has formed a respiratory sac linked to the external environment by an orifice, the pneumostome. The shell can show all degrees of development in the gastropods, from strong, heavy shells (*Nerinea, Turritella, Cypraea*) to no shell at all (*Arion*, nudibranchs). The shell can also be represented by a thin layer covered over by the mantle (*Limax, Aplysia*). In species with shells, the foot and the head can be pulled inside by a single powerful pedal retractor muscle. In the primitive condition, the foot carries a hardened or calcified plate, the operculum, that is used to close off the aperture of the shell when the animal retracts inside.

Some unique derived features

■ The gastropods are characterized by a 180° torsion of the visceral mass during a particular stage of their development. In certain gastropod groups, there can be a secondary detorsion. Fig. 1 summarizes this torsion of the visceral mass on representative trochophore and veliger larvae (1a: ventral view of an early-stage trochophore larva; *mo:* mouth, *ptt:* prototroch, *apt:* apical tuft; 1b: right lateral view of a late-stage trochophore larva; *sh:* shell, *shg:* shell gland; 1c: right lateral view of a veliger larva after bending, but before torsion has occurred—the visceral mass undergoes a bending movement as indicated by the arrow; *a:* anus, *pc:* pallial cavity; 1d: a veliger larva after torsion, with the arrow indicating the direction of the movement; *o:* operculum, *ft:* foot, *sta:* statocyst, *te:* tentacle). Fig. 2, an apical view of an adult gastropod, displays the subsequent torsion of the digestive tube.

Ecology

These species are marine, freshwater or terrestrial. We find gastropods in all environments, even in polar waters and at great marine depths: from the ocean floor to the intertidal zone, at high sea, where many species lead an entirely pelagic existence, in lakes, rivers, and terrestrial environments ranging from the most humid to the driest and most arid (for example in Nevada, Mexico, northern Sahara or Namibia). In the Negev desert, *Sphincterochila* is able to tolerate temperatures of 65°C. We also find gastropods in the mountains, at altitudes of several thousand meters in the Rockies and Himalayas, for example.

The gastropods also include all types of feeding regimes and are therefore present at all links of the marine ecological food chain. They can even parasitize other species: *Thyca* exploits starfish, *Sacculus* is a parasite of ascideans, *Odostomia* is a bivalve parasite. The eulimids are highly modified to exploit echinoderms; some are permanent internal parasites of sea cucumbers. There are even species that lead a tubicolous life style: the vermetids or worm shells. Their shells are tube-shaped and they

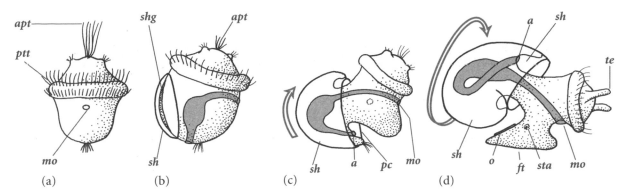

Figure 1

capture suspended particles with the help of long mucous filaments on which microorganisms become trapped. Other species live on algae, within the sediment, in the rocks, etc.

The main problem for terrestrial gastropods is to limit water loss. Those species that live in temperate climates overwinter by hibernating in the soil and spend dry summers in aestivation, that is, in lethargy, buried in the soil, silt or other habitats protected from the sun. Certain species of the genus *Helicella* group together in the early hours of the day on the tips of plant branches. In this way, the individuals escape the elevated temperatures at ground level during the hottest hours of the day. They only come down once

evening comes. It is therefore mainly by their behavior that terrestrial gastropods regulate their water balance, something that has enabled them to colonize all zones across the earth's surface.

The gastropods either have separate sexes or are hermaphroditic. Eggs are laid in the form of clusters, in cocoons or individually. Marine species often have free-swimming trochophore and veliger larvae. Terrestrial and freshwater gastropods produce eggs rich in vitellus; completely formed young hatch from these eggs. Some species brood their eggs, others are viviparous. The pulmonates of the genus *Helix* are consumed and even raised by humans (*Helix pomatia*, Roman snail).

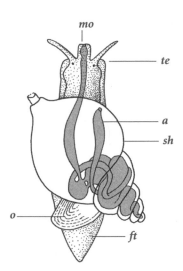

Figure 2

Number of species: 103,000.
Oldest known fossils: the oldest gastropod fossils date from the upper Cambrian. These fossils include the hyperstrophic onychochilid gastropod *Kobayashiella circe* from Shantung, China, the maclurite gastropod *Macluritella walcotti* from Nevada, the sinuitid bellerophont *Sinuella minuta* from Texas, along with the "archaeogastropods" *Sinuopea sweeti* from Wisconsin, *Schizopea typica* from Missouri and *Prohelicotoma uniangularia* from New York (510 MYA). The gastropods have left behind a rich fossil record including 315 families.
Current distribution: worldwide.

Examples

Common limpet: *Patella vulgata; Gibbula adriatica;* common ormer: *Haliotis lamellosa;* common periwinkle: *Littorina littorea;* panther cowry: *Cypraea pantherina;* common whelk: *Buccinum undatum; Murex tenuispina;* Mediterranean cone snail: *Conus ventricosus;* pink-banded shelled sea slug: *Acteon tornatilis; Clio pyramidata;* leopard slug: *Limax maximus;* European red slug: *Arion rufus;* European brown garden snail: *Helix aspersa;* heath snail: *Helicella itala;* sea hare: *Aplysia fasciata; Doris verrucosa; Atlanta peroni; Gasterosiphon deimatis; Entocolax schiemenzi; Thyca entoconcha; Odostomia rissoides; Eulima equestris; Vermetus adansoni.*

Cephalopoda

General Description

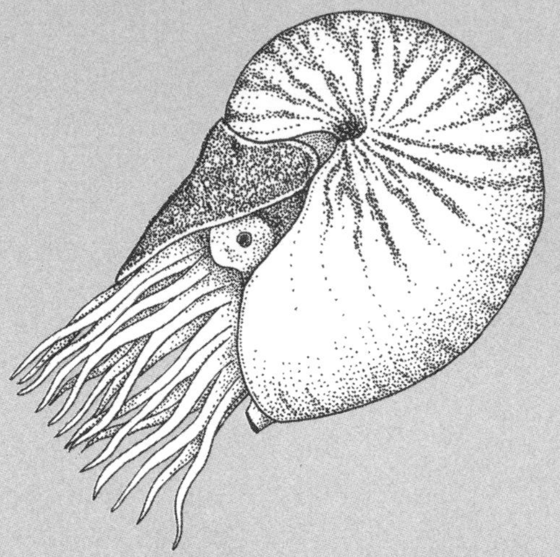

Chambered nautilus
Nautilus macromphalus
up to 20 cm

Common octopus
Octopus vulgaris
up to 300 cm

Paper nautilus
Argonauta argo
10 cm

The cephalopods are elongated marine organisms of large size (from 1 cm to 22 m). In these organisms, the anterior part of the foot has been annexed to the head to form a variable number of pre-oral tentacles. The movement of foot towards the head region along the sagittal plane and the endogastric bending brings all that was in a posterior position in other mollusks (monoplacophora, polyplacophora, solenogastres, caudofoveata) to a ventral position and conserves the bilateral symmetry of these organisms. It is thus that the pallial cavity and its associated complexes—gills, anus, excretory pores, genital pores—find themselves on the ventral side of the animal, opening toward the anterior region, just behind the head. The pallial silt encircles almost the entire body and allows water to move into the pallial cavity; this silt is closed in the median area by the siphon, the posterior end of the foot that has been transformed into a muscular funnel used to expel water from the cavity. The cephalopods are therefore completely enveloped in their mantle, as though in a bag with the opening towards the head, and swim using backward movements by rapidly expelling the water from the pallial cavity. In the primitive state, the mantle secretes a conical shell with its apex oriented towards the posterior end. In the nautiloids, this shell is coiled and divided

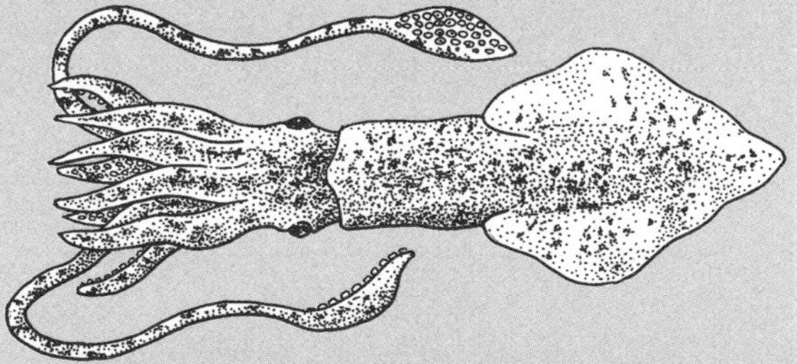

Squid
Loligo vulgaris
20 to 50 cm

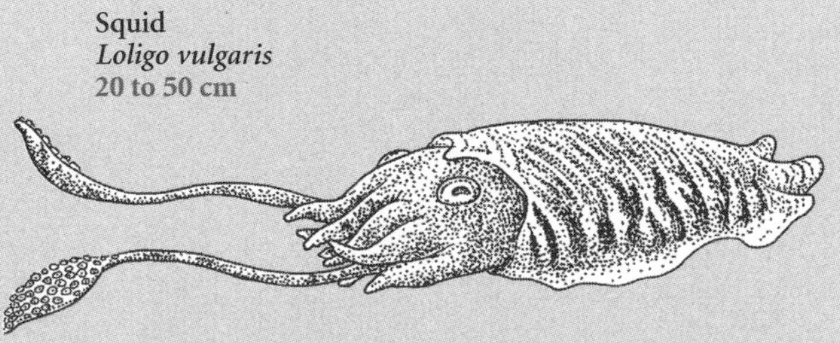

Common cuttlefish
Sepia officinalis
up to 65 cm (of which 30 cm is tentacles)

into chambers of decreasing size. The animal occupies only the last chamber (the largest), but remains in contact with the first chamber by a dorsal tissue cord (siphuncle). In most extant species, the shell either is covered over by the mantle or occurs in a vestigial state only. In the octopi, it has disappeared altogether. In the squid (teuthoids) and cuttlefish (sepioids), the mantle extends to form lateral swimming fins. The cephalopods have large, highly developed lateral eyes. The mouth, found at the center of the tentacles, includes a double lip (typically circular), beak-like jaws characteristic of the group, and a radula. The digestive tube, folded into a U-shape, possesses three parts and is equipped with a paired digestive gland. The coelom is highly developed and includes a genital cavity, a pericardial cavity and two symmetrical renal cavities, which open into the pallial cavity by two renal and two genital coelomoducts. The genital tracts have diverse accessory glands. The nerve ganglia are concentrated in the anterior region and form a "brain" protected by a cartilaginous "skull." Females possess nidimental glands under the skin that empty into the pallial cavity. These glands secrete the material used to form protective egg shells. In the primitive state, there are two pairs of ctenidia, whereas there is only a single pair in the derived state. The nautiloids (nautilus) and fossil ammonoids are classified as "tetrabranchia," whereas the decapods (fossil belemnoids, teuthoids—squid, sepioids—cuttlefish) and the octopods are grouped as "dibranchia." The coloration of the cephalopods is variable. Many species are capable of homochromy, that is, they can take on the color of the surrounding environment. We also find deep-sea squid that are completely translucent, detectable only by different-colored luminescent photophores: *Lycoteuthis diadema* possesses 22 of these organs including ten different structural types.

Some unique derived features

- Tentacles: the anterior end of the foot has been modified into multiple prehensile tentacles. Fig. 1 portrays a longitudinal section of a female squid (*Loligo*) showing half of the tentacles (*te*); *a:* anus, *dc:* digestive caecum, *gd:* genital duct (oviduct), *gc:* genital cavity, *bh:* branchial heart, *syh:* systemic heart, *peric:* pericardial cavity, *sh:* shell, *sk:* skull, *rd:* renal duct (nephridiopore), *ctn:* ctenidium, *pc:* pallial cavity, *st:* stomach, *hp:* hepatopancreas, *dmn:* dorsal mandible, *vmn:* ventral mandible, *ova:* ovary, *is:* ink sac, *rd:* radula, *si:* siphon, *sta:* statocyst, *nsy:* central nervous system, *man:* mantle.

- Siphon: The posterior end of the foot forms the siphon (fig. 1), a muscular funnel made up of two rolled-up lobes. This organ expels water from the pallial cavity.

- The circulatory system, while open in other mollusks, is closed in the cephalopods.

- The nervous system (fig. 1) includes ganglia in most mollusks but becomes highly centralized in the cephalopods, with a type of "brain" protected by a cartilaginous "skull-like" capsule.

- "Ink" sac: most animals have a reservoir of "ink" (fig. 1), a dark fluid released when the animal is alarmed.

- These animals have two skeletal mandibles that form a "beak" (fig. 1). These pieces are represented in left lateral views in fig. 2.

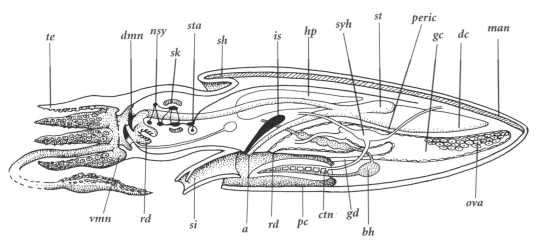

Figure 1

Ecology

The cephalopods are marine predators. They are either pelagic (decapodiformes) or sedentary, living close to ocean floor (octopodiformes). We find them in all of the seas and oceans of the world, up to depths of more than 6000 m. The greatest diversity is found in warm, salty seas, like the Mediterranean or the seas surrounding the Sunda Islands, Indonesia. There are, however, two exceptions; neither the Black Sea nor the Baltic Sea has sufficient salinity levels to support cephalopods. In the cold seas, we mostly find solitary animals. Their isolated way of life is interrupted only for breeding or during the massive gatherings that take place for spawning (notably in *Loligo*). However, grouped movements have also been observed when, for example, several cephalopod species follow schools of fish. The cephalopods feed especially on teleost fish, crustaceans and other mollusks. Some highly derived species specialize on phytoplankton and zooplankton.

The cephalopods are prey to odontocete whales, a few sharks, hake (a gadid teleost fish), the conger eel (another teleost), seals, petrels, and penguins. When frightened, they camouflage their escape by emitting a cloud of ink that blinds, startles or upsets their enemies, if they aren't too large. From tests of aquarium octopi, the cephalopods are known for their intelligence and their highly developed learning abilities. The cuttlefish and the octopus are more sedentary than the squid. It is likely for this reason that squid, tireless swimmers requiring extremely large hunting grounds, are difficult to raise in aquarium settings. Many cephalopods have a surprising ability to change color. They use this ability to mimic their environment to protect themselves or for hunting. However, it is also used in the nuptial parade.

The sexes are separate. After the nuptial parade, the spermatophores offered by the males are

Figure 2

transported by a specialized tentacle (hectocotyle) to the pallial cavity of the female. The eggs are laid onto some kind of supporting surface (coral, sponges, algae, etc.). Telolecithic eggs undergo incomplete cleavage with the persistence of the vitelline reserves. Development is direct, that is, without a separate larval stage; at birth the embryos already greatly resemble small adults. Growth is unlimited in the cephalopods and they can live for many years. The examination of the stomach contents of sperm whales has indicated that the great ocean depths are inhabited by squid of much greater size than any other existing mollusk. Indeed, suckers of 25 cm in diameter and eyes of 40 cm in diameter (the largest of the animal kingdom) have been found. This leaves one to imagine a squid with a body length of 10 m (25 m with the tentacles). This isn't surprising given that the size record for a cephalopod, found in its entirety, is held by *Architeuthis princeps*, captured close to New Zealand in 1933; the body measured 8 m long and the tentacles 14 m, to give a total length of 22 m and a weight of close to 4 tons. Many species of squid and cuttlefish are eaten by humans.

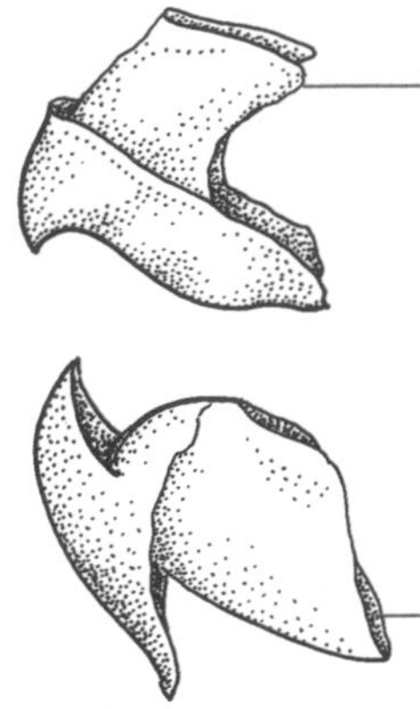

Number of species: 730.
Oldest known fossils: the oldest known cephalopod fossils are nautiloids from the late Cambrian (505 MYA) from China (*Plectronoceras cambria, Theskeloceras benxiense, Eburnoceras pissinum, Acaroceras primordium, Huaiheceras longicollum, Xiaoshanoceras jini, Wanwanoceras exiguum, Yanheceras anhuiense, Aetheloxoceras suxianense, Archendoceras conipartitum, Oonendoceras sinicum*) and the American nautilioid *Balkoceras gracile* from Texas. The cephalopods have left a rich fossil record: 134 nautiloid families, 271 ammonoid families and 45 dibranchial families of which 16 are belemnoids.
Current distribution: worldwide, except the Black Sea.

Examples

Nautilus: *Nautilus pompilius;* squid: *Loligo vulgaris; Lycoteuthis diadema;* giant squid: *Architeuthis princeps;* vampire squid: *Vampyroteuthis infernalis;* common cuttlefish: *Sepia officinalis;* little cuttlefish: *Sepiola atlantica;* common octopus: *Octopus vulgaris;* musky octopus: *Eledone moschata;* paper nautilus: *Argonauta argo.*

Diasoma

A few representatives

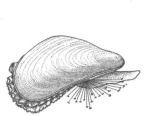

Blue mussel
Mytilus edulis
8 cm
node 13

Razor clam
Solen marginatus
12 cm
node 13

Dentalium entalis
4 cm
node 14

Pulsellum lofotense
5 cm
node 14

Some unique derived features

■ The head has disappeared and the nervous system is secondarily decentralized.

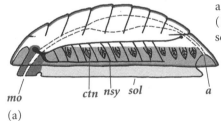

(a)

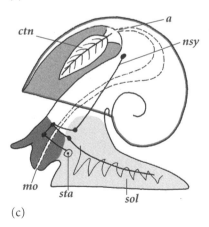

(c)

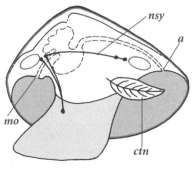

(e)

Fig. 1 shows left lateral views of a polyplacophoran (1a), a monoplacophoran (1b), a gastropod (1c), a cephalopod (1d), a bivalve (1e) and a scaphopod (1f) (*a*: anus, *mo*: mouth, *ctn*: ctenidium, *sol*:

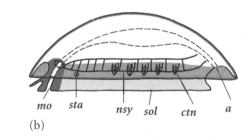

(b)

(d)

■ head ■ foot ■ pallial cavity — mantle margin

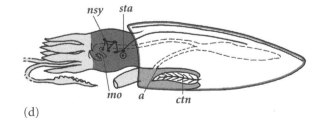

Figure 1

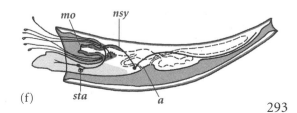

(f)

pedal sole, *sta*: statocyst, *nsy*: nervous system). In the non-ganglioneurans (1a, 1b), the head is almost totally indistinct, but still present, whereas in the visceroconchs (1c, 1d) it is prominent and the nerve ganglia are concentrated anteriorly in the cephalic region. The head is reduced, or even absent in the diasomans (1e, 1f).

- Mantle: in the diasomans, also known as loboconchs, the mantle envelopes the entire animal. On fig. 1 we can see that the mantle protects the entire body except the pedal sole in the non-ganglioneurans (1a, 1b), but only part of the visceral mass of the viscero-chonchs (1c, 1d). The mantle of the diasomans forms two enveloping lobes, as shown in fig. 2, a mussel *(Mytilus edulis)* with its shell removed (top: anterior end of the animal, bottom: posterior end, right: dorsal side, left, ventral side, *by:* byssal threads, *man:* mantle, *aam:* anterior adductor muscle, *pam:* posterior adductor muscle, *exo:* exhalant orifice, *ft:* foot). As a consequence, the

diasoman shell protects the entire animal.

- The foot (fig. 3 and fig. 4) is of variable volume, fluctuating under the control of the hemolymph flow and the retractor muscles. Generally, it is spatula-shaped and blade-like, particularly in burrowing species. In sessile animals (like the mussel), the foot takes the form of a finger. Fig. 3 represents a bivalve of the genus *Tagelus* buried in the sediment. Fig. 4 shows a burrowing bivalve of the genus *Mya,* on which we see the two

siphons originating from the same tube (*exs:* exhalant siphon, *ins:* inhalant siphon, *rv:* right valve).

- The shell of the veliger larva goes through a bivalve stage. The shell of the trochophore larva is primitively univalve. Due to calcification along the medial-dorsal line, the shell takes on a bivalve arrange-ment. This shell remains the same in the adult bivalves. In the veliger larvae of the scaphopods, the two mantle lobes meet and fuse ventrally along the sagittal line. This also occurs in the two valves of the shell that take on a tubular form. In certain scaphopod species, this fusion remains incomplete, such that the shell has a silt (for example, in *Dentalium stenoschizum*), or several aperatures (in *Fissi-dentalium plurifissuratum*).

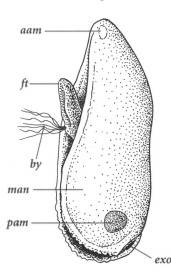

Figure 2

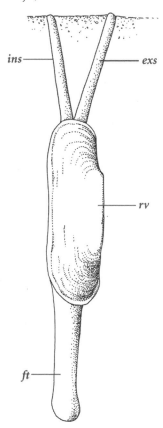

Figure 3

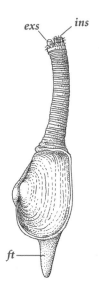

Figure 4

Number of species: 12,400.
Oldest known fossils: the diasomans (or loboconchs) are known from the early Cambrian (530 MYA) from the praenuculid bivalve *Projetaia runnegari* from South Australia and China, and the ribeiriid rostroconchs (a fossil group, closely related to bivalves) *Heraultipegma varensalense* from the Herault département in France and *Watsonella crosbyi* from Massachusetts, United States.
Current distribution: worldwide.

Examples

- BIVALVIA: nut shell: *Nucula nucleus;* blue mussel: *Mytilus edulis;* blacklip pearl oyster: *Pinctada margaritifera;* great scallop: *Pecten maximus;* zebra mussel: *Dreissena polymorpha;* razor clam: *Solen marginatus; Cuspidaria typus.*
- SCAPHOPODA: common tusk shell: *Dentalium vulgare; D. stenoschizum; Entalina quinquangularis; Siphonodentalium galatheae; Fissidentalium plurifissuratum.*

Bivalvia

Blue mussel
Mytilus edulis
8 cm

Noah's ark shell
Arca noae
6 cm

Razor clam
Solen marginatus
12 cm

Chlamys opercularis
15 cm

The bivalves, or lamellibranchs, are laterally compressed mollusks that have conserved their bilateral symmetry. Their shell consists of two articulated valves. Primitively, the two valves are symmetric, but one of the two may have specialized secondarily in some species. The visceral mass lies ventrally and there is a hinge in the mid-dorsal region where the two valves meet and articulate. The shape and color of the valves vary enormously with species and their size ranges from 0.2 to 150 cm with a corresponding weight of up to 250 kg. The shells can be shaped like a hammer, shower head, knife, saber, or pod.

There is no head, pharynx, radula or salivary glands. The body is completely enveloped in the mantle, which has formed two sheets; the external surfaces of these sheets secrete the shell. The movement of the gill cilia creates a current of water. Water enters by the inhalant aperture and exits by the exhalant aperture. These openings are sometimes lengthened by the presence of siphons that, in certain burrowing species, can be extremely long. The two valves are attached together by an elastic dorsal abductor ligament and by one or two adductor muscles that function to close the shell when they contract. The mantle is connected to the foot by a variable number (up to seven) of muscle pairs. This flexible foot is blade-shaped. At its base, a specialized gland is sometimes present, the byssus gland, which produces the byssal threads (tough protein threads used for attachment). The balance organ (statocyst) is housed in the foot.

The mantle margin is rich in sensory organs and, in some species, has even developed colored lobes or tentacles. The eyes can vary in complexity depending on the species: simple pigment cups, organs composed of 70 to 250 photoreceptors, or even complex eyes with retina in a few species. The pallial cavity contains a pair of well-developed gills attached to the mantle on one side and to the foot on the other. The structure of these ctenidia is complex and varies according to the species. The majority of bivalves have gills in the form of filaments, nets or sheets that are used for respiration, but also for capturing suspended food particles in the water. The mouth is armed with four ciliated buccal lobes. The stomach contains a long posterior caecum where the crystalline style and two digestive glands are found. It is also linked to a relatively large hepatopancreas. The heart lies within the pericardium and includes two auricles and a ventricle that pump the blood into the two aortas. The circulatory system is not closed. Two U-shaped nephridia empty into the pericardium. The two genital glands lie either in front of or below the pericardium. The nervous system has an intermediate level of ganglionic fusion.

Some unique derived features

- The shell is made up of two laterally compressed valves in the adult stage. These valves are linked by a medio-dorsal hinge and a non-calcified ligament. Fig. 1 shows the internal view of the right valve of a bivalve (1a; the posterior end of the animal is on the right, the anterior end on the left, and the dorsal side at the top), and an external dorsal view of the same animal (1b; the anterior end of the animal at the top, *hg:* hinge, *l:* ligament, *lu:* lunule, *aam:* anterior adductor muscle, *pam:* posterior adductor muscle, *ps:* pallial sinus, zone where siphons exit, *um:* umbo, *lz:* ligament zone).

- The radula is absent.

- A special gland found at the base of the foot, the byssus gland, produces byssal threads. These threads are adhesive protein filaments that stick to a substrate and harden when they contact water. The byssal threads are important for sedentary animals that anchor to substrates. These threads are illustrated in fig. 2 on the mussel, *Mytilus edulis* (*by:* byssal threads, *sh:* shell, *ft:* foot, *exs:* exhalant siphon).

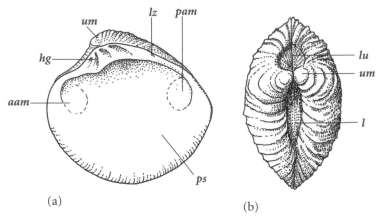

(a)   (b)

Figure 1

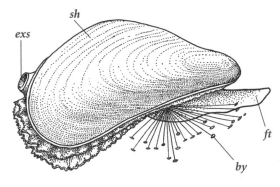

Figure 2

Ecology

The bivalves are marine or freshwater mollusks that include filter-feeders, burrowers and attached surface-dwellers. They are capable of colonizing all substrates. The inhabitants of hard surfaces (oysters, tropical jewel boxes, thorny oysters) cement one valve to the support or attach themselves by means of byssal threads (mussels, pintadines). The surface of the valves is often covered with grooves, ridges, or spines. The burrowing bivalves (clams, soft-shelled clams) often have smooth, flatten shells. Other bivalves are borers, able to perforate wood, calcareous rock, sandstone, coral or clay, either mechanically (pholades, shipworms) or with the help of acidic substances (date mussel), and live in the bored holes. Freshwater bivalves are even present at high altitudes and in Arctic regions. The pill clam (*Pisidium casertanum*), a small bivalve only 4 mm long, lives on the bottom of high mountain lakes, even when the surface of the lake is frozen over.

Driven by the beating of millions of gill cilia, water follows a trajectory that brings oxygen and food particles to the gills and then evacuates through the pallial cavity in the area where the renal aperture and anus empty. Food particles that remain attached to the gill surface will be sent towards the mouth by a mucus filament where the cilia of the four buccal lobes will then help transport them toward the digestive tube. The gills can directly absorb small organic particles by pinocytosis. For example, the greenish color of oysters is due the absorption at the gills of a green lipoprotein pigment that is released by diatoms living at the surface of the silt. The stomach is flanked by a posterior caecum that contains a crystalline style. This style is made up of concentric layers of mucoproteins, including enzymatic proteins. The anterior end of the style projects into the stomach, where it comes up against a toothed cuticular plate. Most often, the style turns itself via the action of vibratile cilia; this action dislodges food from the toothed plate at the same time that it wears the tip of the style to release the enzymes. Even if large particles are ejected via the exhalant siphon, the stomach sorts other particles: small particles, the size of sand grains, go directly toward the digestive tube, whereas larger food particles and enzymes go through two digestive glands where digestion occurs. By filtering water, these micro-phagous animals contribute to the deposition of silt. A Portuguese oyster (*Crassostrea angulata*) produces 1.075 g of silt every 24 hours. The bivalves are prey to boring gastropods, starfish, and many malacophagous marine vertebrates with specialized teeth.

The sexes are normally separate, often with successive hermaphrodism (they are male and female successively in time). The gametes are released directly into the water with the exhalant current. Fertilization takes place in the water, outside the animal or within the pallial cavity. In certain species, the young are protected by special incubating organs or between the gills in certain freshwater species. The larvae follow two possible developmental pathways. The dominant pathway—tro-chophore, veliger—includes a pelagic phase, whereas the second pathway—glochidium, la-sidium—is found only in freshwater bivalves, whose larvae are parasitic, particularly on the gills of teleost fishes. In either case, the trochophore or glochidium both have a dorsal shell that, by default, acquires a typical bivalve arrangement by calcification along the mid-dorsal line. Many bivalve species are consumed and farmed by humans (oysters, giant scallops, mussels, hard- and soft-shelled clams, etc). The history of human culture and commerce is marked by the use of bivalve shells as money and the exploitation of bivalves for pearl production. Indeed, certain bivalve species react to the presence of a parasite or foreign object by the concentric production of periostracum, followed by a calcareous shell and then by a shining layer of nacre that forms the external layer of the "pearl."

Number of species: 12,000.
Oldest known fossils: the oldest bivalve fossil known is that of the praenuculid *Projetaia runnegari* from the early Cambrian from South Australia and China (530 MYA). The bivalves have left a rich fossil record with 215 families.
Current distribution: all the seas of the world, including the polar seas. We find freshwater bivalves in all rivers and lakes, including high mountain lakes at altitudes over 2500 m.

Examples

Nut shell: *Nucula nucleus;* blue mussel: *Mytilus edulis;* date mussel: *Lithophaga mytiloides;* blacklip pearl oyster: *Pinctada margaritifera;* hammer oyster: *Malleus malleus;* pen shell: *Pinna nobilis;* black scallop: *Chlamys varius;* European thorny oyster: *Spondylus gaederopus;* great scallop: *Pecten maximus;* Portuguese oyster: *Crassostrea angulata;* common river mussel: *Unio crassus;* European fingernail clam: *Sphaerium corneum;* heart cockle: *Glossus rubicundus;* zebra mussel: *Dreissena polymorpha;* pill clam: *Pisidium casertanum;* jewel box: *Chama gryphoides;* common cockle: *Cardium edule;* warty Venus shell: *Venus verrucosa;* razor clam: *Solen marginatus;* soft-shelled clam: *Mya arenaria;* common piddock: *Pholas dactylus;* shipworm: *Teredo navalis; Pandora inaequivalvis; Poromya granulata; Lyonsella abyssicola; Cuspidaria typus.*

Scaphopoda

General Description

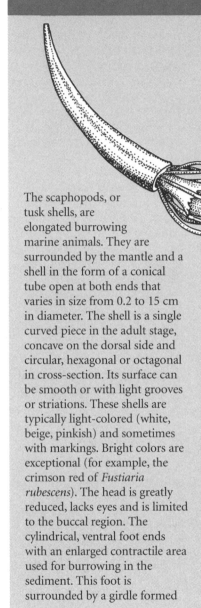

*Pulsellum
lofotense*
5 cm

Tusk shell
Dentalium entalis
4 cm

The scaphopods, or tusk shells, are elongated burrowing marine animals. They are surrounded by the mantle and a shell in the form of a conical tube open at both ends that varies in size from 0.2 to 15 cm in diameter. The shell is a single curved piece in the adult stage, concave on the dorsal side and circular, hexagonal or octagonal in cross-section. Its surface can be smooth or with light grooves or striations. These shells are typically light-colored (white, beige, pinkish) and sometimes with markings. Bright colors are exceptional (for example, the crimson red of *Fustiaria rubescens*). The head is greatly reduced, lacks eyes and is limited to the buccal region. The cylindrical, ventral foot ends with an enlarged contractile area used for burrowing in the sediment. This foot is surrounded by a girdle formed by the lateral lobes of the foot. The mouth opens onto a type of proboscis; it is surrounded by four buccal palps and, on its dorsal side, two flattened lobes elongated by numerous captacules (long ciliated contractile filaments with knobs at their extremities). These structures secrete a mucus which captures micro-organisms in the sediment. The visceral mass is symmetric and stretched dorso-ventrally. The posterior aperture allows water to enter the pallial cavity. The larger anterior aperture enables the foot and the captacules to move in and out of the shell. The non-protractile buccal bulb has a radula and a subradular organ used for tasting. The esophagus is short. The stomach is flanked by two hepatopancreatic lobes with tubular branches. The long intestine ends at the anus which then empties into the pallial cavity. The circulatory and respiratory systems are rudimentary. Gills are absent. Scaphopod respiration is assured by the internal ciliated surface of the mantle. The excretory system includes two latero-ventral nephridia that empty into the pallial cavity. The right nephridiopore is also a genital pore. The lobes of the single gonad occupy the posterior-dorsal region of the body. This gonad becomes associated with the right nephridium once sexual maturity is reached. The nervous system is not centralized.

Some unique derived features

■ Captacules: Two bundles of captacules are found close to the mouth (there can be up to 130 per bundle in some species). These are long, contractile, ciliated filaments with knobs at their extremities that are armed with adhesive glands, which aid in the capture of food. A current created by the cilia also favors the movement of particles towards the mouth. Fig. 1 shows the general organization of a scaphopod with its captacules (*a:* anus, *mo:* mouth, *cpt:* captacules, *pc:* pallial cavity, *st:* stomach, *cg:* cerebral ganglion, *dg:* digestive gland with a hepatopancreatic function, *adg:* adhesive gland, *gn:* gonad, *pg:* pedal ganglion, *vg:* visceral ganglion, *ftl:* foot lobe, *nr:* nephridium, *gea:* genital and excretory aperture, *pa:* posterior aperture of the pallial cavity, *ft:* foot, *rd:* radula, *sta:* statocyst, *nsy:* nervous system, *rz:* respiratory zone of the mantle).

■ The shell has a conical tube-like shape with openings at both ends and is formed by a single curved shell piece. This shell is concave along its dorsal surface, convex on the ventral surface and circular, hexagonal or octagonal in cross-section. Fig. 2 shows the lateral right view of a scaphopod shell.

Ecology

The scaphopods live in the sediments all of seas, from the littoral zone to depths of up to 7000 m (such as *Siphonodentalium galatheae*). These animals burrow obliquely into sandy or silty bottoms, with the posterior end of the shell remaining outside the substrate. Burrowing is done with the help of a club-shaped (or bladelike) foot; the volume of the foot varies through hydrostatic action. The scaphopods are microphagous animals; the captacules are armed with adhesive glands that enable them to capture food particles and bring them to the mouth. They search for their prey in the interstitial spaces of the sediment grains, where the waters are rich in micro-organisms like diatoms and foraminiferans. The scaphopods are prey of polychaetes, shell-boring gastropods, and teleost fishes.

The scaphopods are oviparous animals with separate sexes. The gametes are released into the sea water, where fertilization takes place. Cleavage is total, unequal and spiral. A typical trochophore larva, shaped like a top, hatches from the egg. This larva carries an apical tuft and three ciliated crowns. The veliger larva develops two dorsal mantle lobes that secrete the embryonic bivalve shell. These two lobes meet ventrally along the anterior-posterior plane and fuse the shell valves into their tubular form. In certain species, this fusion remains incomplete such that the shell carries a slit (in *Dentalium stenoschizum*, for example) or several apertures (as in *Fissidentalium plurifissuratum*).

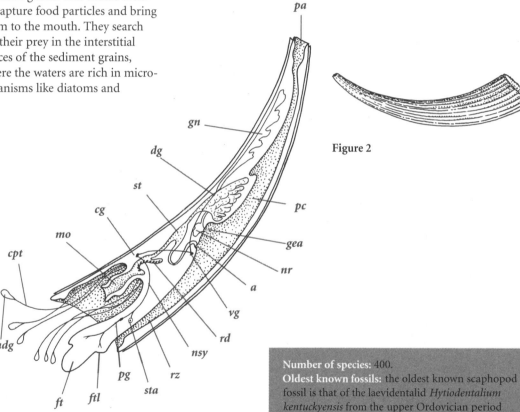

Figure 2

Figure 1

Number of species: 400.
Oldest known fossils: the oldest known scaphopod fossil is that of the laevidentalid *Hytiodentalium kentuckyensis* from the upper Ordovician period from Kentucky, United States (450 MYA). The scaphopods have left a modest fossil record that includes only four families.
Current distribution: all the seas of the world, including polar seas.

Examples

Common tusk shell: *Dentalium vulgare; D. stenoschizum; Fustiaria rubescens; Cadulus quadridentatus; Entalina quinquangularis; Siphonodentalium galatheae; Fissidentalium plurifissuratum; Pulsellum lofotense.*

Chapter 10

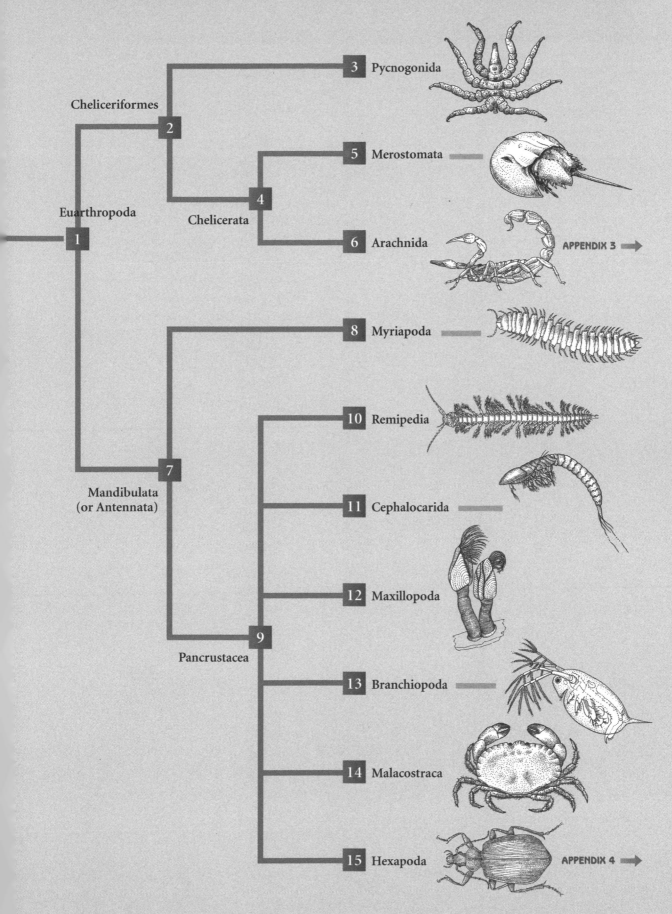

Euarthropoda

Cheliceriformes — 2

Euarthropoda — 1

3 Pycnogonida

Chelicerata — 4

5 Merostomata

6 Arachnida — APPENDIX 3 ➡

Mandibulata (or Antennata) — 7

8 Myriapoda

10 Remipedia

11 Cephalocarida

12 Maxillopoda

Pancrustacea — 9

13 Branchiopoda

14 Malacostraca

15 Hexapoda — APPENDIX 4 ➡

Euarthropoda

This is the animal phylum with the greatest biodiversity. The insects alone have been estimated to include several million species, most of which have yet to be discovered. We find euarthropods from the deepest ocean depths to the top of the highest mountains and in all climates.

We have called this phylum *euarthropoda* and not *arthropoda* here because the latter name is reserved for the taxon that includes all the fossil arthropods. These species, for the most part, appeared at the start of the Cambrian period and classifying them is a delicate matter. For certain scientists (like S. J. Gould in *Wonderful Life: The Burgess Shale and the Nature of History*), these animals show unique body plans and should be placed in their own phylum. For other paleontologists (like D. Briggs or S. Conway-Morris), these organisms are undeniably arthropods. It is this latter vision that we adopt here. The euarthropods therefore correspond to a monophyletic taxon within the arthropods, which is itself found within the monophyletic panarthropod taxon (see 8.25). The trilobites form their own taxon within the euarthropods.

We have adopted a modern view here that does not retain the old division of the unirama and birama. To the contrary, the insects are now placed with the crustaceans in a taxon that we have named "pancrustacea." However, it is very likely that the insects should actually be included within the malacostraca. In the future, it would be interesting to have phylogenetic data reliable enough to reconstruct a likely scenario for the origin of the insects.

Cheliceriformes

A few representatives

Pycnogonum littorale
15 mm
node 3

Horseshoe crab
Limulus polyphemus
30 cm
node 5

Scorpion
Androctonus australis
9 cm
node 6

Cross orb weaver
Araneus diadematus
2.5 cm
node 6

Some unique derived features

- The body is divided into the prosoma and the opisthosoma. The prosoma is composed of the acron and six segments, whereas the opisthosoma has a maximum of twelve segments and a telson.

- The first pair of characteristic appendages are the chelicerae; they are homologous to the second pair of antennae in the mandibulata (fig. 1: dorsal view of the amblypygid *Stygophrynus dammermani*; *che*: chelicera, *ppd*: pedipalp).

- Pedipalps: the second pair of appendages is a pair of palps or jaw-legs called pedipalps (fig. 1).

- The animal has four pairs of walking legs (fig. 1).

- The deutocerebrum is absent (fig. 2: a. the brain of a mandibulate arthropod in the ancestral state; b. cheliceriforme brain; *dc*: deutocerebrum, *agg*: antennal ganglion, *ogg*: optic ganglion, *prc*: protocerebrum, *ttc*: tritocerebrum, *es*: esophagus).

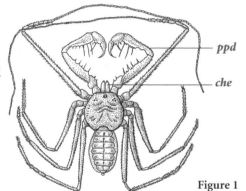

Figure 1

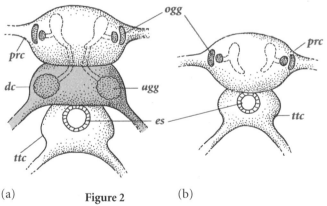

(a)   **Figure 2**   (b)

- In the majority of species, nutrition is obtained by sucking up pre-digested food.

- In the initial stages of embryogenesis, four post-oral segments are formed. On fig. 3, we see three pairs of appendages on a protonymphon larva of a pycnogonid—the chelicera, the palps and the ovigers; the outline of the first pair of legs is visible only on the ventral view (*ov*: oviger, *pb*: proboscis).

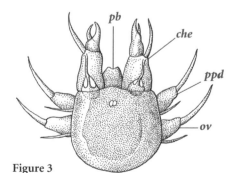

Figure 3

Number of species: 75,450.
Oldest known fossils: *Sanctacaris uncata* from the Burgess Shale was found in Canada and dates from the middle Cambrian (520 MYA).
Current distribution: worldwide.

Examples

- PYCNOGONIDA: *Endeis spinosa; Nymphon gracile; Pycnogonum littorale.*
- MEROSTOMATA: horseshoe crab: *Limulus polyphemus.*
- ARACHNIDA: *Androctonus australis; Buthus occitanus;* cross orb weaver: *Araneus diadematus;* lesser house spider: *Tegenaria domestica; Sarcoptes scabiei* (causes scabies); harvest mite: *Thrombicula autumnalis.*

Pycnogonida

General Description

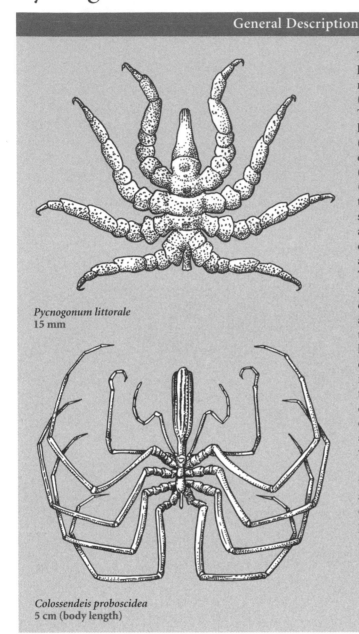

Pycnogonum littorale
15 mm

Colossendeis proboscidea
5 cm (body length)

The pycnogonids (from Greek *pyc:* knobby, and *gonida:* knees) are marine animals that resemble spiders (they are also known as sea spiders). The body is often spindly with long legs. These animals are relatively small (1 to 10 mm), but some polar species or those found at great depths (7000 m) can reach respectable sizes; *Colossendeis* can have a leg span of up to 40 cm; leg span may reach 70 cm in some other species. The pycnogonids are most often dull in color, but some coastal species are green or purple and some deep-sea species are red. The body is very straight with obvious segments. The head (or cephalon) carries a forward-facing proboscis and four eyes found on an ocular tubercle on the dorsal surface. The legs articulate by lateral expansions of the segments. The opisthosoma is reduced. From front to back, the appendages include: a pair of chelicerae, a pair of palps, a pair of ovigers (always present in the male), and, normally, four pairs of long walking legs.

The relatedness of this taxon to the chelicerates seems well established at present. Nevertheless, the homologies of the diverse body parts of pycnogonids and the chelicerates still require some discussion.

Some unique derived features

- Proboscis: the animal has a single pre-oral proboscis (fig. 1: ventral view of *Nymphon rubrum; cho:* chelifore; *eg:* eggs, *ov:* oviger, *ppd:* palp, *pr:* proboscis).

- Ovigers: the male (and sometimes the female) has a pair of ovigerous legs (fig. 1).

- The opisthosoma is reduced or absent and without appendages (fig. 2: posterior end of

Pycnogonum littorale; gp: gonopore, *opi:* opisthosoma).

- The walking legs are made up of nine segments, including the terminal claw (fig. 1).

- The gonads have numerous genital openings, typically found on the second coxa of some or all of the walking legs (fig. 2).

Ecology

Most pycnogonids are carnivorous, feeding on cnidarians (corals, sea anemones), ectoprocts, small polychaetes, or sponges. They place their proboscis directly on their prey and suck up the tissues. Certain species feed on algae, microorganisms or detritus. There are no respiratory or excretory organs; this may be correlated with the body form of the animal, which provides a vast surface for exchange with the external environment. Parasitic forms principally infest mollusks.

The sexes are separate. The simple gonads have diverticulae that extend into the femurs of the legs; they have numerous orifices that are generally found on the legs. The ovules are fertilized as soon as they are released. The males receive them and carry them on their ovigers until hatching. The protonymphon larvae have the first three pairs of appendages (the chelicerae, the palps and the ovigers) and the outline of the first pair of legs.

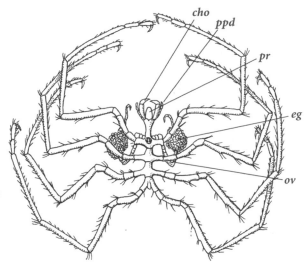

Figure 1

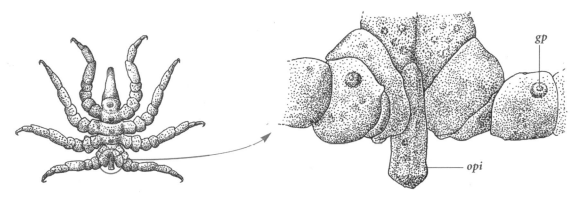

Figure 2

Number of species: 1,000.
Oldest known fossils: an unnamed larval form from the upper Cambrian (510 MYA) was found in Sweden. The most ancient adult forms are *Palaeoisopus problematicus, Palaeopantopus maucheri* and *Palaeothea devonica* from the lower Devonian from Germany (395 MYA).
Current distribution: worldwide, principally in cold waters.

Examples

Achelia echinata; Colossendeis proboscidea; C. scotti; Dodecolopoda mawsoni; Endeis spinosa; Nymphon gracile; N. rubrum; Pycnogonum littorale.

Chelicerata

A few representatives

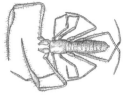

Horseshoe crab
Limulus polyphemus
30 cm
node 5

Scorpion
Androctonus australis
9 cm
node 6

Cross orb weaver
Araneus diadematus
2.5 cm
node 6

Pseudoscorpion
Neobisium tuzeti
4 mm
node 6

Some unique derived features

- The prosoma carries a shield-like carapace (fig. 1: dorsal view of *Ricinoides sjöstedti,* size 20 mm; *sd:* shield).

- Genital somite: The first or second segment of the opisthosoma is modified into a genital somite

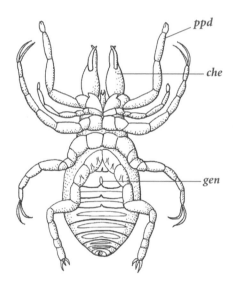

Figure 2

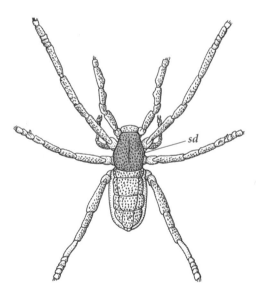

Figure 1

(fig. 2: ventral view of the solifuge *Mossamedessa abnormis* showing the genital opening on the second segment of the opisthosoma; *che:* chelicera, *gen:* genital orifice, *ppd:* pedipalp).

Number of species: 74,450.
Oldest known fossils: the most ancient arthropod with chelicerate characteristics was *Sanctacaris uncata* from the middle Cambrian (520 MYA) from Canada. It is the sister group of all other known chelicerates.
Current distribution: worldwide.

Examples

- MEROSTOMATA: *Limulus polyphemus.*
- ARACHNIDA: *Androctonus australis; Buthus occitanus;* cross orb weaver spider: *Araneus diadematus;* lesser house spider: *Tegenaria domestica; Copidognathopsis kerguelenensis; Sarcoptes scabiei* (causes scabies); harvest mite: *Thrombicula autumnalis.*

Merostomata

General Description

The merostomates include an exclusively fossil taxon, the eurypterids, and the xiphosurans (or horseshoe crabs). As the name implies, this latter group has an easily recognizable horseshoe-shaped carapace. These animals have a pair of chelicerae and five pairs of walking legs made up of seven joints each. The first pair of walking legs is homologous to the pedipalps of other cheliceriformes. The first joint of the first four pairs of legs has been modified into an inward pointing spine that macerates food as it is moved toward the mouth.

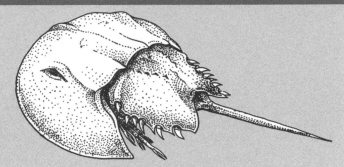

Horseshoe crab
Limulus polyphemus
30 cm

The opisthosoma has six pairs of appendages. The first pair is fused and forms a genital operculum with the two genital pores on the ventral side. The other pairs are modified into membranous gills. These gills are constantly moving and help in swimming. Finally, the opisthosoma has a spine-shaped telson, characteristic of the taxon.

Some unique derived features

- Gills: the appendages of the opisthosoma have been modified into gills (fig. 1: ventral view of *Limulus polyphemus*; g: gills, che: chelicera, go: genital operculum, ts: telson).

- The telson is long and pointed (fig. 1).

Ecology

The horseshoe crabs are bottom-dwellers and generally live on the sandy bottom of shallow marine waters. The shape of the carapace helps the animal to dig in the sediment. They are predators or necrophagic omnivores, feeding on mollusks (principally bivalves), worms and dead animals.

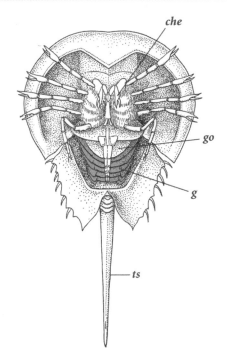

Figure 1

Number of species: 5.
Oldest known fossils: the most ancient xiphosuran is *Eolimulus alatus* from the middle Cambrian from Sweden (520 MYA). The oldest eurypterids (*Waeringopterus priscus, Eocarcinosoma ruedemanni, Dolichopterus antiquus, Pterygotus deepkillensis*) are all from the lower Ordovician from New York (USA, 470 MYA).
Current distribution: northwest coast of the Atlantic, Asian coastal waters from Korea and Japan to the east of India and the Philippines.

Examples

Horseshoe crab: *Limulus polyphemus; Tachypleus tridentatus.*

➡ *see appendix 3, p. 522*

Arachnida

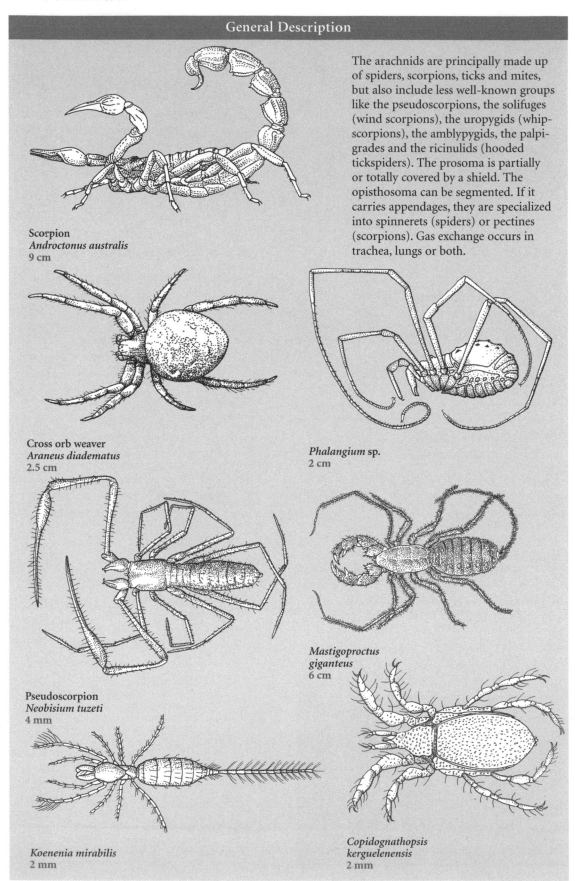

General Description

The arachnids are principally made up of spiders, scorpions, ticks and mites, but also include less well-known groups like the pseudoscorpions, the solifuges (wind scorpions), the uropygids (whip-scorpions), the amblypygids, the palpi-grades and the ricinulids (hooded tickspiders). The prosoma is partially or totally covered by a shield. The opisthosoma can be segmented. If it carries appendages, they are specialized into spinnerets (spiders) or pectines (scorpions). Gas exchange occurs in trachea, lungs or both.

Scorpion
Androctonus australis
9 cm

Cross orb weaver
Araneus diadematus
2.5 cm

Phalangium sp.
2 cm

Pseudoscorpion
Neobisium tuzeti
4 mm

Mastigoproctus giganteus
6 cm

Koenenia mirabilis
2 mm

Copidognathopsis kerguelenensis
2 mm

Some unique derived features

- Opisthosoma: the appendages of the opisthosoma have been lost, reduced or modified into spinnerets or pectines (combs) (fig. 1: ventral view of a spider showing the posterior spinnerets and the two openings of the lungs; *che:* chelicera, *spin:* spinnerets, *stg:* stigmata, the opening of a lung).

- Respiration: the animal breathes by tracheae, lungs or both at the same time (fig. 1).

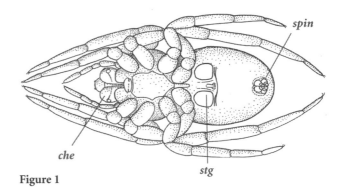

Figure 1

Ecology

The arachnids are almost all terrestrial. There are a few marine and freshwater mites and some rare spiders that live in fresh water, like the water spider (*Argyroneta aquatica*). However, from an evolutionary point of view, this corresponds to a return to water. The scorpions are viviparous animals famous for their metamerized abdomen that ends in a venomous apparatus. A sting from this apparatus is deadly for the scorpion's prey and can sometimes be dangerous for humans. The venom includes two types of active substances: neurotoxins that act on the nervous system and hemorrhagins that inhibit blood coagulation and cause local tissue necrosis.

In the spiders, the poison glands are found on the tip of the chelicerae. Spigots on the spinnerets produce the silk threads that are used to weave a web, to construct a shell lodge, to make the egg cocoons, etc. Spiders are carnivorous animals that feed principally on insects. However, certain large tarantulas can capture small rodents or birds. The acari include both ticks and mites. Most free-living mites are predators of small animals living in soil litter. However, they can also be symbionts of other arthropods (millipedes, ants, beetles). The majority of parasitic acari are ticks, hematophagous vertebrate parasites. Some acari are responsible for well-known diseases: *Sarcoptes scabiei*, the causal agent of human scabies, constructs galleries under the epidermis; in North America, *Dermacentor andersoni* is the vector of the *Rickettsia* that causes Rocky Mountain spotted fever; a bite from a harvest mite (*Thrombicula autumnalis*) can provoke an itching sensation.

Number of species: 74,445 (of which 1,200 are scorpions, 35,000 are spiders, and 30,000 are ticks and mites).
Oldest known fossils: the most ancient fossil is that of the scorpion *Dolichophonus loudonensis* from the lower Silurian from Scotland (435 MYA). We can also cite: the mites *Protospeleorchestes pseudoprotacarus, Protacarus crani, Pseudoprotacarus scoticus, Palaeotydeus devonicus, Paraprotacarus hirsti* from the lower Devonian from Scotland (395 MYA); the pseudoscorpion *Dracochela deprehando* from the middle Devonian from New York (USA, 380 MYA); the opilians (unnamed) from the lower Carboniferous from Scotland (330 MYA), the phalangiotarbids *Goniotarbus tuberculatus, Mesotarbus intemedius* and *Leptotarbus torpedo* from the upper Carboniferous from England (315 MYA); and the solifuge *Protosolpuga carbonaria* from the upper Carboniferous from Illinois (310 MYA). The oldest spider is *Attercopus filmbriunguis* from the lower Devonian from New York (USA, 395 MYA).
Current distribution: worldwide.

Examples

- SCORPIONES: *Androctonus australis; Buthus occitanus.*
- PSEUDOSCORPIONES: *Neobisium tuzeti.*
- UROPYGI: *Mastigoprotus giganteus.*
- AMBLYPYGI: *Stegophrynus dammermani; Phrynictus lunatus.*
- ARANEAE: cross orb weaver spider: *Araneus diadematus;* black widow: *Latrocdectus mactans;* lesser house spider: *Tegenaria domestica.*
- OPILIONES: *Phalangium* sp.
- ACARI: *Copidognathopsis kerguelenensis; Demodex folliculorum* (parasite of sebaceous glands); *Sarcoptes scabiei* (causal agent of scabies); harvest mite: *Thrombicula autumnalis.*
- PALPIGRADI: *Koenenia mirabilis.*

Mandibulata (or Antennata)

A few representatives

Polydesmus sp.
15 mm
node **8**

Goose barnacle
Lepas anatifera
5 cm
node **12**

Edible crab
Cancer pagurus
25 cm
node **14**

Swallowtail
Papilio machaon
8 cm
node **15**

Some unique derived features

- Mandibles: the head has mandibles, that is, appendages with few joints, often hard and used for scraping, cutting or grinding (fig. 1: posterior view of a mandible of the grasshopper *Dissosteira carolina;* we can easily distinguish the apodemes where the muscles attach, the outer hinge, and the heavily sclerotized incisive process (shown in color); *apo:* apodeme, *hin:* hinge, *inp:* incisive process).

- The 18S rRNA gene sequences support the monophyly of this taxon.

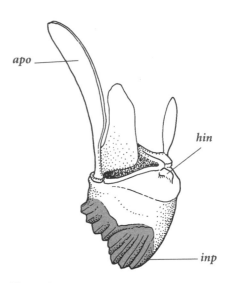

Figure 1

Number of species: 880,964.
Oldest known fossils: six ostracod families were already present at the start of the Cambrian period (540 MYA). We can cite: *Bradoria scrutator* and *Hipponicharion eos* from Canada, *Cambria sibirica* from east Siberia, *Kunmingella maxima* from Yunnan, China, *Uskarella prisca* from Kazakhstan, *Hesslandona* sp. and *Strenuella* sp. from England. Some phyllocarids date from the same period, like *Hymenocaris* sp. from from Europe, North America, Australia and New Zealand (540 MYA) or *Perspicaris* sp. from Yunnan (China, 540 MYA).
Current distribution: worldwide.

Examples

- MYRIAPODA: millipede: *Lulus scandinavius; Polydesmus* sp.; centipede: *Geophilus electricus*
- PANCRUSTACEA:—REMIPEDIA: *Speleonectes lucayensis.* —CEPHALOCARIDA: *Hutchinsoniella macracantha.* —MAXILLOPODA: *Amphiascus* sp.; *Tigriopus fulvus;* goose barnacle: *Lepas anatifera.* —BRANCHIOPODA: brine shrimp: *Artemia salina; Daphnia pulex.* —MALACOSTRACA: common prawn: *Palaemon serratus;* edible crab: *Cancer pagurus; Oniscus asellus;* scud: *Gammarus pulex.* —HEXAPODA: *Ctenosta bastardi;* swallowtail: *Papilio machaon; Drosophila melanogaster; Scarcophaga* sp.

Myriapoda

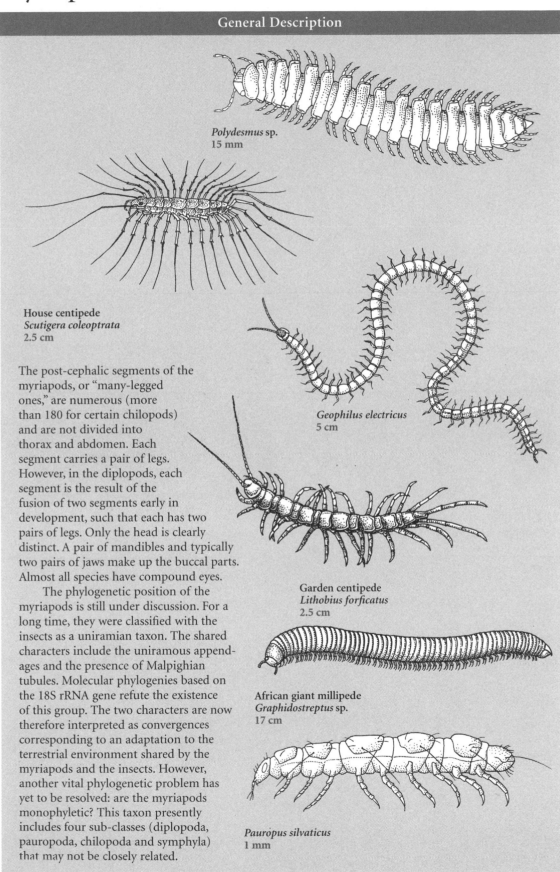

Polydesmus sp.
15 mm

House centipede
Scutigera coleoptrata
2.5 cm

Geophilus electricus
5 cm

The post-cephalic segments of the myriapods, or "many-legged ones," are numerous (more than 180 for certain chilopods) and are not divided into thorax and abdomen. Each segment carries a pair of legs. However, in the diplopods, each segment is the result of the fusion of two segments early in development, such that each has two pairs of legs. Only the head is clearly distinct. A pair of mandibles and typically two pairs of jaws make up the buccal parts. Almost all species have compound eyes.

The phylogenetic position of the myriapods is still under discussion. For a long time, they were classified with the insects as a uniramian taxon. The shared characters include the uniramous appendages and the presence of Malpighian tubules. Molecular phylogenies based on the 18S rRNA gene refute the existence of this group. The two characters are now therefore interpreted as convergences corresponding to an adaptation to the terrestrial environment shared by the myriapods and the insects. However, another vital phylogenetic problem has yet to be resolved: are the myriapods monophyletic? This taxon presently includes four sub-classes (diplopoda, pauropoda, chilopoda and symphyla) that may not be closely related.

Garden centipede
Lithobius forficatus
2.5 cm

African giant millipede
Graphidostreptus sp.
17 cm

Pauropus silvaticus
1 mm

Some unique derived features

- Tömösvary's organ: there are specialized sensory organs of unknown function called Tömösvary's organs. These organs are composed of clusters of epidermal sensory cells controlled by the protocerebrum.

- The ancestral bifurcating appendages have lost a branch (fig. 1: ventral view of the symphylid *Scutigerella immaculata* showing the antennae and the uniramous legs; *aa*: antenna, *l*: leg). This is why for a long time the myriapods were classified in the unirama.

- Tracheae: gas exchange occurs via a tracheal system. This character is a convergence with the hexapods.

- Two Malpighian tubules assure excretion.

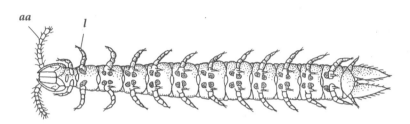

Figure 1

Ecology

The diplopods are almost exclusively vegetarians, living in the humus or under tree bark. They can cause much agricultural damage (beets, potatoes, strawberries). The chilopods are predators. The appendages of the first trunk segment have been transformed into powerful hooks containing venom, the forcipules. As these organisms prefer humid, dark environments, we find them under rocks, in mosses, leaf litter, etc. Certain diplopods have glands that secrete toxic, repulsive substances. Development is direct and growth continues after hatching. The growth zone lies just anterior to the telson.

Number of species: 12,050 (8,000 diplopods, 400 pauropods, 150 symphylids, 3,500 chilopods).
Oldest known fossils: many unnamed species from the upper Silurian are without doubt myriapods, particularly a diplopod from Scotland (410 MYA). There may also be *Cambropodus gracilis* from the middle Cambrian from Utah (United States, 515 MYA), although it is not certain that this species is a true myriapod.
Current distribution: worldwide.

Examples

- DIPLOPODA: millipede: *Lulus scandinavius*; pill millipede: *Glomeris marginata*; *Graphidostreptus* sp.: *Polydesmus* sp.
- CHILOPODA: centipede: *Geophilus carpophagus*; *G. electricus*; stone centipede: *Lithobius forficatus*; Megarian banded centipede: *Scolopendra cingulata*; *S. subspinipes*; house centipede: *Scutigera coleoptrata*.
- PAUROPODA: *Decapauropus cuenoti*; *Pauropus silvaticus*.
- SYMPHYLA: *Scutigerella immaculata*.

Pancrustacea

A few representatives

Hutchinosoniella macracantha
2.5 mm
node 11

Goose barnacle
Lepas anatifera
5 cm
node 12

Edible crab
Cancer pagurus
25 cm
node 14

Ctenosta bastardi
4 cm
node 15

Some unique derived features

- They have nauplius-type larvae. This larva is not segmented and has three pairs of swimming appendages, respectively from front to back, the antennules, the antennae and the mandibles (fig. 1: nauplius larva of a cirriped crustacean *Balanus* sp. (300 μm); *a1:* antennule, *a2:* antenna, *ma:* mandible).

- The 18S rRNA gene sequences support the monophyly of this taxon.

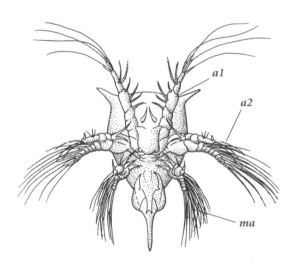

Figure 1

Number of species: 868,914.
Oldest known fossils: six ostracod families were already present at the start of the Cambrian period (540 MYA). We can cite: *Bradoria scrutator* and *Hipponicharion eos* from Canada, *Cambria sibirica* from east Siberia, *Kunmingella maxima* from Yunnan, China, *Uskarella prisca* from Kazakhstan, *Hesslandona* sp. and *Strenuella* sp. from England. Some phyllocarids date from the same period, like *Hymenocaris* sp. from from Europe, North America, Australia and New Zealand (540 MYA) or *Perspicaris* sp. from Yunnan (China, 540 MYA).
Current distribution: worldwide.

Examples

- REMIPEDA: *Speleonectes lucayensis.*
- CEPHALOCARIDA: *Hutchinsoniella macracantha.*
- MAXILLOPODA: *Amphiascus* sp.; *Tigriopus fulvus;* goose barnacle: *Lepas anatifera.*
- BRANCHIOPODA: brine shrimp: *Artemia salina; Triops cancriformis; Daphnia pulex.*
- MALACOSTRACA: *Nebalia bipes;* common prawn: *Palaemon serratus;* lobster: *Homarus gammarus;* hermit crab: *Eupagurus bernhardus;* edible crab: *Cancer pagurus;* common sea slater: *Ligia oceanica; Oniscus asellus;* aquatic sowbug: *Idotea basteri;* amphipod: *Gammarus pulex; Orchestia gammarellus;* sand hopper: *Talitrus saltator.*
- HEXAPODA: *Ctenosta bastardi;* swallowtail: *Papilio machaon;* migratory locust: *Locusta migratoria; Drosophila melanogaster; Sarcophaga* sp.

Remipedia

General Description

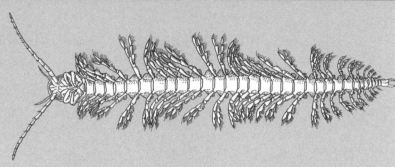

The remipedes were only recently discovered by J. Yager in the cavernous, subterranean waters of the Bahamas. They are small, transparent animals that live in the water column; observing them in these obscure waters was therefore difficult. As all of their post-cephalic metameres are identical, they could be considered to represent the most basal branch of the pancrustacean tree. This is further supported

Lasionectes entrichoma
3 cm

by the fact that the antennules and antennae are branched. However, for the moment, there is no solid proof for this hypothesis.

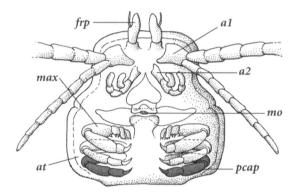

Figure 1

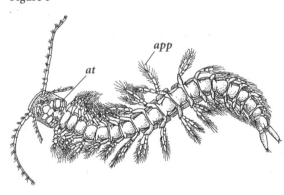

Figure 2

Number of species: 11.
Oldest known fossils: *Tesnusocaris goldichi*, from the lower Carboniferous period from Texas (United States, 340 MYA), belongs to the enantiopoda, a group of fossil remipedes.
Current distribution: underwater caves of the Gulf of Mexico (Cuba, Bahamas, Yucatan) and the Canary Islands.

Some unique derived features

■ The buccal appendages are found in a ventral depression of the cephalon called the atrium (fig. 1: ventral view of the anterior end of *Speleonectes lucayensis* showing the atrium; *a1:* antennule, *a2:* antenna, *pcap:* post-cephalic appendage, *at:* atrium, *mo:* mouth, *frp:* frontal process).

■ The first appendage, which from an anatomical point of view should be post-cephalic, is a "maxilliped" (*max*) integrated into the cephalon (fig. 1).

■ The trunk appendages are all laterally oriented (fig. 2: ventral view of *Speleonectes* sp. showing the lateral appendages; *app:* appendage).

■ The gonopores have unusual positions; they are found on the seventh post-cephalic segment of females and on the fourteenth segment of males.

Ecology

The remipedes are strictly cave-dwelling organisms, found in waters that lead to the sea. We find them in the halocline, that is, in the water mass that results from the contact between fresh and sea water. We know little of their biology, and particularly that of their reproduction and development. Their feeding method is debated; some believe they are carnivorous animals, whereas others feel they are microphagous filter-feeders. The presence of numerous setae on the buccal region tends to support the latter hypothesis.

Examples

Lasionectes entrichoma; Speleonectes lucayensis; S. ondinae.

Cephalocarida

The cephalocarids are small, marine animals that were only recently discovered. In 1955, H. L. Sanders discovered this group in the shallow waters of the Northeastern United States. For the moment, only *Hutchinsoniella macracantha* has been studied alive. The body is elongated and subdivided into head, thorax and abdomen. A shield surrounds the head and forms a lateral folded edge. The nine thoracic segments have lateral folds of the same size. The abdomen, made up of eleven segments, lacks appendages. Their morphology, anatomy, biology and development have led certain authors to suggest

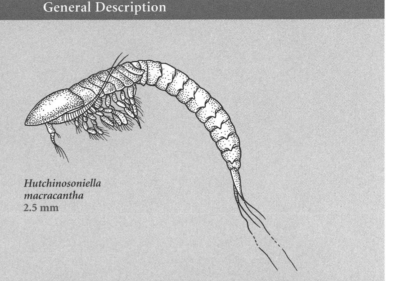

Hutchinosoniella macracantha
2.5 mm

that the cephalocarids are the most basal group of the euarthropoda. Nothing really confirms this hypothesis, despite a strong resemblance between the members of this group and some crustaceans from the upper Cambrian from Sweden.

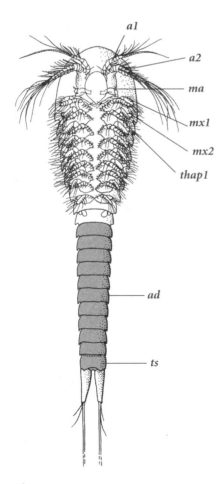

Figure 1

Some unique derived features

- There is a loss of the "abdominal" appendages.
- Abdominal segments: the number of segments is reduced to eleven (fig. 1: ventral view of *H. macracantha* showing the eleven abdominal segments, including the telson; *a1*: antennule, *a2*: antenna, *ad*: abdomen, *ma*: mandible, *mx1*: first maxilla, *mx2*: second maxilla, *thap1*: first thoracic appendage, *ts*: telson).

Ecology

The cephalocarids are part of the meiofauna, that is, they live in the superficial layers of the bottom sediments (gravel or silt). Here, they filter-feed on small particles from the sediment. The development is progressive; in *H. macracantha*, the first stage is a metanauplius and is followed by 18 other stages.

Number of species: 9.
Oldest known fossils: *Lepidocaris rhyniensis* from the lower Devonian from Scotland (395 MYA).
Current distribution: worldwide, except on the European coasts.

Examples

Hutchinsoniella macracantha; *Lightiella incisa*; *Sandersiella acuminata*.

315

Maxillopoda

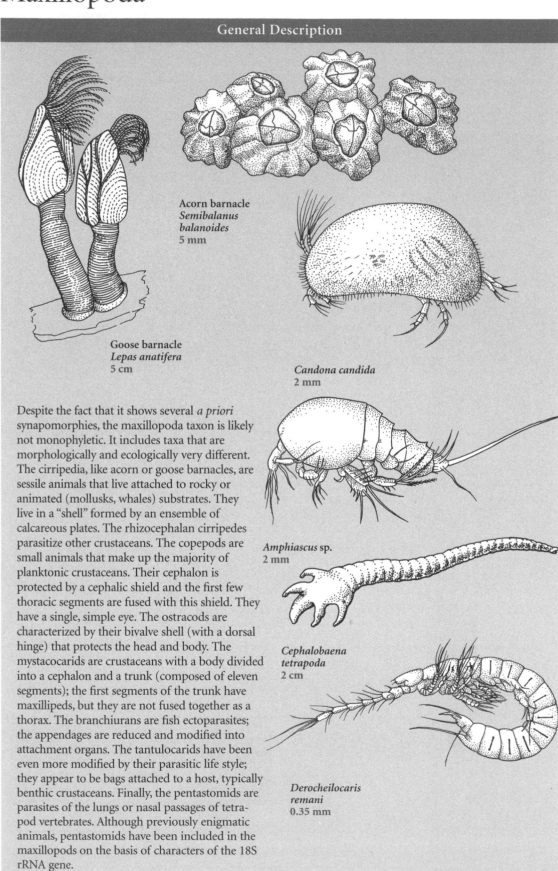

Acorn barnacle
*Semibalanus
balanoides*
5 mm

Goose barnacle
Lepas anatifera
5 cm

Candona candida
2 mm

Amphiascus sp.
2 mm

*Cephalobaena
tetrapoda*
2 cm

*Derocheilocaris
remani*
0.35 mm

Despite the fact that it shows several *a priori* synapomorphies, the maxillopoda taxon is likely not monophyletic. It includes taxa that are morphologically and ecologically very different. The cirripedia, like acorn or goose barnacles, are sessile animals that live attached to rocky or animated (mollusks, whales) substrates. They live in a "shell" formed by an ensemble of calcareous plates. The rhizocephalan cirripedes parasitize other crustaceans. The copepods are small animals that make up the majority of planktonic crustaceans. Their cephalon is protected by a cephalic shield and the first few thoracic segments are fused with this shield. They have a single, simple eye. The ostracods are characterized by their bivalve shell (with a dorsal hinge) that protects the head and body. The mystacocarids are crustaceans with a body divided into a cephalon and a trunk (composed of eleven segments); the first segments of the trunk have maxillipeds, but they are not fused together as a thorax. The branchiurans are fish ectoparasites; the appendages are reduced and modified into attachment organs. The tantulocarids have been even more modified by their parasitic life style; they appear to be bags attached to a host, typically benthic crustaceans. Finally, the pentastomids are parasites of the lungs or nasal passages of tetrapod vertebrates. Although previously enigmatic animals, pentastomids have been included in the maxillopods on the basis of characters of the 18S rRNA gene.

Some unique derived features

- The abdominal appendages have been lost secondarily.

- The thorax has a maximum of six segments.

- The abdomen has a maximum of four segments.

- The carapace is reduced.

- The genital appendages are carried on the first abdominal somite and are associated with the male gonopore.

- The eye of the nauplius larva has some particular characteristics.

Ecology

The ecology of the maxillopods is diverse. The copepods are planktonic animals for the most part. Given their position in the marine food chain, they are of considerable ecological importance. The ostracods may also be planktonic, but many species of this group are benthic animals. We can find them as readily in fresh water, mosses, or humus as at great ocean depths (up to 7000 m). The cirripede crustaceans are generally filter-feeders; they are widespread on rocky substrates of the intertidal zone.

They are very interesting animals from a morphological perspective because they have lost their abdomen secondarily and attach themselves to a substrate by their head. Historically, it was Charles Darwin that conducted the first modern revision of this group. The mystacocarids are marine animals of the meiofauna, living in the sands of the littoral zone. Many species of the maxillopoda taxon (cirripedes, copepods, branchiurans, pentastomids) are parasites.

Number of species: 15,214 (ostracoda: 5,650; mystacocarida: 10; copepoda: 8,405; branchiura: 150; cirripedia: 900; tantulocarida: 4; pentastomida: 95).

Oldest known fossils: the oldest maxillopods are ostracods; six families were already present at the start of the Cambrian period (540 MYA). We can cite: *Bradoria scrutator* and *Hipponicharion eos* from Canada, *Cambria sibirica* from east Siberia, *Kunmingella maxima* from Yunnan, China, *Uskarella prisca* from Kazakhstan, *Hesslandona* sp. and *Strenuella* sp. from England. The oldest copepod is *Kabatarina pattersoni* from the lower Cretaceous from Brazil (110 MYA). The oldest cirripede is *Priscansermarinus barnetti* from the middle Cambrian from British Columbia, Canada (520 MYA). The oldest pentastomid is an unnamed marine form from the start of the Ordovician from Sweden (490 MYA).

Current distribution: worldwide.

Examples

- COPEPODA: *Amphiascus* sp.; *Canthocamptus* sp.; *Cyclops* sp.; *Tigriopus fulvus.*

- CIRRIPEDIA: goose barnacle: *Lepas anatifera*; acorn barnacle: *Semibalanus balanoides*; Poli's stellate barnacle: *Chthamalus stellatus*; *Elminius modestus*; parasitic barnacle (parasite of the green crab *Carcinus maenas*): *Sacculina carcini.*

- OSTRACODA: *Candona candida; Cypris* sp.; *Cythereis* sp.; *Conchoecia* sp.

- MYSTACOCARIDA: *Derocheilocaris remanei.*

- TANTULOCARIDA: *Basipodella atlantica.*

- BRANCHIURA: *Argulus foliaceus.*

- PENTASTOMIDA: *Cephalobaena tetrapoda.*

Branchiopoda

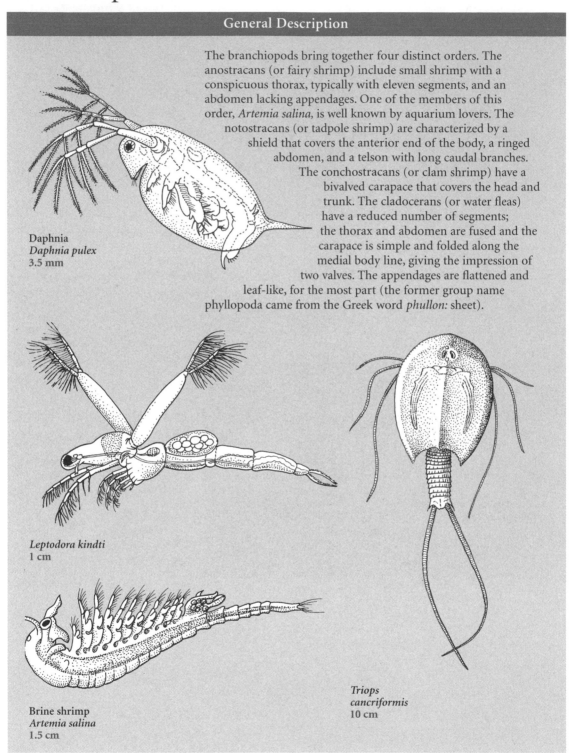

General Description

The branchiopods bring together four distinct orders. The anostracans (or fairy shrimp) include small shrimp with a conspicuous thorax, typically with eleven segments, and an abdomen lacking appendages. One of the members of this order, *Artemia salina*, is well known by aquarium lovers. The notostracans (or tadpole shrimp) are characterized by a shield that covers the anterior end of the body, a ringed abdomen, and a telson with long caudal branches. The conchostracans (or clam shrimp) have a bivalved carapace that covers the head and trunk. The cladocerans (or water fleas) have a reduced number of segments; the thorax and abdomen are fused and the carapace is simple and folded along the medial body line, giving the impression of two valves. The appendages are flattened and leaf-like, for the most part (the former group name phyllopoda came from the Greek word *phullon*: sheet).

Daphnia
Daphnia pulex
3.5 mm

Leptodora kindti
1 cm

Brine shrimp
Artemia salina
1.5 cm

Triops cancriformis
10 cm

Some unique derived features

- The abdominal appendages have been lost secondarily.
- The maxillae are reduced or absent. There are no maxillipeds.
- The genital appendages are found on the first abdominal segment and are associated with the male gonopores.

318

Ecology

The branchiopods are mostly freshwater animals adapted to life in temporary environments. They can tolerate a large range of salinities and often have resistant eggs that enable them to conquer these ephemeral habitats. Curiously, the anostracans swim ventral side up. They are an important food resource for birds. The notostracans, although represented by few species, have a widespread distribution. We find them in temporary environments like rice paddies. The conchostracans also have a worldwide distribution (except Antarctica). Most species are benthic. The cladocerans include a few marine species. Many species are benthic; planktonic species swim via the movements of their antennae.

Number of species: 934 (anostraca: 275; notostraca: 9; conchostraca: 200; cladocera: 450).
Oldest known fossils: the conchostracans *Cyzicus* sp. and *Asmussia membranacea,* from the lower Devonian, were widespread throughout the world (408 MYA).
Current distribution: worldwide.

Examples

- ANOSTRACA: brine shrimp: *Artemia salina; Linderiella massaliensis.*
- NOTOSTRACA: *Triops cancriformis; Lepidurus apus.*
- CONCHOSTRACA: *Cyzicus bucheti; Lynceus gracilicornis.*
- CLADOCERA: daphnia: *Daphnia magna; D. pulex; Leptodora kindti.*

Malacostraca

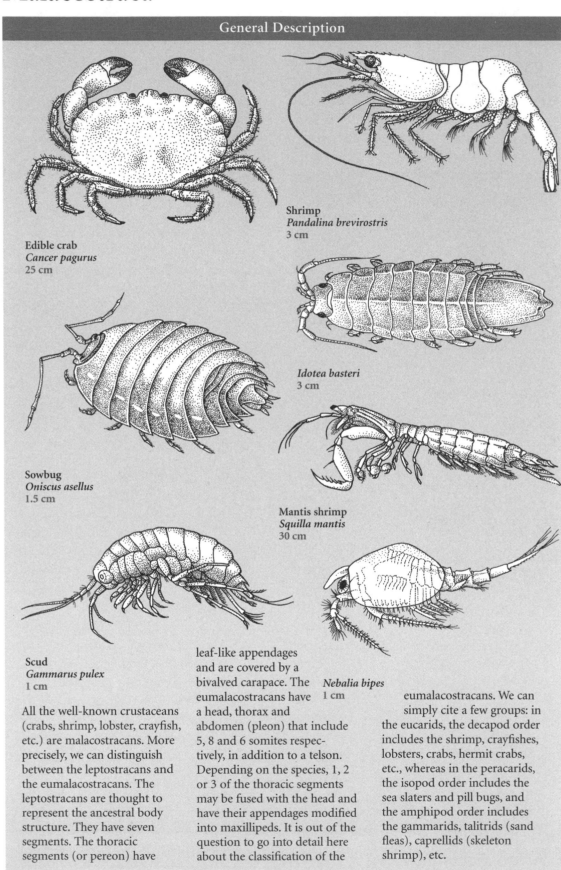

Edible crab
Cancer pagurus
25 cm

Shrimp
Pandalina brevirostris
3 cm

Idotea basteri
3 cm

Sowbug
Oniscus asellus
1.5 cm

Mantis shrimp
Squilla mantis
30 cm

Scud
Gammarus pulex
1 cm

Nebalia bipes
1 cm

All the well-known crustaceans (crabs, shrimp, lobster, crayfish, etc.) are malacostracans. More precisely, we can distinguish between the leptostracans and the eumalacostracans. The leptostracans are thought to represent the ancestral body structure. They have seven segments. The thoracic segments (or pereon) have leaf-like appendages and are covered by a bivalved carapace. The eumalacostracans have a head, thorax and abdomen (pleon) that include 5, 8 and 6 somites respectively, in addition to a telson. Depending on the species, 1, 2 or 3 of the thoracic segments may be fused with the head and have their appendages modified into maxillipeds. It is out of the question to go into detail here about the classification of the eumalacostracans. We can simply cite a few groups: in the eucarids, the decapod order includes the shrimp, crayfishes, lobsters, crabs, hermit crabs, etc., whereas in the peracarids, the isopod order includes the sea slaters and pill bugs, and the amphipod order includes the gammarids, talitrids (sand fleas), caprellids (skeleton shrimp), etc.

Some unique derived features

- Cephalo-thorax: the anterior end of the body is structured into a cephalo-thorax.

- The cephalon has a maximum of three pairs of maxillipeds.

- The thorax has eight segments, and the abdomen 6 or 7 segments (not including the telson).

- The abdomen has five pairs of branched pleopods along with a pair of branched uropods.

- The male gonopores are found on the eighth thoracic segment, whereas the female gonopores are on the sixth thoracic segment.

- The visual system has a double chiasma (fig. 1: a. the hexapod and malacostracan system, with a double optic chiasma between the different ganglia of the gray matter, the lamina, the medulla and the lobula, b. the system for all other crustaceans, without the lobula or an optic chiasma; *br:* brain, *lam:* lamina, *lob:* lobula, *med:* medulla, *oma:* ommatidium, *x1:* first chiasma, *x2:* second chiasma). If this double chiasma of the malacostracans is homologous to that of the hexapods, the hexapods and the malacostracans should be united into a single taxon.

- The eye of the nauplius larva has characteristics unique to the malacostracans.

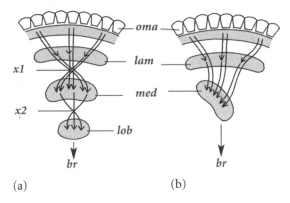

(a)   (b)

Figure 1

Ecology

The leptostracans live primarily in low-oxygen environments where they feed on suspended organic debris. The eumalacostracans have conquered all possible aquatic habitats: swimming animals like shrimps, benthic animals like crabs, freshwater animals such as crayfish, interstitial fauna like the tanaids. The pill bugs are terrestrial crustaceans. Finally, many isopods are parasitic. The malacostracans have a major ecological role. For example, coastal crabs are very efficient detritivores and scavengers. Without them, our beaches would have a disagreeable odor. We should note that many malacostracans, especially crabs and shrimps, have been found in deep-sea hydrothermal vents.

Number of species: 22,671 (20 leptostraca, 350 hoplocarida, 11,150 peracarida and, in the eucarida, 90 euphausiacea and the remaining decapoda—the most diverse order).

Oldest known fossils: the oldest malacostracans are leptostracans like *Hymenocaris* sp. from the lower Cambrian from Europe, North America, Australia and New Zealand (540 MYA) or *Perspicaris* sp. from the lower Cambrian from Yunnan, China (540 MYA).

Current distribution: worldwide.

Examples

- LEPTOSTRACA: *Nebalia bipes.*
- EUMALACOSTRACA:—HOPLOCARIDA: mantis shrimp: *Squilla mantis.*—EUCARIDA, EUPHAUSIACEA: krill: *Euphausia superba.*—EUCARIDA, DECAPODA: white shrimp: *Penaeus schmitti;* brown shrimp: *Crangon crangon;* common prawn: *Palaemon serratus; P. elegans; Pandalina brevirostris,* Norway lobster: *Nephrops norvegicus;* common lobster: *Homarus gammarus;* hermit crab: *Eupagurus bernhardus;* edible crab: *Cancer pagurus;* green crab: *Carcinus maenas;* Risso's crab: *Xantho pilipes,* spider crab: *Maja squinado.*—PERACARIDA, ISOPODA: *Armadillidium* sp.; sea slater: *Ligia oceanica; Oniscus asellus;* aquatic sowbug: *Idotea basteri;* sowbug: *Asellus* sp.; bopyrid (decapod parasite): *Bopyrus* sp.—PERACARIDA, AMPHIPODA: *Gammarus locusta; G. pulex; Orchestia gammarellus;* sand hopper: *Talitrus saltator;* skeleton shrimp: *Caprella linearis.*

➡ *see appendix 4, p. 523*

Hexapoda

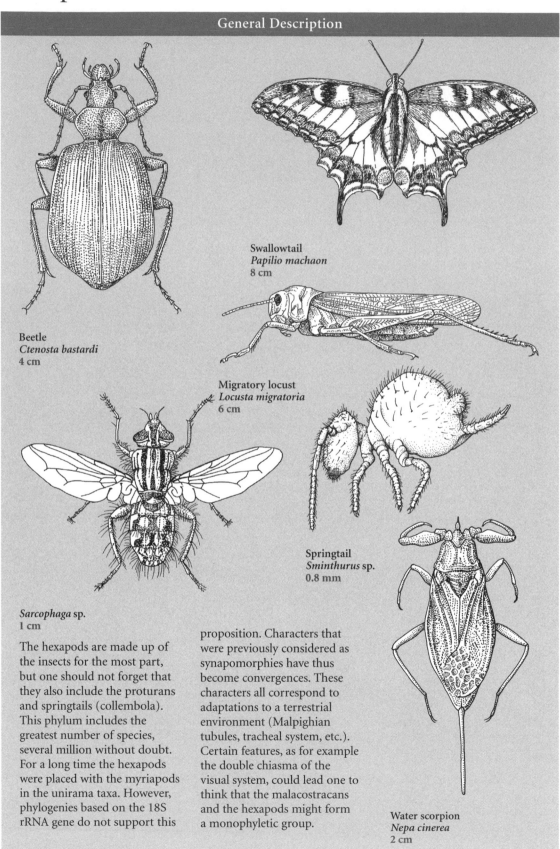

General Description

Beetle
Ctenosta bastardi
4 cm

Swallowtail
Papilio machaon
8 cm

Migratory locust
Locusta migratoria
6 cm

Springtail
Sminthurus sp.
0.8 mm

Sarcophaga sp.
1 cm

Water scorpion
Nepa cinerea
2 cm

The hexapods are made up of the insects for the most part, but one should not forget that they also include the proturans and springtails (collembola). This phylum includes the greatest number of species, several million without doubt. For a long time the hexapods were placed with the myriapods in the unirama taxa. However, phylogenies based on the 18S rRNA gene do not support this proposition. Characters that were previously considered as synapomorphies have thus become convergences. These characters all correspond to adaptations to a terrestrial environment (Malpighian tubules, tracheal system, etc.). Certain features, as for example the double chiasma of the visual system, could lead one to think that the malacostracans and the hexapods might form a monophyletic group.

Some unique derived features

- Labium: The second pair of jaws are fused into a lower lip, or labium (fig. 1: lower lip of an orthopteran; *glo:* glossae, *men:* mentum, *pgl:* paraglossae, *lbp:* labial palps, *sb:* submentum). There is a structural convergence of the labium with the symphyla, a sub-class of myriapods.

- The second pair of antennae has been lost.

- Thorax: in the post-cephalic region, there is a distinct thorax made up of three segments and followed by an abdomen.

- The organisms have three pairs of legs, hence the name hexapod.

- Abdominal segments: there is a maximum of 11.

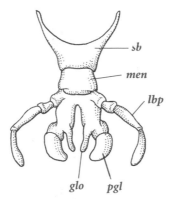

Figure 1

- There has been a loss of the abdominal appendages. However, some persist, such as the styli of the diplurans, the furca of the collembolans, and the genitalia.

- Trachea: gas exchange takes place via a tracheal system. This is a convergence with the myriapods.

- Malpighian tubules: Malpighian tubules, derived from the ectoderm, are used for excretion. This is a convergence with the myriapods.

- The visual pathway has a double chiasma (see 10.14, fig. 1). This is a convergence with the malacostracans. However, if the double chiasmas are in fact homologous, they would be a synapomorphy uniting the malacostraca and the hexapoda.

Ecology

The hexapods have conquered all terrestrial biotopes. Some species have returned secondarily to the water, like the water beetles, the water scorpions, and the water boatmen. Their biodiversity is prodigious and the number of species defies our imagination. For example, there are more than 9,000 species of ants alone. The development of the hexapods includes a series of molts. We distinguish three molting patterns, depending on the type of transformation that occurs. In the ametabola, the animal develops without changing body form. However, the majority of insects undergo a metamorphosis. The hemimetabola have an "incomplete" metamorphosis, that is, the young strongly resemble the adult (as in the cricket, for example). On the other hand, the larvae of the holometabola undergo a complete metamorphosis; the larvae differ from the adult forms (the imagoes) both morphologically and ecologically. For example, a maggot does not at all resemble a fly. It is for this reason that we can find aquatic larvae and terrestrial imagoes, free-living larvae and parasitic imagoes, etc. It is futile to describe the ecological impact of the insects. Their medical importance as parasites (fleas, lice, etc.) or disease vectors (mosquitoes, blackflies, tsetse flies, bugs, etc.) is enormous; their agricultural role is equally important both as pollinators and pests.

Number of species: 830,075 by an approximate order-by-order count (protura: 100; collembola: 2,000; diplura: 100; thysanura and archaeognatha: 700; ephemeroptera: 2,100; odonata: 5,500; blattaria: 3,700; mantodea: 1,800; isoptera: 2,000; plecoptera: 1,600; orthoptera: 20,000; dermaptera: 1,100; phasmida: 2,500; zoraptera: 25; psocoptera: 2,600; hemiptera: 35,000; thysanoptera: 4,000; anoplura: 500; mallophaga: 2,800; homoptera: 33,000; coleoptera: 300,000; neuroptera: 4,700; hymenoptera: 125,000; mecoptera: 500; siphonaptera: 1,750; diptera: 150,000; trichoptera: 7,000; lepidoptera: 120,000). These numbers correspond to the approximate number of described species. The actual number is certainly much greater (some estimate up to 30 million insect species in the tropical forests alone). **Oldest known fossils:** the oldest hexapod fossil is a springtail *Rhyniella praecursor* from the lower Devonian from Scotland (395 MYA). **Current distribution:** worldwide.

Examples

- COLLEMBOLA: *Anurida maritima.*
- INSECTA: seven-spotted ladybeetle: *Coccinella septempunctata;* cockchafer: *Melolontha melolontha;* swallowtail: *Papilio machaon;* migratory locust: *Locusta migratoria;* bee: *Apis mellifera; Drosophila melanogaster; Sarcophaga* sp.; body louse: *Pediculus humanus.*

Chapter 11

Deuterostomia

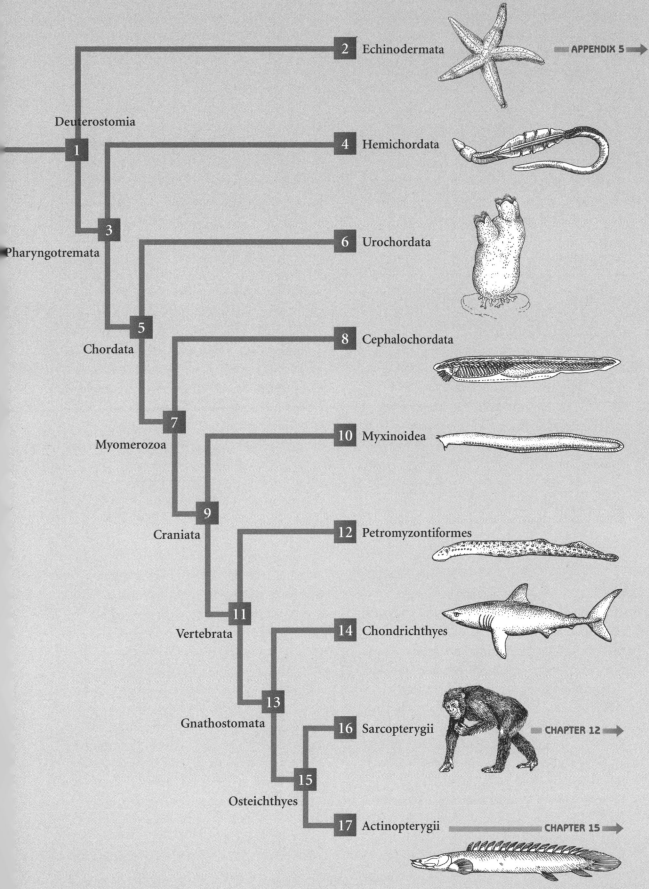

Deuterostomia

1

Pharyngotremata

3

Chordata

5

Myomerozoa

7

Craniata

9

Vertebrata

11

Gnathostomata

13

Osteichthyes

15

2 Echinodermata — APPENDIX 5 ➡

4 Hemichordata

6 Urochordata

8 Cephalochordata

10 Myxinoidea

12 Petromyzontiformes

14 Chondrichthyes

16 Sarcopterygii — CHAPTER 12 ➡

17 Actinopterygii — CHAPTER 15 ➡

The deuterostomian phylogeny, as it is presented here, contains two main areas of controversy: the phylogenetic position and monophyly of the hemichordates (enteropneusta and pterobranchia), and the position of the myxinoids (hagfish) and petromyzontids (lampreys) in relation to the gnathostomes (jawed deuterostomes).

The monophyly of the hemichordates has been called into question. Indeed, the enteropneusts share several features with the chordates that are not shared by the pterobranchs: the organization of the gill slits, the dorsal nerve cord and several structures of the digestive tube. Jefferies has also suggested that the hemichordates are paraphyletic, but with the pterobranchs and then the enteropneusts as the most basal lineages of the bilaterally symmetrical deuterostomes. Indeed, he has proposed a deuterostomian phylogeny that is completely different from the one shown here. It is based on the interpretation of strange calcichordate fossils that occurred from the Cambrian until the Devonian. These animals were flattened and completely asymmetric, as though they had been laid on their right side and covered in calcareous plates similar to those of the echinoderms. Jefferies brings the calcichordates and all other deuterostomes that, according to him, correspond to a sort of hemichordate that has lain down on its side, into the *Dexiothetica* clade. Within this fundamentally asymmetrical clade of deuterostomes, the echinoderms are the sister group of the chordates and the cephalochordates are the sister group of a clade that includes the urochordates and the craniates. Each *Dexiothetica* group would have a calcichordate origin, that is, would have an asymmetric, bottom-lying organism as an ancestor. Adult bilateral symmetry would therefore correspond to a secondary arrangement. A more recent reanalysis of the calcichordate anatomy has suggested that these organisms are actually bizarre echinoderms. This interpretation is further supported by the fact that Jefferies's theory would imply a number of significant secondary losses, notably that of the calcareous plates.

Within the craniates, the phylogenetic position of the lampreys has led to significant ink flow. This is partly because the consequences of the polemic go beyond just the lampreys to affirming phylogenetic systematics in general. The hagfish and lampreys were traditionally placed within the "agnatha," a paraphyletic group. However, the analysis of extant agnathan characters using Hennig's method has revealed more than a dozen synapomorphies common to lampreys and vertebrates. The paraphyly of this group was even more flagrant when fossils were included in the analysis. However, in order to protect the extant agnathan group, the "cyclostomes," some conservative authors have suggested that several cartilaginous structures are homologies: the lingual piston, the dentary plates, the gill pockets, and the single naso-hypophyseal canal. They even go as far as to say that the hagfishes must have possessed the common characters of the lampreys and the gnathostomes, but lost them secondarily as a result of their ectoparasitic life style. This argument is not based on any embryological observations that might attest to the presumed regression, because we know next to nothing about the development of the hagfishes. In terms of molecular phylogenies, no reliable data enable us to treat this question. Indeed, the closest extant outgroups are the cephalochordates and the urochordates; both of these are already too distant from the craniates in terms of time since divergence. Because of long-branch attraction, these different available outgroup combinations lead to instability in the position of the basal craniate groups. The position of the lampreys must therefore be addressed using separate analyses of numerous independent genes.

➡ *see appendix 5, p. 524*

Echinodermata

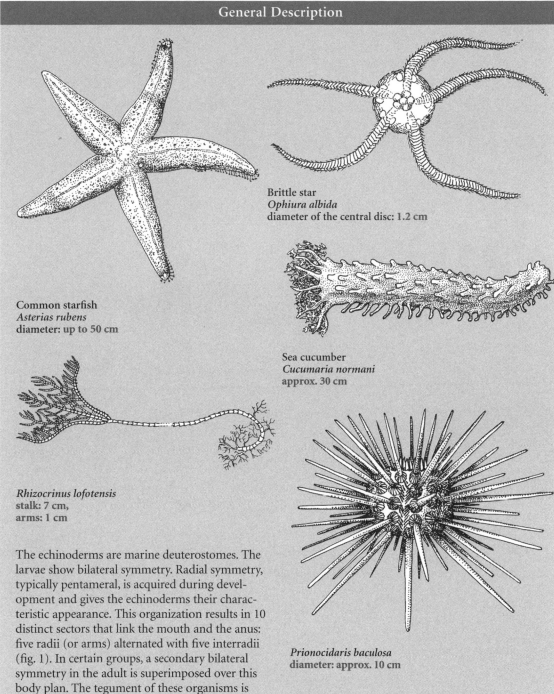

General Description

Brittle star
Ophiura albida
diameter of the central disc: 1.2 cm

Common starfish
Asterias rubens
diameter: up to 50 cm

Sea cucumber
Cucumaria normani
approx. 30 cm

Rhizocrinus lofotensis
stalk: 7 cm,
arms: 1 cm

Prionocidaris baculosa
diameter: approx. 10 cm

The echinoderms are marine deuterostomes. The larvae show bilateral symmetry. Radial symmetry, typically pentameral, is acquired during development and gives the echinoderms their characteristic appearance. This organization results in 10 distinct sectors that link the mouth and the anus: five radii (or arms) alternated with five interradii (fig. 1). In certain groups, a secondary bilateral symmetry in the adult is superimposed over this body plan. The tegument of these organisms is rough or spiny because it is covered in sharp spines or reinforced by internal discontinuous (spicules) or continuous (plates) skeletal formations of dermal origin. The echinoderms are divided into two subphyla. The pelmatozoans unite sessile organisms attached to substrates by a stalk in the larval state and typically in the adult state as well. The mouth is on the same side of the animal as the anus, at the center of a circle of tentacular arms. This group includes the extant crinoids and

several fossil groups. The other subphylum, the eleutherozoans, appeared at the end of the Cambrian (500 MYA) and contain the free-living echinoderms; in this group, the anus is on the opposite side of the body to the mouth. It includes four classes: the echinoidea (sea urchins), the asteroidea (starfishes), the ophiuroidea (brittle stars in which the anus is secondarily blocked) and the holothuroidea (sea cucumbers).

Some unique derived features

- Skeleton: the organic and crystalline microstructure of the skeleton is very characteristic. The internal dermal skeleton (fig. 1: sea urchin shell; *A* to *E*: radius, *I*: interradius) is composed of units of calcite (calcium carbonate) mono-crystals.

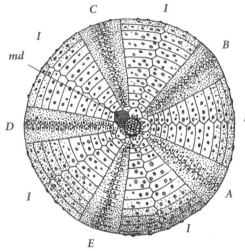

Figure 1

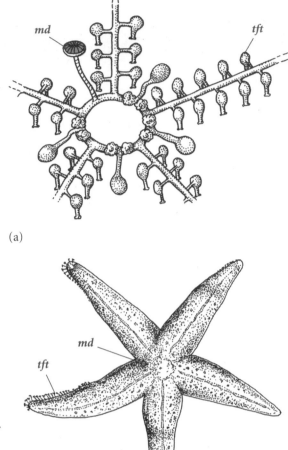

(a)

(b)

Figure 2

- Within the body, there is a system of seawater-filled chambers, the water vascular system (fig. 2a). This system communicates with the external environment via one or several water pores across a "perforated plate" or madreporite (*md*, fig. 2). Using differential water pressure, this system enables the animal to control its tube feet (*tft*, fig. 2a and 2b). These feet often end with suckers that assist with movement.

Ecology

The echinoderms are present in all seas, including the deepest and coldest waters. They are most often benthic animals. Feeding modes vary depending on the class. They can be suspension feeders, filtering suspended particles from the seawater (sea lilies and feather stars, sea cucumbers), burrowing animals that feed on organic particles found in the sediments (certain sea cucumbers, sand dollars), scavengers (brittle stars and starfishes), active predators (starfishes, certain sea urchins) or rock grazers (urchins). Typically the sexes are separate, although there are a few hermaphroditic species. Fertilization is external and takes place in the seawater. Most species have a free-swimming pelagic larva. The adult form develops from a bud on the left side of the larva.

> **Number of species:** 6,000.
> **Oldest known fossils:** *Helicoplacus* from Nevada, United States, dates from the start of the Cambrian (580 MYA).
> **Current distribution:** worldwide.

Examples

- ASTEROIDEA: common starfish: *Asterias rubens;* beaded sea star: *Astropecten aurantiacus.*
- OPHIUROIDEA: *Ophiothrix fragilis; Amphipholis squamata; Ophiura albida.*
- HOLOTHUROIDEA: northern sea cucumber: *Cucumaria frondosa; C. normani; Leptosynapta inhaerens.*
- ECHINOIDEA: *Sphaerechinus granularis;* European purple sea urchin: *Paracentrotus lividus;* green sea urchin: *Psammechinus miliaris; Prionocidaris baculosa;* purple heart urchin: *Spatangus purpureus.*
- CRINOIDEA: rosy feather star: *Antedon bifida; Rhizocrinus lofotensis.*

Pharyngotremata

A few representatives

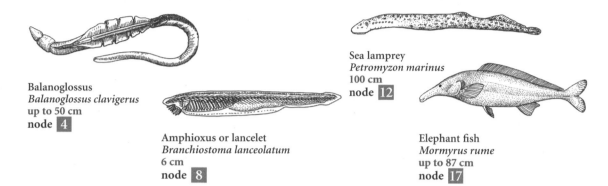

Balanoglossus
Balanoglossus clavigerus
up to 50 cm
node **4**

Amphioxus or lancelet
Branchiostoma lanceolatum
6 cm
node **8**

Sea lamprey
Petromyzon marinus
100 cm
node **12**

Elephant fish
Mormyrus rume
up to 87 cm
node **17**

Some unique derived features

- Ciliated pharyngeal (or branchial) slits open laterally. They result from the fusion of two of the three embryonic tissue layers, the endoderm and the ectoderm. These slits are shown in fig. 1 on a hemichordate and in fig. 2 on a chordate larva (*a:* anus, *at:* atrium, *atp:* atriopore, *co:* collar, *cc:* collar coelom, *pc:* proboscis coelom, *tc:* trunk coelom, *bs:* branchial slit, *p:* proboscis, *bp:* branchial pouch, *ph:* pharynx, *bpo:* branchial pore, *t:* trunk, *dt:* digestive tube, *nc:* nerve cord). The arrows indicate the direction of water flow.

- The pharyngeal slits are supported by a skeleton composed, at least primitively, of cartilaginous gill bars.

- The dorsal hollow nerve cord is formed by an invagination of the neuroectoderm.

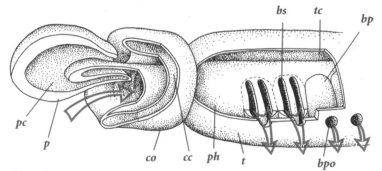

Figure 1

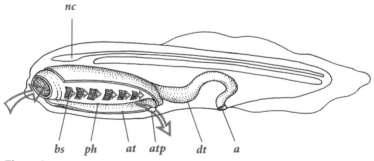

Figure 2

Number of species: 52,331.
Oldest known fossils: *Pikaia* is a cephalochordate from the lower Cambrian (530 MYA). Older still in the Cambrian (540 MYA), but more enigmatic, is an unnamed appendicularian (urochordata) from China.
Current distribution: worldwide.

Examples

- HEMICHORDATA: balanoglossus: *Balanoglossus clavigerus; Rhabdopleura normani.*
- CHORDATA: sea squirt: *Ascidia mentula;* salp: *Salpa maxima;* amphioxus or lancelet: *Branchiostoma lanceolatum;* crested newt: *Triturus cristatus;* human: *Homo sapiens.*

Hemichordata

General Description

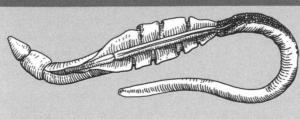

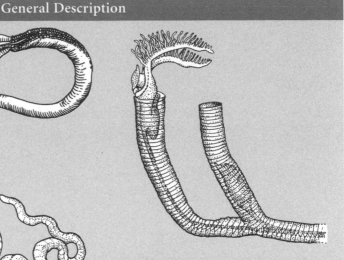

Balanoglossus
Balanoglossus clavigerus
up to 50 cm

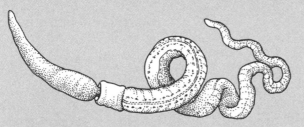

Saccoglossus cambrensis
approx. 15 cm

Rhabdopleura normani
0.03 cm without the stalk,
length of zooidal tube: 0.3 cm

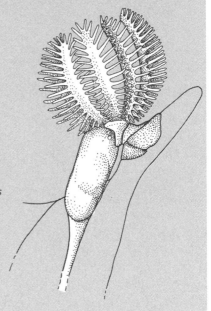

The hemichordates bring together two classes of worm-like, marine deuterostomes: the enteropneusta (acorn worms) and the pterobranchia (sea angels). Their bodies are soft, cylindrical, and divided into three parts: a conical anterior end (the proboscis or protosome), followed by a collar (mesosome) and then a long trunk (metasome). The proboscis is a muscular organ that enables the acorn worms to burrow into sand using contractile movements. The collar is a small cylinder that carries the ventral mouth and, in the pterobranchs, a pair of tentacles. A hard rod found within the collar is similar to a rudimentary notochord. At the anterior end of the trunk, the branchial slits open on each side, crossing the walls of the pharynx and trunk. This arrangement brings the hemichordates closer to the chordates within the pharyngotremata clade. The trunk makes up most of the body and contains the gonads. The enteropneusts look like large worms, ranging in size from 9 to 45 cm long. A Brazilian species, *Balanoglossus gigas*, measures more than 2.5 m. The pterobranchs are colonial, tube-dwelling hemichordates that carry a pair of tentacles on the collar used for capturing suspended organic particles from the seawater. The graptolites, similar to the pterobranchs, are a fossil group of colonial, marine hemichordates that were very abundant in the Silurian (415 MYA).

Cephalodiscus dodecalophus
Body length of one individual: 0.2 cm

Some unique derived features

■ Proboscis or glans: the anterior end of the animal is made up of a proboscis (*p*). In fact, the hemichordates are well characterized by the division of their bodies into three parts, as shown in fig. 1 in an enteropneust (1a, *Saccoglossus*) and a pterobranch (1b, *Cephalodiscus*): the proboscis or protosome is followed by a collar (*co*) or mesosome, and then by a long trunk (*t*, metasome).

Ecology

The pterobranchs are especially abundant in the seas of the Southern hemisphere. They live in deep water within tubes that they secrete. The enteropneusts typically live in shallow marine waters; they are found along all shorelines. Sixty percent of these species are found in warm seas. Some species live at great depths (*Grandiceps abyssicola*). Most of them dig U-shaped galleries in the sand or silt in the marine intertidal zone. They ingest sediment from which they extract organic particles. Their feces are found on the surface of the sand, in a small pile at the posterior opening of the gallery. The hemichordates move by contracting the proboscis. They all feed on organic particles that they find in the sediments (enteropneusta) or in suspension (pterobranchia). In both instances, the proboscis is used to trap the food. The vibratile cilia that cover the proboscis lead these particles toward the mouth. The sexes are separate. Fertilization is external and development can be direct or indirect. In the latter case, the larva is called a "tornaria" and strongly resembles that of the echinoderms.

Number of species: 85.
Oldest known fossils: *Rhabdotubus johanssoni* is a rhabdopleuran pterobranch from the middle Cambrian from Sweden (515 MYA). *Yunnanozoon lividum,* from Chengjiang, China (530 MYA), may be a hemichordate closely related to balanoglossus.
Current distribution: worldwide.

Examples

Balanoglossus: *Balanoglossus clavigerus; Saccoglossus cambrensis; Rhabdopleura normani; Cephalodiscus dodecalophus.*

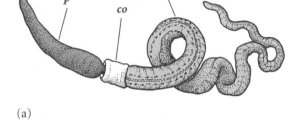

(a)

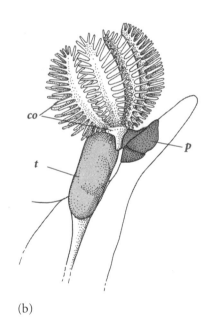

(b)

Figure 1

Chordata

A few representatives

Sea peach
Halocynthia papillosa
10 cm
node 6

Amphioxus or lancelet
Branchiostoma lanceolatum
6 cm
node 8

Garganey
Anas querquedula
approx. 37 cm
node 16

Elephant fish
Mormyrus rume
up to 87 cm
node 17

Some unique derived features

■ Larva (fig. 1): at least initially, the chordates have an elongated, swimming larva with bilateral symmetry and a tail (*t*). In the more recent chordates (amniotes), this larval form doesn't exist *per se* because of development within the amniote egg.

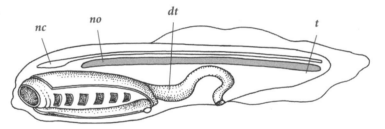

Figure 1

■ Notochord: a rigid dorsal rod, the notochord (*no*), supports the organism, at least in the larval form. This chord is not made of bone, but rather of a fibrous tissue.

■ Body plan (fig. 1): it is characteristic and includes (from dorsal to ventral): the dorsal hollow nerve chord (*nc*) that forms from the neuroectoderm, the notochord and the ventral digestive tube (*dt*) of endodermal origin.

Number of species: 52,246.
Oldest known fossils: *Pikaia* is a cephalochordate from the lower Cambrian (530 MYA). Older still in the Cambrian (540 MYA), but more enigmatic, is an unnamed appendicularian (urochordata) from China.
Current distribution: worldwide.

Examples

■ UROCHORDATA: *Oikopleura albicans;* salp: *Salpa maxima;* sea squirt: *Ascidia mentula;* star ascidian: *Botryllus schlosseri.*

■ MYOMEROZOA: amphioxus or lancelet: *Branchiostoma lanceolatum;* perch: *Perca fluviatilis;* crested newt: *Triturus cristatus;* cat: *Felis catus.*

Urochordata

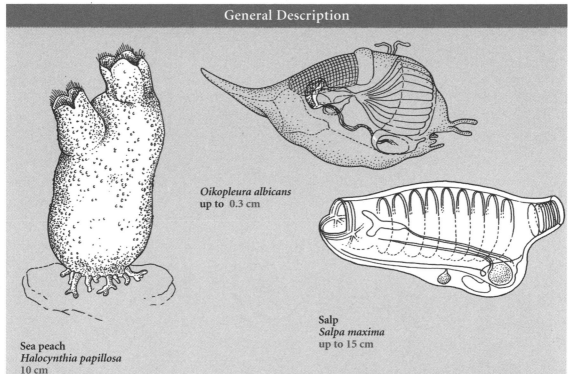

Oikopleura albicans
up to 0.3 cm

Salp
Salpa maxima
up to 15 cm

Sea peach
Halocynthia papillosa
10 cm

The urochordates (or tunicates) are deuterostomes that have a notochord only in the tail during the larval stage. There are three classes that at first sight do not resemble each other. The ascidians make up most of the group. They are urochordates with a sessile adult stage and a free-swimming larva. From the outside, the adult looks like a bag placed on a substrate with two pores or siphons, one inhalant (buccal) and one exhalant (atrial) (see arrows on fig. 1). Their bodies are covered in a tunic that is more or less thick depending on the species. The thaliaceans are transparent, free-swimming species shaped like small barrels. In this group, the inhalant siphon is found on the opposite side of the body to the exhalant siphon. The tail is found only in the larval stage. There are both solitary (*Salpa maxima*) and pelagic colonial species (*Thalia democratica*). Colonial species share a common tunic (*Pyrosoma*). The appendicularians may be transparent, neotenic urochordates with a long tail that remains in the adult state along with the notochord. These species are enveloped in a delicate jelly (that is not exactly a tunic). In certain species, this jelly covers the entire body. These animals of 3 mm in length swim among and filter-feed on marine plankton.

Some unique derived features

- A sessile adult stage (at least primitively). The metamorphosis of a swimming chordate larva into a sessile adult is shown in fig. 1 (*tu:* tunic, *bb:* branchial basket).

- The presence of a tunic (fig. 1) that contains a polysaccharide similar to cellulose, tunicin.

Ecology

In the urochordates, water enters by the inhalant siphon, crosses the pharyngeal walls and exits by the exhalant siphon (fig. 1). The pharynx is highly specialized for filtering seawater and forms a pharyngeal or branchial "basket" (*bb*, fig. 1) that secretes mucus. This basket traps suspended food particles that are then progressively moved toward the digestive tube by the abundant pharyngeal ciliature. The sessile ascidians can be colonial with a thick tunic and a single common exhalant siphon. They are simultaneous hermaphrodites. Fertilization can take place internally within a cavity (the spermatozoids enter by the inhalant siphon). In this case, tiny larvae with a tail and notochord leave from the exhalant siphon. Fertilization can also be external. After a pelagic existence, the "tadpole" settles head-first onto a substrate (fig. 1). The thaliaceans are free-swimming, planktonic animals. Muscular bands encircle their bodies. The contraction of these muscles expels water from the exhalant siphon, found on the opposite side of the body to the inhalant siphon. The created water current therefore serves for both the propulsion of the animal and for water filtration. The life cycle of the thaliaceans includes an alternation of sexual and asexual generations. The larvae have notochords. The ascidians and the thaliaceans both include colonial forms that can measure several meters in length. The appendicularians are planktonic, swimming organisms that filter-feed on small phytoplankton (nanoplankton). They live within a protective mucus coating that they produce themselves and that serves as a receptacle for filtering water.

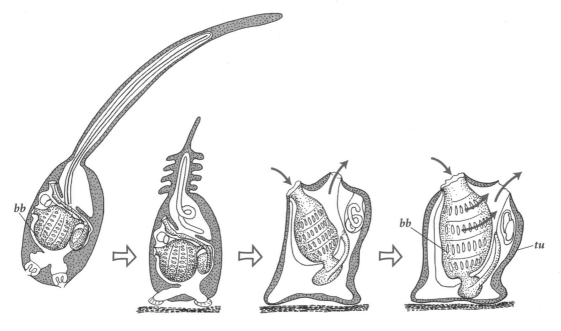

Figure 1

Number of species: 1,300.
Oldest known fossils: *Paleooikopleura* is an appendicularian from the lower Cambrian from China (530 MYA). Older still in the Cambrian (540 MYA), but more enigmatic, is an unnamed appendicularian from China.
Current distribution: worldwide.

Examples

■ THALIACEA: salp: *Salpa maxima;* colonial salp: *S. democratica; Thalia democratica;* atlantic pyrosome: *Pyrosoma atlanticum; Doliolum denticulatum.*
■ ASCIDIACEA: star ascidian: *Botryllus schlosseri;* sea squirt: *Ascidia mentula;* sea peach: *Halocynthia papillosa; Ciona intestinalis; Phallusia mammillata; Stolonica aggregata.*
■ APPENDICULARIA: *Oikopleura albicans.*

Myomerozoa

A few representatives

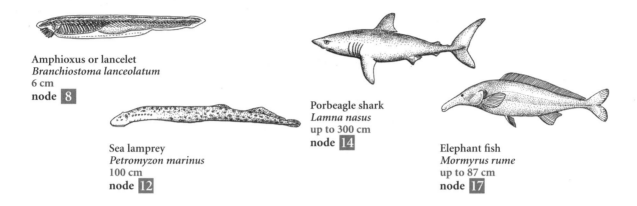

Amphioxus or lancelet
Branchiostoma lanceolatum
6 cm
node 8

Sea lamprey
Petromyzon marinus
100 cm
node 12

Porbeagle shark
Lamna nasus
up to 300 cm
node 14

Elephant fish
Mormyrus rume
up to 87 cm
node 17

Some unique derived features

■ Somites: the intermediate embryonic tissue layer (mesoderm) is segmented along the anterior-posterior axis on either side of the notochord (*no*) and hollow nerve cord (*nc*), that is, this tissue is subdivided into small successive blocks called somites (*so*) (fig. 1: transparent embryo on which we see the mesoderm). Each block will give rise to one or more organs, principally muscular formations; these blocks are therefore called myotomes.

■ The adult shows bilateral symmetry.

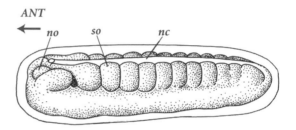

Figure 1

Number of species: 50,946.
Oldest known fossils: *Pikaia* is from the lower Cambrian (530 MYA).
Current distribution: worldwide.

Examples

■ CEPHALOCHORDATA: amphioxus or lancelet: *Branchiostoma lanceolatum.*
■ CRANIATA: hagfish: *Myxine glutinosa*; perch: *Perca fluviatilis*; crested newt: *Triturus cristatus*; Hermann's tortoise: *Testudo hermanni*; cat: *Felis catus*; domestic chicken: *Gallus gallus.*

Cephalochordata

General Description

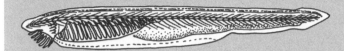

Amphioxus or lancelet
Branchiostoma lanceolatum
6 cm

The cephalochordates are marine organisms that resemble small fishes. These animals are tapered to a point at both extremities and are laterally compressed. They range in size from 5 to 7.5 cm long, 6 to 8 mm high and 3 to 4 mm thick. They are slightly off-white in color, almost colorless. There is no head, skull or fins. A cutaneous fold on the posterior end plays the role of the caudal fin. The mouth is surrounded by large rigid finger-like projections (cirri). The buccal cavity is short and leads to a large branchial pharynx that takes up almost the entire anterior end of the body. The branchial slits open to the exterior in the larval form. In the adult, the body wall forms a vertical fold that covers the branchial region; this fold results in the formation of a peribranchial cavity (atrium) that communicates with the external environment via a common unpaired orifice, the branchial pore (atriopore). The notochord is strong, covers the entire length of the animal and is present throughout its life. On either side of the notochord, the musculature is broken up into successive segments (the myomeres).

Some unique derived features

- The notochord (*no*, fig. 1) extends anteriorly beyond the mouth (*mo*).

- There is a segmented musculature within the notochord (fig. 2). Discoid muscle plates are vertically aligned along the notochord. Each plate consists of one or two muscle cells (*mc*) that contain contractile filaments and have a cytoplasmic projection (*cp*) that makes contact with the overlying hollow neural cord (*nc*) via small holes in the notochordal sheath.

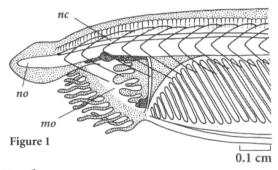

Figure 1

0.1 cm

Ecology

The cephalochordates live at depths of less than 50 m. They are able to swim for brief periods via lateral undulations of the body. However, we typically find them lying on their sides or buried obliquely in the sand, with only the anterior region emerging into the water. These organisms filter-feed on organic particles suspended in the seawater. Ciliature on the branchial slits creates a water current that brings water in through the mouth. The water then crosses the branchial pharynx via the slits, ending in the peribranchial chamber (atrium) where it then exits the animal by the branchial pore (atriopore). This water serves as much for respiration as it does for filter-feeding. An organ called the endostyle secretes mucus that traps food particles. These particles are then led toward the digestive tube. The sexes are separate. Fertilization is external; it takes place in the seawater.

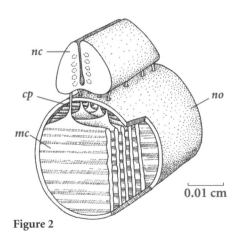

Figure 2

0.01 cm

Examples

- Amphioxus or lancelet: *Branchiostoma lanceolatum; Asymmetron lucayanum.*

Number of species: 13.
Oldest known fossils: *Pikaia* (lower Cambrian, 530 MYA); *Palaeo-branchiostoma hamatotergum* (lower Permian, 280 MYA), South Africa.
Current distribution: in all the seas of the world, particularly in tropical regions and in the shallow waters of the littoral zones.

Craniata

A few representatives

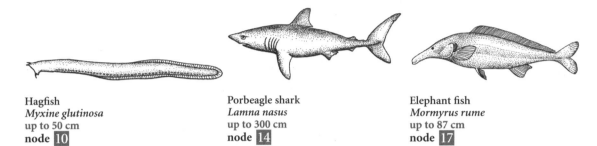

Hagfish
Myxine glutinosa
up to 50 cm
node **10**

Porbeagle shark
Lamna nasus
up to 300 cm
node **14**

Elephant fish
Mormyrus rume
up to 87 cm
node **17**

Some unique derived features

■ The presence of a skull, primitively composed of cartilaginous arches and fibrous plates (neurocranium). The primitive organization of the braincase is shown in fig. 1 (*boc:* basioccipital, *bsp:* basisphenoid, *nac:* nasal capsule, *otc:* otic capsule, *et:* ethmoid, *ex:* exoccipital, *set:* sphenethmoid, *soc:* supraoccipital).

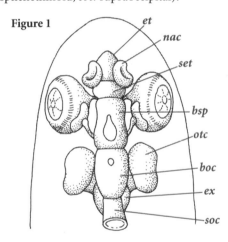

Figure 1

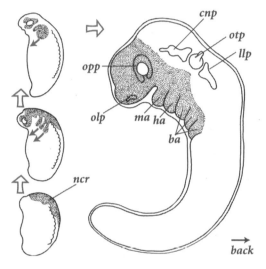

Figure 2

■ During the development of the central and axial nervous systems, dorsal ridges form longitudinally. Migratory cells with diverse fates originate from this neural crest. These "neural crest cells" (*ncr*) are shown by the stippled area in fig. 2.

■ Sensory organs: olfactory, visual, and otic organs develop from small plates on the embryo's surface. These epidermal placodes are shown in fig. 2 (*ba:* branchial arches, *ha:* hyoid arch, *ma:* mandibular arch, *llp:* dorsolateral placode from which the lateral lines originate, *cnp:* dorso-lateral placode from which the cranial nerve originates, *olp:* olfactive placode, *opp:* optic placode, *otp:* otic placode).

■ The mineralization of the skeleton involves calcium phosphate.

Examples

■ Myxinoidea: hagfish: *Myxine glutinosa.*
■ Vertebrata: river lamprey: *Lampetra fluviatilis;* thorny skate: *Raja radiata;* perch: *Perca fluviatilis;* green frog: *Rana esculenta;* green lizard: *Lacerta viridis;* house mouse: *Mus musculus;* chimpanzee: *Pan troglodytes.*

Number of species: 51,066.
Oldest known fossils: *Myllokunmingia fengjiaoa* and *Haikouichthys ercaicunensis,* from the lower Cambrian from Yunnan, China (530 MYA) may already be vertebrates. The conodonts (fig. 3) are craniates, the oldest of which are *Eoconodontus* and *Proconodontus* from the upper Cambrian from Texas (500 MYA). A bone fragment from an arandaspid, an incontestable craniate, is known from the base of the Ordovician (480 MYA). **Current distribution:** worldwide.

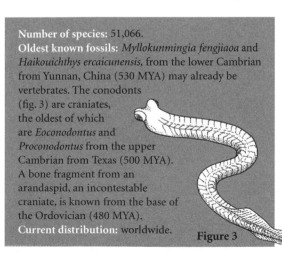

Figure 3

Myxinoidea

General Description

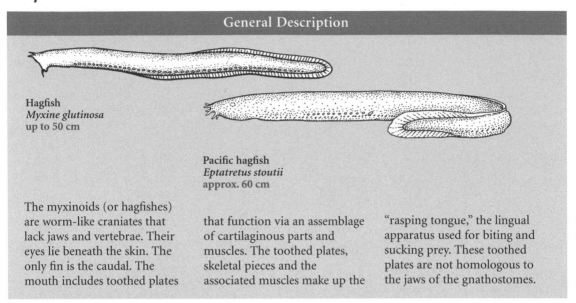

Hagfish
Myxine glutinosa
up to 50 cm

Pacific hagfish
Eptatretus stoutii
approx. 60 cm

The myxinoids (or hagfishes) are worm-like craniates that lack jaws and vertebrae. Their eyes lie beneath the skin. The only fin is the caudal. The mouth includes toothed plates that function via an assemblage of cartilaginous parts and muscles. The toothed plates, skeletal pieces and the associated muscles make up the "rasping tongue," the lingual apparatus used for biting and sucking prey. These toothed plates are not homologous to the jaws of the gnathostomes.

Some unique derived features

■ The branchial apparatus is very different from that of the vertebrates: it is made up of numerous small onion-shaped pouches containing the gills. A dorsal view of this apparatus is shown in fig. 1. in *Myxine* (1a), where water is evacuated by a single branchial opening, and in *Eptatretus* (1b), where it leaves by the same number of openings as there are pouches (*ANT:* anterior end). The number of branchial openings varies with the species: 1 in *Myxine glutinosa*, 6 in *Paramyxine atami*, 8 in *Eptatretus okinoseanum*, and 12 in *E. stoutii*.

■ The head ends in a single nasopharyngeal opening (*npo*, fig. 2 and fig. 3a) that leads into the pharynx (*ph:* fig. 3a). Six tactile tentacles (*te*, fig. 2)

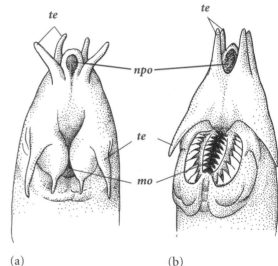

(a)   (b)

Figure 2

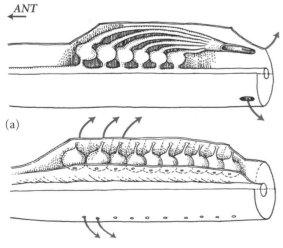

ANT

(a)

(b)

Figure 1

surround the mouth (*mo:* mouth, *no:* notochord, *top:* toothed plate, *nc:* hollow nerve cord).

■ An esophago-cutaneous duct of unknown function leads from the esophagus to the exterior on the left side of the animal.

■ Lingual apparatus: the toothed plates function via the action of a "piston" (made up of cartilaginous pieces and muscles arranged in a characteristic fashion). Fig. 3b and fig. 3c show how this lingual apparatus, or piston, functions (simplified version) (*alc:* anterior lingual cartilage, *mlc:* medial lingual cartilage, *plc:* posterior lingual cartilage). When the inferior muscle contracts and the superior muscle extends, the toothed plate is projected towards the exterior (fig. 2).

Ecology

The myxinoids are found only in marine environments. They generally bury themselves in the silt, only leaving it to feed on sick or dead fishes that they skin by biting and sucking. Reproduction and development are poorly known in this group.

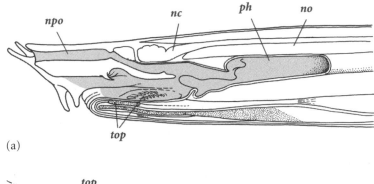

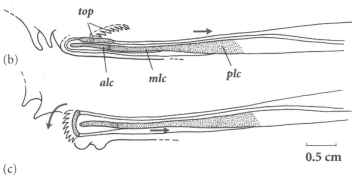

(a)

(b)

(c)

Figure 3

Number of species: 22.
Oldest known fossils: *Myxineides gonorum* from Monceaux-les-Mines, France, dates from the upper Carboniferous (300 MYA). The myxinid lineage dates from before the middle Cambrian (540 MYA).
Current distribution: cold and temperate seas of the Northern and Southern hemispheres. A few species can be found in tropical seas, but only in deep waters (Venezuela, Brazil, Philippines, etc.).

Examples

Hagfish: *Myxine glutinosa:* Pacific hagfish: *Eptatretus stoutii; Paramyxine atami.*

Vertebrata

A few representatives

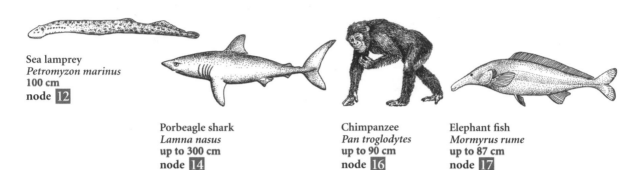

Sea lamprey
Petromyzon marinus
100 cm
node 12

Porbeagle shark
Lamna nasus
up to 300 cm
node 14

Chimpanzee
Pan troglodytes
up to 90 cm
node 16

Elephant fish
Mormyrus rume
up to 87 cm
node 17

Some unique derived features

- Vertebrae: endoskeletal elements surround the notochord and run in succession along an anterior-posterior line: these are the vertebrae. On fig. 1 the notochord (*no*) and the rudimentary vertebrae (*rv*) are shown in the lamprey. In more recent vertebrates, the notochord has largely been replaced by the vertebrae (*v*); only relic pieces found between the vertebrae remain (fig. 2, a portion of the vertebral column of a teleost fish *Lampanyctus leucopsarus*).

- There are two semicircular canals in the inner ear (fig. 3b,

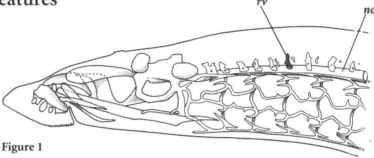

Figure 1

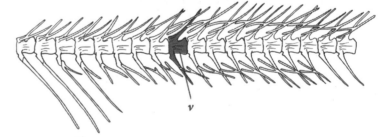

Figure 2

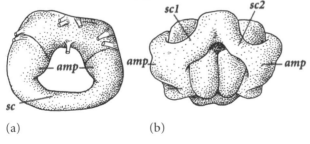

(a)   (b)

Figure 3

sc1, sc2) used for orientation (there was only one in the hagfishes, fig. 3a; *amp*: ampulla, *sc*: semicircular canal).

- There are extrinsic eye muscles.

- There is nervous regulation of the heart.

- There is a system of lateral sensory lines, containing neuromasts.

- We see the appearance of an individual spleen and pancreas.

Examples

- PETROMYZONTIFORMES: river lamprey: *Lampetra fluviatilis*.
- GNATHOSTOMATA: thorny skate: *Raja radiata*; Atlantic salmon: *Salmo salar*; green frog: *Rana esculenta*; grass snake: *Natrix natrix*; Hermann's tortoise: *Testudo hermanni*; domestic chicken: *Gallus gallus*; house mouse: *Mus musculus*.

Number of species: 50,911.
Oldest known fossils: *Myllokunmingia fengjiaoa* and *Haikouichthys eraicunensis* from Yunnan, China, date from the lower Cambrian (530 MYA) and were most likely vertebrates. The first incontestable vertebrate is represented by fragments of the arandapsid *Porophoraspis* from the lower Ordovician from Australia. *Sacabambaspis janvieri*, from the middle Ordovician from Bolivia (470 MYA), is the most complete fossilized vertebrate of this time period.
Current distribution: worldwide.

Petromyzontiformes

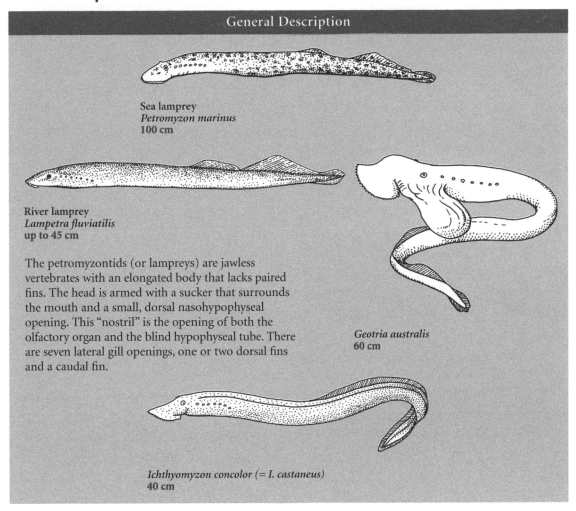

General Description

Sea lamprey
Petromyzon marinus
100 cm

River lamprey
Lampetra fluviatilis
up to 45 cm

The petromyzontids (or lampreys) are jawless
vertebrates with an elongated body that lacks paired
fins. The head is armed with a sucker that surrounds
the mouth and a small, dorsal nasohypophyseal
opening. This "nostril" is the opening of both the
olfactory organ and the blind hypophyseal tube. There
are seven lateral gill openings, one or two dorsal fins
and a caudal fin.

Geotria australis
60 cm

Ichthyomyzon concolor (= *I. castaneus*)
40 cm

Some unique derived features

- There is a specialized branchial
  skeleton called the "branchial
  basket" (*bb*, fig. 1; *ac*: annular
  cartilage, *dc*: dentary cartilage,
  nac: nasal capsule, *otc*: otic
  capsule, *pic*: piston cartilage,
  cw: cranial wall, *sty*: stylet).
  Skeletal pieces support the
  branchial system; the branchial
  arches (*ba*, fig. 2) are external
  to the gill pouches (*gp*).

- An olfactory organ opens into
  a blind tube, the hypophyseal
  tube (fig. 2; *no*: notochord, *ph*:
  pharynx, *dt*: digestive tube, *ht*:
  hypophyseal tube, *nc*: hollow
  nerve cord).

- Mouth: the mouth is
  surrounded by a sucker (fig. 3)
  and includes a "tongue" with

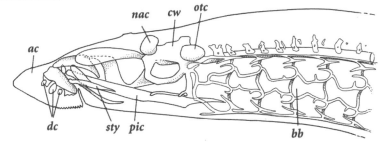

Figure 1

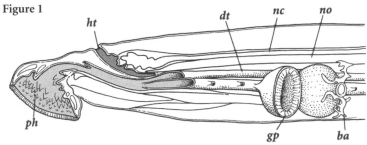

Figure 2

horny teeth fixed to cartilaginous plates (annular cartilage and dentary cartilage; fig. 1). These pieces pivot around a long cartilaginous piston (fig. 1). The anterior-posterior movements of this piston enable the animal to suck the blood of its victims.

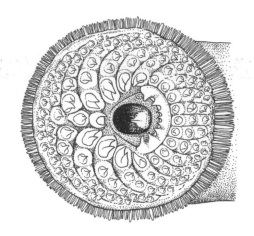

Figure 3

Number of species: 38.
Oldest known fossils: *Haikouichthys eraicunensis*, from the lower Cambrian from Yunnan, China (530 MYA), may have been closely related to the lampreys. An indubitable lamprey, *Hardistiella montanensis*, dates from the upper Carboniferous from Montana, United States (325 MYA).
Current distribution: these animals are absent from waters between 30° N and 30° S. Four species are present in the Southern hemisphere (genera *Mordacia* and *Geotria*). There are 34 species in the Northern hemisphere (Eurasia and North America, Atlantic and Pacific Oceans).

Ecology

Most adult petromyzontids live in marine environments, whereas the larvae live in fresh water. The adult forms are typically ectoparasites that feed on the blood of fishes or cetaceans. The lampreys pass through a larval stage and undergo metamorphosis. For a long time, the larvae were considered to be a separate species from the adults to which the name ammocoetes was given. This name is still used to refer to the larvae.

Examples

River lamprey: *Lampetra fluviatilis;* Pacific lamprey: *Entosphenus tridentatus;* sea lamprey: *Petromyzon marinus; Geotria australis; Ichthyomyzon unicuspis; Mordacia mordax.*

Gnathostomata

A few representatives

Porbeagle shark
Lamna nasus
up to 300 cm
node 14

Chimpanzee
Pan troglodytes
90 cm
node 16

Garganey
Anas querquedula
37 cm
node 16

Elephant fish
Mormyrus rume
up to 87 cm
node 17

Some unique derived features

■ Jaws: have an upper bilateral mandible (fig. 1) made up of the pterygo-palatoquadrate cartilage (*pqc*) and a lower bilateral mandible made up of Meckel's cartilage (*Mc*). These mandibles are produced from a modification to the most anterior branchial arch. Just posterior to the jaw, the associated branchial slit forms the spiracle or vent (*v*), an opening linked to the pharynx via a canal (spiracular canal).

■ There is a third semicircular canal in the inner ear (fig. 3; *amp:* ampulla, *ec:* endolymphatic canal, *sc1* and *sc2:* vertical semicircular canals, *sc3:* horizontal semicircular canal). These canals are used for orientation.

■ The nerve fibers are covered by myelin sheaths (fig. 4: *ax:* axon, *Sc:* Schwann cell, *ms:* myelin sheaths, *Rn:* Ranvier node).

■ Hemoglobin: this molecule has two types of amino acid chains, an α chain and a β chain. The molecule is tetrameric, with 2 α and 2 β chains.

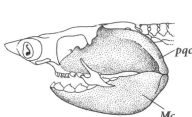

Figure 1

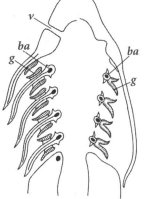

Figure 2

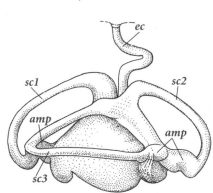

Figure 3

■ The branchial skeleton (fig. 2: *ba:* branchial arches) is internal to the gills (*g*). On fig. 2 the arrangement of the branchial arches is shown (dorsal view) for a chondrichthyan (*left*) and an osteichthyan (*right*).

Figure 4

Number of species: 50,873.
Oldest known fossils: acanthodian scales have been found from the end of the Ordovician in Siberia (430 MYA).
Current distribution: worldwide.

Examples

■ CHONDRICHTHYES: thorny skate: *Raja radiata*.
■ OSTEICHTHYES: Atlantic salmon: *Salmo salar;* green frog: *Rana esculenta;* grass snake: *Natrix natrix;* green lizard: *Lacerta viridis;* Hermann's tortoise: *Testudo hermanni;* domestic chicken: *Gallus gallus;* house mouse: *Mus musculus;* common dolphin: *Delphinus delphis*.

343

Chondrichthyes

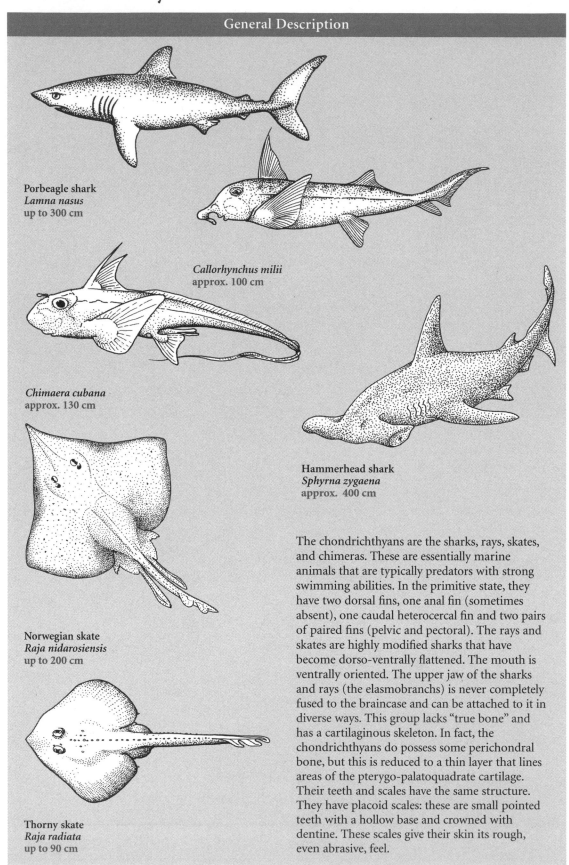

General Description

Porbeagle shark
Lamna nasus
up to 300 cm

Callorhynchus milii
approx. 100 cm

Chimaera cubana
approx. 130 cm

Hammerhead shark
Sphyrna zygaena
approx. 400 cm

Norwegian skate
Raja nidarosiensis
up to 200 cm

Thorny skate
Raja radiata
up to 90 cm

The chondrichthyans are the sharks, rays, skates, and chimeras. These are essentially marine animals that are typically predators with strong swimming abilities. In the primitive state, they have two dorsal fins, one anal fin (sometimes absent), one caudal heterocercal fin and two pairs of paired fins (pelvic and pectoral). The rays and skates are highly modified sharks that have become dorso-ventrally flattened. The mouth is ventrally oriented. The upper jaw of the sharks and rays (the elasmobranchs) is never completely fused to the braincase and can be attached to it in diverse ways. This group lacks "true bone" and has a cartilaginous skeleton. In fact, the chondrichthyans do possess some perichondral bone, but this is reduced to a thin layer that lines areas of the pterygo-palatoquadrate cartilage. Their teeth and scales have the same structure. They have placoid scales: these are small pointed teeth with a hollow base and crowned with dentine. These scales give their skin its rough, even abrasive, feel.

Some unique derived features

- A layer of prismatic calcified cartilage of characteristic structure lines the cartilage of the endoskeleton. A cross-section of a skeletal element of a chondrichthyan is shown in fig. 1a. The doted area indicates the cartilage and the dashes along its surface, the prismatic structures. These structures are arranged on the surface as

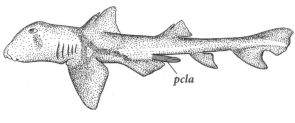

Figure 2

shown in fig. 1b (*hyc*: hyaline cartilage, *prc*: prismatic calcified cartilage).

- In the males, the pelvic fins have a special apparatus used in copulation; the pelvic claspers (*pcla*) are shown in fig. 2 on a 65 cm long Port Jackson shark *Heterodontus portusjacksoni*.

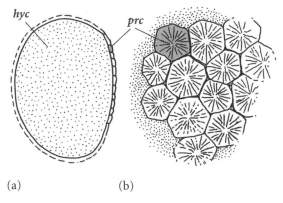

(a)   (b)

Figure 1

Number of species: 846.
Oldest known fossils: isolated scales date from the upper Silurian from central Asia (410 MYA).
Current distribution: worldwide.

Ecology

The chondrichthyans are typically marine predators that feed on other fishes. They may be pelagic or live on the ocean floor. There are a few freshwater species. They do not have a swim bladder (buoyancy organ) and therefore continuously swim to prevent sinking. Some scientists think that certain fossil gnathosomes, like the placoderms, show signs of having had lungs and that the presence of lungs may be the primitive state in the gnathosomes. If this is true, the chondrichthyans have lost their lungs secondarily. Certain species feed on plankton (basking shark) and others on mollusks (certain skates). This group includes oviparous, ovoviparous and even viviparous species, with some cases of feeding and even cannibalism *in utero*.

Examples

Lesser spotted dogfish: *Scyliorhinus canicula;* great white shark: *Carcharodon carcharias;* gummy shark: *Mustelus antarcticus;* spiny dogfish: *Squalus acanthias;* tope shark: *Galeorhinus galeus;* Port Jackson shark: *Heterodontus portusjacksoni;* longnose sawshark: *Pristiophorus cirratus;* common torpedo: *Torpedo torpedo;* thorny skate: *Raja radiata;* rabbitfish: *Chimaera monstrosa.*

Osteichthyes

A few representatives

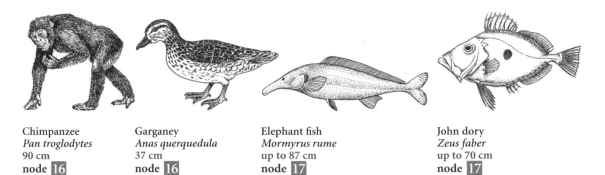

Chimpanzee
Pan troglodytes
90 cm
node 16

Garganey
Anas querquedula
37 cm
node 16

Elephant fish
Mormyrus rume
up to 87 cm
node 17

John dory
Zeus faber
up to 70 cm
node 17

Some unique derived features

■ Bone: there is a new type of bone, the endochondral bone. The osteichthyans therefore have two types of "true" bone. Endochondral bone results from the destruction of the embryonic cartilage and the reconstruction of the piece in bone. This process is at the origin of the spongy bones and the bones of the endoskeleton. In contrast, "dermal" bone forms by another process and has a different origin; it is formed from the dermis.

■ Assemblages of bony parts of dermal origin appear, notably the maxillary, premaxillary and dentary bones that all carry teeth. The bones of dermal origin are shown in white on fig. 1 in *Amia calva* (*ang:* angular, *ant:* antorbital, *cl:* cleithrum, *de:* dentary, *dsp:* dermosphenotic, *esc:* extrascapular, *fr:* frontal, *io:* infraorbitals, *iop:* interopercular, *la:* lacrimal, *mx:* maxilla, *na:* nasal, *op:* opercular, *pa:* parietal, *pcl:* post-cleithrum, *pmx:* premaxilla, *po:* post-orbital, *pro:* preopercular, *pt:* post-temporal, *pto:* pterotic, *san:* surangular, *scl:* supracleithrum, *smx:* supramaxilla, *sop:* subopercular).

■ The pectoral girdle (*pec*), the ensemble of bones that join the anterior limbs to the vertebral column, also contains dermal bones (indicated in fig. 1).

■ Branchial arches 1 and 2 articulate on the same bony piece found ventrally, the basibranchial.

■ Air sacs are connected to the digestive tube. In the primitive state, these sacs are lungs but can act as an air bladder (or swim bladder) used for buoyancy. Certain authors believe that fossil vertebrates, like the placoderms from the Devonian, must have also possessed this type of sac. If this is the case, these sacs would be an original trait of the gnathosomes and not of the osteichthyans; the chondrichthyans would have lost their lungs secondarily.

■ The fin membranes are supported by dermal fin rays, the lepidotrichia.

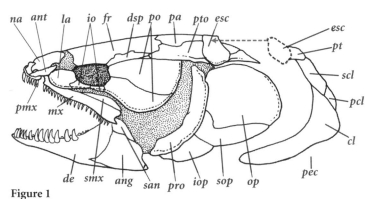

Figure 1

Number of species: 50,027.
Oldest known fossils: *Andreolepis hedei* from Sweden dates from the upper Silurian (420 MYA).
Current distribution: worldwide.

Examples

■ SARCOPTERYGII: African lungfish: *Protopterus dolloi;* green frog: *Rana esculenta;* green lizard: *Lacerta viridis;* house mouse: *Mus musculus;* cat: *Felis catus.*
■ ACTINOPTERYGII: common eel: *Anguilla anguilla;* perch: *Perca fluviatilis;* Atlantic salmon: *Salmo salar.*

➡ *see chap. 12, p. 350*

Sarcopterygii

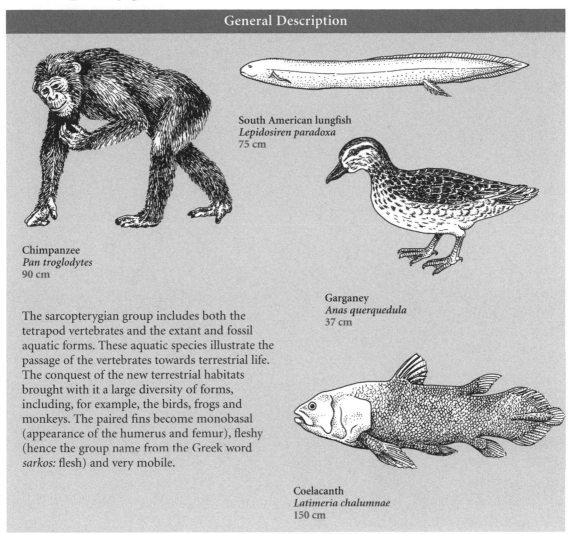

General Description

Chimpanzee
Pan troglodytes
90 cm

South American lungfish
Lepidosiren paradoxa
75 cm

Garganey
Anas querquedula
37 cm

Coelacanth
Latimeria chalumnae
150 cm

The sarcopterygian group includes both the tetrapod vertebrates and the extant and fossil aquatic forms. These aquatic species illustrate the passage of the vertebrates towards terrestrial life. The conquest of the new terrestrial habitats brought with it a large diversity of forms, including, for example, the birds, frogs and monkeys. The paired fins become monobasal (appearance of the humerus and femur), fleshy (hence the group name from the Greek word *sarkos:* flesh) and very mobile.

Some unique derived features

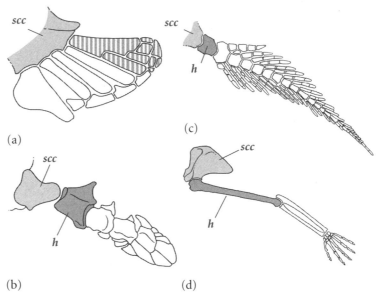

■ The endoskeleton of the fins (or limbs) is monobasal, meaning that they are attached to the girdles by a single articulation. The girdles, like the "pelvis" for example, are skeletal assemblages that link the limbs to the vertebral column. The endoskeletal elements of the pectoral limbs of several sarcopterygians are shown in fig. 1: in an actinistian (coelacanth, 1b), a dipnoan (*Neoceratodus*, 1c), a tetrapod (human, 1d; *h:* first segment or humerus, *scc:* scapulocoracoid or scapula). In a non-sarcopterygian, such as a

Figure 1

347

sturgeon (1a), the striped part of the endoskeleton of the pectoral fin is the meta-pterygium; this part is assumed to be homologous to the monobasal pectoral fin of the sarcopterygii.

■ There is true enamel on the teeth.

■ Branchial arches: the last branchial arch (fifth, *ba5*) is attached ventrally to the preceding arch (fourth, *ba4*). The branchial arches are a vertical succession of boney "ribs" arranged side by side in a dorso-ventral direction that provide skeletal support to the gills. Originally, there were five branchial arches in the sarcopterygii. A dorsal view of the branchial skeleton is shown in fig. 2 (*ANT*: anterior end) for a fossil actinopterygian (*Pteronisculus,* 2a) and for two sarcopterygians, the coelacanth (*Latimeria,* 2b) and a dipnoan (*Neoceratodus,* 2c). In the actinopterygian, the fifth branchial arch does not articulate on the fourth. In the two sarcopterygians, the ventral junction between the fourth and fifth branchial arches is indicated.

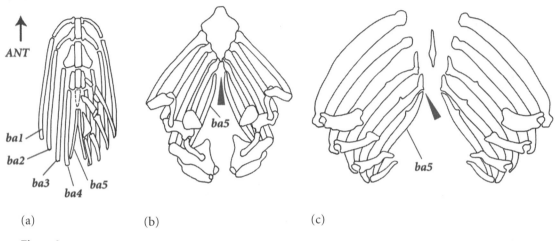

Figure 2

Number of species: 26,315.
Oldest known fossils: *Psarolepis romeri* from China and Vietnam dates from the upper Silurian (420 MYA).
Current distribution: worldwide.

Examples

Coelacanth: *Latimeria chalumnae;* African lungfish: *Protopterus dolloi;* green frog: *Rana esculenta;* green lizard: *Lacerta viridis;* grass snake: *Natrix natrix;* Hermann's tortoise: *Testudo hermanni;* dog: *Canis familiaris;* house mouse: *Mus musculus;* cat: *Felis catus;* domestic chicken: *Gallus gallus.*

Ecology

The current forms are mostly terrestrial; they are found in all habitats and at all latitudes. There are many aquatic fossil species. The extant species limited to aquatic habitats are principally the dipnoans (tropical freshwater), the coelacanth (coasts of the Comoro Islands, South Africa and the Sulawesi Islands) and the cetaceans (found in all seas; these correspond to a "return to the water" for the mammals). We can also mention that some terrestrial fossil sarcopterygians underwent a readaptation to water as well (ichthyosaurs, placodonts, sauropterygians, marine snakes, mosasaurs). In this case, aerial respiration remained, as with the cetaceans.

⟶ *see chap. 15, p. 502*

Actinopterygii

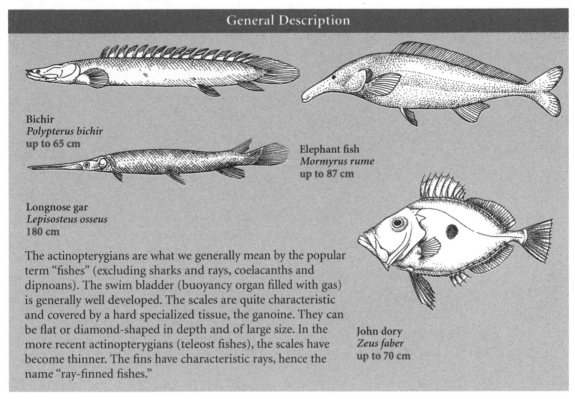

General Description

Bichir
Polypterus bichir
up to 65 cm

Elephant fish
Mormyrus rume
up to 87 cm

Longnose gar
Lepisosteus osseus
180 cm

The actinopterygians are what we generally mean by the popular term "fishes" (excluding sharks and rays, coelacanths and dipnoans). The swim bladder (buoyancy organ filled with gas) is generally well developed. The scales are quite characteristic and covered by a hard specialized tissue, the ganoine. They can be flat or diamond-shaped in depth and of large size. In the more recent actinopterygians (teleost fishes), the scales have become thinner. The fins have characteristic rays, hence the name "ray-finned fishes."

John dory
Zeus faber
up to 70 cm

Some unique derived features

■ Each tooth has a small cap of a special mineralized tissue: acrodin (fig. 1).

■ The anterior dorsal fin has been lost (fig. 2); the fin that remains is homologous to the

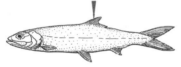

Figure 2

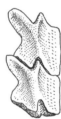

Figure 3

posterior dorsal fin of the chondrichthyans. This fin can be secondarily subdivided.

■ The scales articulate against each other via a system of peg and socket joints (fig. 3).

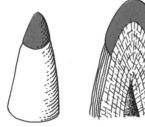

Figure 1

Number of species: 23,712.
Oldest known fossils: from scales: *Andreolepis* and *Lophosteus* from Europe and *Naxilepis* from China all date from the upper Silurian (420 MYA). Fossilized animals: *Cheirolepis* from Scotland and Canada and *Stegotrachelus*, *Moythomasia* and *Orvikuina* from Scotland, Germany and Estonia all date from the middle Devonian (380 MYA).
Current distribution: worldwide.

Ecology

The actinopterygians make up almost half of all vertebrate species. They have colonized all aquatic habitats: from −11,000 m to +4,500 m, from hot springs (43°C) to very cold waters (−1.8°C). They are extremely diversified in morphology, ecology and behavior.

Examples

Senegal bichir: *Polypterus senegalus*; longnose gar: *Lepisosteus osseus*; Siberian sturgeon: *Acipenser baeri*; common eel: *Anguilla anguilla*; Atlantic salmon: *Salmo salar*; bluefin tuna: *Thunnus thynnus*; John dory: *Zeus faber*; perch: *Perca fluviatilis*.

Chapter 12

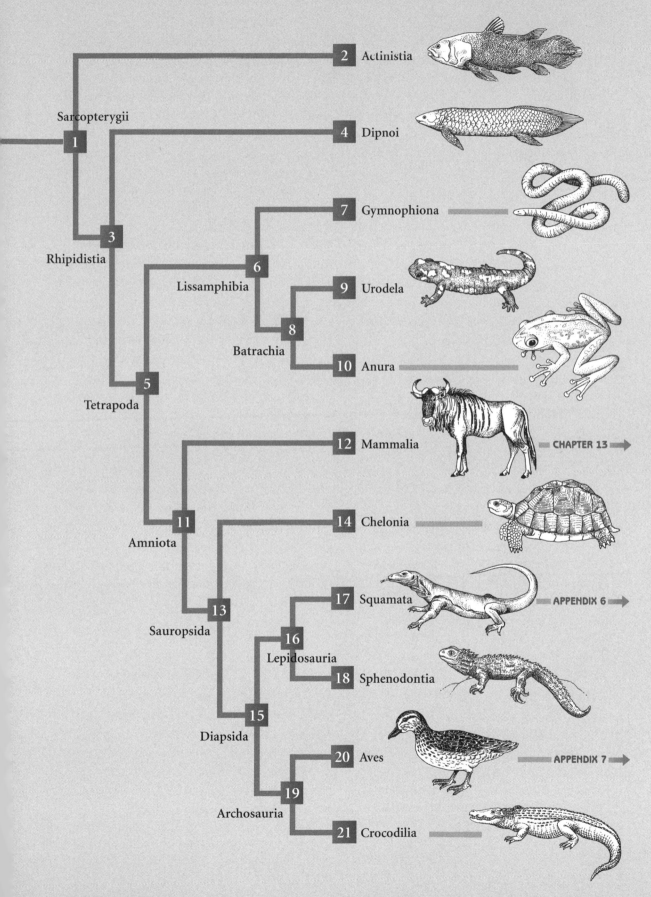

1 Sarcopterygii

2 Actinistia

3 Rhipidistia

4 Dipnoi

6 Lissamphibia

7 Gymnophiona

5 Tetrapoda

9 Urodela

8 Batrachia

10 Anura

12 Mammalia   CHAPTER 13 ➡

11 Amniota

14 Chelonia

17 Squamata   APPENDIX 6 ➡

13 Sauropsida

16 Lepidosauria

18 Sphenodontia

15 Diapsida

20 Aves   APPENDIX 7 ➡

19 Archosauria

21 Crocodilia

For a long time now we have debated the phylogenetic affinities of the actinistians, a group for which fossils closely related to the extant coelacanths were already present in the Devonian. Indeed, they have been proposed to be the sister group of the chondrichthyes, the sister group of the actinopterygians, the sister group of the tetrapods, the sister group of the dipnoans, and the sister group of the clade containing dipnoans + tetrapods. The actinistians are strange animals with fat-filled lungs, a high level of blood urea, unique rostral organs, ventral kidneys, a liquid-filled noto-chord, and a reduced brain. To this mosaic of relic sarcopterygian and specialized actinistian characters, one must add a certain number of specializations linked to life in deep, cold water and juvenile traits like those associated with the heart structure and a reduced ossification of the endoskeleton. Given all this, it is not surprising that, even though we know the anatomy of the extant *Latimeria* relatively well, the evolutionary interpretation of its characters produces contrasting phylogenetic results.

Amphibians in the classic sense, that is, including the fossils, make up a grade of non-amniotic tetrapods. It is therefore a para-phyletic group. Indeed, certain fossil clades from the Carboniferous and the Permian, like the anthracosaurians, the seymouriamorphs and the diadectomorphs, are more closely related to the amniotes than to other batra-chomorph amphibians. These three fossilized amphibian groups, along with the amniotes, constitute the reptiliomorph clade.

The phylogeny of the amniotes has undergone several revisions during the last twenty years. In the 1980s, there was a polemic over the phylogenetic position of the birds (see introduction) and, more recently, a certain number of studies have called into question the traditional position of the turtles (chelonia). The tree presented here is the traditional cladogram of the amniotes in which the birds are archosaurs and the chelonians are the sister group of the diapsids. During the first polemic, the sister group relationship between birds and mammals was debated. However, recent work, including several molecular phylogenies, all agree that, among extant amniotes, birds and crocodiles should be sister groups. In addition, the recent discovery of some spectacular fossils has confirmed that birds originated from the dinosaurs.

In contrast, the current discussions on the position of the chelonians may well change the image that we have of these animals. Traditionally, the turtles were considered as the most basal sauropsid lineage. They have no temporal fenestrae, whereas their sister group, the diapsids, is characterized by a double temporal fenestra behind the orbit. In 1996, two paleontologists showed by the cladistic analysis of anatomical characters of extant and fossil amniotes that chelonians are diapsids and, more precisely, that they are the sister group of the sauropterygii (nothosaurs, plesiosaurs, elasmosaurs). This position was confirmed by additional analyses in 1999. However, if we leave only extant species in this phylogenetic tree, we see that the turtles are the sister group of the lepidosaurs. The absence of the temporal openings in the turtles would therefore correspond to a secondary loss. Several molecular studies confirm that the turtles are diapsids. However, they do not appear as the sister group of the lepidosaurs, but rather as either the sister group of the birds + crocodiles group (mitochondrial genome, simultaneous analysis of 11 nuclear proteins), or the sister group of the crocodiles alone (simultaneous analysis of 18 genes, analyses of the LDHA, LDHB, α-enolase, and myoglobin genes). Numerous paleontologists continue to publish cladistic analyses where the so-called classic tree presented in this book is upheld. Therefore, although the exact sister group of the chelonians has yet to be clearly identified, in the absence of further information we have decided to present the result of the classic cladistic analyses here.

Actinistia

General Description

The coelacanth is the only extant species known from the actinistia, a group for which there are many fossil species. We find fossils, identical in form to the coelacanth, that are up to 70 million years old. The coelacanth is gray-blue in color with white spots; it can measure more than 1.5 m in length. It is easily recognizable by its fins. Indeed, the fleshy basal lobe of the anterior dorsal fin is reduced and armed with hollow ptergiophores (hence the name "coelacanth" meaning "hollow spine").

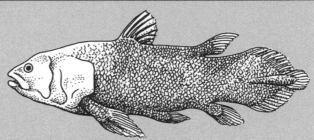

Coelacanth
Latimeria chalumnae
150 cm

The paired fins have a long fleshy lobe and are very mobile. The coordination of their movements is that of a four-legged animal (tetrapod). The caudal fin is very characteristic with three distinct lobes.

Some unique derived features

■ Caudal fin: there is a ventral and dorsal lobe, along with a small median lobe. An extension of the notochord runs into this median lobe (fig. 1).

■ The first dorsal fin (fig. 1) is armed with large, hollow spines.

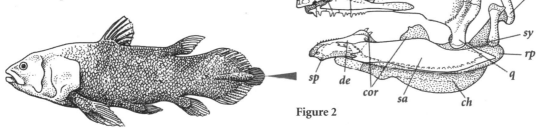

Figure 1

Figure 2

■ The maxillary bone is lost. On the bony head of a coelacanth (fig. 2; *apa:* autopalatine, *q:* quadrate, *ch:* ceratohyal, *cor:* coronoids, *de:* dentary, *hm:* hyomandibular, *mt:* metapterygoid (epipterygoid), *rp:* retroarticular process, *pg:* ptergyoid, *sa:* splenio-angular, *sp:* splenial, *sy:* symplectic), we see that the teeth are carried on the ectopterygoid (*ecp*), the palatines (*pal*), and the parasphenoid (*psp*, see fig. 3).

■ Rostral organ: the snout encloses a special electroreceptor organ, the rostral organ. The posterior openings of this organ are indicated on fig. 3 (*ro*).

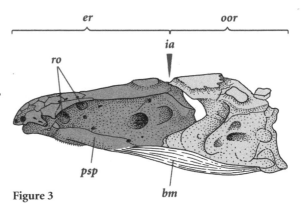

Figure 3

■ There is an intracranial articulation (*ia*, fig. 3) between the ethmoid region (*er*) and the otico-occipital region (*oor*) that enables novel movements of the head skeleton. This articulation is actually a primitive feature of the sarcopterygii that was lost independently in the tetrapods and dipnoans. It is, however, unique among extant fauna and is accompanied by a very long basicranial muscle (*bm*) in the most recent coelacanths (fossil *Macropoma* and extant *Latimeria*). This muscle runs over 50% of the length of the parasphenoid (*psp*).

Ecology

The coelacanth lives at depths of between 70 and 400 m. It is a predator, feeding especially on lanternfish. In Indonesia, this species was captured at a depth of 100 to 150 m, above a volcanic slope. The ragged surface and underwater caves of this slope were most likely used as shelter by the coelacanth. Reproduction in this species is poorly known. We also know nothing about the size of the populations, their age structure, or the age at sexual maturity. The species is ovoviviparous; five fetuses were found in the genital tract of a female and took up almost the entire oviduct. The young develop here until a very advanced stage. Gestation is certainly longer than one year. These animals can live for at least 25 years.

Number of species: 1.
Oldest known fossils: *Euporosteus eifeliensis* from Germany dates from the Devonian (380 MYA).
Current distribution: the coelacanth lives off the coast of the Comoro Islands. The first specimen was found in 1938 off the east coast of South Africa. It was studied by M. Courtenay-Latimer (hence the name *Latimeria*). In 1998, other coelacanths (brown-colored) were found close to the volcanic islands north of Sulawesi Island in Indonesia.

Examples

Coelacanth: *Latimeria chalumnae.*

Rhipidistia

A few representatives

Australian lungfish
Neoceratodus forsteri
180 cm
node 4

Common toad
Bufo bufo
up to 15 cm
node 10

Chimpanzee
Pan troglodytes
90 cm
node 12

Indian gharial
Gavialis gangeticus
700 cm
node 21

Some unique derived features

- The functional lung has alveoli. A dorsal view of the lungs of the dipnoan *Lepidosiren* are shown in fig. 1 (*al*: alveoli, *g*: glottis, *lu*: lung, *dt*: digestive tube).

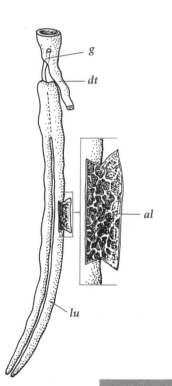

Figure 1

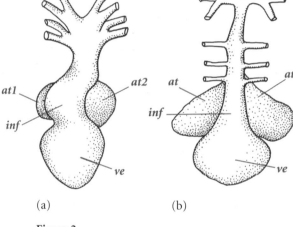

(a)   (b)

Figure 2

- The infundibulum or pulmonary trunk of the heart is convoluted. Fig 2a shows a ventral view of the heart of the lungfish *Neoceratodus*. Compare this with a ventral view of a dogfish shark (*Cetorhinus*) heart (fig. 2b); the infundibulum is on the right (*at*: atrium, *inf*: infundibulum, *ve*: ventricle).

- There are two atria in the heart (fig. 2a; *at1* and *at2*), whereas there is only a single atrium in non-rhipidistians (fig. 2b).

- There is a glottis (*g*; fig. 1), the meeting point of the larynx and digestive tube.

- The larva has a ciliated epidermis.

Number of species: 26,314.
Oldest known fossils: *Youngolepis praecursor* and *Diabolepis speratus* from Yunnan, China, date from the lower Devonian (410 MYA); *Powichthys thorsteinssoni* from the Canadian Arctic dates from this same period.
Current distribution: worldwide.

Examples

- DIPNOI: Australian lungfish: *Neoceratodus forsteri*; South American lungfish: *Lepidosiren paradoxa*.
- TETRAPODA: green frog: *Rana esculenta*; domestic chicken: *Gallus gallus*; gorilla: *Gorilla gorilla*.

Dipnoi

General Description

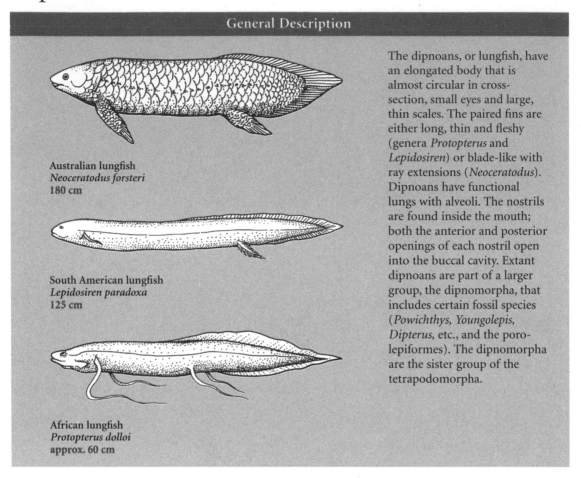

Australian lungfish
Neoceratodus forsteri
180 cm

South American lungfish
Lepidosiren paradoxa
125 cm

African lungfish
Protopterus dolloi
approx. 60 cm

The dipnoans, or lungfish, have an elongated body that is almost circular in cross-section, small eyes and large, thin scales. The paired fins are either long, thin and fleshy (genera *Protopterus* and *Lepidosiren*) or blade-like with ray extensions (*Neoceratodus*). Dipnoans have functional lungs with alveoli. The nostrils are found inside the mouth; both the anterior and posterior openings of each nostril open into the buccal cavity. Extant dipnoans are part of a larger group, the dipnomorpha, that includes certain fossil species (*Powichthys, Youngolepis, Dipterus*, etc., and the porolepiformes). The dipnomorpha are the sister group of the tetrapodomorpha.

Some unique derived features

■ The mouth contains crushing dental plates with fan-shaped ridges. These plates have different shapes depending on the species (fig. 1a: pterygoidian dental plate of *Neoceratodus forsteri,* 1b: that of the fossil *Protopterus protopteroides*). The upper plates are carried on the pterygoid and prearticular bones (pterygoidian dental plate), the lower plates on the coronoid bones (coronoidian dental plate). A ventral view of the skull of *Neoceratodus* is shown in fig. 2 (*cho:* choana, *hm:* hyomandibular, *pvt:* prevomerine teeth, *ptp:* pterygoidian dental plate, *pg:* pterygoid, *psp:* parasphenoid, *o:* operculum, *cob:* circumorbital bone).

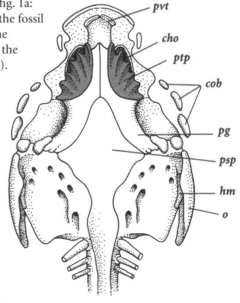

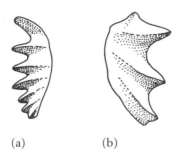

(a)   (b)

Figure 1

Figure 2

■ Certain jaw bones (maxilla and premaxilla) have disappeared. A lateral view of the skull of *Neoceratodus* is shown in fig. 3 (*q:* quadrate, *pvt:* prevomerine teeth, *it:* internasal, *cob:* circumorbital bone, *dp:* dental plate, *ats:* anterior splenial, *sa:* spleno-angular).

NOTE: these characters are those of the dipnoiformes that include the extant dipnoans and all fossil dipnoan species except *Powichthys* and *Youngolepis*.

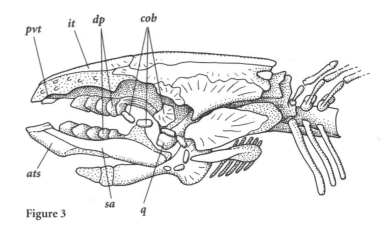

Figure 3

Ecology

The dipnoans live in slow-moving fresh water, in muddy waters, and sometimes in small, ephemeral water bodies. They feed on crustaceans, mollusks and small fishes. They breathe using their gills if there is sufficient oxygen in the water. If this is not the case, they are also able to breathe air into their lungs at the water's surface. The protopterans and the lepidosirens dig a chamber in the mud during droughts. They live in this "cocoon" (surrounded by a mucus matrix), connected to the external environment by only a breathing hole, until the water returns and liberates them from the matrix. During the rainy season, their breeding season, it is the male that builds the tunnel-shaped nest in which the female will lay her eggs. The male then fertilizes these eggs and protects them until hatching.

Number of species: 6.
Oldest known fossils: *Youngolepis praecursor* and *Diabolepis speratus* from Yunnan, China, date from the lower Devonian (410 MYA); *Powichthys thorsteinssoni* from the Canadian Arctic dates from this same period.
Current distribution: the genus *Lepidosiren* is found in the Amazon basin of South America, the genus *Protopterus* in tropical Africa, and the genus *Neoceratodus* in Australia (Queensland).

Examples

African lungfish: *Protopterus dolloi;* South American lungfish: *Lepidosiren paradoxa;* Austrialian lungfish: *Neoceratodus forsteri*.

Tetrapoda

A few representatives

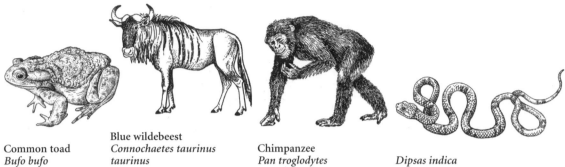

Common toad
Bufo bufo
up to 15 cm
node **10**

Blue wildebeest
Connochaetes taurinus taurinus
up to 200 cm
node **12**

Chimpanzee
Pan troglodytes
90 cm
node **12**

Dipsas indica
70 cm
node **17**

Some unique derived features

■ There is a lacrimal duct between the eye and the nasal sac. In the actinopterygians, the inhalant (*inn*) and exhalant (*exn*) nostrils are external (fig. 1a). In the tetrapods, the exhalant nostril empties into the buccal cavity; it is called the choana (*cho*, fig. 1b). The nasal sac (*ns*) is linked to the eye by the lacrimal duct (*lcd*).

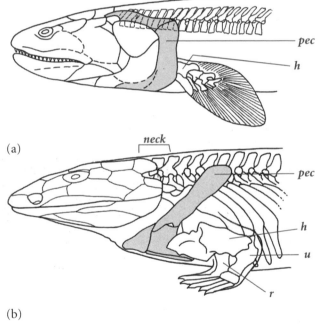

(a)

(b)

Figure 2

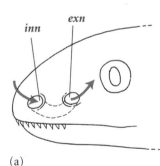

(a)

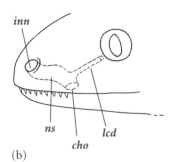

(b)

Figure 1

■ Appearance of the neck: the pectoral girdle becomes separated from the head. In *Eusthenopteron,* a non-tetrapod sarcopterygian (fig. 2a), the pectoral girdle (*pec*) is attached directly to the dermal skull roof. In a tetrapod (*Ichthyostega,* fig. 2b), it is separated by the neck.

■ Limbs: the tetrapods are characterized by their paired locomotory limbs with digits. The skeleton of these paired limbs is derived from that of the paired fins of certain sarcopterygian fossil species. A limb is most often made up of five digits (initially from 6 to 8) along with carpal bones, a radius/ulna (fig. 2b: *r/u*) and the humerus (*h*) for the upper limb, or tarsal bones, a tibia/fibula and a femur for the lower limbs.

- The hyomandibular bone, previously used for supporting the jaw, takes on an auditory function: it conducts sound from the tympanum (*ty*) to the oval window (*ow*). It is now called the columella or stapes (*st*) and is shown in a ventral view on an anuran skull (bullfrog: *Rana catesbeiana*) in fig. 3a and in a transverse section of a frog's head in fig. 3b.

- The first cervical vertebra, called the atlas, has become specialized.

NOTE: certain fossil groups (panderichthyids, osteolepiformes) and the tetrapods are included in the choanata clade, a clade defined by the presence of a choana, the exhalant nostril that empties internally into the buccal cavity. With the inclusion of their fossil sister group, the rhizodontiformes, the choanates make up the larger tetrapodomorpha clade. The tetrapodomorphs are the sister group of the dipnomorphs, which include the porolepiformes (fossil group) and the dipnoans.

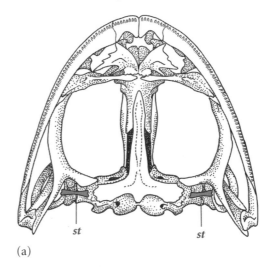

(a)

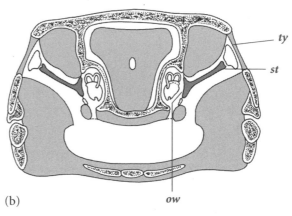

(b)   *ow*

Figure 3

Number of species: 26,308.
Oldest known fossils: *Elginerpeton pancheni* from Scotland dates from the upper Devonian (368 MYA); *Obruchevichthys* sp. from Russia dates from the same period. The mandible of *Livoniana multidentata* from Estonia dates from the middle Devonian (375 MYA). Traces of a leg from an unknown species were found in Australia (390 MYA).
Current distribution: worldwide.

Examples

- Lɪssᴀᴍᴘʜɪʙɪᴀ: aquatic caecilian: *Typhionectes compressicaudus;* crested newt: *Triturus cristatus;* green frog: *Rana esculenta;* green lizard: *Lacerta viridis;* grass snake: *Natrix natrix;*

- Aᴍɴɪᴏᴛᴀ: common dolphin: *Delphinus delphis;* chimpanzee: *Pan troglodytes;* leatherback turtle: *Dermochelys coriacea;* domestic chicken: *Gallus gallus.*

Lissamphibia

A few representatives

Caecilia sp.
up to 61 cm
node **7**

Fire salamander
Salamandra salamandra
25 cm
node **9**

Common toad
Bufo bufo
up to 15 cm
node **10**

Some unique derived features

- Teeth: the lissamphibians have either bicuspid (fig. 1, for extant species) or multicuspid teeth that hinge on a pedicel. A ventral section of a bicuspid pedicellate tooth is shown for a gymnophionan (fig. 2a), a urodele (fig. 2b) and an anuran (fig. 2c). We can see that the articulated crown (*acr*) is attached to the pedicel (*ped*) and the jaw bone (*jb*) by a ligament (*l*).

- Loss of the jugal bone: on a lateral view of an anuran skull, we can see that the jugal bone, which normally lies between the maxilla (*mx*) and the quadratojugal (*qj*), has been lost in the lissamphibia (fig. 3; *sq*: squamosal).

- The labyrinth, or inner ear, of the lissamphibia has a unique sensory plate called the *papilla amphibiorum.*

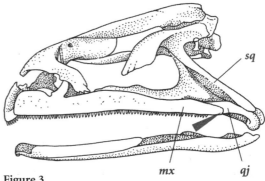

Figure 1

Figure 3

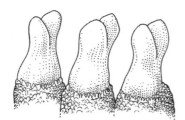

(a) (b) (c)

Figure 2

- Ribs are short, or nonexistent: as the thoracic cage is not closed, respiratory movements are impossible. It is for this reason that the lungs are inflated by swallowing gulps of air. Nonetheless, most oxygen is obtained through the skin.

- Visual focusing in the lissamphibians is performed by the backward and forward movements of the lens. This movement is controlled by an intrinsic muscle of mesodermal origin unique to the lissamphibians, the muscle *protractor lentis*. There is one such muscle in the urodeles and two in the anurans.

Number of species: 4,975.
Oldest known fossils: *Triadobatrachus massinoti* from Madagascar dates from the lower Triassic (240 MYA). The lissamphibians belong to a larger clade that includes some fossil species (microsaurs, nectridians, eryopids, dissorophids, etc.), the batrachomorpha. This clade dates from the lower Carboniferous. The batrachomorphs are the sister group of the reptiliomorphs. Along with the amniotes, the reptiliomorphs include certain fossils that were previously found in the amphibian grade (such as anthracosaurs, seymouriamorphs and diadectomorphs).
Current distribution: worldwide, except for the poles and in marine environments. Their presence is most often linked to that of freshwater or humid environments. The greatest species richness is found in tropical rain forests.

Examples

- GYMNOPHIONA: Greater yellow-banded caecilian: *Ichthyophis glutinosus;* aquatic caecilian: *Typhionectes compressicaudus; Dermophis oaxacae.*
- BATRACHIA: alpine salamander: *Salamandra atra;* crested newt: *Triturus cristatus;* tiger salamander: *Ambystoma tigrinum;* Spanish ribbed newt: *Pleurodeles waltli;* European tree frog: *Hyla arborea;* bullfrog: *Rana catesbeiana;* common toad: *Bufo bufo;* green frog: *Rana esculenta;* African clawed frog: *Xenopus laevis.*

Gymnophiona

General Description

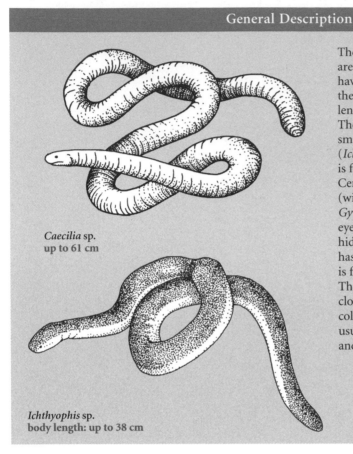

Caecilia sp.
up to 61 cm

Ichthyophis sp.
body length: up to 38 cm

The gymnophionans (or caecilians) are burrowing lissamphibians that have secondarily lost their limbs; they therefore resemble large worms. Their length varies between 6 and 140 cm. The skin is smooth and slimy with small scales embedded in the dermis (*Ichthyophis, Caecilia, Hypogeophis*) and is folded so that it forms transverse rings. Certain genera have no dermal scales (without doubt a secondary loss: *Gymnophis, Siphonops, Typhlonectes*). The eyes have atrophied and are sometimes hidden below the skin. The tympanum has been lost. A special sensory tentacle is found between the eye and the nostril. The caudal region is so short that the cloaca is almost at the terminal end. The color of these species is variable, but usually dark: brown, olive-brown, black and even midnight blue.

Some unique derived features

- Complete loss of the limbs and girdles. This loss is certainly very old because there are no longer even embryonic traces of the girdles.

- There is a protractile tentacle rich in tactile buds. The position of this organ is indicated in fig. 1 for the greater yellow-banded caecilian (*Ichthyophis glutinosus*). The details of this tentacle and its sac are shown in fig. 2. The tentacle (*te*) is housed in a cutaneous fold (*cf*), all of which is then found in a tentacular sac (*tes*) delimited by a connective tissue sheath (*cts*). The retractor muscle, *retractor tentaculi*, inserts on the tentacular sac and serves to elongate and moisten (via mucus excretion) the tentacle when it contracts. This mucus mixes with the aromatic molecules of the environment. The back and forth movements of the tentacle brings these molecules to the nasal-vomerine organ lined with an olfactive epithelium.

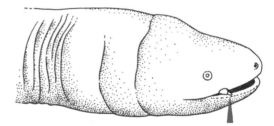

Figure 1

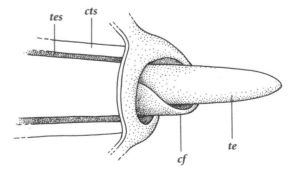

Figure 2

362

Ecology

All gymnophionans live in moist humus or in the mud of marshes, except for the genus *Typhlonectes*, which is an aquatic group. They feed on worms and small arthropods. They are hunted by snakes, but other vertebrates seem to avoid them as prey. This is likely due to their nauseating and toxic cutaneous secretions. Males possess a copulatory organ and fertilization is internal. No gymnophionans lay their eggs in water. For example, *Ichthyophis* sp. lays a dozen large eggs filled with vitellus in moist burrows close to water. The mother rolls herself around these eggs. During embryogenesis, the eggs double in size by absorbing water. At hatching the newborn has external gills and will move directly towards a water body, where it will live until metamorphosis. Several gymnophionan genera are ovoviviparous and release young into the environment that already have the adult morphology. In certain species, the oviducts play the role of the uterus and the larvae feed on the "uterine milk" secreted by glands found in the oviduct wall.

Number of species: 165.
Oldest known fossils: an unidentified gymnophionan from Arizona, United States, dates from the lower Jurassic (200 MYA). However, the group is certainly much older.
Current distribution: tropical forests—from southern Mexico to northern Argentina, Central Africa, East African forests, Seychelles, southern India, Sri Lanka, Southeast Asia, Indomalaya. They are absent from Madagascar, Australia and the Antilles.

Examples

Greater yellow-banded caecilian: *Ichthyophis glutinosus;* aquatic caecilian: *Typhlonectes compressicaudus; Siphonops annulatus;* caecilia: *Caecilia tentaculata; Dermophis oaxacae; Geotrypetes seraphini; Schistometopum gregorii.*

Batrachia

A few representatives

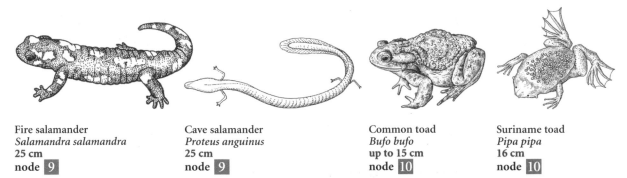

Fire salamander
Salamandra salamandra
25 cm
node **9**

Cave salamander
Proteus anguinus
25 cm
node **9**

Common toad
Bufo bufo
up to 15 cm
node **10**

Suriname toad
Pipa pipa
16 cm
node **10**

Some unique derived features

■ Inner ear: two bones lie in the oval window (*fenestra ovalis*) (*ow*, fig. 1) of the inner ear: the stapes or columella (*st*) and the operculum (*op*). A transverse section of an anuran head (fig. 1) shows the location of the operculum (*sc:* semi-circular canals, *ec:* extracolumella, *tym:* tympanic membrane, *ie:* inner ear, *me:* middle ear, *ph:* pharynx, *sac:* sacculus, *Et:* Eustachian tube linking the cavity of the middle ear to the pharynx). The oper-culum is connected to the pectoral girdle (*ssc:* supra-scapular) by a muscular arrangement. This arrangement is shown in fig. 2 on a salam-ander. Vibrations are trans-mitted from the legs to the pectoral girdle (*ssc*) and then from there to the operculum via the opercularis muscle (*om*). The operculum then transmits these vibrations to the inner ear. We should note that, unlike the frogs, the salamanders have lost the tympanum. The stapes is therefore linked to the squa-mosal (*sq*) by a ligament (*l*). Vibrations are also transmitted from the mandible to the oval window via the squamosal, the ligament and the stapes.

■ The development of the choanae is unique in the

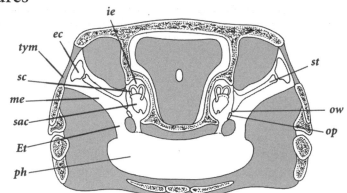

Figure 1

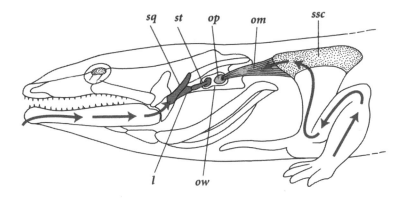

Figure 2

batrachians and results in endodermal choanae. Fig. 3 shows the nasal sac (*ns*) of a frog with the nasolacrimal duct (*nld*), the external nare or nostril (*en*), and the internal nare or choana (*cho*) that empties into the buccal cavity. Fig. 4 shows the formation of

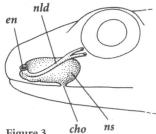

Figure 3

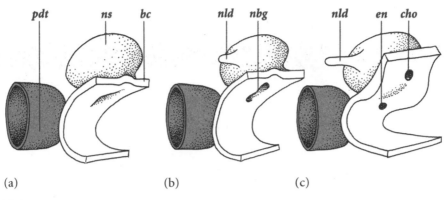

Figure 4

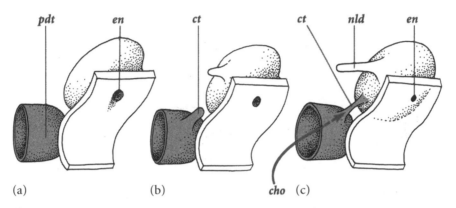

Figure 5

the choanae in non-batrachians like the gymnophionans and the amniotes; fig. 5 shows this same process in the batrachians (anurans and urodeles). In non-batrachians (fig. 4), the wall of the buccal cavity (*bc*) and the nasal sac (*ns*) are both of ectodermal orgins and meet to form the naso-buccal groove (*nbg*) (fig. 4a and fig. 4b). This groove will eventually form two orifices, the external nare and the choana (fig. 4c), while at the same time the nasolacrimal

duct will develop. The choana (*cho*) is therefore derived from the ectoderm. In the batrachians (fig. 5), the nasal sac and wall of the buccal cavity give rise to a single orifice, the external nare (ectodermal origin; fig. 5a). The anterior end of the primitive digestive tube (*pdt*, of endodermal orgin) sends out an extension that joins the nasal sac to form the choanal tube (*ct*, fig. 5c). The nasal sac is therefore connected to

the digestive tube and the choana (*cho*) is of endodermal orgin.

- Pigment: there is a green pigment in the rods of the retina.

Examples

- URODELA: fire salamander: *Salamandra salamandra;* Chinese giant salamander: *Andrias davidianus;* tiger salamander: *Ambystoma tigrinum;* crested newt: *Triturus cristatus;* cave salamander: *Proteus anguinus; Plethodon cinereus.*
- ANURA: Hochstetter's frog: *Leiopelma hochstetteri;* African clawed frog: *Xenopus laevis;* common spadefoot toad: *Pelobates fuscus;* common toad: *Bufo bufo;* European tree frog: *Hyla arborea;* green frog: *Rana esculenta.*

Number of species: 4,810.
Oldest known fossils: *Triadobatrachus massinoti* from Madagascar dates from the lower Triassic (240 MYA).
Current distribution: worldwide, in all climates and on almost all continents; they are absent from both Arctic (above the Arctic Circle) and Antarctic regions.

Urodela

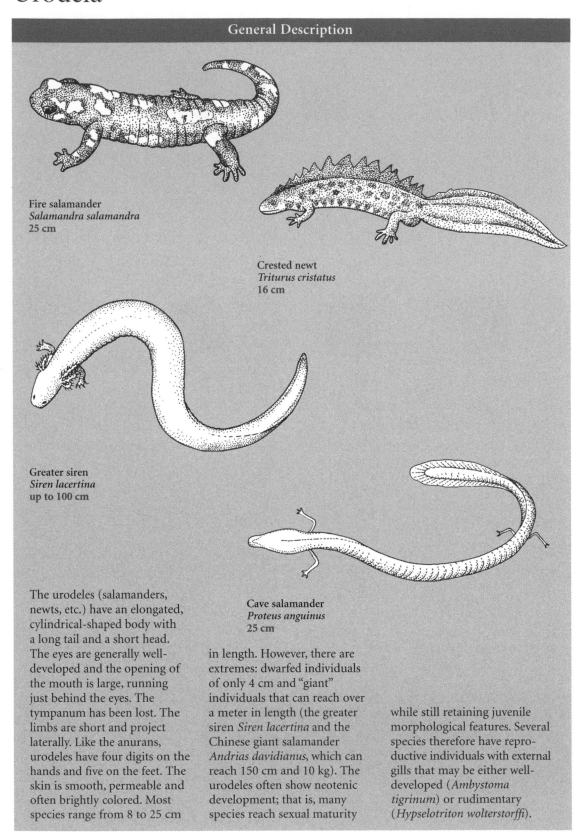

General Description

Fire salamander
Salamandra salamandra
25 cm

Crested newt
Triturus cristatus
16 cm

Greater siren
Siren lacertina
up to 100 cm

Cave salamander
Proteus anguinus
25 cm

The urodeles (salamanders, newts, etc.) have an elongated, cylindrical-shaped body with a long tail and a short head. The eyes are generally well-developed and the opening of the mouth is large, running just behind the eyes. The tympanum has been lost. The limbs are short and project laterally. Like the anurans, urodeles have four digits on the hands and five on the feet. The skin is smooth, permeable and often brightly colored. Most species range from 8 to 25 cm in length. However, there are extremes: dwarfed individuals of only 4 cm and "giant" individuals that can reach over a meter in length (the greater siren *Siren lacertina* and the Chinese giant salamander *Andrias davidianus*, which can reach 150 cm and 10 kg). The urodeles often show neotenic development; that is, many species reach sexual maturity while still retaining juvenile morphological features. Several species therefore have reproductive individuals with external gills that may be either well-developed (*Ambystoma tigrinum*) or rudimentary (*Hypselotriton wolterstorffi*).

Some unique derived features

■ The monophyly of the urodeles is not based on any morphological or anatomical derived characters. The external signs that distinguish them within the lissamphibia (long tail, short limbs) are in fact only primitive characters of the amphibians and are already present in other extant and fossil tetrapods. It is said that the urodeles may be paraphyletic and constitute the stem group of anurans. However, the monophyly of this group seems to be supported by phylogenies of four different mitochondrial genes.

Ecology

The urodeles are even more sensitive to the absence of water than the anurans, explaining their fragmented and limited distribution. Most species divide their life cycle between aquatic (where at least reproduction and larval development occurs) and terrestrial habitats; some species are arboreal, others are cavernous, a few are strictly aquatic and finally still others are strictly terrestrial. We do not find them above 4500 m altitude. They are widely distributed in temperate zones, with one extreme case, the Siberian salamander (*Hynobius keyserlingii*), whose distribution reaches the Arctic Circle. The urodeles feed on worms, larval and adult insects, crustaceans, mollusks, eggs and anuran tadpoles. They are hunted by birds, rodents, canid and mustelid carnivores and by aquatic snakes. They reproduce without copulation, regardless of whether fertilization is internal or external. After a nuptial parade, the male deposits his spermatophores, small cones that contain the spermatozoids, in front of the female. She grabs them using her cloacal lips and stores them in a spermatheca. Most urodeles are oviparous. After fertilization, the eggs are deposited in water or moist earth. At hatching, the larvae possess three pairs of well-developed external gills and will take on an aquatic existence until metamorphosis. The alpine salamander (*Salamandra atra*) is an example of a viviparous urodele: the young develop and metamorphose within the maternal oviduct and are born in the adult form. Gestation lasts two to four years. This unusual characteristic is thought to be an adaptation to cool mountain temperatures.

Number of species: 429.
Oldest known fossils: *Albanerpeton inexpectatum* from the Aveyron region (France) dates from the middle Jurassic (170 MYA).
Current distribution: in the Americas, the urodeles are found from southern Canada to northern Bolivia. They are also present throughout Europe, Turkey and the Middle East. In Asia, the distribution is fragmented: we find them in Siberia, Manchuria, the Korean peninsula, Japan and the southwestern region of China. In Africa, they are present only in the extreme north of the Maghreb region (Morocco, Algeria and Tunisia). They are thus absent from almost all of Africa, from Madagascar and from the Arabian Peninsula. They are also absent from India, Southeast Asia, Australia and the Pacific Islands, the southern half of South America and the Arctic and Antarctic zones.

Examples

Fire salamander: *Salamandra salamandra;* alpine salamander: *Salamandra atra;* Siberian salamander: *Hynobius keyserlingii;* Chinese giant salamander: *Andrias davidianus;* tiger salamander: *Ambystoma tigrinum; Hypselotriton wolterstorffi;* crested newt: *Triturus cristatus;* Spanish ribbed newt: *Pleurodeles waltli;* cave salamander: *Proteus anguinus; Plethodon cinereus; Amphiuma means.*

Anura

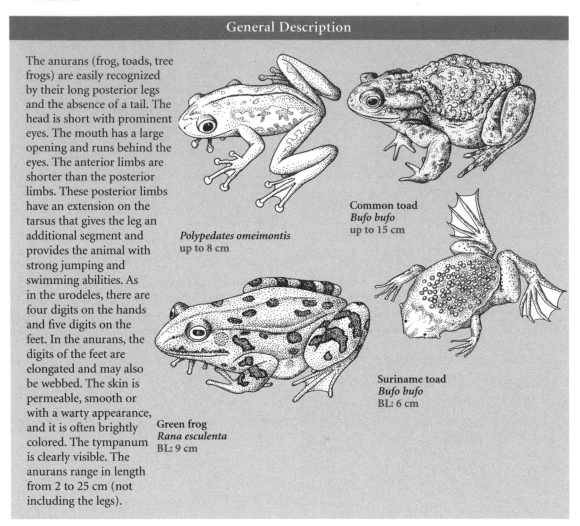

The anurans (frog, toads, tree frogs) are easily recognized by their long posterior legs and the absence of a tail. The head is short with prominent eyes. The mouth has a large opening and runs behind the eyes. The anterior limbs are shorter than the posterior limbs. These posterior limbs have an extension on the tarsus that gives the leg an additional segment and provides the animal with strong jumping and swimming abilities. As in the urodeles, there are four digits on the hands and five digits on the feet. In the anurans, the digits of the feet are elongated and may also be webbed. The skin is permeable, smooth or with a warty appearance, and it is often brightly colored. The tympanum is clearly visible. The anurans range in length from 2 to 25 cm (not including the legs).

Polypedates omeimontis
up to 8 cm

Common toad
Bufo bufo
up to 15 cm

Suriname toad
Bufo bufo
BL: 6 cm

Green frog
Rana esculenta
BL: 9 cm

Some unique derived features

Within the salientians (that is, *Triadobatrachus* and the anurans):

■ The salientians have a characteristic elongation of the ilium (*il*) toward the anterior end of the body. This is seen in the skeleton of a common toad (*Bufo bufo*) in fig. 3, on an anuran fossil species *Vieraella* (fig. 1) and on the fossil salientian *Triadobatrachus* (fig. 2; *atl:* atlas, *as:* astragalus, *cal:* calcaneum, *ri:* rib, *f:* femur, *fp:* frontoparietal, *h:* humerus, *mtt:* metatarsals, *mx:* maxilla, *na:* nasal, *pmx:* premaxilla, *prt:* prootic, *qj:* quadratojugal, *rau:* radioulna (fused radius and ulna), *sm:* sacrum, *sq:* squamosal, *ssc:* suprascapula, *ur:* urostyle).

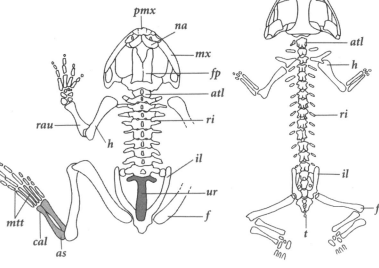

Figure 1

Figure 2

■ There is an extreme reduction (fig. 1, fig. 2) or complete disappearance (fig. 3) of the ribs. In *Vieraella* (fig. 1) we can see some residual ribs on the anterior vertebrae.

In the modern salientians (that is, the anurans):

■ The postsacral vertebrae are transformed into a rigid central rod, the urostyle. This is readily seen in the common toad (fig. 3). In *Triadobatrachus*, however, there is still a residual tail (*t*, fig. 2).

■ The number of vertebrae is reduced (fig. 3; 8 vertebrae in the toad).

■ Two of the ankle bones, the astragalus and calcaneum, are greatly elongated and form an additional segment in the posterior legs.

Ecology

More so than in the urodeles, there is an increasing independence from the aquatic environment in the anurans. The anurans inhabit a large diversity of biotopes: equatorial forests, deserts, tundra, and mountains up to the permanent snow line. They are absent from the marine environment, although a couple of species can tolerate brackish waters. The anurans feed principally on arthropods and protect themselves from their numerous predators via toxic cutaneous secretions. Contrary to the urodeles and gymnophionans, which make only short, sharp cries, the anurans emit powerful, specific and varied vocalizations. These songs have important roles in social interactions, territorial defense and reproduction. Correlated with this singing ability is a well-developed tympanum that ensures strong auditory capacities. One of the most interesting aspects of this group lies in their diverse reproductive mechanisms. With only the rare exception, fertilization is external. The nuptial parade results in a type of false copulation called an amplexus: the male uses his front legs to clasp the female under the forearms or around the abdomen. Classically, the fertilized eggs develop in the aquatic

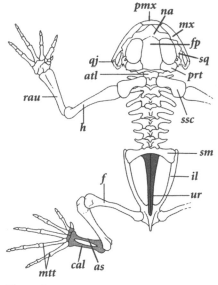

Figure 3

environment and produce legless larvae with external gills and a tail (tadpoles). The metamorphosis into the adult form is spectacular. Predation on the egg clutches and larvae is very heavy in the anurans. For this reason, many species have acquired ways to protect their offspring. Certain pipid species house their eggs within the skin on their backs. Some tree frogs (genus *Gastrotheca*) keep their eggs in a "marsupial"-like pouch found on the back until the eggs reach an advanced developmental stage (or up to the final stage). In other species, like *Assa darlingtoni*, males have incubating pouches above the groin. The clutch is therefore terrestrial; after hatching, these pouches will continue to hold the larvae until development is completed. Other species transport their tadpoles on their backs. Finally, certain mountainous toads (genus *Nectophrynoides*) are viviparous: larval development takes place in the female genital tract during a nine-month period. The young are then released to the world.

Examples

Hochstetter's frog: *Leiopelma hochstetteri*; tailed frog: *Ascaphus truei*; midwife toad: *Alytes obstetricans*; Suriname toad: *Pipa pipa*; African clawed frog: *Xenopus laevis*; common spadefoot toad: *Pelobates fuscus*; parsley frog: *Pelodytes punctatus*; *Ceratophrys ornata*; common toad: *Bufo bufo*; European tree frog: *Hyla arborea*; green frog: *Rana esculenta*; bullfrog: *R. catesbeiana*.

Number of species: 4,381.
Oldest known fossils: *Vieraella herbstii* (fig. 1) from the lower Jurassic from Argentina (205 MYA) is the first true frog known (with a urostyle). *Triadobatrachus* (fig. 2), from the lower Triassic from Madagascar, is not a true frog: it differs from the modern anurans in that it lacks a urostyle and elongated posterior legs. *Triadobatrachus* together with the anurans form the salientian clade.
Current distribution: worldwide, in all climates and on almost all continents; they are absent from Arctic (beyond the Arctic Circle) and Antarctic regions.

Amniota

A few representatives

Chimpanzee
Pan troglodytes
90 cm
node 12

European pond turtle
Emys orbicularis
14 cm
node 14

Dipsas indica
70 cm
node 17

African grey parrot
Psittacus erithacus
33 cm
node 20

Some unique derived features

- Amnion: a membrane surrounds the embryo and forms a liquid-filled sac where the young develops (fig. 1; *all:* allantois, *ci:* chorion, *cic:* chorionic cavity, *ys:* yolk sac). This membrane is the amnion (*an*) and the liquid that fills it, the amniotic fluid (*af*). The amniotes are in fact tetrapods that have freed themselves from an aquatic environment for reproduction. However, the embryo (*em*), protected by the hard-shelled egg, still develops within an aquatic milieu within the amnion. In therian mammals, development is *in utero,* but the amnion is still present.

- The occipital condyle is convex. In fig. 2, there are two posterior views (fig. 2a and fig. 2b) and two ventral views (fig. 2c and fig. 2d) of four different

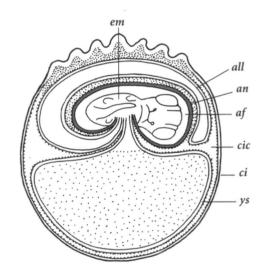

Figure 1

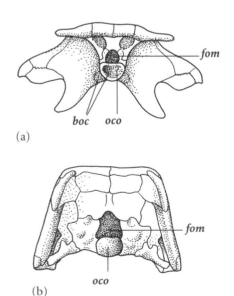

(a)

(b)

Figure 2

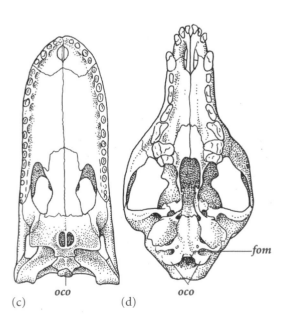

(c)

(d)

skulls. In non-amniote tetrapods from the Carboniferous, like *Palaeogyrinus* (fig. 2a), the occipital condyle (*oco*) is concave, whereas it is convex in the amniote *Paleothyris* from the same period (fig. 2b), as it is in two extant amniotes, a crocodile (fig. 2c) and a canid mammal (fig. 2d; *boc:* basioccipital, *fom:* foramen magnum).

■ There has been a loss of one of the bones from the skull roof, the intertemporal. Fig. 3 shows the skull roof of a non-amniote tetrapod from the Carboniferous, *Palaeogyrinus* (fig. 3a), and an amniote from the same period, *Paleothyris* (fig. 3b). The intertemporal (*itt*) is present in the former and missing in the latter (*q:* quadrate, *pfo:* parietal foramen, *j:* jugal, *la:* lacrimal, *mx:* maxilla, *na:* nasal, *en:* external nares, *pa:* parietal, *pf:* prefrontal, *pmx:* premaxilla, *po:* post-orbital, *pof:* post-frontal, *pp:* post-parietal, *qj:* quadratojugal, *smx:* septomaxilla, *sq:* squamosal, *sut:* supratemporal, *tab:* tabular).

■ On the amniote skull roof, the frontal bone (*fr*) borders the orbit (*or,* fig. 3).

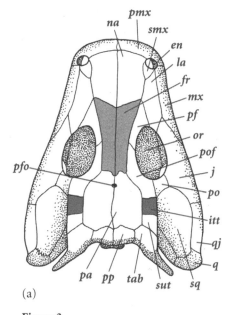

(a)

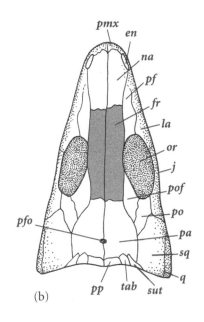

(b)

Figure 3

Number of species: 21,333.
Oldest known fossils: *Westlothania lizziae* from Scotland dates from the lower Carboniferous (340 MYA). The amniote clade, along with some fossil groups (diadectomorpha, seymouriamorpha, anthracosauria), are part of the larger reptilomorpha clade. The reptilomorpha are the sister group of the batrachomorpha, a separate clade that includes the fossil and extant "amphibians" (lissamphibia).
Current distribution: worldwide, at all latitudes and altitudes, except at great ocean depths.

Examples

■ Mammalia: cat: *Felis catus;* human: *Homo sapiens.*
■ Sauropsida: leatherback turtle: *Dermochelys coriacea;* Hermann's tortoise: *Testudo hermanni;* green lizard: *Lacerta viridis;* grass snake: *Natrix natrix;* tuatara: *Sphenodon punctatus;* Nile crocodile: *Crocodylus niloticus;* domestic chicken: *Gallus gallus.*

→ *see chap. 13, p. 388*

Mammalia

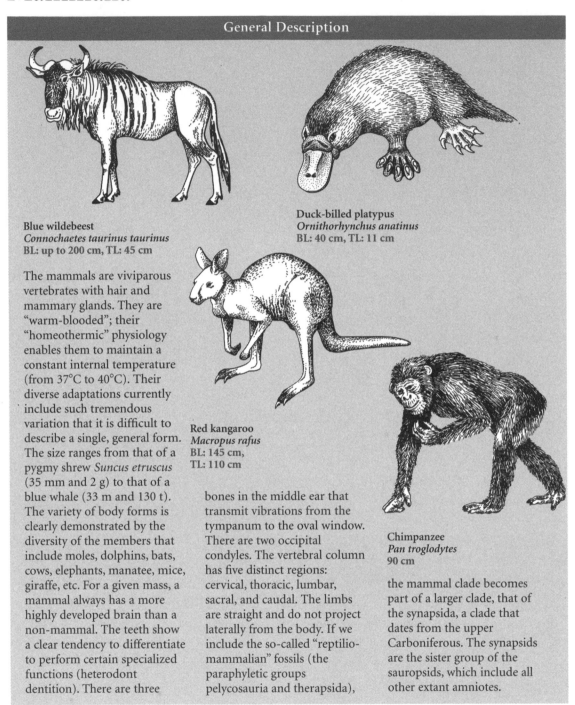

General Description

Blue wildebeest
Connochaetes taurinus taurinus
BL: up to 200 cm, TL: 45 cm

Duck-billed platypus
Ornithorhynchus anatinus
BL: 40 cm, TL: 11 cm

Red kangaroo
Macropus rufus
BL: 145 cm,
TL: 110 cm

Chimpanzee
Pan troglodytes
90 cm

The mammals are viviparous vertebrates with hair and mammary glands. They are "warm-blooded"; their "homeothermic" physiology enables them to maintain a constant internal temperature (from 37°C to 40°C). Their diverse adaptations currently include such tremendous variation that it is difficult to describe a single, general form. The size ranges from that of a pygmy shrew *Suncus etruscus* (35 mm and 2 g) to that of a blue whale (33 m and 130 t). The variety of body forms is clearly demonstrated by the diversity of the members that include moles, dolphins, bats, cows, elephants, manatee, mice, giraffe, etc. For a given mass, a mammal always has a more highly developed brain than a non-mammal. The teeth show a clear tendency to differentiate to perform certain specialized functions (heterodont dentition). There are three bones in the middle ear that transmit vibrations from the tympanum to the oval window. There are two occipital condyles. The vertebral column has five distinct regions: cervical, thoracic, lumbar, sacral, and caudal. The limbs are straight and do not project laterally from the body. If we include the so-called "reptilio-mammalian" fossils (the paraphyletic groups pelycosauria and therapsida), the mammal clade becomes part of a larger clade, that of the synapsida, a clade that dates from the upper Carboniferous. The synapsids are the sister group of the sauropsids, which include all other extant amniotes.

Some unique derived features

- There is dentary-squamosal jaw articulation. Figs. 1, 2 and 3 show lateral views of three synapsid skulls. In some non-mammalian synapsids, like the pelycosaur *Ophiacodon* from the lower Permian (fig. 1), there is an articular (*ar*)–quadrate (*q*) jaw articulation. In other non-mammalian synapsids, like the cynodont *Probainognathus* (fig. 2), there is a double articulation: articular-quadrate and dentary-squamosal (the dentary has a posterior blade that connects with the squamosal). Finally, in extant mammals (*Canis*, fig. 3) only the dentary and squamosal take part in the jaw articulation (*ang:* angular, *q:* quadrate, *cor:* coronoid, *de:* dentary, *epg:* epipterygoid, *fr:* frontal, *tf:* temporal fenestra, *j:* jugal, *la:* lacrimal, *mx:* maxilla, *na:* nasal, *occ:* occipital, *or:* orbit, *pa:* parietal, *pf:* prefrontal, *pg:* pterygoid, *pmx:* premaxilla, *po:* post-orbital, *qj:* quadrato-jugal, *sq:* squamosal).

- The jugal teeth have two roots.

- The internal wall that separates the orbits consists of the orbitosphenoid bone and the ascending process of the palatine.

- Hair, the hair erector muscles (*arrector pili* muscles), and the associated sebaceous glands are all characteristic of the mammals. In the dermis, there are also sweat glands that play important behavioral and thermoregulatory roles.

- Mammary glands are characteristic of mammals. These glands synthesize milk, the only food of newborns. The mammary glands are arranged in pairs along two thoracic-abdominal lines, but the number vary from species to species.

Ecology

Because of their homeothermy, the mammals have been able to conquer the coldest environments (Arctic and Antarctic poles), along with diverse habitats of more temperate and arid regions. The chiropterans (bats) have invaded the air (particularly after sunset) and the cetaceans (dolphins, whales, etc.) the sea. The mammals are even present underground (moles, shrews, mole rat, etc.). There are many factors that have likely contributed to the extreme success of this group: they are viviparous (for the most part), independent of water for reproduction, and homeothermic, there is a strong bond between mother and offspring, and they have complex behaviors, strong social interactions and a learning capacity superior to that of other vertebrate groups (for the most part).

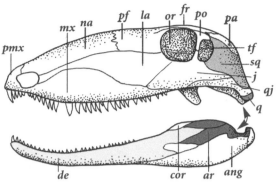

Figure 1

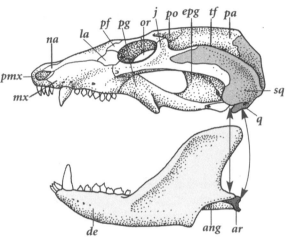

Figure 2

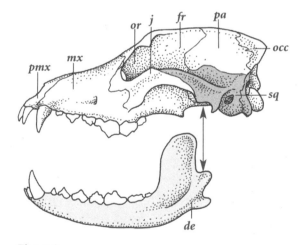

Figure 3

Number of species: 4,496.
Oldest known fossils: *Haramiya* from England dates from the upper Triassic (Norian, 220 MYA); *Morganucodon* and *Eozostrodon* from Europe and *Sinoconodon* from Yunnan, China, also date from the upper Triassic (Rhetian, 210 MYA).
Current distribution: worldwide.

Examples

Duck-billed platypus: *Ornithorhynchus anatinus;* red kangaroo: *Macropus rufus;* three-toed sloth: *Bradypus tridactylus;* European hedgehog: *Erinaceus europaeus;* cat: *Felis catus;* house mouse: *Mus musculus;* gorilla: *Gorilla gorilla;* human: *Homo sapiens;* African elephant: *Loxodonta africana;* domestic horse: *Equus przewalskii caballus;* common dolphin: *Delphinus delphis;* red bat: *Lasiurus borealis;* giant pangolin: *Manis gigantea.*

Sauropsida

A few representatives

European pond turtle
Emys orbicularis
14 cm
node **14**

Dipsas indica
70 cm
node **17**

African grey parrot
Psittacus erithacus
33 cm
node **20**

Indian gharial
Gavialis gangeticus
700 cm
node **21**

Some unique derived features

- There is a ventral blade on the cervical vertebrae called the hypapophysis (*hy*). In a synapsid from the upper Permian (*Ophiacodon*, fig. 1), the cervical vertebrae lack the hypapophysis. Four cervical vertebrae are shown in a tuatara (fig. 2), in the lizard *Ophisaurus apodus* (fig. 3), and in the monitor lizard *Varanus monitor* (fig. 4). On fig. 5, we see the two first cervical vertebrae and an anterior dorsal vertebra of the Balkan whip snake *Coluber gemonensis* (*nar*: neural arch, *cy*: condyle, *di*: diapophysis, *pap*: parapophysis, *ce*: centrum, *ply*: pleurapophysis, *pra*: proatlas, *pz*: postzygapophysis, *z*: zygosphene).

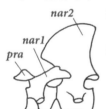

- The iris of the eye contains striated muscles.

- The sauropsids produce ornithuric acid.

Figure 1

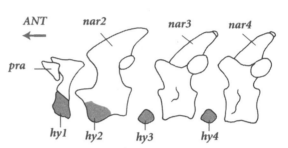

Figure 2

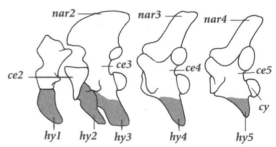

Figure 3

> **Number of species:** 16,837.
> **Oldest known fossils:** *Hylonomus lyelli* is from Nova Scotia, Canada, and dates from the upper Carboniferous (315 MYA). A yet unnamed sauropsid was discovered in Scotland from the late Carboniferous (340 MYA).
> **Current distribution:** worldwide.

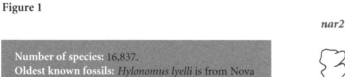

Figure 4

Examples

- CHELONIA: Hermann's tortoise: *Testudo hermanni*; leatherback turtle: *Dermochelys coriacea*.
- DIAPSIDA: green lizard: *Lacerta viridis*; grass snake: *Natrix natrix*; red worm lizard: *Amphisbaena alba*; tuatara: *Sphenodon punctatus*; domestic chicken: *Gallus gallus*; ostrich: *Struthio camelus*; great cormorant: *Phalacrocorax carbo*; snowy owl: *Nyctea scandiaca*; Nile crocodile: *Crocodylus niloticus*.

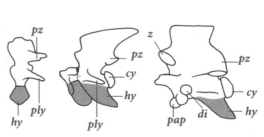

Figure 5

Chelonia

General Description

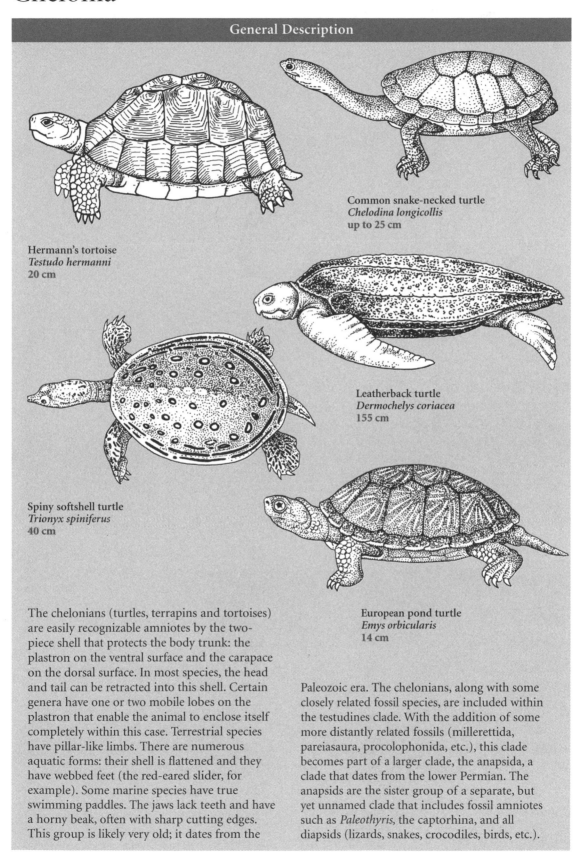

Hermann's tortoise
Testudo hermanni
20 cm

Common snake-necked turtle
Chelodina longicollis
up to 25 cm

Leatherback turtle
Dermochelys coriacea
155 cm

Spiny softshell turtle
Trionyx spiniferus
40 cm

European pond turtle
Emys orbicularis
14 cm

The chelonians (turtles, terrapins and tortoises) are easily recognizable amniotes by the two-piece shell that protects the body trunk: the plastron on the ventral surface and the carapace on the dorsal surface. In most species, the head and tail can be retracted into this shell. Certain genera have one or two mobile lobes on the plastron that enable the animal to enclose itself completely within this case. Terrestrial species have pillar-like limbs. There are numerous aquatic forms: their shell is flattened and they have webbed feet (the red-eared slider, for example). Some marine species have true swimming paddles. The jaws lack teeth and have a horny beak, often with sharp cutting edges. This group is likely very old; it dates from the Paleozoic era. The chelonians, along with some closely related fossil species, are included within the testudines clade. With the addition of some more distantly related fossils (millerettida, pareiasaura, procolophonida, etc.), this clade becomes part of a larger clade, the anapsida, a clade that dates from the lower Permian. The anapsids are the sister group of a separate, but yet unnamed clade that includes fossil amniotes such as *Paleothyris*, the captorhina, and all diapsids (lizards, snakes, crocodiles, birds, etc.).

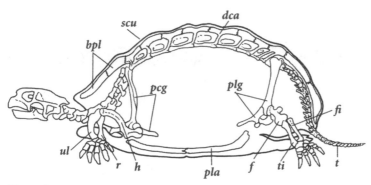

Figure 1

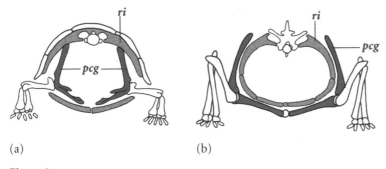

(a)   (b)

Figure 2

tween the dermal badges are not the same as those between the bony plates.

- The ventral side of the body is protected by the plastron shell (fig. 1).

- The girdles lie within the rib cage. Two cross-sections of the pectoral girdle are shown in fig. 2 for a turtle (2a) and an alligator (2b). In the turtle, the pectoral girdle is enclosed within the ribs (*ri*: rib), whereas in the alligator, this girdle remains exterior to the rib cage.

- The maxillary, premaxillary and dentary bones all lack teeth. The skull of *Lepidochelys* is shown in fig. 3 (*de*: dentary, *mx*: maxilla, *pmx*: premaxilla)

- The extant chelonians have a horny beak.

Some unique derived features

- The dorsal carapace is composed of bony plates (*bpl*) that are fused together and covered by thick scales, the scutes (*scu*) or dermal badges. The bony plates are attached to the ribs and vertebrae. A longitudinal section of the skeleton of *Geochelone elephantopus* is shown in fig. 1 (*dca*: dorsal carapace, *plg*: pelvic girdle, *pcg*: pectoral girdle, *ul*: ulna, *f*: femur, *h*: humerus, *fi*: fibula, *pla*: plastron, *t*: tail, *r*: radius, *ti*: tibia). The sutures be-

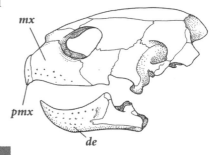

Figure 3

Number of species: 290.
Oldest known fossils: *Proganochelys quenstedtii* from Germany and Thailand dates from the upper Triassic (210 MYA).
Current distribution: worldwide, in warm and temperate zones.

Ecology

The chelonians are mostly either carnivores or omnivores. They can be terrestrial, living in temperate and warm climates, and can even be found in scorching deserts. In temperate climates, they hibernate during the cold season. Many species are aquatic. There are also a few species that live in the open ocean like the leatherback turtle. In all cases, the chelonians are oviparous; the marine species return to land to lay their eggs on beaches. The Galápagos Islands are home to giant tortoises. These animals have an incredible lifespan: certain species can live more than 120 years.

Examples

Galápagos tortoise: *Geochelone elephantopus*; giant tortoise: *Testudo gigantea*; Hermann's tortoise: *Testudo hermanni*; common snapping turtle: *Chelydra serpentina*; European pond turtle: *Emys orbicularis*; red-eared slider: *Pseudemys scripta elegans*; common map turtle: *Graptemys geographica*; green sea turtle: *Chelonia mydas*; loggerhead turtle: *Caretta caretta*; leatherback turtle: *Dermochelys coriacea*; Florida softshell turtle: *Trionyx ferox*; matamata: *Chelus fimbriatus*.

Diapsida

A few representatives

Dipsas indica
70 cm
node 17

Tuatara
Sphenodon punctatus
up to 70 cm
node 18

African grey parrot
Psittacus erithacus
33 cm
node 20

Indian gharial
Gavialis gangeticus
700 cm
node 21

Some unique derived features

- There are two temporal openings (fenestrae) behind the orbit. The skull of an extant sphenodont (*Sphenodon*) is shown in fig. 1 (*ltf:* lower temporal fenestra, *utf:* upper temporal fenestra, *or:* orbit).

- There is a large sub-orbital fenestra that opens on the palate. Dorsal views of two extant diapsids are

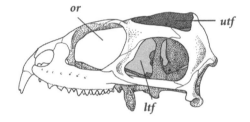

Figure 1

shown in fig. 2, that of a lepidosaur (the lizard *Tupinambis nigropunctatus,* fig. 2a) and that of an archosaur (the crocodile *Alligator mississipiensis,* fig. 2b; *sof:* sub-orbital fenestra).

- There is a ridge-and-groove ankle joint (tarsal bones) between the tibia and the astragalus.

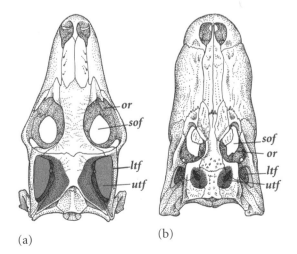

(a)   (b)

Figure 2

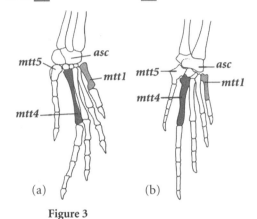

(a)   (b)

Figure 3

- The first metatarsal is only half the length of the fourth. Fig. 3 shows the left foot of a fossil sphenodont *Homoeosaurus* from the Jurassic (3a) and that of an extant lizard *Tupinambis* (3b; *asc:* fused astragalus and calcaneum, *mtt1:* first metatarsal).

Number of species: 16,547.
Oldest known fossils: *Petrolacosaurus kansensis* from Kansas, United States, dates from the upper Carboniferous (300 MYA). This clade, along with some fossil species (*Paleothyris,* captorhinids), is part of a larger, unnamed group that is characterized by the loss of the connection between the post-orbital and supratemporal bones. This clade is the sister group of a separate clade (the anapsida) that includes the chelonians and several fossil species (millerettids, pareiasaurs, etc.).
Current distribution: worldwide in terrestrial environments. They are absent from open sea water; crocodiles and marine snakes are coastal.

Examples

- LEPIDOSAURIA: Komodo dragon: *Varanus komodoensis;* green lizard: *Lacerta viridis;* grass snake: *Natrix natrix;* red worm lizard: *Amphisbaena alba;* tuatara: *Sphenodon punctatus.*
- ARCHOSAURIA: ostrich: *Struthio camelus;* albatross: *Diomedea exulans;* peregrine falcon: *Falco peregrinus;* great cormorant: *Phalacrocorax carbo;* Nile crocodile: *Crocodylus niloticus.*

Lepidosauria

A few representatives

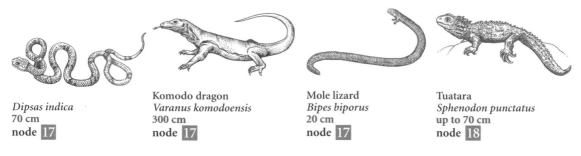

Dipsas indica
70 cm
node 17

Komodo dragon
Varanus komodoensis
300 cm
node 17

Mole lizard
Bipes biporus
20 cm
node 17

Tuatara
Sphenodon punctatus
up to 70 cm
node 18

Some unique derived features

■ The astragalus and calcaneum of the ankle skeleton are fused. The left foot of a sphenodont fossil *Homoeosaurus* from the Jurassic is shown in fig. 1, whereas that of the extant sphenodont *Sphenodon* is shown in fig. 2. A posterior leg of a lizard (*Tupinambis*) is shown in fig. 3. Finally, the primitive arrangement is outlined in fig. 4 by the foot of a pelycosaur from the Permian, *Ophiacodon* (synapsid lineage) (*as*: astragalus, *asc*: fused astragalus-calcaneum, *cal*: calcaneum, *cen*: centralia, *mtt1*: first metatarsal, *mtt5*: bent fifth metatarsal, *dta1*: first distal tarsal, *dta4+5*: fused fourth and fifth distal tarsals).

■ In the ankle bones, there is a loss of the centralia and the first and fifth distal tarsals (figs. 1 to 3). We see these bones in non-lepidosaurs like *Ophiacodon* (fig. 4).

■ The fifth metatarsal is bent or hooked (figs. 1 to 3).

■ Along with some fossils species (sauropterygians, ichthyosaurs), this clade is part of a larger clade, the lepidosauromorpha. The lepidosauromorphs are the sister group of the archosauromorphs, a distinct clade that includes the crocodiles and birds. In the lepidosauromorphs, the lamina of the quadrate bone supports the tympanum. There is also a loss of the postparietal and supratemporal bones (two bones of the skull roof) and locomotion is performed by body undulations.

Examples

■ SPHENODONTIA: tuatara: *Sphenodon punctatus*.
■ SQUAMATA: tokay gecko: *Gekko gecko*; common iguana: *Iguana iguana*; chameleon: *Chamaeleo chamaeleon*; Komodo dragon: *Varanus komodoensis*; slow worm: *Anguis fragilis*; grass snake: *Natrix natrix*; green anaconda: *Eunectes murinus*; desert horned viper: *Cerastes cerastes*; red worm lizard: *Amphisbaena alba*.

Figure 1 **Figure 2**

Number of species: 6,852.
Oldest known fossils: *Saurosternon bainii* from South Africa (upper Permian, 250 MYA).
Current distribution: they are found worldwide in terrestrial environments, except at the poles. Most marine lepidosaurs are coastal (for example, the iguana of the Galápagos Islands and marine snakes).

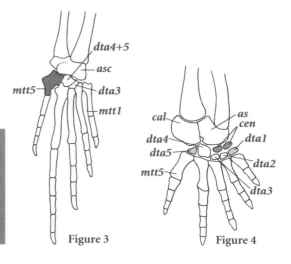

Figure 3 **Figure 4**

see appendix 6, p. 525

Squamata

General Description

The squamate group includes vertebrates with elongated bodies, long tails and horny scales. They move about by either quadruped or rectilinear locomotion (snakes, worm lizards and certain lizards have lost their limbs secondarily). As diapsids, they have two fenestrae in the temporal region, one above the other. These openings are separated by a bar of bone. The lizards have lost the part of the bottom bar that delimits the lower fenestra such that the lower surface remains open. The snakes show this same pattern, except they have also lost the intermediate bar that separates the upper and lower fenestrae. Their temples are therefore completely open, leaving the neurocranium exposed. This arrangement allows a maximal degree of jaw and skull flexibility, enabling the snakes to ingest large prey.

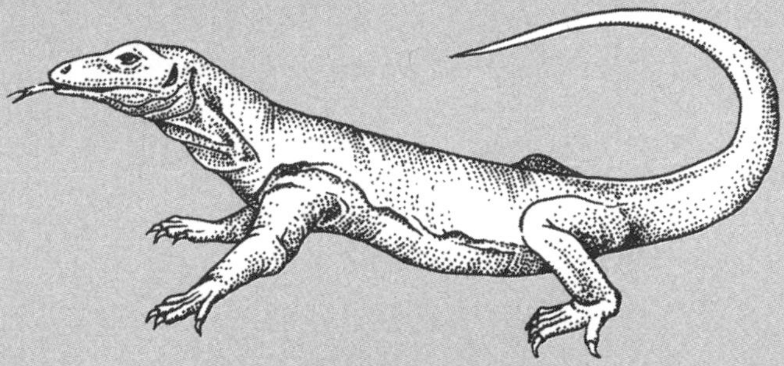

Komodo dragon
Varanus komodoensis
300 cm

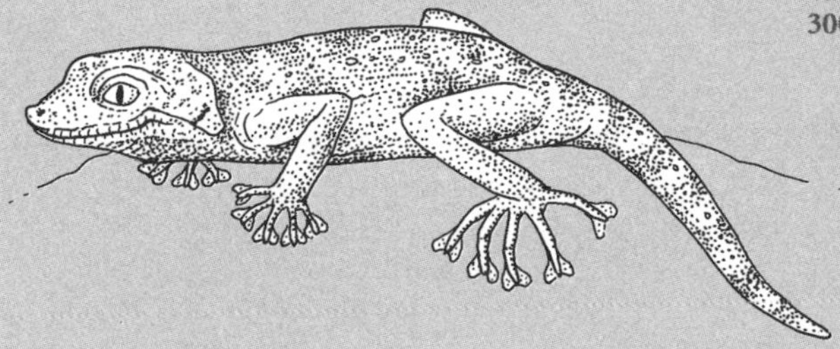

Fan-footed gecko
Ptyodactylus hasselquistii
15 cm

Dipsas indica
70 cm

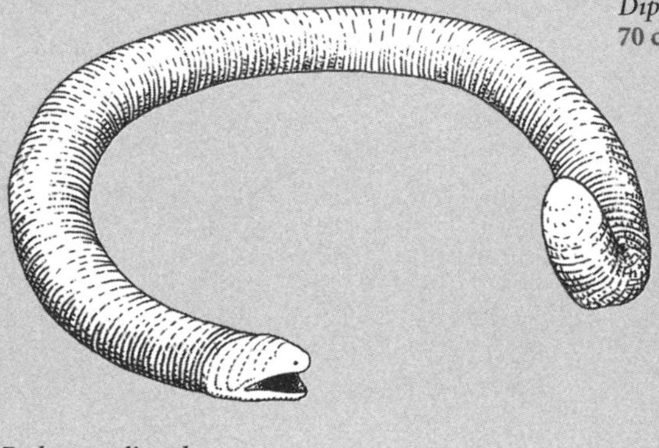

Red worm lizard
Amphisbaena alba
61 cm

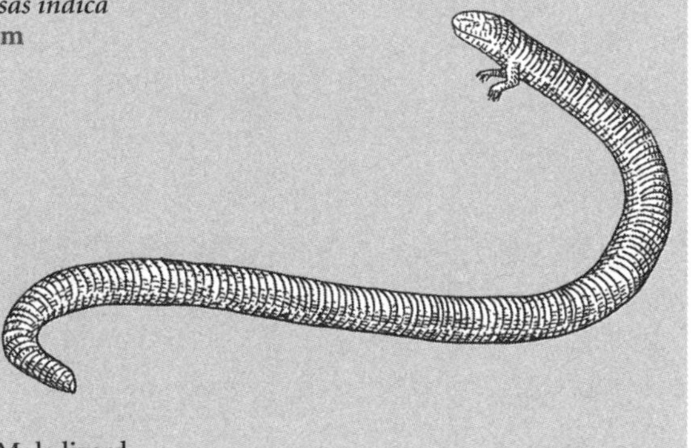

Mole lizard
Bipes biporus
20 cm

Some unique derived features

■ The squamates have lost the lower temporal bar. A non-squamate skull (*Sphenodon*) is shown in fig. 1. We note the two temporal fenestrae; the upper fenestra (*utf*) is closed along the lower edge by the post-orbital-squamosal bar, whereas the lower fenestra (*ltf*) is closed off by the jugal-quadratojugal bar. In a squamate, such as the lizard *Tupinambis nigropunctatus* (fig. 2), the lower bar has disappeared and the lower temporal fenestra is therefore open along the bottom (*q:* quadrate, *j:* jugal, *or:* orbit, *pa:* parietal, *po:* post-orbital, *qj:* quadratojugal, *sq:* squamosal).

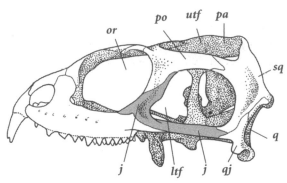

Figure 1

■ The quadrate bone is mobile. In the squamates, as in many tetrapods, the jaw articulation is carried out by the articular (*ma:* mandible) and quadrate (otic region) bones. The squamates have a mobile quadrate bone (we say that it is streptostylic) enabling the mouth to open very widely (fig. 3).

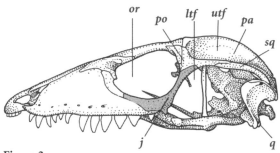

Figure 2

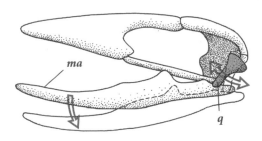

Figure 3

■ The quadratojugal (fig. 1), the bone that lies between the cheek and the otic region, has been lost. This bone is found in non-squamates such as *Sphenodon* (fig. 1), but is no longer present in squamates such as *Tupinambis* (fig. 2).

Ecology

In our time, the squamates represent an extremely adaptable group in full expansion. They are insectivores or carnivores, oviparous or viviparous. In temperate and cold climates, they carry out hibernation. They have the unusual ability of "sniffing" with the help of a forked tongue that they constantly flick into the air. This action samples air particles and brings them to the Jacobson's organ, a type of "internal nose" found within an opening of the palate. The snakes have developed a very specialized feeding mode; they ingest large prey items without chewing them. The amphisbaenians (worm lizards) are legless, fossorial squamates; they are neither lizards nor snakes.

Number of species: 6,850.
Oldest known fossils: *Becklesius hoffstetteri*, *Saurillodon proraformis*, and *Dorsetisaurus purbeckensis* from Portugal date from the upper Jurassic (150 MYA).
Current distribution: in terrestrial environments, they are found worldwide (except at the poles). Most marine squamates are coastal animals (for example, the iguana of the Galápagos Islands and marine snakes).

Examples

Tokay gecko: *Gekko gecko;* common iguana: *Iguana iguana;* chameleon: *Chamaeleo chamaeleon;* Komodo dragon: *Varanus komodoensis;* slow worm: *Anguis fragilis;* green lizard: *Lacerta viridis;* grass snake: *Natrix natrix;* green anaconda: *Eunectes murinus;* desert horned viper: *Cerastes cerastes;* red worm lizard: *Amphisbaena alba.*

Sphenodontia

General Description

The extant sphenodonts (or rhynchocephalans) externally resemble large lizards. They can reach up to 70 cm in length. There currently remains only two, closely related species, the ultimate living fossils of a group that prospered approximately 200 million year ago. The body is stocky and laterally compressed with large limbs. Each limb has five clawed digits linked together by a short membrane at the base. The head and back are covered in small tubercles and the ventral surface in transverse bony plates. A clear dorsal ridge is formed by spiny tubercles that start above the head and continue to the end of the tail. The eyes are large and dark with a "second eyelid," the nictitating membrane (a convergence with the archosaurs). The tympanum has been lost. There is a "third eye"

Tuatara
Sphenodon punctatus
up to 70 cm

on the top of the head; this eye, found in the pineal foramen, is 5 mm in diameter with a retina, lens and optic nerve. The scale that covers this eye is transparent in the young and becomes darker with age. The sphenodonts are generally brown but have the capacity to change color.

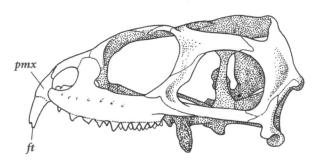

Figure 1

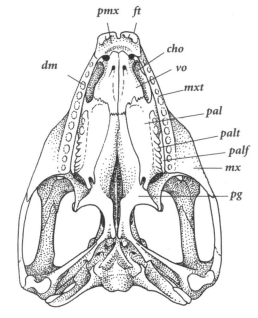

Figure 2

Some unique derived features

- The premaxilla carries two or three teeth that are fused to the bone. A lateral view of the skull of *Sphenondon punctatus* is shown in fig. 1 (*ft*: fused teeth, *pmx*: premaxilla). These fused teeth appear convergently in the Triassic rhynchosaurs.

- Teeth: there is a double row of teeth on the upper jaw, an external row of maxillary teeth and an internal row of palatine teeth. A ventral view of the skull of *Sphenodon punctatus* is shown in fig. 2 (*mxt*: maxillary teeth, *palt*: palatine teeth, *mx*: maxilla, *cho*: internal nare or choana, *pal*: palatine, *pg*: pterygoid, *vo*: vomer). The palatine bones (*pal*) are enlarged and partially close off the palatine fenestra (*palf*), or sub-orbital fenestra.

- They have acrodont teeth, that is, the teeth are attached by their base to the bone by ankylosis (fig. 3a). Another type of ankylosed implantation occurs with pleurodonty (fig. 3b), where the lateral surface of the teeth are fixed in a groove that runs along the external surface (labial) of the supporting bone. Finally, teeth that are attached in alveoli are called thecodont teeth (fig. 3c). Schematic cross-sections of the mandible of a sphenodont (3a), a lizard (3b) and a crocodile (3c) are shown in fig. 3. Acrodont teeth are also found by convergence in certain lizards like the agamids, the chameleons, the amphisbaenians, and the trogonophinans. Pleurodont teeth are found in other lizards and thecodont teeth in the archosaurs (i.e., crocodiles).

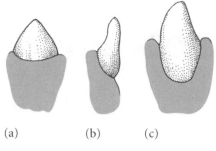

(a)　　　　(b)　　　(c)

Figure 3

Number of species: 2.
Oldest known fossils: *Planocephalosaurus* from Virgina, United States, and *Brachyrhinodon taylori* from Scotland both date from the middle Triassic (230 MYA). **Current distribution:** littoral zones of small islands off the coast of New Zealand.

Examples

Tuatara: *Sphenodon punctatus;* Brother's Island tuatara: *S. guntheri.*

Ecology

The tuatara (*Sphenodon punctatus*) is a placid terrestrial animal. It is crepuscular or nocturnal and lives in heaths and herbaceous areas of the littoral zones of small islands off the coast of New Zealand. They feed principally on earthworms, arthropods and snails, bird eggs and chicks, and lizards. They are territorial and live alone in burrows that they dig themselves. They can also take over abandoned burrows of shearwaters and, in some instances, may even live in cohabitation with the bird in the same burrow. There are no copulatory organs and copulation occurs by direct cloacal contact. The spermatozoids are stored by the female from one summer to the next. She will eventually lay 9 to 12 eggs in a hole dug in the ground. Hatching takes place 13 to 15 months later (this is a record for the length of an incubation period). The tuatara is the cold-blooded animal with the lowest heat requirement; its preferred temperature is 12°C and can remain active to temperatures as low as 7°C. Its metabolism and growth are extremely slow; sexual maturity is reached at 20 years of age and growth continues until 50 years of age. The lifespan is approximately 100 years.

Archosauria

A few representatives

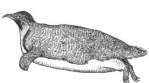

Kiwi
Apteryx australis
70 cm
node 20

King penguin
Aptenodytes patagonica
95 cm
node 20

African grey parrot
Psittacus erithacus
33 cm
node 20

Indian gharial
Gavialis gangeticus
700 cm
node 21

Some unique derived features

■ Antorbital fenestra: the archosaur skull has a lateral fenestra (opening in the cranial wall) in front of the orbit; it is clearly visible in all non-avian dinosaurs. However, it has been lost secondarily in the crocodiles and has fused with the orbit in the birds. Fig. 1 shows the skull of a ceratosaur (Jurassic fossil) (*af*: antorbital fenestra, *mf*: mandibular fenestra, *ltf*: lower temporal fenestra, *n*: nostril, *or*: orbit).

■ Mandibular fenestra: on the mandible there is a posterior-lateral fenestra (*mf*; fig. 1). It is clearly visible in the crocodiles, but not always in certain extant birds.

■ There is a muscular area of the stomach, the gizzard.

■ Nictitating membrane: the archosaurs have a transparent membrane that protects the eye called the nictitating membrane.

■ Teeth are laterally compressed. A tooth, 5 cm high, from the lower right jaw of a herbivorous dinosaur (*Iguanodon*) from the Jurassic is shown in fig. 2, in an internal view (lingual, 2a) and a posterior view (2b). The width of the tooth in 2b is smaller than that of 2a. In the crocodiles, carnivorous dinosaurs, and even in the toothed birds (*Archaeopteryx*, archaeornithes), the teeth are always laterally compressed in this way.

(a)   (b)

Figure 2

■ In the archosauromorphs, all the cervical vertebrae have double-headed ribs (*cr*), as in a crocodile (fig. 3; *nar1*: first neural arch, *ic1*: first intercentrum, *ce1*: first centrum, *pra*: proatlas).

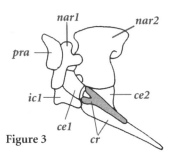

Figure 3

Figure 1

Number of species: 9,695.
Oldest known fossils: *Proterosuchus* from South Africa and *Archosaurus rossicus* from Russia both date from the end of the Permian (250 MYA). Along with certain fossil species (trilophosauridae, rhynchosauria, prolacertiformes, etc.), the archosaurs are part of a larger group that dates from the upper Permian, the archosauromorpha. The archosauromorphs are the sister group of the lepidosauromorphs, a distinct clade that includes lizards and snakes. Other than birds and crocodiles, the archosaurs contain a large number of fossil species, called dinosaurs. The crocodiles separated very early from the dinosaur lineage (in the lower Triassic). However, the birds are directly descended from this lineage and are the only extant dinosaurs.
Current distribution: worldwide.

Examples

■ CROCODILIA: American alligator: *Alligator mississippiensis;* black caiman: *Melanosuchus niger;* Nile crocodile: *Crocodylus niloticus;* Indian gharial: *Gavialis gangeticus.*

■ AVES: domestic chicken: *Gallus gallus;* albatross: *Diomedea exulans;* peregrine falcon: *Falco peregrinus;* green violet-eared hummingbird: *Colibri thalassinus,* razorbill: *Alca torda.*

➡ *see appendix 7, p. 526*

Aves

General Description

Garganey
Anas querquedula
37 cm

Kiwi
Apteryx australis
70 cm

King penguin
Aptenodytes patagonica
95 cm

Rock dove
Columba livia
34 cm

African grey parrot
Psittacus erithacus
33 cm

Griffon vulture
Gyps fulvus
100 cm

The birds are archosaurs capable of powered flight, supported and propulsed by their anterior limbs. All extant birds have feathers and a hard case that covers the mandibles, the beak (fig. 1). The body is light, largely due to the reduction of the bone mass. The sternum (fig. 1) extends into a keel that provides a large surface area for the attachment of the alar muscles. The dorsal vertebrae (lower thoracic, lumbar and sacral) are more or less fused together, providing a certain rigidity to the rib cage. The number of vertebrae, particularly cervical vertebrae, is highly variable (25 cervical vertebrae in the swan, 16 in the blackbird). The tail has atrophied into the pygostyle. The hand bones help support the wing, but the carpal and metacarpal bones have fused and the digits are reduced (digits II, III, IV still remain). The birds are variable in size, often with a pronounced sexual dimorphism. Indeed, their size varies from that of the male ostrich (250 cm and 130 kg) to that of the hummingbird *Mellisuga helenae* (6 cm and 2 g). Along with certain fossil species (pterosaurs and all dinosaurs), this group is part of a larger clade that dates from the lower Triassic, the ornithodira. The ornithodira is the sister group of the crurotarsi, a group that includes the crocodiles.

Some unique derived features

■ Remiges (or flight feathers) provide the surface area required for flight. Other dermal structures homologous to feathers have been found in certain fossil theropod dinosaurs, but these structures would not have enabled powered flight.

■ The first toe, or hallux (*ha*, fig. 1), is directed backwards with the claw turned toward the three other toes. Fig. 1 shows the skeleton of a domestic chicken (*be*: beak, *to2*: second toe, *fu*: furcula or wishbone, *h*: humerus, *s*: sternum or keel).

■ The anterior limbs have undergone a backward rotation, enabling them to fold along the body (fig. 1).

NOTE: the following character is unique to extant avifauna, but we can also find it in certain fossilized theropod dinosaurs.

■ The clavicles meet on the ventral side and form an original bone, the furcula or wishbone (fig. 1).

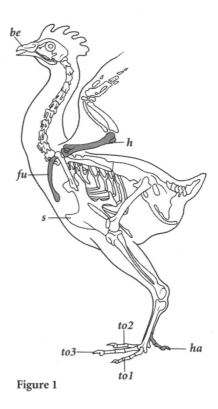

Figure 1

Ecology

The birds are present in all habitats, including the coldest, but are not found at great ocean depths. It is a group in full expansion. Their success is likely due to a combination of their colonization of the air, their independence from water for reproduction, and their homeothermy. It is likely for energetic reasons that they are most often predaceous carnivores (insects or vertebrates) or granivores. They are only rarely herbivores, except for geese and the extraordinary case of the hoatzin, *Opisthocomus hoazin*. Vision is always excellent. Some birds have lost their ability to fly and have specialized in running (ostriches) or swimming in the open sea (penguins). They are always oviparous. Their highly complex behaviors are the subject of much research. They can carry out some of the farthest seasonal migrations found in the animal world.

Number of species: 9,672.
Oldest known fossils: *Archaeopteryx lithographica* from Germany dates from the upper Jurassic (150 MYA).
Current distribution: worldwide, including the coldest continent (penguins from Antarctica).

Examples

Domestic chicken: *Gallus gallus*; albatross: *Diomedea exulans*; peregrine falcon: *Falco peregrinus*; green violet-eared hummingbird: *Colibri thalassinus*, razorbill: *Alca torda*; emperor penguin: *Aptenodytes forsteri*; brown pelican: *Pelecanus occidentalis*; African grey parrot: *Psittacus erithacus*; ostrich: *Struthio camelus*; greater flamingo: *Phoenicopterus ruber*; garganey: *Anas querquedula*; common raven: *Corvus corax*; blue tit: *Parus caeruleus*.

385

Crocodilia

General Description

Nile crocodile
Crocodylus niloticus
480 cm

Cuvier's dwarf caiman
Palaeosuchus palpebrosus
145 cm

At first glance, the crocodiles look like large lizards of up to 10 m in length. They have a bulky, elongated head with the nostrils and eyes in an anterior position, a massive body trunk that is wide and flat, and a laterally compressed tail. The body is covered in large horny plates supported by pieces of dermal bone, the osteoderms. The

internal anatomy of the crocodiles is very different from that of the lizards and is much closer to the dinosaurs and birds than to the lepidosaurs. In particular, they possess a four-chambered heart and a cerebral cortex that is more highly developed than that of the lizards and snakes, and their legs are more vertical while walking than those of the lizards. In fact, crocodilian ancestors that lived during the Triassic period were bipedal with vertical legs, as were certain terrestrial fossil crocodiles. By specializing on an amphibious lifestyle, the modern crocodiles have somewhat lost this arrangement. Along with certain fossil species (ornithosuchids, phytosaurs, etc.), this clade is part of a larger clade that dates from the lower Triassic, the crurotarsi. The crurotarsi are the sister group of the ornithodirans, a distinct group that includes the pterosaurs and the dinosaurs (and thus the birds).

Indian gharial
Gavialis gangeticus
700 cm

Some unique derived features

- A secondary palate is formed by ventral extensions of the maxilla (*mx*), palatines (*pal*) and the pterygoids (*pg*); these extensions push the internal nares (or choanae, *cho*) posteriorly into the occipital region. The choanae now open on the pterygoids. A ventral view of an alligator skull is shown in fig. 1 (*boc:* basi-occipital, *q:* quadrate, *cy:* occipital condyle, *ecp:* ectopterygoid, *palf:* palatine fenestra, *pmx:* premaxilla, *qj:* quadratojugal).

- The lower temporal fenestra is triangular in shape. A dorsal view of an alligator skull is shown in fig. 2 (*fr:* fused frontals, *ltf:* lower temporal fenestra, *utf:* upper temporal fenestra, *j:* jugal, *or:* orbit, *pa:* fused parietals, *po:* post-orbital, *sq:* squamosal).

- Ankle: there has been a functional rotation in the ankle of the crurotarsi where the calcaneum (*cal*)

functions as part of the foot and the astragalus (*as*) as part of the tibia (*ti*). In this way, the plane of the ankle flexion (*fl*) forms a diagonal, contrary to the horizontal flexion plane of the basal archosaurs like *Proterosuchus* or the ornithodirans (pterosaurs, dinosaurs and thus birds).
A schematic view of the foot of *Proterosuchus* is shown in fig. 3a, that of an ornithodiran (here, a dinosaur) in fig. 3b and fig. 4b, and that of a crocodile in fig. 3c and fig. 4a. In the crocodile, the line of flexion passes between the proximal tarsal bones (astragalus and calcaneum); this is therefore called an intratarsal or crurotarsal joint (fig. 4a: crocodile leg). In the ornithodirans, this line passes between the proximal and the distal tarsal bones and is thus called a mesotarsal joint (fig. 4b: dinosaur leg; *mtt:* metatarsal, *fi:* fibula, *dta:* distal tarsal).

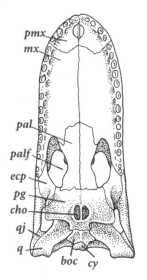

Figure 1

Ecology

The crocodiles are aquatic predators (there is only a single marine species), good swimmers, carnivorous and necrophagous, with largely crepuscular and noctural activity. Their lungs have an enormous air capacity; after closing the valves of the nostrils and choanae, folding back the dermal fold on the tympanum and protecting the eyes using the nictitating membrane, they can remain underwater for approximately one hour. The choanae (internal nares) empty far back into the throat such that inspired air does not pass through the mouth. This means that the mouth can remain open underwater without blocking respiration. The crocodiles can therefore hold their prey below the water while still breathing by their nostrils, which remain above the water. They are oviparous; the females lay their eggs in nests close to the water.

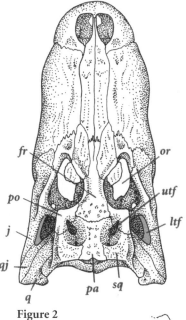

Figure 2

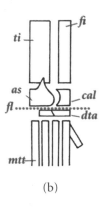 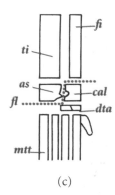

(a) (b) (c)

Figure 3

Number of species: 23.
Oldest known fossils: *Protosuchus*, a sphenosuchid from South Africa, dates from the upper Triassic (230 MYA). *Hesperosuchus agilis* from Arizona, United States, dates from this same period.
Current distribution: the crocodiles live in the rivers of tropical and subtropical zones and are found on all the corresponding continents, except Europe.

Examples

American alligator: *Alligator mississippiensis;* black caiman: *Melanosuchus niger;* common caiman: *Caiman crocodilus;* saltwater crocodile: *Crocodylus porosus;* Nile crocodile: *C. niloticus;* false gharial: *Tomistoma schlegelii;* Indian gharial: *Gavialis gangeticus.*

(a) (b)

Figure 4

Chapter 13

Mammalia

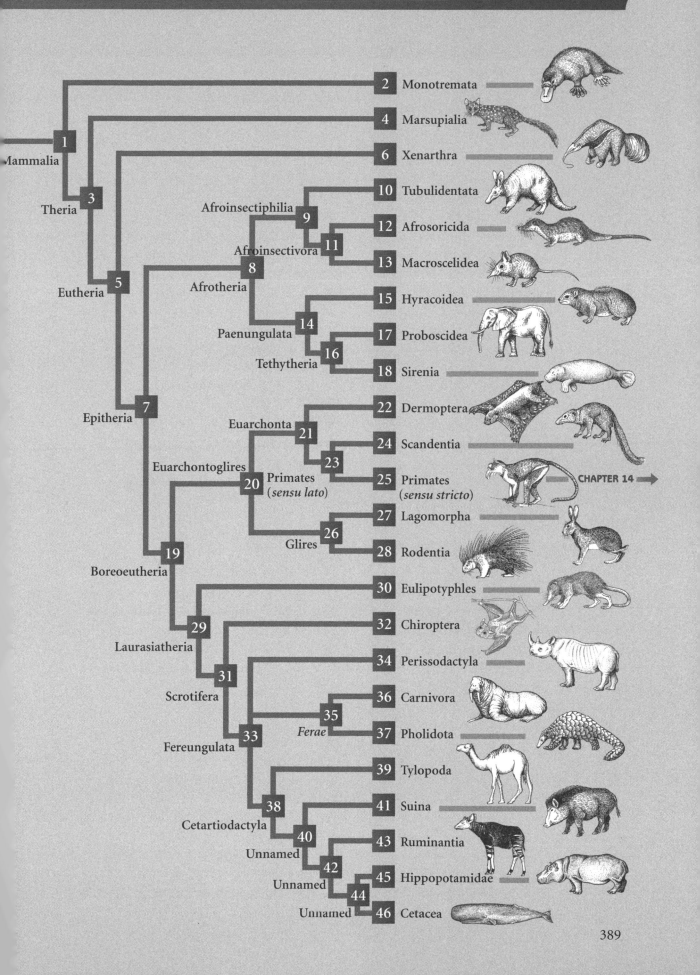

1 Mammalia

2 Monotremata

3 Theria

4 Marsupialia

5 Eutheria

6 Xenarthra

Afroinsectiphilia 9

10 Tubulidentata

11

12 Afrosoricida

Afroinsectivora
8

13 Macroscelidea

Afrotheria

15 Hyracoidea

14

17 Proboscidea

Paenungulata

16

Tethytheria

18 Sirenia

7 Epitheria

Euarchonta

21

22 Dermoptera

24 Scandentia

23

Euarchontoglires

Primates
(*sensu lato*)

20

25 Primates
(*sensu stricto*)   CHAPTER 14 →

27 Lagomorpha

26

Glires

28 Rodentia

19 Boreoeutheria

30 Eulipotyphles

29

Laurasiatheria

32 Chiroptera

31

34 Perissodactyla

Scrotifera

35

36 Carnivora

Ferae

37 Pholidota

33 Fereungulata

39 Tylopoda

38

Cetartiodactyla

41 Suina

40

Unnamed

43 Ruminantia

42

Unnamed

45 Hippopotamidae

44

Unnamed   46 Cetacea

mammalia

The classification of the mammals is particular in that the "orders" (marsupials, rodents, primates, carnivores, etc.) have been relatively stable for a long time, but the relationships among these orders has undergone a complete revision over the last few years. The elucidation of these relationships, based on anatomical characters, has always been difficult. Indeed, these characters are rarely stable: they are subject to convergences and reversions such that there is a strong degree of conflict among them, even when the fossil species are taken into account.

Two independent molecular phylogenies, each based on several genes and incorporating all the mammalian orders, have found the same inter-ordinal relationships. These relationships are also logical in biogeographical terms. The two research teams have since merged their DNA sequence data and the information that has emerged, from 19 nuclear and 3 mitochondrial genes, has become more precise and has permitted them to date the divergences using sophisticated analytical tools. The orders of placental mammals differentiated from each other during the Cretaceous period, between 100 and 70 MYA, well before the disappearance of the non-avian dinosaurs. Sixty-five million years ago, during the Cretaceous-Tertiary extinction event, the orders that we find today had already formed their individual lineages. It was therefore necessary to abandon the idea that the mammalian radiation occurred after the disappearance of the dinosaurs, a scenario that was presented in all textbooks. Another team of researchers have published results based on 7 nuclear markers, all corroborating the new relationships. These results were further supported by structural genomic markers, like the chromosome structure and the position of transposons. This degree of correspondence could not be found by chance alone. This modern version of the relationships among the eutherian orders continues to be supported by new studies that make use of the diverse range of molecular techniques now available. The

anatomical data that was used to support the previous view are so inconsistent that they support this new tree as much as they did the old one.

One hundred million years ago, when South America separated from Africa and the Atlantic Ocean was born, the ancestors of the placental mammals were divided between two isolated continents, separated by the sea levels of the time. The afrotherians are the descendants of the African fauna, whereas the xenarthrians are the descendants of part of the South American fauna. Ninety-five million years ago, some placental mammals moved into the northern continents and gave rise to the boreoeutherians. Consistent with this scenario, molecular studies have invalidated or modified two aspects of the traditional orders. First, the artiodactyla have been shown to include the cetaceans and we must therefore refer to this group as the Cetartiodactyla. Second, as previously suspected, the insectivores are polyphyletic. The afrosoricids (tenrecids, chrysochlorids, potamogalids), limited to Africa and Madagasar, are afrotherians, whereas the eulipotyphles (talpids, soricids, erinaceids, solenodontids) are laurasiatherians.

By examining their phylogeny, and therefore their classification, we can find some remarkable examples of morphological and ecological convergences in the placental mammals. The afrotherians and laurasia-therians independently gave rise to moles (golden moles in the first and talpid moles in the second), hedgehogs (the hedgehog tenrecs and true hedgehogs, respectively), shrews (shrew tenrecs and soricid shrews, respectively), permanent swimmers (sirenians and cetaceans, respectively), and anteaters (aardvark and pangolins, respectively). This vast anatomical plasticity was previously recognized by the famous similarities between the marsupial mammals, found in South America and Australia, and the placental mammals of the other continents.

Monotremata

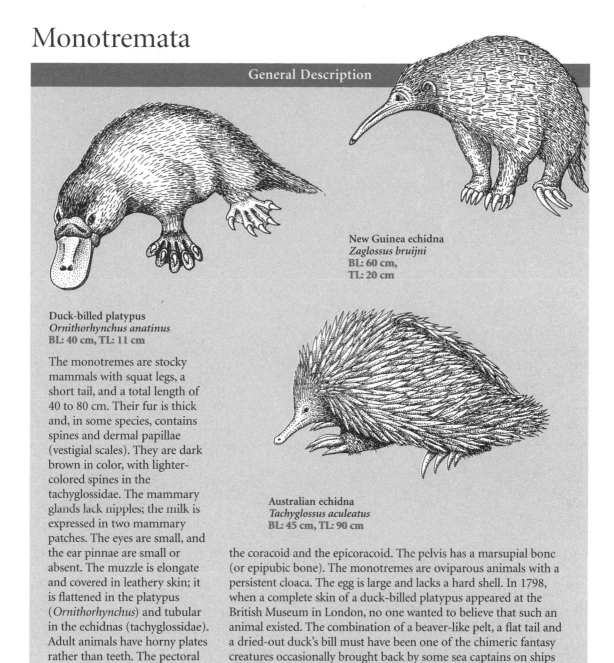

General Description

New Guinea echidna
Zaglossus bruijni
BL: 60 cm,
TL: 20 cm

Duck-billed platypus
Ornithorhynchus anatinus
BL: 40 cm, TL: 11 cm

Australian echidna
Tachyglossus aculeatus
BL: 45 cm, TL: 90 cm

The monotremes are stocky mammals with squat legs, a short tail, and a total length of 40 to 80 cm. Their fur is thick and, in some species, contains spines and dermal papillae (vestigial scales). They are dark brown in color, with lighter-colored spines in the tachyglossidae. The mammary glands lack nipples; the milk is expressed in two mammary patches. The eyes are small, and the ear pinnae are small or absent. The muzzle is elongate and covered in leathery skin; it is flattened in the platypus (*Ornithorhynchus*) and tubular in the echidnas (tachyglossidae). Adult animals have horny plates rather than teeth. The pectoral girdle shows some primitive characteristics and is made up of three elements: the scapula, the coracoid and the epicoracoid. The pelvis has a marsupial bone (or epipubic bone). The monotremes are oviparous animals with a persistent cloaca. The egg is large and lacks a hard shell. In 1798, when a complete skin of a duck-billed platypus appeared at the British Museum in London, no one wanted to believe that such an animal existed. The combination of a beaver-like pelt, a flat tail and a dried-out duck's bill must have been one of the chimeric fantasy creatures occasionally brought back by some sea captains on ships coming from oriental markets. It took nearly a century and numerous long discussions, founded on fragmented data, finally to place the monotremes among the mammals.

Some unique derived features

- The monotremes have a cornified spur on each hind foot. This spur is venemous in males. Fig. 1 shows the right foot pad of the male Australian echidna (*Tachyglossus aculeatus*, 1a), the New Guinea echidna (*Zaglossus bruijni*, 1b), and the duck-billed platypus (*Ornithorhynchus anatinus*, 1c), indicating the position of the spur (*spr*).

- The monotremes lose their teeth at adulthood. It is worth noting that this same pattern is also observed in the pholidota and in some xenarthrans such as anteaters.

Ecology

The monotremata draw together a semi-aquatic and semi-fossorial species, the duck-billed platypus (*Ornithorhynchus anatinus*) that frequents warm water, and two terrestrial species, the tachyglossids that we find in open forests and semi-arid regions. The platypus specializes in hunting small aquatic animals (worms, crustaceans, and even small fish) that it captures under water with the help of the wrinkles, furrows and filtering transverse ridges of its beak. Its large webbed fore and hind feet, as well as its flat tail, make it well-equipped for swimming. It is also a fossorial animal that excavates complex burrows with numerous branches. This subterranean refuge includes an incubation chamber, found thirty centimeters above the water's surface, where the female lays between 1 and 3 soft eggs of about 1.7 cm in size. The female does not have a ventral pouch and must brood the eggs. The young hatch from the eggs after approximately 10 days, 2.5 cm long, naked and blind. They suckle the milk that oozes from their mother's mammary patch for four months, and then, at roughly 35 cm long, they leave the burrow to begin hunting. Although the spurs on the hind feet of females are temporary, they are permanent and become venomous in males. This venom is dangerous, even to humans. The phenomenal daily consumption of worms and crustaceans by a single platypus, in addition to the requirement of a large water source and soil embankments, makes this species particularly difficult to keep in captivity.

The echidnas or tachyglossids are specialized in hunting ants, termites, and other insects that they capture using their long tongues. With their strong claws, they can bury themselves very quickly when threatened. However, even if only half-buried, they are still impossible to dislodge because of the sharp spines on their backs and the fact that they anchor themselves using their claws. These animals are also remarkably strong. The echidnas can go without food for more than a month. The females lay their

(a)

(b)

(c)

Figure 1

eggs into a temporary ventral pouch where they are carried for approximately 10 days. At hatching, the young measure no more than 12 mm in length. They lap their mother's thick milk from tufts of her fur. Six to 8 weeks later, once they have grown to 10 cm in length and have developed spines, the young leave the pouch for a den. At one year of age, they become sexually mature. Strangely, some male echidnas in a zoo in Prague were observed to develop a ventral pouch every 28 days. The tachyglossids are among the few mammals that can live over 50 years.

Number of species: 3.
Oldest known fossils: *Steropodon* dates from the start of the Cretaceous from Australia (115 MYA).
Current distribution: Australia, Tasmania, New Guinea.

Examples

Duck-billed platypus: *Ornithorhynchus anatinus*; Australian echidna: *Tachyglossus aculeatus*; New Guinea echidna: *Zaglossus bruijni*.

Theria

A few representatives

Red kangaroo
Macropus rufus
BL: 145 cm
node 4

Pink fairy armadillo
Chlamydophorus truncatus
BL: 15 cm
node 6

European rabbit
Oryctolagus cuniculus
BL: 40 cm
node 27

Sperm whale
Physeter catodon
14 m
node 46

Some unique derived features

■ There is a supraspinatous fossa on the scapula (shoulder blade). The change from a transversally positioned limb to one that is perfectly vertical (parasagittal) has led to major modifications of the musculature of the pectoral girdle. Almost all of the muscles that were inserted on the coracoid (*cor*) are now on the scapula (*sc*). The coracoid is no longer a distinct bone in the theria but has been reduced to a process at the edge of the glenoid cavity (*gc*), the socket that receives the head of the humerus. Fig. 1 shows a left lateral view of the pectoral girdle of an echidna (1a) and that of a cat's scapula (1b), both oriented so that the front of the animal is to the left. In the therians (1b), the external surface of the scapula has a thick ridge, the spine (*sp*), above which lies the supraspinatous fossa (*spf*) and below which is found the subspinatous fossa (*sbf*). In non-therians, the external surface of the scapula is smooth (1a). We also note that the therians have lost the procoracoid (*pcor*) and the interclavicle (*icv*), bones that are still present in the monotremes (fig. 1a). The therian girdle is therefore essentially made up of the scapula and the clavicle (*clv*, not shown on the cat).

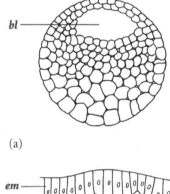

(a)

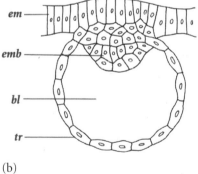

(b)

Figure 2

■ The blastocyst has a single cell layer. Two embryos in the blastula stage are shown in fig. 2 (*emb:* embryoblast, *em:* endometrium), one of a frog (2a) and one of a human during implantation on the 6th day (2b). In the frog, as in other non-therian amniotes, the cavity of the blastocyst, or blastocoel (*bl*), is surrounded by a wall composed of numerous cell layers. In therians,

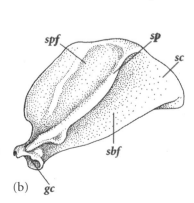

(a)

(b)

Figure 1

most of the blastocoel wall is made up of a single cell layer, the trophoblast (*tr*).

■ The upper molars are tribosphenic, that is, they have an additional cusp on the lingual (or internal) side of the tooth. This cusp interacts with a basin on the corresponding molar on the lower jaw, forming a mortise and tenon occlusion. Upper and lower molars of a fossil non-therian mammal from the lower Jurassic, *Kuehneotherium* (3a), and a fossil eutherian mammal from the upper Cretaceous, *Gypsonictops* (3b), are

illustrated in fig. 3. On the diagram of the upper molars, the front is to the left, the external edge is above (*ext*), and the internal (lingual) edge is below. On the lower molars, the internal edge is above and the external is below (*ext*). The dots on 3a show the three occlusion surfaces of the upper and lower teeth in the non-therian. In the case of tribosphenic molars (3b), there are six such surfaces, of which only two are shown here; these are two new ones that have appeared with the presence of the new cusp, the protocone (*prc*).

■ The prefrontal and postfrontal bones have been lost.

■ The mammary gland has a nipple.

■ The cochlea is spiral-shaped with at least one turn.

■ The egg is microlecithal, meaning that it is small (from 60 to 250 μm in diameter) and lacks vitelline reserves. Complete cell division divides the entire egg into two daughter cells, then four, etc. The ancestral condition found outside of the therians is a telolecithal egg; a large egg (up to 9 cm in diameter) with abundant vitelline reserves. In these eggs, the cellular cytoplasm separates from the vitellus and unites with the nucleus of the animal pole to form a flattened disc, the germinal disc. The division of the egg is therefore partial, leaving the vitellus undivided. This arrangement is found in the myxines, chondrichthyans, teleost fishes, actinistians, sauropsids and monotremes.

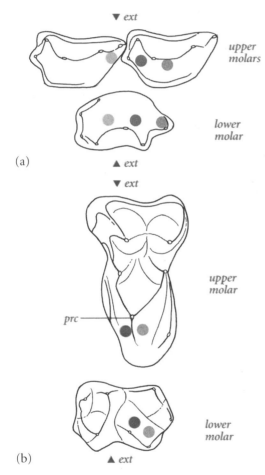

(a)

(b)

Figure 3

NOTE: The sister group of the monotremes is called *Therimorpha* (the therimorphs) and, of course, includes all extant marsupials and eutherians (the therians), along with several fossil groups whose branches lie between those of the monotremes and therians. These mammals of the Mesozoic include the triconodonts, the multituberculates, *Kuehneotherium*, *Trinodon*, *Shuotherium*, the symmetrodonts, the dryolestoids, *Amphitherium*, the peramurids, *Vincelestes*, *Aegialodon* and finally *Pappotherium*, the sister group of the therians. The synapomorphies described above are all present in extant therians, but may not all be present in the different fossil groups. For example, the tribosphenic molar is found only in the tribosphenidians (*Aegialodon*, *Pappotherium*, and the therians).

Examples

■ Marsupialia: southern opossum: *Didelphis marsupialis;* red kangaroo: *Macropus rufus.*

■ Eutheria: three-toed sloth: *Bradypus tridactylus;* Indian pangolin: *Manis crassicaudata;* European hedgehog: *Erinaceus europaeus;* common tree shrew: *Tupaia glis;* Eurasian badger: *Meles meles;* mountain zebra: *Equus zebra.*

Marsupialia

General Description

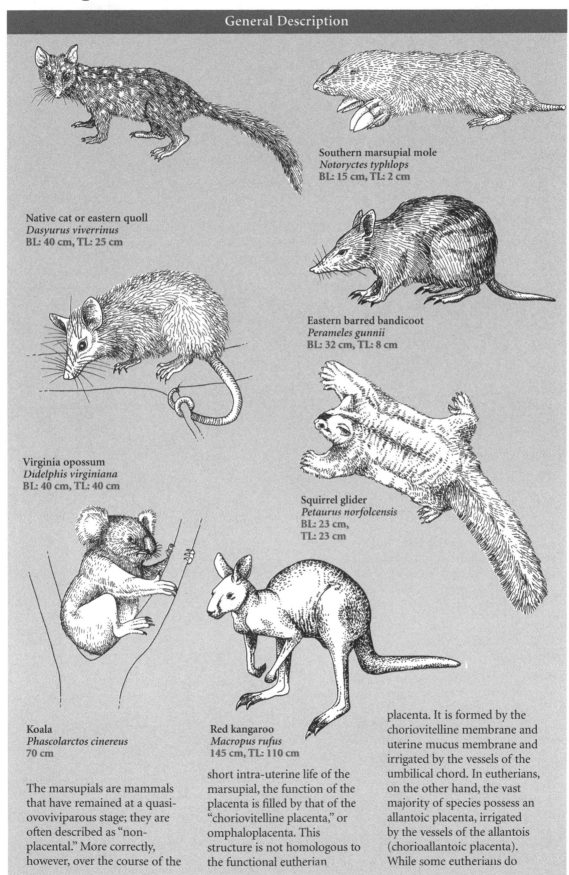

Native cat or eastern quoll
Dasyurus viverrinus
BL: 40 cm, TL: 25 cm

Southern marsupial mole
Notoryctes typhlops
BL: 15 cm, TL: 2 cm

Eastern barred bandicoot
Perameles gunnii
BL: 32 cm, TL: 8 cm

Virginia opossum
Didelphis virginiana
BL: 40 cm, TL: 40 cm

Squirrel glider
Petaurus norfolcensis
BL: 23 cm,
TL: 23 cm

Koala
Phascolarctos cinereus
70 cm

Red kangaroo
Macropus rufus
145 cm, TL: 110 cm

The marsupials are mammals that have remained at a quasi-ovoviviparous stage; they are often described as "non-placental." More correctly, however, over the course of the short intra-uterine life of the marsupial, the function of the placenta is filled by that of the "choriovitelline placenta," or omphaloplacenta. This structure is not homologous to the functional eutherian placenta. It is formed by the choriovitelline membrane and uterine mucus membrane and irrigated by the vessels of the umbilical chord. In eutherians, on the other hand, the vast majority of species possess an allantoic placenta, irrigated by the vessels of the allantois (chorioallantoic placenta). While some eutherians do develop an omphaloplacenta (eulipotyphles, rodents), it is never exclusively used and either disappears early in development (except in the hedgehogs and moles) or does not touch the uterine wall (inverse omphaloplacenta of rodents). In parallel, the allantoic placenta is always present. Among the marsupials, *Perameles* is the exception, having both an omphalo-placenta and an allantoic placenta. However, what really distinguishes the marsupials from the eutherians is not the absence of a placenta, but rather the short intra-uterine stage and the fact that they are, in a sense, born twice. Indeed, the young marsupial is born in a very precocial stage and completes its development in the marsupial pouch. For these reasons, we have often thought of the Mar-supialia as imperfect mammals. In fact, the marsupials are the result of a unique radiation, 120 million years ago, that led to a diversity of forms and adapta-tions virtually equivalent to that found among all other extant mammals. We find marsupials of all shapes and sizes (from 10 to 200 cm long, not including the tail), from all habitats (other than marine), and of all lifestyles and diets. If fossil marsupials are also included, this wealth of diversity is still more impressive.

Marsupials generally have thick fur. The nipples, num-bering between 2 and 27, are well developed and equipped with a mammary muscle that injects milk into the mouth of the poorly developed young. The nipples lie within a fold of skin called the marsupial pouch, the appearance and position of which is highly variable among species. Each marsupial young completes its development attached to a nipple. Epipubic bones are present, having disappeared only in the Tasmanian wolf (*Thylacinus cynocephalus*). Marsupials have between 18 and 56 teeth, with 4 or 5 incisors per jaw quadrant. In fact, the dental formula varies considerably among species. Interestingly, the articular-quadrate articulation of the jaw with the skull (which could be described as "reptilian") is functional for the first 25 days after birth. The bony palate is fenestrated. The brain is simple and the cortex lacks convo-lutions. The core body temperature is lower than in the placental mammals (34 to 36 °C) and is more dependent on the ambient temperature. The penis is contained within a pouch that lies posterior to the scrotum. Ancestrally, there were 2 separate vaginae and uteri, but in modern forms some fusion has occurred with a diversity of intermediate forms. The anus and genital tracts open into a cloaca that is equipped with a sphincter.

Some unique derived features

■ A strip of jugal bone extends the length of the zygomatic arch, up to the anteriolateral edge of the glenoid fossa. In fig. 1, the lateral views of three skulls are represented: a eutherien (*Canis lupus*, 1a) and two marsupials (*Thylacinus cynocephalus*, 1b, *Didelphis marsupialis*, 1c). We see that the jugal bone (*j*) does not extend to the glenoid fossa (*gf*) in the first skull, but it does in the second and third.

■ The zygomatic process of the squamosal bone (fig. 1: *sq*) is generally robust dorsally and laterally.

■ Teeth: on both the upper and lower jaws, the teeth are milk teeth with the exception of the last premolar. In modern marsupials, dental modification is thought to be linked to the mode of lactation. When the newborn is attached to the mother's nipple, its buccal cavity is complelely filled by its tongue and the nipple: there is no space for tooth development. The development of the first generation of anterior premolars is delayed and only the last premolar (*Pm4* if we consider that *Pm2* has disappeared) has a replace-ment tooth. Most marsupials have 3 premolars

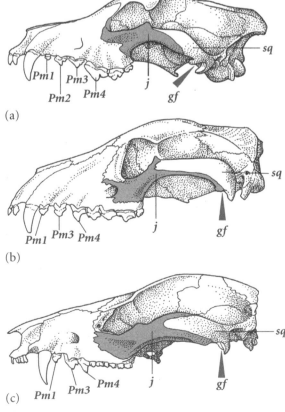

Figure 1

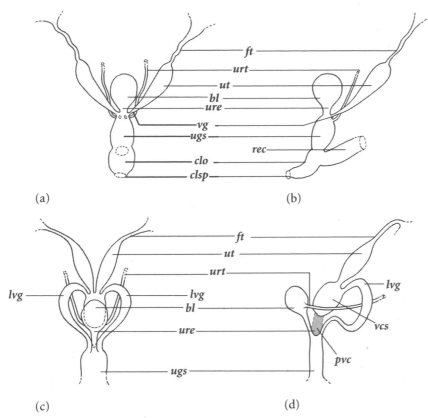

(a)

(b)

(c)

(d)

Figure 2

form a single structure, the median vaginal cul-de-sac. In some species, this structure can extend to meet the urogenital sinus. In the kangaroo, it even opens into this sinus and forms a permanent and functional median vagina through which the young are born. In most kangaroo species, this connection is established during the first birth and then remains. In fact, the lateral vaginae are too narrow for the passage of the fetuses. In other marsupials, the lower part of the vaginal cul-de-sac is only reabsorbed at birth and, via the action of certain hormones, a pseudovaginal canal is formed in the neighboring connective tissue through which the fetus moves from the uterus to the urogenital sinus.

and 4 molars, whereas most placental mammals have 4 premolars and 3 molars. The direct correspondence between the functional milk dentation of the marsupials and that of the eutherians (except *Pm4*, of course) has been highly debated.

■ The female gential tract is composed of two long lateral vaginae (*lvg*) connected to a double uterus (*ut*) and a pseudovaginal canal (*pvc*). Ventral (2ac) and left lateral (2bd) views of the female urogential tracts of an echidna (2ab) and an opossum (2cd) are represented in fig. 2 (*clo:* cloaca, *rec:* rectum, *clsp:* cloacal sphincter, *ure:* urethra, *urt:* ureter, *bl:* bladder). In mammals, the fallopian tubes (*ft*), the uterus and the vagina (*vg*) are formed from the Mullerian ducts. Initally, all of these structures are paired, but over the course of mammalian evolution, a progressive fusion of the two vaginae and two uteri has occurred. In the monotremes, the vaginae (fig. 2ab) are poorly developed and it is the urogenital sinus (*ugs*) that plays the role of the vagina during mating. In marsupials, the two Mullerian ducts meet sagittally at their vaginal section and may fuse together. In highly developed systems, there are long curved lateral vaginae and vaginal cul-de-sacs (*vcs*) that connect with the corresponding cervix. The vaginal cul-de-sacs meet medially and

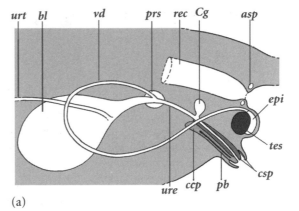

(a)

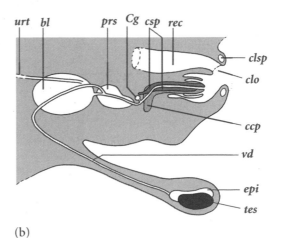

(b)

Figure 3

- In the male genital tract, the scrotum lies anterior to the penis. Sagittal sections of the male urogenital system of a cat (a) and a kangaroo (b) are shown in fig. 3 (anterior end on the left, posterior on the right; *ccp:* corpora cavernosa of the penis, *csp:* corpus spongiosum of the penis, *vd:* vas deferens, *epi:* epididymis, *Cg:* Cowper's gland, *pb:* penis bone, *prs:* prostate gland, *asp:* anal sphincter, tes: testicles; *clsp:* cloacal sphincter).

- Choriovitelline placenta: there is no development of the chorioallantoic membrane, thus giving rise to a choriovitelline placenta.

Ecology

The marsupials can be diurnal or nocturnal, terrestrial or arboreal, gliders or diggers, insectivores, carnivores, herbivores, omnivores or nectarivores. They are found in a diverse range of terrestrial environments such as tropical forests, savannas, arid and desert regions, and prairies. The variety of ecological niches occupied by this group is comparable to that of all placental mammals. Indeed, the comparison of these two groups reveals multiple spectacular examples of convergent evolution between unrelated species that have adopted equivalent life styles. The South American marsupial fossil *Thylacosmilus* shows a stunning resemblance to the saber-toothed tigers of North America (carnivores). The marsupial mole *Notoryctes* lacks nothing when compared to placental moles (eulipotyphles and afrosoricids). The marsupial glider *Petaurus* strongly resembles the flying squirrels (rodents). The thylacine or Tasmanian wolf (recently extinct) was very wolf-like. There are also marsupial mice (*Sminthopsis*), rats (*Antechinus*), cats (*Dasyurus*), "anteaters" (*Myrmecobius*), etc.

The marsupials have a very short gestation period (8 to 42 days depending on the species) and, in a sense, a double birth. During the early stages of development, the embryo is fed by a "choriovitelline placenta." Young marsupials are therefore born for the first time in a "premature" state compared with placental mammals, with the forelimbs more strongly developed than the hind limbs. At a length of only 0.5 to 3 cm, they leave the mother's urogenital sinus naked and blind. By grasping the mother's fur, they migrate along a hairline, moistened prior to birth by the mother. Despite the fact that the mother does not help during this climb, the young reaches the marsupial pouch within a few minutes. It then attaches to a nipple and will remain attached for several weeks. When the young leaves this pouch, it is "born" for a second time.

Examples

Western barred bandicoot: *Perameles fasciata;* southern opossum: *Didelphis marsupialis;* native cat or eastern quoll: *Dasyurus viverrinus;* Tasmanian devil: *Sarcophilus harrisi;* thylacine or Tasmanian wolf: *Thylacinus cynocephalus;* yellow-bellied glider: *Petaurus australis;* red kangaroo: *Macropus rufus;* long-nosed potoroo: *Potorous tridactylus;* southern marsupial mole: *Notoryctes typhlops;* numbat: *Myrmecobius fasciatus;* spotted cuscus: *Phalanger maculatus;* koala: *Phascolarctos cinereus;* coarse-haired wombat: *Vombatus ursinus;* silky shrew opossum: *Caenolestes fuliginosus.*

Number of species: 272.
Oldest known fossils: *Holoclemensia* from Texas (USA) dates from the middle Cretaceous (100 MYA). *Kokopellia* from Utah from the lower Cretaceous is older and may be a marsupial. These fossils confirm the North American origin of the marsupials during the Cretaceous. Certain authors argue for a South American origin, where a Paleocene radiation of a great diversity of forms (ten families) would support an earlier diversification. Fossils of a common South American family that date from the late Eocene were found on the Antarctic Peninsula, supporting the hypothesis of an early dispersal towards Australia via the Antarctic. In Australia, marsupial fossils are known from the lower Eocene and start to become abundant and already highly diversified in the late Oligocene. They reach Europe and North Africa during the Eocene and remain there until the Miocene. The placental mammals appeared very early in Asia, but the marsupials were considered quasi-absent from this continent up until 1998, when *Asiatherium* from Mongolia (upper Cretaceous) was discovered. South America had marsupials from the start of the Tertiary, remained isolated for a long time, and then became reconnected with North America during the Pleistocene. During this time many families of North American placental mammals moved into South America. A replacement of the fauna then occurred, leading to the extinction of many marsupial families on this continent.
Current distribution: Australia, Tasmania, New Guinea, Sulawesi Islands, South America, Central America, southern North America. Their current distribution may seem to be residual and complementary to that of the placental mammals. Indeed, if we exclude the placental mammals introduced by humans to Australia, this isolated continent seems in the Quaternary to be that of the marsupials. In South America, only two families (Didelphidae and Caenolestidae), resembling mice and rats, still exist and only a few opossums re-established in North America, around 3 MYA.

Eutheria

A few representatives

Giant anteater
Myrmecophaga tridactyla
BL: 110 cm
node 6

African savanna elephant
Loxodonta africana africana
BL: 690 cm
node 17

Garden dormouse
Eliomys quercinus
BL: 14 cm
node 28

Giant pangolin
Manis gigantea
BL: 80 cm
node 37

Some unique derived features

■ The epipubic bones have been lost. A left lateral view of the pelvic girdle of a cat, where the epipubic bones are absent, is represented in fig. 1a. In contrast, fig. 1b shows the position of this bone (*epu*) on a left lateral view of the pelvic girdle of a platypus. This bone is also shown on the full skeleton (fig. 1c; *ac:* acetabulum, *cvr1:* first

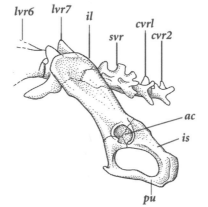

(a)

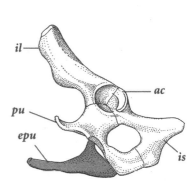

(b)

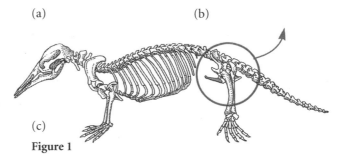

(c)

Figure 1

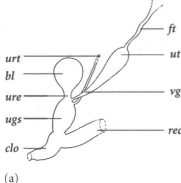

(a)

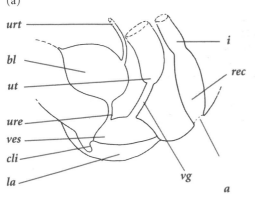

(b)

Figure 2

caudal vertebra, *il:* ilium, *is:* ischium, *lvr6:* sixth lumbar vertebra, *pu:* pubis, *svr:* sacral vertebrae).

■ The cloaca has disappeared, that is, the openings of the urogenital and digestive tracts are henceforth separated. Fig. 2 (*cli:* clitoris, *la:* labia, *i:* intestine, *ft:* fallopian tubes, *ut:* uterus, *ure:* urethra, *urt:* ureter, *ves:* vestibule, *vg:* vagina, *bl:* bladder) shows left lateral views of the end points of the digestive and urogenital tracts of a monotreme, echidna (2a), and a therian, human (2b). In the nontherian (2a), the rectum (*rec*) and the urogenital sinus (*ugs*) empty into the same cavity, the cloaca (*clo*), which has a single external opening. In the therians, the digestive tube has a separate opening, the anus (*a*).

399

The vagina is median and permanent and results from the fusion of part of the Mullerian ducts. Ventral views of the female urogenital tracts of an echidna (3a, monotremata), an opossum (3b, marsupialia) and a mouse (3c, eutheria) are represented in fig. 3. In all mammals, the Mullerian ducts give rise to the fallopian tubes, the uterus and the vagina. Initially, these structures are paired, but over the course of mammalian evolution, a progressive fusion of the two vaginae and two uteri has occurred. However, in the monotremes and the marsupials, there is no true permanent median vagina. In the eutherians (fig. 3c), there is only one, median and permanent vagina (see 13.4).

NOTE: the bladder of the mouse (fig. 3c) has been moved to the left.

- The ureters (fig. 3c) pass laterally to the derivatives of the Mullerian ducts (fallopian tubes, uterus and vagina). The primitive condition, found in the monotremes and marsupials, is the reverse (fig. 3a; *clsp*: cloacal sphincter); the derivatives of the Mullerian ducts pass laterally to the ureters. Indeed, in the monotremes, both the uteri and the fallopian tubes are lateral (fig. 3a) and in the marsupials, the vaginae are lateral (*lvg*) and connect at a vaginal cul-de-sac (*vcs*, fig. 3b).

- The intra-uterine life is prolonged by anatomical and hormonal changes.

- The optic foramen is widely separated from the sphenorbital fissure. Posterodorsal views of the cranial floor (the skull roof has been removed) of a marsupial (*Perameles*, bandicoot) and a eutherian (*Bradypus*, sloth) are represented in fig. 4 (*ofo*: optic foramen, *ovf*: oval foramen, the passage of nerve V3, *sf*: sphenorbital fissure).

- The corpus callosum connects the two cerebral hemispheres of the brain.

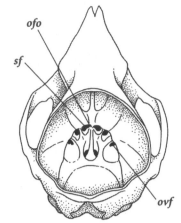

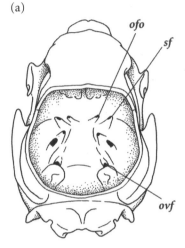

(a)

(b)

Figure 4

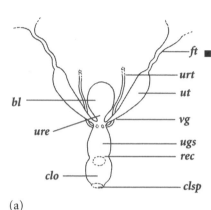

(a)

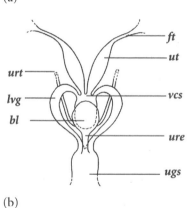

(b)

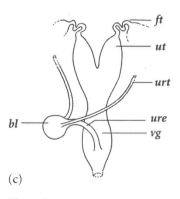

(c)

Figure 3

Examples

- XENARTHRA: three-toed sloth: *Bradypus tridactylus*; nine-banded armadillo: *Dasypus novemcinctus*.
- EPITHERIA: Indian pangolin: *Manis crassicaudata*; Eurasian badger: *Meles meles*; wild cat: *Felis sylvestris*; mountain zebra: *Equus zebra*.

Xenarthra

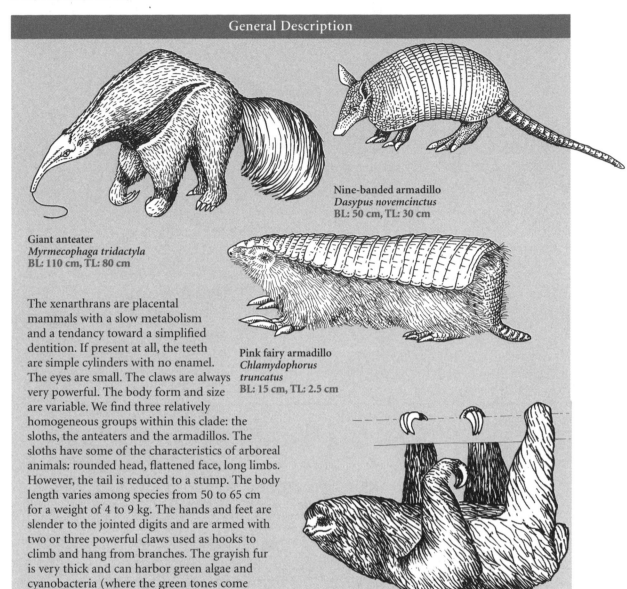

Giant anteater
Myrmecophaga tridactyla
BL: 110 cm, TL: 80 cm

Nine-banded armadillo
Dasypus novemcinctus
BL: 50 cm, TL: 30 cm

Pink fairy armadillo
Chlamydophorus truncatus
BL: 15 cm, TL: 2.5 cm

Three-toed sloth
Bradypus tridactylus
BL: 60 cm, TL: 7 cm

The xenarthrans are placental mammals with a slow metabolism and a tendancy toward a simplified dentition. If present at all, the teeth are simple cylinders with no enamel. The eyes are small. The claws are always very powerful. The body form and size are variable. We find three relatively homogeneous groups within this clade: the sloths, the anteaters and the armadillos. The sloths have some of the characteristics of arboreal animals: rounded head, flattened face, long limbs. However, the tail is reduced to a stump. The body length varies among species from 50 to 65 cm for a weight of 4 to 9 kg. The hands and feet are slender to the jointed digits and are armed with two or three powerful claws used as hooks to climb and hang from branches. The grayish fur is very thick and can harbor green algae and cyanobacteria (where the green tones come from).

In contrast to the sloths, the anteaters have a long, almost tubular, muzzle with a narrow mouth at the tip. They use their long, sticky tongue to enter into the openings and galleries of ant hills and termite mounds. Teeth are absent. The hands are strongly modified such that they are supported by the external part of the last phalange and have the claws turned inward towards the body. The tail is long, scaly and prehensile in arboreal species, like tamandua, and very fluffy in the giant anteater. The length and color of the fur is variable. The body length from the tip of the muzzle to the base of the tail varies from 16 to 120 cm. The tail length ranges from 18 to 90 cm and the overall weight from 0.5 to 35 kg.

The armadillos are more lively animals than the sloths or anteaters. The dorsal side of the body, the tail and the external surfaces of the limbs are all covered in epidermally derived boney scales. The ensemble of scales forms a type of "carapace" with scattered hairs. They are stocky animals, generally brown in color with a long muzzle and long ear pinnae. The fossorial genus *Chlamydophorus* are the only members of the group with white fur and reduced eyes and ears.

Some unique derived features

- In the dorsal region, the ischium (*is*) is fused with the sacrum (*sm*). Left lateral views of the pelvic girdle of a cat (fig. 1a) and giant anteater (fig. 1b; *ac:* acetabulum, *cvr1:* first caudal vertebra, *il:* ilium, *lvr6:* sixth lumbar vertebra, *pu:* pubis) are represented in fig. 1. In the xenarthrans (with the exception of *Cyclopes*), a large part of the dorsal surface of the ischium is in contact with the transverse processes of the sacrum (for example *Myrmecophaga:* giant anteater) or those of the pseudosacral vertebrae (for example, *Bradypus:* three-toed sloth). This bone is fused to 2 to 4 vertebrae, forming a sacro-ischium suture (*sis*), and, at the anterior side of the suture, a sacro-ischium foramen (*sif*).

- There are additional processes and articulations on the posterior thoracic and lumbar vertebrae. A cranial view of the third lumbar vertebra of a human is shown in fig. 2a: the only articular surface that we see besides that of the centrum (*vrc*) is the normal prezygapophysis (*prz*). In the xenarthra, the articulation between vertebrae is more complex with two accessory postzygapophyses (*apz*) nested within two accessory prezygapophyses (*aprz*) in addition to

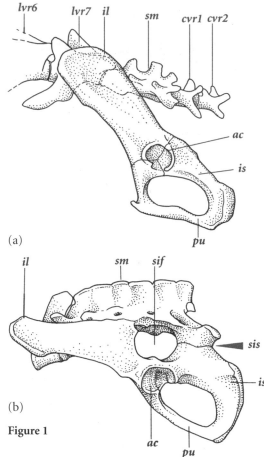

Figure 1

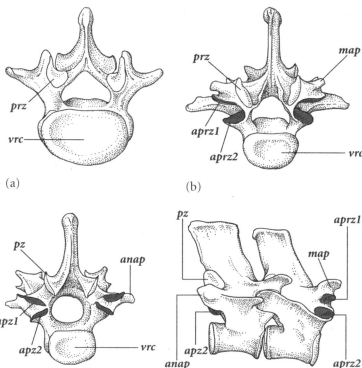

Figure 2

the normal post- (*pz*) and prezygapophyses. These new articular surfaces are the sites of what is called xenarthrous articulations. Fig. 2b and 2c show cranial and caudal views respectively of the second lumbar vertebra of a giant anteater (*Myrmecophaga*). Fig. 2d shows a right lateral view of the second and third lumbar vertebrae of this species. The additional articulations are indicated: *aprz1*, *aprz2*, *apz1*, *apz2* (*anap:* anapophysis, *map:* metapophysis).

- There is a characteristic deletion of three amino acids in the alpha-crystallin protein.

- Molecular phylogenies based on the sequences of 19 nuclear genes and 3 mitochondrial genes bring together the xenarthrans (anteaters, armadillos and sloths). These genes include hemoglobin, myoglobin, alpha-crystallin, cytochrome c, and ribonuclease genes along with mitochondrial ribosomal genes. In addition, independent analyses of other nuclear (a different fragment of *RAG1*, *c-myc* oncogene, growth hormone) and mitochondrial (*ND6*) genes support this clade.

Ecology

The xenarthrans are terrestrial or arboreal animals found in forests, savannas and prairies. The sloths are hanging arboreal species that feed exclusively on leaves and fruit. Fermentation takes place in their large stomachs, where leaf cellulose is transformed by the action of anaerobic bacteria into fatty acids capable of passing into the blood. This type of fermentation is performed, by convergence, in other parts of the digestive system (*stomach, rumen, caecum*) in several other mammalian groups: the ruminants, equids, herbivorous marsupials of the genus *Setonix*, leaf-eating monkeys of the genus *Colobus*, hyracoids. Mating in the sloths takes place in the trees, male and female face to face and hanging by their arms. Gestation lasts 5 months and 3 weeks. Birth also takes place in the trees, with the mother suspended by her arms. The fully formed young is born head first with no fetal membrane. It breathes immediately and then helps with the rest of the birth, holding onto the mother as it comes out. The sloths move very slowly and sleep or rest approximately 20 hours per day (hence the name). This stationarity in addition to their camouflage in the leaves means that the populations are not threatened. They were domesticated as pets by the Indians of the Amazonian forest.

The armadillos are primarily nocturnal, fossorial omnivores with relatively poor vision and an extremely well developed sense of smell. The number of mammae varies from one species to another, and the number of teeth can vary within a single species. Mating generally takes place in July. After fertilization and the early developmental stages, there is a pause of 14 weeks before implantation occurs. Gestation then lasts around four months such that the young are born in February or March. The nine-banded armadillo systematically gives birth to quadruplets. In the seven-banded armadillo this polyembryony can give rise to 4, 8 or 12 genetically identical young. In the southern United States, the principal enemies of the armadillos are no longer coyotes, wolves and puma but rather automobile drivers. Indeed, we see many dead armadillos along the sides of American roads in the early hours of the morning. When these animals feel threatened, they can bury themselves very quickly. Some species, like the three-banded armadillo, can roll itself into a closed ball. The genus *Chlamydophorus* contains fossorial armadillos that rarely leave their underground galleries.

The anteaters are exclusive insectivores. The giant anteater is diurnal and terrestrial whereas the tamandua is nocturnal and both terrestrial and arboreal. The silky anteater is strictly nocturnal and arboreal. When threatened, they stand up on their hind legs and show their powerful claws. Both parents of the silky anteater take care the young and regurgitate grounded-up insects. The young can attach itself to the father's fur.

Number of species: 29.
Oldest known fossils: *Protegotherium* from Brazil dates from the Paleocene (60 MYA). The xenarthrans have a rich fossil record. The sloths reached their peak relatively recently, during the Pleistocene, with giant species like *Megatherium* (the size of an elephant), *Mylodon* and *Megalonyx*. The giant armadillo *Glyptodon*, 4 m in length with a rigid shield-like carapace, was still alive during the Pleistocene.
Current distribution: South America, Central America, and southern North America.

Examples

Nine-banded armadillo: *Dasypus novemcinctus;* seven-banded armadillo: *D. septemcinctus;* three-banded armadillo: *Tolypeutes matacus,* giant armadillo: *Priodontes giganteus;* pink fairy armadillo: *Chlamydophorus truncatus;* three-toed sloth: *Bradypus tridactylus;* giant anteater: *Myrmecophaga tridactyla;* tamandua: *Tamandua tetradactyla;* silky anteater: *Cyclopes didactylus.*

Epitheria

A few representatives

African savanna elephant
Loxodonta africana africana
BL: 690 cm
node 17

Brown hare
Lepus capensis
BL: 60 cm
node 27

European hedgehog
Erinaceus europaeus
BL: 20 cm
node 30

Bactrian camel
Camelus ferus
BL: 290 cm
node 39

Some unique derived features

- The fibula is not in direct contact with the femur. Anterior views of the tibia (*ti*) and fibula (*fi*) of the right leg of different mammals are shown in fig. 1: 1a: *Phascolarctos* (koala, Marsupialia), 1b: *Elephas* (Asian elephant, Proboscidea), 1c: *Manis* (pangolin, Pholidota), 1d: *Erinaceus* (hedgehog, Eulipotyphles). We can see the reduction in the fibula on external views of the right tibia and fibula of some ungulates: 2a: *Equus* (horse, Perissodactyla), 2b: *Hippopotamus* (hippopotamus, Hippopotamidae). In non-epitherian mammals (here, marsupials, 1a), the fibula reaches the level of the tibia and touches the femur. In all other mammals, the proximal

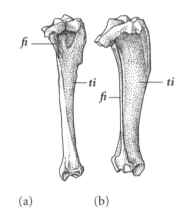

(a)   (b)

Figure 2

head of the fibula is lower than that of the tibia.

- The longitudinal division (and its traces) of the median vagina, the result of the partial fusion of the Mullerian ducts, has completely disappeared.

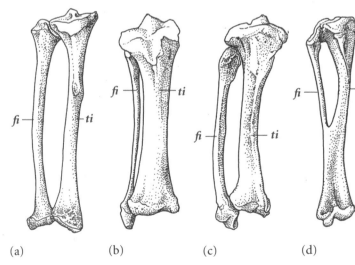

(a)   (b)   (c)   (d)

Figure 1

Number of species: 4,192.
Oldest known fossils: the gypsonictopid *Prokennalestes* (Leptictida) from Mongolia dates from the lower Cretaceous (100 MYA).
Current distribution: worldwide.

Examples

- **AFROTHERIA**: aardvark: *Orycteropus afer;* short-snouted elephant shrew: *Elephantulus brachyrhynchus;* African elephant: *Loxodonta africana*.
- **BOREOEUTHERIA**: Indian pangolin: *Manis crassicaudata;* European hedgehog: *Erinaceus europaeus;* mountain zebra: *Equus zebra*.

Afrotheria

A few representatives

Aardvark
Orycteropus afer
BL: 130 cm
node 10

Giant otter shrew
Potamogale velox
BL: 32 cm
node 12

Rock hyrax
Procavia capensis
50 cm
node 15

African savanna elephant
Loxodonta africana africana
BL: 690 cm
node 17

Some unique derived features

The relatedness of the afrotherians is indicated by numerous molecular traits:

- There are unique chromosomal arrangements. The most notable is that two chromosomes, corresponding to chromosomes 5 and 21 in humans, are physically linked in the afrotherians. This also occurs for chromosomes 1 and 19.

- There are unique amino acid deletions in the sequence coding for apolipoprotein B (amino acids 79 to 82) along with a deletion of 9 base pairs in the *BRCA1* gene (breast and ovarian cancer susceptibility gene 1).

- There are novel "SINE" type transposable elements in the genome whose sequence is exclusive to the afrotherians; they have justly been named "afro-SINEs."

- Molecular phylogenies based on the sequences of 19 nuclear genes and 3 mitochondrial genes all regroup the afrotherians together. These genes include hemoglobin, myoglobin, alpha-crystallin, cytochrome c, and ribonuclease genes along with mitochondrial ribosomal genes. In addition, independent analyses of other nuclear (a fragment of *RAG1*, *c-myc* oncogene, growth hormone) and mitochondrial (*ND6*) genes support this clade.

Number of species: 74.
Oldest known fossils: *Anaptogale, Chianshania, Wanogale, Linnania,* and *Stenanagale* are anagalids of the lower Paleocene from China (65 MYA).
Current distribution: This group represents the old African strain of placental mammals. Today, we find them in Africa, the Middle East, Madagascar and Southeast Asia. We also find them in rivers, in the shallow seas of the coastal intertropical Atlantic zone, in the Red Sea, and in the coastal waters of east Africa, Madagascar, southern India, Sri Lanka, Southeast Asia, New Guinea, northern Australia, New Caledonia and the Caroline, Solomon, and Marshall Islands. Although these "orders" of placental mammals originated in Africa, time and past contact between different continents have enable some to acquire vast distributions. For example, in the second half of the Tertiary period, mammoths were found across all of Eurasia and North America, and the Steller's sea cow lived in the Bering Strait.

Examples

- AFROINSECTIPHILIA: aardvark: *Orycteropus afer;* short-snouted elephant shrew: *Elephantulus brachyrhynchus;* giant otter shrew: *Potamogale velox.*
- PAENUNGULATA: African elephant: *Loxodonta africana;* dugong: *Dugong dugon;* western tree hyrax: *Dendrohyrax dorsalis.*

Afroinsectiphilia

A few representatives

Giant otter shrew
Potamogale velox
BL: 32 cm
node 12

Aardvark
Orycteropus afer
BL: 130 cm
node 10

Cape golden mole
Chrysochloris asiatica
BL: 10 cm
node 12

North African elephant shrew
Elephantulus rozeti
BL: 12 cm
node 13

Some unique derived features

- The equivalents of chromosomes 2 and 8 in humans are physically linked in this group.

- The equivalents of chromosomes 3 and 20 in humans are physically linked in this group. This association has been lost secondarily in the chrysochlorids.

- The equivalents of chromosomes 10 and 17 in humans are physically linked in this group. This association has been lost secondarily in the chrysochlorids.

- Molecular phylogenies based on the sequences of 19 nuclear genes and 3 mitochondrial genes unite the tubulidents, macroscelids and afrosoricids. These genes include hemoglobin, myoglobin, alpha-crystallin, cytochrome c, and ribonuclease genes along with mitochondrial ribosomal genes. In addition, independent analyses of other nuclear (a fragment of *RAG1*, *c-myc* oncogene, growth hormone) and mitochondrial (*ND6*) genes support this clade.

Number of species: 61.
Oldest known fossils: *Anaptogale, Chianshania, Wanogale, Linnania,* and *Stenanagale* are anagalids of the lower Paleocene from China (65 MYA).
Current distribution: Africa, Madagascar.

Examples

Tubulidentata: aardvark: *Orycteropus afer.*
Afroinsectivora: short-snouted elephant shrew: *Elephantulus brachyrhynchus;* giant otter shrew: *Potamogale velox.*

Tubulidentata

General Description

The tubulidentata includes only a single, extant species, the aardvark. It is a medium-sized placental mammal with a body length of between 120 and 170 cm, a tail length of 45 to 60 cm and a weight of 50 to 80 kg. It is easily recognizable by its elongated head, long upright ears, and long mobile snout. The neck is short, the body stocky, the posterior limbs are longer than the anterior limbs, the rump is high, and the tail is muscular and thick. The anterior legs have four digits armed with powerful claws used for digging. These claws cover each phalange and form a type of "hoof." The posterior limbs are digitigrade with 5 toes that also end in powerful claws. The fur is sparse and brownish in color. The zygomatic arch is strong. The teeth are found in high numbers in the fetus, implanted within alveoli. However, in the adult animal, only 5 to 7 jugal teeth are maintained per jaw quadrant (fig. 2). These teeth are column-shaped, rootless, and are covered in a cement rather than enamel. The dentine is structured into parallel, hexagonal prisms, unique to this group. The column-shaped teeth grow

Aardvark
Orycteropus afer
BL: 130 cm, TL: 50 cm

over the entire lifetime of the animal. The tongue is straight, flattened, protractable, and rich in saliva; it serves to capture termites. The stomach is simple, but strongly muscular in the pyloric region, a convergent organization that can also be found in the pangolins (pholidotes), along with other anteaters and exclusive termite-eaters. As in these animals, the salivary glands in the aardvark are extremely well developed.

Some unique derived features

■ The teeth are very characteristic. They are column-shaped, lack enamel, grow continuously, and are composed of numerous hexagonal prisms (1500 prisms for the second molar). The pulp cavity is relatively open at

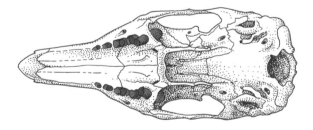

Figure 2

the base of the tooth and sends vertical canals towards the top of the tooth; there is a canal present in the axis of each ivory prism. Fig. 1 shows a drawing of a vertical cross-section of an adult tooth (*pcn*: pulp canal, *iv*: ivory, *oiv*: old ivory, *pri*: prisms). Fig. 2, a ventral view of an aardvark's skull, shows the tubular shape of the jugal teeth.

■ The claws are powerful and used for digging. The claws of the fore and hind feet of an aardvark are shown in fig. 3 (*to1*: first toe).

NOTE: the foot is represented in a plantigrade position. However, the aardvark is actually a digitigrade animal when walking. The plantigrade position is taken only when the animal crouches and uses

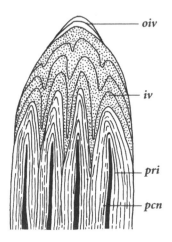

oiv

iv

pri

pcn

Figure 1

its forefeet to dig a burrow or tear open a termite mound. This change in position is permitted by the presence of two foot pads, the metatarsal pad (*mp*) and a calcanean pad (*cp*).

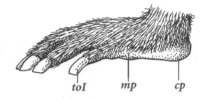

Figure 3

Ecology

The aardvark, the only species in this order, is a solitary, nocturnal animal that digs for its main food source, termites (although it may sometimes eat ants). We find it in savannas and rain forests, mostly in areas where termites of the genera *Macrotermes*, *Cubitermes*, *Trinervitermes* and *Bellicotermes* are present. Olfaction is the dominant sense. The aardvark is an efficient digger, capable of hollowing out a shelter for protection within a few minutes. During the day, it sleeps in its burrow. The animal blocks the entrance to this burrow from the inside to keep out intruders, such as pythons. The burrow is typically 2 to 3 m long with a diameter of 40 cm. It slopes into the soil at a 45° angle and ends in a spherical chamber. Abandoned burrows are used by warthog families to such an extent that, in some regions of Africa, warthogs are present only in areas where there are also aardvarks. The burrows can also be used by porcupines, hedgehogs, rodents, genets, jackals, hyenas, and a bat (*Nycteris thebaica*).

If food is abundant, an aardvark will remain within a restricted territory, follow the same paths, and visit the same termite mounds approximately once a week. It feeds by breaking open the mounds and using its sticky tongue to capture termites. An interval of one week allows the termites to rebuild their nest and the population to recover. The predators of the aardvark are leopards, lions and hyenas.

In different regions of Africa, mating takes place at different periods in the year. Gestation lasts approximately 7 months and typically produces a single young, born pink and naked, that will spend the first weeks of its life in the burrow. At six months old, the young will dig its own burrow close to that of its mother and will continue to search for food with her. During the next rut, male offspring leave their mothers. The lifespan is approximately 15 years.

Number of species: 1.
Oldest known fossils: *Leptomanis* from Quercy, France, dates from the lower Oligocene (30 MYA).
Current distribution: sub-Saharan Africa.

Examples

Aardvark: *Orycteropus afer.*

Afroinsectivora

A few representatives

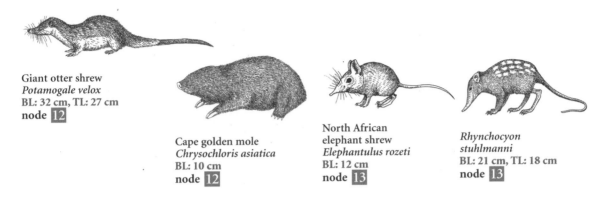

Giant otter shrew
Potamogale velox
BL: 32 cm, TL: 27 cm
node 12

Cape golden mole
Chrysochloris asiatica
BL: 10 cm
node 12

North African
elephant shrew
Elephantulus rozeti
BL: 12 cm
node 13

*Rhynchocyon
stuhlmanni*
BL: 21 cm, TL: 18 cm
node 13

Some unique derived features

■ Molecular phylogenies based on the sequences of 19 nuclear genes and 3 mitochondrial genes unite the macroscelids and the afrosoricids. These genes include hemoglobin, myoglobin, alpha-crystallin, cytochrome c, and ribonuclease genes along with mitochondrial ribosomal genes.

■ The placenta is hemochorial and discoid (fig. 1). With a hemochorial-type placenta, the uterine epithelium disappears and the uterine capillaries spread out and form the maternal blood lacunae that connect to the fetal tissue (in the chorial syncytium). The placental villi are found in a disc-shaped region; we therefore say that the placentation is discoid.

NOTE: In other afrotherians, the placentation is endotheliochorial (except for some hyracoids). In this case, the uterine epithelium also disappears and the uterine capillaries are in direct contact with the fetal tissue, but this time without the formation of blood lacunae. A hemochorial and discoid placenta is also found by convergence in the haplorrhinian primates, the rodents and the chiropterans.

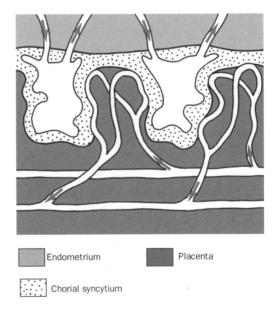

□ Endometrium   ■ Placenta

▦ Chorial syncytium

Figure 1

Number of species: 60
Oldest known fossils: *Anaptogale, Chianshania, Wanogale, Linnania,* and *Stenanagale* are anagalids of the lower Paleocene from China (65 MYA).
Current distribution: Africa, Madagascar.

Examples

■ AFROSORICIDA: giant otter shrew: *Potamogale velox*.
■ MACROSCELIDEA: short-snouted elephant shrew: *Elephantulus brachyrhynchus*.

Afrosoricida

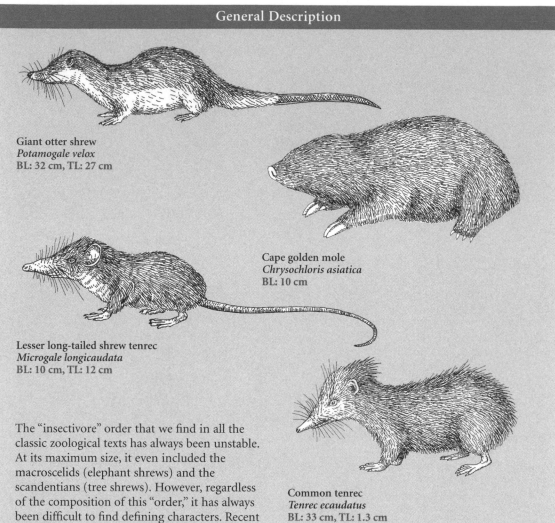

General Description

Giant otter shrew
Potamogale velox
BL: 32 cm, TL: 27 cm

Lesser long-tailed shrew tenrec
Microgale longicaudata
BL: 10 cm, TL: 12 cm

Cape golden mole
Chrysochloris asiatica
BL: 10 cm

Common tenrec
Tenrec ecaudatus
BL: 33 cm, TL: 1.3 cm

The "insectivore" order that we find in all the classic zoological texts has always been unstable. At its maximum size, it even included the macroscelids (elephant shrews) and the scandentians (tree shrews). However, regardless of the composition of this "order," it has always been difficult to find defining characters. Recent genomic approaches have shown that the insectivores are polyphyletic and that the strictly African families (tenrecidae, potamogalidae, chrysochloridae) are not related to the holarctic families. The afrosoricida derives its name from the geographic limits of the three families it contains. It comprises small animals (body length from 7 to 39 cm), with dense fur, short legs and powerful claws. The tail is short, except in the otter shrews. The eyes are also small. Each of the three families possesses physical traits that distinguish them from each other. The tenrecidae (tenrecs) have spines scattered throughout their fur, or only spines (most notably in the hedgehog tenrec that greatly resembles a hedgehog). Their muzzle is conical and their tail is very short. The potamogalids (otter shrews) look like small otters. The body is elongated. The head is long and flattened with an enlarged upper lip, long stiff bristles, and nostrils found behind a horny plate that closes like a valve while diving. The legs

are short, the feet are unwebbed, and the tail is laterally compressed. The chrysochlorids (golden moles) greatly resemble true moles. They have a cylindrical body that tapers anteriorly. The nose is naked and horny. The eyes are reduced, hidden below the fur. The ear pinnae have disappeared. The anterior legs are short and strong, armed with four fingers, two of which have particularly powerful claws. The fur is short, dense and soft. It is interesting to note that over their evolution the mammals have produced three mole-like groups: once in Africa with the chrysochlorids, once in the holarctic region with the talpids, and once in Australia with the notoryctid marsupials. The "hedgehog" form has also evolved twice: once in the holarctic region with the erinaceids and once in Madagascar with the hedgehog tenrecs (*Setifer* and *Echinops* genera).

Some unique derived features

■ Molecular phylogenies based on the sequences of 19 nuclear genes and 3 mitochondrial genes regroup the afrosoricids. These genes include hemoglobin, myoglobin, alpha-crystallin, cytochrome c, and ribonuclease genes along with mitochondrial ribosomal genes. In addition, independent analyses of other nuclear (a fragment of *RAG1*, *c-myc* oncogene, growth hormone) and mitochondrial (*ND6*) genes support this clade.

Number of species: 45.
Oldest known fossils: *Prochrysochloris*, a chrysochlorid from the Kenya, dates from the lower Miocene (20 MYA).
Current distribution: sub-Saharan Africa, Madagascar. The tenrecs are limited to Madagascar, but they have been introduced to Maurice Island, Reunion Island, Mayotte in the Comoro Islands, and Mahé Island in the Seychelles. The potamogalids are found in humid regions of central Africa and the chrysochlorids are found across the entire southern half of Africa.

Ecology

They are terrestrial animals that live in forests, scrub regions, semi-arid and steppe zones, but they also can be found in swampy areas. All the species are fossorial, and some are completely subterranean, like the golden moles. However, the hedgehog tenrecs can climb trees and, although they have their burrows along rivers and streams, the aquatic tenrecs and otter shrews are aquatic species that spend a large part of their time in the water. These species are omnivores, both diurnal and nocturnal. In the afrosoricids, we find the record number of young per litter; the common tenrec can have up to 31 young. Litter sizes of 12 to 15 are common for this species and the female has 12 pair of nipples.

Examples

Giant otter shrew: *Potamogale velox;* Nimba otter shrew: *Micropotamogale lamottei;* common tenrec: *Tenrec ecaudatus;* greater hedgehog tenrec: *Setifer setosus;* aquatic tenrec: *Limnogale mergulus;* lesser long-tailed shrew tenrec: *Microgale longicaudata;* golden mole: *Chrysochloris congicus;* giant golden mole: *Chrysospalax trevelyani;* Hottentot golden mole: *Amblysomus hottentotus.*

Macroscelidea

The macroscelids are the "elephant shrews," small placental mammals with an elongated, flexible muzzle. The nostrils are found at the end of this "trunk." Their eyes and ear pinnae are large. The body is short, but the tail is long. The tibia and metatarsal bones are elongated such that the posterior limbs are longer than the anterior limbs. These posterior limbs are used for jumping. The fur is thick and soft. It is generally brown with a lighter ventral surface, but the shades vary among species. The trunk is naked at the end and has long whiskers at its base. Scent glands are present at the base of the tail and emit a musky odor, particularly in males. The body size varies from 9.5 cm to 31.5 cm, and the tail from 8 to 26.5 cm. The macroscelids were previously placed in the insectivore order. However, they differ greatly from the other members of this

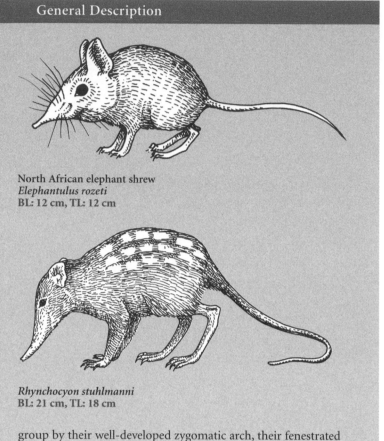

North African elephant shrew
Elephantulus rozeti
BL: 12 cm, TL: 12 cm

Rhynchocyon stuhlmanni
BL: 21 cm, TL: 18 cm

group by their well-developed zygomatic arch, their fenestrated palate (except in *Rhynchocyon*), and their intestinal caecums.

Some unique derived features

- The posterior limb (*pl*) is longer than the anterior limb and has an elongated foot (fig. 1).

- The muzzle is elongated into a short and flexible trunk (*tr*) (fig. 1).

- The upper canine is shaped like a premolar. A lateral left view of a macroscelid skull (*Elephantulus brachyrhynchus*) is represented in fig. 2a (*I*: incisor, *M*: molar, *Pm*: premolar), showing the small, premolar-like canine (*C*). This tooth is also visible on the ventral view of the same skull (fig. 2b). The unique shape of this tooth is obvious when we compare it with the canine of a carnivore (fig. 2c).

- The palate is strongly fenestrated (*f*, fig. 2b), a rare attribute in the eutherians.

Ecology

The macroscelids are diurnal, nocturnal or crepuscular and live in prairies, savannas and rocky,

arid zones. We also find them in forested areas, but no species is strictly limited to this habitat type. They feed on insects, sometimes mollusks or eggs and, for the larger species of the *Rhynchocyon* genus, even small vertebrates. The trunk plays a predominant role in the detection of prey. Hunting is active and fast, but the flight ability is even faster; the leaping abilities of these animals enable them to attain speeds that put them out of harm's way from most of their predators. During leaping movements, the long tail is used for balance. They nest in leaf nests, in galleries that they dig themselves or in burrows built by other mammals, but also in tree trunks. Outside the reproductive period, these animals live solitary lives or form loose social groups with no obvious social link among individuals. To delimit their territory, they place excrement along the territory boundary. To defend the territory, they drum the ground with the two posterior legs.

In contrast to the insectivores, the group in which the macroscelids were previously placed, the number

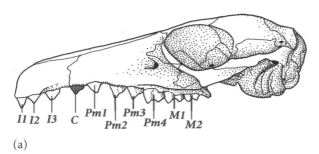

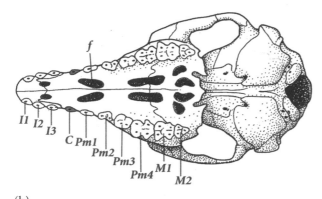

(a)

$I1$ $I2$ $I3$ C $Pm1$ $Pm2$ $Pm3$ $Pm4$ $M1$ $M2$

Figure 1

of young produced per cycle is only one or two, despite the fact that the up to 60 ovules can be released by each ovary during estrus. After a gestation period of approximately 2 months, the female will give birth to completely formed young that are covered in hair, capable of jumping and fleeing from predators. Longevity in this group can reach up to 3 years.

(b)

f

$I1$ $I2$ $I3$ C $Pm1$ $Pm2$ $Pm3$ $Pm4$ $M1$ $M2$

Number of species: 15.
Oldest known fossils: *Anaptogale, Chianshania, Wanogale, Linnania,* and *Stenanagale* are anagalids of the lower Paleocene from China (65 MYA). *Chambius* is a macroscelid from the lower Eocene from Tunisia (50 MYA).
Current distribution: sub-Saharan Africa, with a single species found in North Africa (*Elephantulus rozeti*).

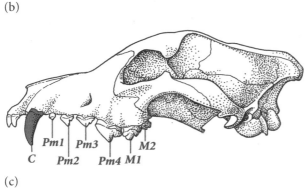

(c)

C $Pm1$ $Pm2$ $Pm3$ $Pm4$ $M1$ $M2$

Figure 2

Examples

Short-snouted elephant shrew: *Elephantulus brachyrhynchus;* North African elephant shrew: *Elephantulus rozeti;* black and rufous elephant shrew: *Rhynchocyon petersi;* four-toed elephant shrew: *Petrodromus tetradactylus;* short-eared elephant shrew: *Macroscelides proboscideus.*

Paenungulata

A few representatives

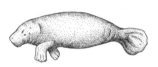

Rock hyrax
Procavia capensis
50 cm
node 15

African savanna elephant
Loxodonta africana africana
BL: 690 cm
node 17

African manatee
Trichechus senegalensis
400 cm
node 18

Some unique derived features

- The mastoid bone is either relegated to the ventro-lateral region of the skull by the posterior extension of the squamosal bone or is covered over by this bone; this situation is referred to as "amastoid." In the hyracoideans and proboscideans, the mastoid bone is no longer exposed in the occipital region (this arrangement is also found convergently in the Cetacea and the Suina). The exposed mastoid

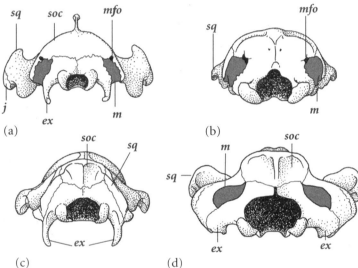

Figure 1

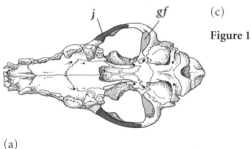

(a)

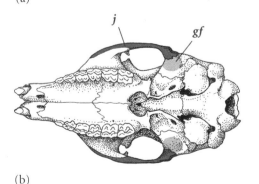

(b)

Figure 2

bone of the sirenians was secondarily acquired from amastoid ancestors. Fig. 1 shows occipital views of non-paenungulate skulls (*Didelphis*, Marsupialia, 1a, and *Orycteropus*, Tubulidentata, 1b) and paenungulate skulls (*Procavia*, Hyracoidea, 1c, and *Trichechus*, Sirenia, 1d); *ex:* exoccipital, *mfo:* mastoid foramen, *j:* jugal, *m:* mastoid, *soc:* supraoccipital, *sq:* squamosal.

Note: the mastoid bone is not a separate bone but rather a dorso-lateral extension of the petrosal bone.

- The jugal bone extends posteriorly to the anterio-lateral edge of the glenoid fossa, the fossa that connects with the mandibular condyle. Fig. 2 shows ventral views of a non-paenungulate skull (Carnivora, *Canis,* 2a) and a paenungulate skull (Hyracoidea, *Procavia capensis,* 2b); *gf:* glenoid fossa, *j:* jugal.

- The carpal bones are dorso-ventrally compressed and serially arranged.

- The lunate, the second bone in the upper row of the carpal bones, normally articulates with the radius proximally, with the unciform and magnum bones distally, with the scaphoid exteriorly, and with the pyramidal interiorly. In the paenungulates, the lunate and the unciform have little to no contact.

- The placenta is zonal, that is, the chorionic villi form a ring and there is a small, independent yolk sac and a large allantois.

- There is a mobile element in the genome, specific to the afrotherians ("afro-SINEs"), that has a 45 base pair deletion in the paenungulates.

- Molecular phylogenies based on the sequences of 19 nuclear genes and 3 mitochondrial genes unite the paenungulates. These genes include hemoglobin, myoglobin, alpha-crystallin, cytochrome c, and ribonuclease genes along with mitochondrial ribosomal genes. In addition, independent analyses of other nuclear (a fragment of *RAG1*, *c-myc* oncogene, growth hormone) and mitochondrial (*ND6*) genes support this clade.

Number of species: 13.
Oldest known fossils: the embrithopod *Phenacolophus* from Mongolia and China dates from the early Paleocene (60 MYA).
Current distribution: Africa (except in the Atlas Mountains and western Sahara), Middle East, and Southeast Asia, in rivers, coastal waters, and shallow coastal waters of intertropical Atlantic Ocean, in the Red Sea, and in coastal areas of East Africa, Madagascar, southern India, Sri Lanka, Southeast Asia, New Guinea, north Australia, New Caledonia, and the Caroline, Solomon and Marshall Islands.

Examples

HYRACOIDEA: western tree hyrax: *Dendrohyrax dorsalis*.
TETHYTHERIA: African elephant: *Loxodonta africana;* dugong: *Dugong dugon*.

Hyracoidea

General Description

The hyraxes are small, stocky mammals that look something like a groundhog. Their size varies from 40 to 50 cm in length for a weight of between 2.5 and 3.5 kg. They are plantigrade ungulates; they have hoof-like nails on their toes and walk on their heels. They have elastic pads on the soles of their feet that are kept moist by specialized sweat glands. There are 4 toes on the forefeet and 3 toes on the hindfeet (II, III, IV). The toenails are flat and black, except for the nail on the second toe of the hindfeet, which carries a long, curved claw used for preening. The hyraxes are tailless. The snout is short, the eyes are large, the ear pinnae are short and rounded. The upper lip is clefted. The upper incisors grow continuously and the lower incisors are proclivous (i.e., angled forward). The canines are absent. The stomach is simple, but a well-developed caecum houses bacteria able to degrade cellulose. Another bicornate caecum, whose function is poorly known, is found near

Rock hyrax
Procavia capensis
50 cm

the colon. The fur is dense and can vary in color from light gray to black, gray-brown, light brown, ochre or sandy brown. A dorsal gland is found in the middle of the back; this glandular zone is long and straight, lacks fur and is surrounded by erectile hairs that vary in color from white to orange.

Some unique derived features

- The dorsal gland is characteristic of the hyraxes. This glandular zone is long and straight, covering approximately 2 cm². It is hairless but is surrounded by erectile hairs. The zone includes a high density of eccrine and apocrine sweat glands and sebaceous glands. During the period of the rut, the glandular secretions are creamy and strong-smelling.

- Plantigrade: the hyraxes show secondary plantigrade feet with elastic pads on the soles of their feet that carry specialized sweat glands. This type of plantigrade foot is considered a secondary acquistion because the ancestors of the hyraxes were non-plantigrade ungulates. Fig. 1 shows a

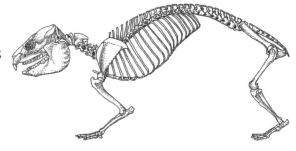

Figure 1

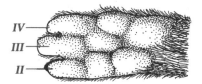

IV —
III —
II —

Figure 2

left lateral view of the skeleton of *Heterohyrax brucei* on which we see that the anterior limb is supported by the heel of the foot. Fig. 2, the bottom of the left hindfoot of *Dendrohyrax*, shows the elastic foot pads.

- Incisors: the hyraxes have a pair of continuously growing upper incisors. They are shown in fig. 3 on a ventral view of the skull of *Procavia capensis* (*I*).

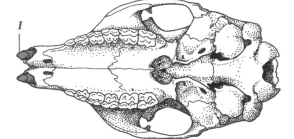

I

Figure 3

Ecology

The hyraxes are principally herbivorous animals and only rarely feed on insects (crickets). They live in almost all terrestrial environments: tropical rain forests, steppes, semi-arid regions, and mountainous regions up to 5000 m in altitude. They are terrestrial, rupicolous and arboreal. They can be solitary or live in couples or colonies. Depending on the species, they can be diurnal or nocturnal (nocturnal in dense forests). In areas with few available trees, the hyraxes may live in crevasses, hollows, scree, and other cavities with small openings. They never dig their own burrows, but when rocky areas are already fully occupied, they may use burrows left by other species (for example, aardvarks or meerkats). The hyraxes are territorial animals that coat local trees or rocks with secretions from their dorsal glands by rubbing up against them. In addition, they protect their territories by screaming, grinding their teeth, showing their dorsal gland, stamping their forefeet and defecating in certain areas of the territory. Their enemies are diurnal and nocturnal carnivores and birds of prey: leopards, panthers, golden cats, servals, crowned eagles, etc. In colonies that are out in the open (rocky areas, for example), sentinels keep watch over the colony. If danger approaches, a shrill whistle alerts the tribe. This whistle is sufficient to interrupt the meal of the entire colony and cause the members to seek shelter in holes or under rocks.

Gestation lasts 7 to 8 months, a long period for a mammal of this size. The young, 1 to 3, are born in a well-developed state and are immediately able to follow their mothers. They start to graze with their mothers very early and quickly adopt a mixed feeding regime; all young are weaned before they are three months old. The hyraxes like contact and will often crowd together. The young of the rock hyrax like to sit on their mother's back. Sexual maturity is reached at 16 months old and the lifespan is approximately 12 years. These animals are easily maintained in captivity.

Number of species: 6.
Oldest known fossils: *Titanohyrax* from Algeria dates from the lower Eocene (50 MYA).
Current distribution: Africa (except in the Atlas Mountains and west of the Sahara Desert), Middle East.

Examples

Western tree hyrax: *Dendrohyrax dorsalis;* Southern tree hyrax: *D. arboreus;* Rock hyrax: *Procavia capensis;* gray hyrax: *Heterohyrax brucei.*

Tethytheria

A few representatives

African savanna elephant
Loxodonta africana africana
BL: 690 cm
node 17

Asian elephant
Elephas maximus
BL: 600 cm
node 17

African manatee
Trichechus senegalensis
400 cm
node 18

Dugong
Dugong dugon
270 cm
node 18

Some unique derived features

- The heart has a bifid apex.

- The orbits (*or*) are moved forward on the anterior-posterior axis of the skull to such an extent that they open before the jugal teeth. Fig. 1 shows left lateral views of some tethytherian skulls (1a: a modern proboscidean, a young Asian elephant; 1b: a fossil proboscidean, *Moeritherium;* 1c: a sirenian, the manatee *Trichechus senegalensis*) and a non-tethytherian (1d: a carnivore, *Canis*). In the *Canis* skull, the orbit opens above the last molars, whereas in the tethytherians, the orbit opens anteriorly to all molars and premolars.

> **Number of species:** 7.
> **Oldest known fossils:** the oldest tethytherian is the embrithopod *Phenacolophus* from Mongolia and China that dates from the early Paleocene (60 MYA).
> **Current distribution:** sub-Saharan Africa and Southeast Asia (from southern India and Sri Lanka to northern Vietnam and in southern China, southern Himalayas, Sumatra and Borneo); rivers, coastal waters, and shallow intertropical waters of the eastern and western Atlantic, the Red Sea, and coastal areas of East Africa, Madagascar, southern India, Sri Lanka, and from South-east Asia to the coasts of southern China, New Guinea, north Australia, New Caledonia, and the Caroline, Solomon and Marshall Islands.

Examples

- PROBOSCIDEA: African elephant: *Loxodonta africana*.
- SIRENIA: African manatee: *Trichechus senegalensis*.

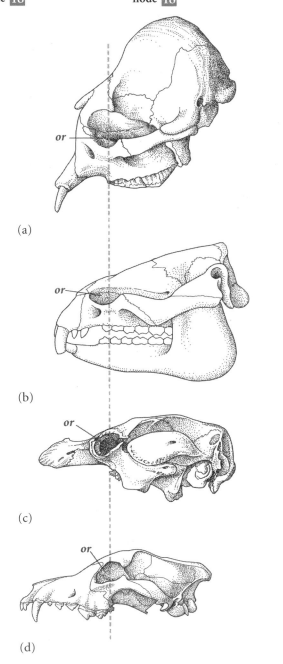

(a)

(b)

(c)

(d)

Figure 1

Proboscidea

General Description

The elephants are the largest extant terrestrial mammals; their size varies from 220 to 400 cm to the top of the shoulders, from 260 to 700 cm in body length and they weigh between 1.5 and 6 tons. The skull is short and high with nasal fossae that lie above the orbits. The upper incisors have unlimited growth and lack enamel; they form the long, ivory tusks that the elephant uses for defense. The canine teeth have been lost. The molars are made up of vertical ridges (laminae) of ivory surrounded by cementum and a layer of enamel. They appear and wear out from back to front. The nose and the upper lip have been transformed into a flexible trunk used for breathing, drinking, smelling, grabbing and hitting. The lower lip is fleshy and pointed. The eyes are small. The ear lobes can be very large.

African savanna elephant
Loxodonta africana africana
BL: 690 cm, TL: 130 cm

African forest elephant
Loxodonta africana cyclotis
BL: 440 cm, TL: 60 cm

Asian elephant
Elephas maximus
BL: 600 cm,
TL: 130 cm

The neck is short. The limbs are column-like, each with 5 toes. The heel puts pressure on a pad of connective tissue on the bottom of the foot that forms the sole. The tail is slender. The skin is gray and covered in sparse hairs.

Some unique derived features

■ The loss of the lower canines. This characteristic is seen on the left lateral view of an adult elephant skull (fig. 1): in addition to the transformation of the third upper incisors into tusks (*ts*), there has been loss of teeth; only the jugal teeth are found on the upper and lower jaws. This can also be seen on a left lateral view of a fossil proboscidean from Egypt and Libya that dates from the Eocene (*Moeritherium*, 37 MYA) (fig. 2; *LlC*: loss of lower canines, *C*: upper canine, *de*: dentary, *fr*: frontal, *I1*: first incisor, *j*: jugal, *mx*: maxillary, *na*: nasal, *or*: orbit, *pa*: parietal, *pmx*: premaxillary, *ptp*: post-tympanic process, *pgs*: post-glenoid part of the squamosal, *soc*:

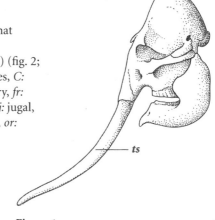

Figure 1

supraoccipital, *sq:* squamosal, *eam:* external auditory meatus).

- The orbit opens on the maxillary (fig. 2), whereas in other mammals it opens behind this bone at the level of the frontal and jugal bones. Fig. 3 shows a left lateral view of a skull of a young Asian elephant where the limits of the maxillary are clearly indicated (*la:* lacrimal, *pg:* pterygoid). In fig. 4, the maxillary bone is shown on a left lateral view of a skull of a non-proboscidean, a carnivore (*Canis*).

- In the temporal region of skull wall, the ascending process of the palatine is reduced.

- The external auditory meatus is delimited by the squamosal bone. It is closed ventrally by the post-glenoid part of the squamosal and by the post-tympanic process (fig. 2). These two parts meet completely in modern species (fig. 3).

NOTE: The characters described above define both extant and fossil proboscideans. The presence of a trunk, nasal orifices higher on the face than the orbits, and the third incisor transformed into a continuously growing tusk are characters that could define the proboscideans, if we limit the description to only extant fauna. However, these characters would not be valid for many fossil species, starting with *Moeritherium*.

Ecology

The elephants are large herbivores of dense forests, savannas, jungles, dry forests, steppes, plains and mountains up to 5000 m. They are absent from arid desserts. They feed on plants, bamboo, roots, bulbs, buds, and fruits. Depending on their size, they can consume between 100 and 200 kg of vegetation per day. In Africa, there are two distinct ecotypes that use different habitats: the savanna elephant uses mostly open habitats, whereas the forest, or dwarf, elephant (*Loxodonta africana cyclotis*), the smaller and less well-known ecotype (shoulder height of 160 to 205 cm and weight of 900 to 1500 kg), frequents the dense forests of central Africa. In Asia, at least four extant sub-species are recognized, along with four others that disappeared before the Middle Ages (Mesopotamia, China, Persia, Java).

Elephants are social animals that live in matriarchal family units. There are also old solitary males and "clubs" of young males. The Asian elephant can form crèches in which a few females will survey all the young while the herd is searching for food. Indeed, most of an elephant's time is dedicated to finding food and to cleaning itself. Elephants do not sleep much. They undergo seasonal migrations along particular routes in order to get to a water source or a zone rich in

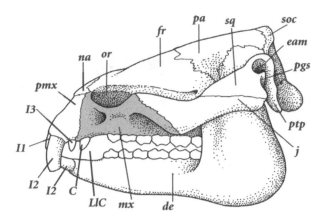

Figure 2

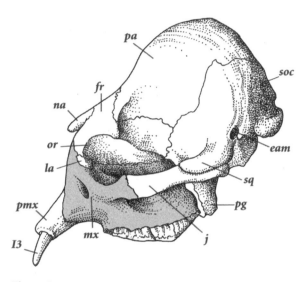

Figure 3

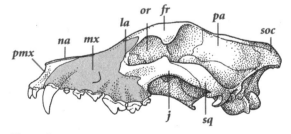

Figure 4

food. For example, the savanna elephant will frequent wet savanna habitats during the dry season, when food is abundant. During the rainy season, it will migrate to the dry savanna or steppes, where it can then take advantage of flowering and rapidly growing vegetation. Along these true migration routes, different families will join together into a large, loosely structured herd containing up to 1000 individuals and running up to 500 km long. However, these migrations are becoming more and more limited as human habitations and activities have expanded. Elephants also look for water sources: they love to bathe and to coat their heads, necks and backs in mud. Once dry, this mud protects them from the sun and insects.

For communication, elephants have the ability to make a rich diversity of sounds that they use in association with the position and movement of their ears and trunk. Their sense of smell, touch and sound are very strong, but their ability to see is mediocre. Their predators include lions, spotted hyenas, wild dogs, and crocodiles, all of which attack the young. When a herd member dies, the others will mourn the death for a certain period and will sometimes cover the corpse with branches. A mother will sometimes pull around the corpse of her young for an entire day. Elephants tend to have few combats. When the number of dominant males is low for the number of females, a secondary male can copulate with the same females without conflict. The gestation period lasts 22 months. During birth, the mother will often be helped by other females. Only a single young is usually produced, twins are very rare. A calf will nurse for 2 years, and may sometimes continue for 4 to 6 years, but typically starts to eat vegetative matter at a very early age. A female will give birth at an interval of 2 to 4 years.

The major growth period is completed by 30 years of age. However, the tusks continue to grow throughout an individual's life. Sexual maturity is reached at 10 to 12 years of age in the savanna elephant, and at 8 to 10 years of age in the forest elephant. A savanna elephant may live 60 years, whereas a forest elephant can live up to 80 years.

Even if a forest elephant will rarely encounter a savanna elephant because they live in different biotypes, the two ecotypes are not genetically isolated. Mitochondrial genetic markers typical to the savanna elephant have been found in the forest elephant. There is thus little justification for considering the forest elephant as a distinct species. The sub-Saharan African elephants were never domesticated. In contrast, the Asian elephant was domesticated in India 2500 years B.C. for use as transport, work and war animals. The Carthaginians domesticated a small, North African sub-species of the savanna elephant that is now extinct. It is with the help of these animals that Hannibal and his army were able to cross the Alps in 220 B.C.

Number of species: 2.
Oldest known fossils: *Phosphatherium* from Morocco dates from the late Paleocene (55 MYA). If we consider the entire sister lineage of the sirenia, which includes the desmostylians (fossils), the proboscideans (extant) and the embrithopods (fossils), the oldest known fossil would be that of the embrithopod *Phenacolophus* from Mongolia and China dating from the early Paleocene (60 MYA).
Current distribution: elephants currently live in sub-Saharan Africa and Southeast Asia (from southern India and Sri Lanka to northern Vietnam, and in southern China, the southern Himalayas, Sumatra and Borneo).

Examples

African elephant: *Loxodonta africana;* Asian elephant: *Elephas maximus.*

Sirenia

General Description

The sirenians are heavy, elongated animals with a rotund form due to a thick fat layer. The body length varies between 250 and 400 cm for a minimum weight of 360 kg. The adult is almost hairless, except for whiskers and sensory hairs on the squarish muzzle. However, the fetus is completely covered in velvety fur, and the young retain a few, sparse hairs. The skull has a muzzle and a ventrally deflected jaw. The mouth is closed by two large upper lips that hang laterally. In the manatees, the upper lip is divided into 3 parts and the nostrils can be sealed off by skin flaps. The eyes are small with a nictitating membrane and lacrimal glands that produce oily secretions. There are no ear pinnae. The tympanic bulla is only loosely attached to the basicranium. The ribs are massive and have undergone pachyostosis, making the animal considerably heavier. The horizontal stability of the body is due to the elongated lungs and an almost horizontal diaphragm. In contrast to the cetaceans, the sirenians are not active divers; they sink without effort by regulating the volume of air in their lungs. The canine

African manatee
Trichechus senegalensis
400 cm

Dugong
Dugong dugon
270 cm

teeth have been lost. In the dugong, the jugal teeth are reduced, but the incisors grow continuously, as in the elephants. This pattern is reversed in the manatee; it loses its incisors before sexual maturity and has extra jugal teeth that are replaced progressively from back to front. The neck is difficult to see externally. The hands have been transformed into five-fingered swimming paddles with no phalangeal extension (hyperphalangy). The sirenians have lost their posterior limbs, but a vestigial pelvis remains. A fold in the skin forms a horizontal tail that differs in shape from that of the cetaceans. It is round with a fleshy medial crest in the manatees. In the dugongs, this tail is bifurcated. The skin color ranges from gray to grayish-brown, the ventral surface being slightly lighter than the dorsal surface.

Some unique derived features

■ The outer tympanic bone (ectotympanic) has a characteristic teardrop shape, with a greatly thickened ventral surface and both anterior and posterior attachment sites. This bone is obliquely oriented to the anterior-posterior body plane, with its lower surface found further below the skull than its upper surface. The tympanic membrane is supported by this framework. Fig. 1 shows a left lateral view of the skull of a manatee *Trichechus senegalensis* with the position of the ectotympanic bone indicated. Fig. 1b shows an enlargement of

the auditory bulla (*to:* tholus, by which the auditory bulla is loosely attached to the squamosal, *ect:* ectotympanic, *ex:* exoccipital, *fr:* frontal, *j:* jugal, *mx:* maxillary, *or:* orbit, *pa:* parietal, *pg:* pterygoid, *pma:* petrosal-mastoid part, *pmx:* premaxillary, *pet:* petrosal, *soc:* supraoccipital, *sq:* squamosal).

■ The premaxillary bones have large posterior processes that surround the reduced nasal bones and that meet up with the frontal bones such that the maxillaries are no longer connected to the frontals. Fig. 2 shows dorsal views of two skulls; 2a is that of the manatee *Trichechus manatus,* a sirenian, and 2b is that of a non-sirenian, a carnivore *Canis.* In non-sirenians, it is normally the nasal (*na*) and the maxillary (*mx*) bones that touch the frontal bones, not the premaxillaries. In certain mammalian groups, the premaxillaries can have long, thin posterior processes that meet up with the frontals (glires, cetaceans) but do not block the contact between the maxillaries and the frontals. This situation is also found by convergence in extant proboscideans, but not in many of the fossil species (see *Moeritherium,* clade 17).

- The posterior limbs have been lost. This loss is found convergently in the cetaceans. Fig. 3 shows a left lateral view of a manatee skeleton (*Trichechus manatus*).

- Hands: they form swimming paddles, but without hyperphalangy (fig. 3).

- Bones: the sirenians have largely acellular bones that are very dense and heavy. These bones have evolved by pachyostosis and osteosclerosis.

- Loss of the ethmoidal foramen. This foramen normally opens in the orbital region of the skull wall, either on the orbitosphenoid, on the frontal, between these two bones, or on the presphenoid.

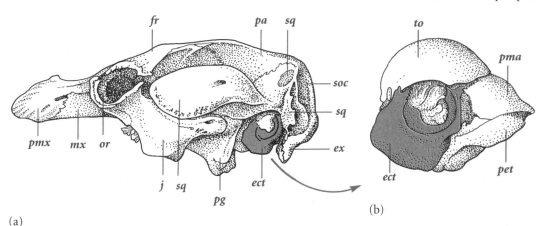

(a)

(b)

Figure 1

Ecology

The sirenians are large aquatic animals that we find in rivers or coastal waters in areas where aquatic vegetation is abundant. They graze in algal beds or on freshwater aquatic plants in shallow waters of 1 to 2 m (sometimes up to 10 m). Their predators are large sharks and humans. They are social animals that live in families or large family groups. For example, family groups of dugongs can include 80 to 100 individuals that all graze together in a single marine prairie. However, we can also find young, solitary males. It is possible that two males may covet the same female, but male-male combats have never been observed. The male approaches the female and touches her muzzle, in a sort of kiss, and then brings himself close to her. Mating takes place in the water by ventral contact. Gestation lasts between 5 and 11 months depending on the species. The birth of the single young, sometimes two, also takes place in the water. Reproduction can occur at any time in the

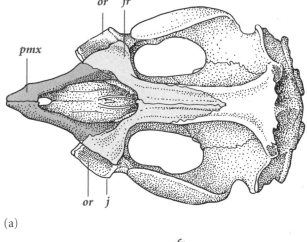

(a)

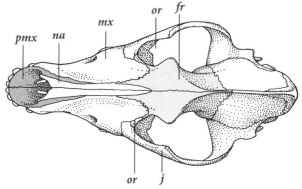

(b)

Figure 2

year. A breeding pair are very close and the male may caress the female even outside the mating period. The female will nurse for 18 months, but the young starts on a mixed diet very early on. Both parents will guard the young and the female may carry it on her back or take it in her arms. Sexual maturity is reached at 3 or 4 years old and an individual may live between 20 and 30 years.

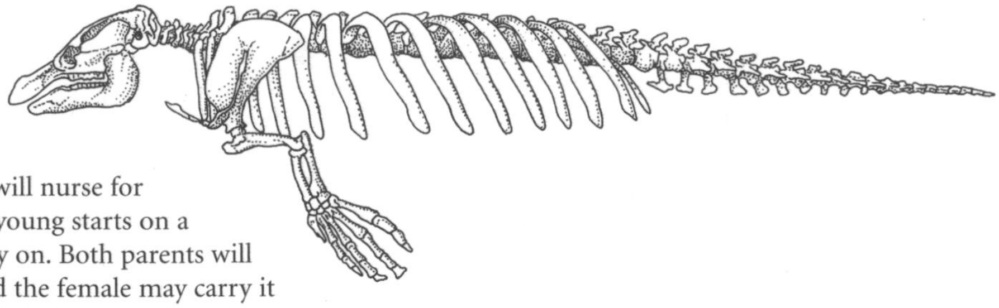

Figure 3

Despite the fact that they are officially protected in many countries, all sirenian species are threatened with extinction. This is to due to a combination of the destruction and pollution of their aquatic habitats and hunting. More than 200 years ago, humans caused the extinction of the largest representative of this order, the Steller's sea cow (*Rhytina gigas*), a marine animal 7 m long and weighing 4 tons. The last known individual was killed in 1768, only 27 years after its discovery.

Number of species: 5.
Oldest known fossils: *Prorastomus* from Jamaica and *Eotheroides* from Egypt both date from the Eocene (40 MYA).
Current distribution: rivers, coastal waters, and shallow tropical seas. The manatees live in the intertropical waters of the eastern and western Atlantic, in the Caribbean Sea, the Amazon basin and many rivers of western Africa and Florida. The dugong is found in the Red Sea, along the coastal areas of east African, Madagascar, southern India, Sri Lanka, from Southeast Asia to southern China, New Guinea, north Australia, New Caledonia, and the Caroline, Solomon and Marshall Islands.

Examples

Dugong: *Dugong dugon;* Amazonian manatee: *Trichechus inunguis;* West Indian manatee: *T. manatus;* African manatee: *T. senegalensis.*

Boreoeutheria

A few representatives

Guinea baboon
Papio papio
BL: 70 cm
node 25

Black rat
Rattus rattus
BL: 20 cm
node 28

European hedgehog
Erinaceus europaeus
BL: 20 cm
node 30

European wild cat
Felis sylvestris
BL: 60 cm
node 36

Some unique derived features

■ Molecular phylogenies based on the sequences of 19 nuclear genes and 3 mitochondrial genes regroup the boreoeutherians. These genes include hemoglobin, myoglobin, alpha-crystallin, cytochrome c, and ribonuclease genes along with mitochondrial ribosomal genes. In addition, independent analyses of other nuclear (a fragment of *RAG1*, *c-myc* oncogene, growth hormone) and mitochondrial (*ND6*) genes support this clade.

Number of species: 4,118.
Oldest known fossils: if the leptictid *Prokennalestes* of the gypsonictopid family is indeed a boreoeutherian, it would be the oldest (early Cretaceous from Mongolia, 100 MYA). However, this relationship seems to be questionable for the moment. Evidence from molecular dating indicates that the boreoeutherian lineage became clearly distinct only 95 MYA. The oldest indisputable boreoeutherians are therefore the cimolestans *Otlestes* from Uzbekistan and *Deccanolestes* from India (Middle Cretaceous, 95 MYA).
Current distribution: worldwide.

Examples

■ EUARCHONTOGLIRES: Philippine flying lemur: *Cynocephalus volans;* common tree shrew: *Tupaia glis;* crested gibbon: *Hylobates concolor;* European hare: *Lepus europaeus;* black rat: *Rattus rattus.*

■ LAURASIATHERIA: European hedgehog: *Erinaceus europaeus;* red fox: *Vulpes vulpes;* giant pangolin: *Manis gigantea;* common long-eared bat: *Plecotus auritus;* mountain zebra: *Equus zebra;* Bactrian camel: *Camelus ferus;* desert warthog: *Phacochoerus aethiopicus;* elk: *Cervus elaphus;* bottlenose dolphin: *Tursiops truncatus;* hippopotamus: *Hippopotamus amphibius.*

Euarchontoglires

A few representatives

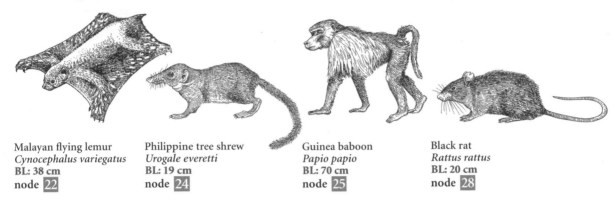

Malayan flying lemur
Cynocephalus variegatus
BL: 38 cm
node 22

Philippine tree shrew
Urogale everetti
BL: 19 cm
node 24

Guinea baboon
Papio papio
BL: 70 cm
node 25

Black rat
Rattus rattus
BL: 20 cm
node 28

Some unique derived features

- There is a 6 base pair deletion in the prion protein gene (*prnB* gene).
- There is a characteristic deletion of 18 amino acids in the SCA1 protein (spinocerebellar ataxia type 1).
- There is a characteristic deletion of 54 base pairs in the *ATXN1* gene (Ataxin-1 gene).
- There is a characteristic deletion of 3 base pairs in the *TNF* gene (tumor necrosis factor gene).
- There are three characteristic insertion sites for "mariner" type transposable elements (LINE MLT1A0) in the *CFTR* gene (the Cystic Fibrosis Transmembrane Conductance Regulator gene that provokes cystic fibrosis when modified).
- There is typically a hemochorial placenta. In a hemochorial placenta (fig. 1b), the uterine epithelium disappears and the uterine capillaries spread, creating maternal blood lacunae that form the connection with the fetal tissue (with the chorionic epithelium). An epitheliochorial placenta is shown in fig. 1a for comparison; here, the uterine epithelium is retained.

NOTE: There are several exceptions to this feature. In the strepsirrhinian primates (lemurs, loris), there is an epithelio-chorial placenta (fig. 1a). In the Scandentia (tree shrews), there is an endotheliochorial placenta; the uterine epithelium disappears and the uterine capillaries connect with the fetal tissue, but without the formation of the blood lacunae. Hemochorial placentas are found by convergence in the anteaters, armadillos, hyraxes, macroscelids, afrosoricids, certain bats, certain carnivores, and certain eulipotyphles. This trait is therefore very labile and therefore of little phylogenetic value.

- Molecular phylogenies based on the sequences of 19 nuclear genes and 3 mitochondrial genes bring together the euarchontoglires. These genes include hemoglobin, myoglobin, alpha-crystallin, cytochrome c, and ribonuclease genes along with mitochondrial ribosomal genes. In addition, independent analyses of other nuclear (a fragment of *RAG1, c-myc* oncogene, growth hormone) and mitochondrial (*ND6*) genes support this clade.

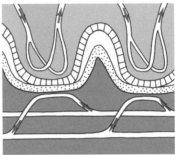

Figure 1a

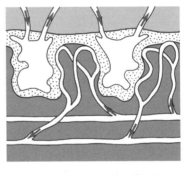

|||| Uterine epithelium   ⬚ Chorionic epithelium
▨ Endometrium   ■ Placenta

Figure 1b

Number of species: 2,304.
Oldest known fossils: *Purgatorius* from Montana (Paleocene, 65 MYA).
Current distribution: worldwide; in Antarctica, Australia and New Zealand, they were introduced by humans.

Examples

EUARCHONTA: Philippine flying lemur: *Cynocephalus volans*; common tree shrew: *Tupaia glis*; crested gibbon: *Hylobates concolor*.
GLIRES: European hare: *Lepus europaeus*; black rat: *Rattus rattus*.

Euarchonta

A few representatives

Malayan flying lemur
Cynocephalus variegatus
BL: 38 cm
node 22

Philippine tree shrew
Urogale everetti
BL: 19 cm
node 24

Guinea baboon
Papio papio
BL: 70 cm
node 25

Some unique derived features

■ Molecular phylogenies based on the sequences of 19 nuclear genes and 3 mitochondrial genes unite the euarchontans. These genes include hemoglobin, myoglobin, alpha-crystallin, cytochrome c, and ribonuclease genes along with mitochondrial ribosomal genes.

Number of species: 203.
Oldest known fossils: *Purgatorius* from Montana dates from the start of the Paleocene (65 MYA).
Current distribution: worldwide (including humans). Without humans, the euarchontans are found in the intertropical regions of all continents except Australia.

Examples

DERMOPTERA: Philippine flying lemur: *Cynocephalus volans.*
PRIMATES: common tree shrew: *Tupaia glis;* crested gibbon: *Hylobates concolor.*

Dermoptera

General Description

The colugos or flying lemurs are gliding and climbing mammals about the size of a cat. They cannot stand upright and are extremely handicapped on land. Their body, neck, tail and limbs are connected to a large, thick membrane that allows them to glide long distances. This membrane is covered in a soft fur and is extremely variable in color. The dorsal surface is generally gray-brown with scattered pale yellow spots, whereas the underside is shiny with tones of orange, red-brown or yellow. The head is dog-like (origin of the genus name *Cynocephalus*), the neck is long and very mobile, the body is slender, and the limbs are elongate. It should be noted that the gliding membrane has appeared several times in unrelated mammalian families:

Malayan flying lemur
Cynocephalus variegatus
BL: 38 cm, TL: 22 cm

the dermopterans; among the marsupials, in the pygmy gliders (*Acrobates*), squirrel gliders (*Petaurus*), and great gliders (*Petauroides*); among the rodents, in the scaly-tailed squirrels (*Anomalurus, Idiurus*) and flying squirrels (*Glaucomys, Petaurista*). However, the uropatagium (caudal gliding membrane between the hind legs and tail) is present only in the colugos (and in the chiropterans, but these animals have powered flight).

Some unique derived features

- The first and second lower incisors are large and prominent. They are flattened and comb-like with the anterior edge divided by 6 to 20 longitudinal grooves. Fig. 1 shows a dorsal view of the front part of a colugo's mandible.

- The gliding membrane (or patagium) is completely covered in fur. It extends from the neck to the fingers and from the fingers to the toes and then continues to the tip of the tail.

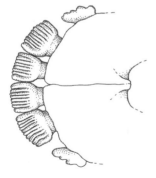

Figure 1

Examples

Philippine flying lemur: *Cynocephalus volans;* Malayan flying lemur: *Cynocephalus variegatus.*

Number of species: 2.
Oldest known fossils: *Elpidophorus, Dracontolestes,* and *Navajovius* from North America (Montana, Utah and Colorado, respectively) date from the lower Paleocene (60 MYA). *Berruvius* from France also dates from this period.
Current distribution: Southeast Asia: Indochinese Peninsula, Myanmar, and the Malay Archipelago including Java, Sumatra, Borneo and the Philippines.

Ecology

The colugos lead a nocturnal existence and are found in the tropical forests of Southeast Asia. They hang on the underside of tree branches and move easily within the trees. They go from tree to tree by gliding and can cover a distance of up to 70 m. During the day, they rest, hanging, often within a hollowed tree, but with their heads upturned, much like the sloths. They are strict vegetarians eating leaves, buds, flowers, and sometimes fruit. The females have only a single young at a time; it is born in a somewhat undeveloped state, only slightly more developed than a young marsupial. It will spend the first few weeks of its life suckling, firmly attached to one of the mother's two nipples. The young will continue to cling to the mother's fur for a long period of time, even after it has its own fur.

Primates (*sensu lato*)

A few representatives

Northern tree shrew
Tupala belangeri
BL: 18 cm
node 24

Pen-tailed tree shrew
Ptilocercus lowii
BL: 13 cm
node 24

Diana monkey
Cercopithecus diana
BL: 50 cm
node 25

Indri
Indri indri
BL: 65 cm
node 25

Some unique derived features

- The petrosal bone forms the auditory bulla. The auditory bulla (or tympanic bulla, *tb*) is a very hard bony bulb at the base of the skull that encloses the middle ear (or tympanic cavity) including the three ossicles that carry sound vibrations from the tympanum to the oval window. In the primates and scandentians, this bulla (fig. 1, shown on a tree shrew) is completely formed by the petrosal (*ptr*, colored), the periotic or the petrous part of the temporal bone. In non-primates, it is typically composed of four bones, including the petrosal (according to the group: alisphenoid (*asp*), tympanic, squamosal on the anterior side, basisphenoid (*bsp*), petrosal, entotympanic (*ent*) on the posterior side). Fig. 2 (posterior side facing upward and anterior side facing downward, *ec*: ectotympanic, *sao*: orifice for the passage of the stapedial artery, *to*: tubal orifice, *cf*: carotid foramen, *smf*: stylomastoid foramen)

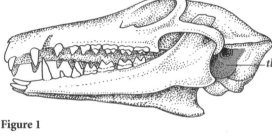

Figure 1

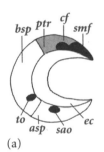 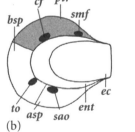 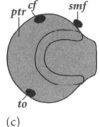

(a)   (b)   (c)

Figure 2

shows schematic ventral views and the bone composition of the auditory bulla of a eulipotyphle (2a, *Erinaceus*), a macroscelid (2b, *Macroscelides*) and a scandentian (2c, *Tupaia*, tree shrew).

- The brain has a well-developed occipital lobe (visual control center), indicating the importance of vision over the other senses. There is a calcarine sulcus.

Examples

- SCANDENTIA: common tree shrew: *Toupaia glis*; Elliot tree shrew: *Anathana ellioti*; pen-tailed tree shrew: *Ptilocercus lowii*.
- PRIMATES (*sensu stricto*): ruffed lemur: *Lemur variegatus*; indri: *Indri indri*; aye-aye: *Daubentonia madagascariensis*; slender loris: *Loris tardigradus*; Philippine tarsier: *Tarsius syrichta*; common marmoset: *Callithrix jacchus*; yellow baboon: *Papio cynocephalus*; human: *Homo sapiens*.

> **Number of species:** 201.
> **Oldest known fossils:** *Purgatorius* from Montana dates from the start of the Paleocene (65 MYA) and was for a long time considered to be the oldest known primate. However, this species is closer to the dermopterans than to the primates. The oldest incontestable primates are older than the oldest known scandentians: they are the Ypresian adaptids *Donrusselia* (France) and *Cantius* (Belgium, France, Wyoming, 55 MYA).
> **Current distribution:** worldwide (including humans). Without humans, the primates and scandentians are found in the intertropical regions of all continents except Australia.

Scandentia

General Description

Northern tree shrew
Tupaia belangerl
BL: 18 cm, TL: 18 cm

Philippine tree shrew
Urogale everetti
BL: 19 cm,
TL: 15 cm

Pen-tailed tree shrew
Ptilocercus lowii
BL: 13 cm, TL: 17 cm

The scandentians, or tree shrews, are small arboreal placental mammals that are very squirrel-like in appearance, with long, bushy tails. The body length varies between 15 and 20 cm and the tail length is the same, or even longer in the pen-tailed tree shrew. The ears and eyes are large, the muzzle is somewhat elongated and the orbits are turned slightly forward. There are claws on all toes. The big toes are not opposable; although the big toes of the hindfeet can be moved away from the other toes, they cannot reach the degree of complete opposability of the true primates. These characteristics have sometimes been considered as indicators of the relatedness of this group to the true primates, but other times they have been seen as convergences linked to an arboreal lifestyle. The fur is thick and soft and quite variable in color among species, with a predominance of dark reds (but we also see black, gray and brown). The tail of most species is covered in bushy hairs but is more scaly in the pen-tailed tree shrews. The tree shrews have been a veritable puzzle for zoologists; they are primate-like by their brain, skull anatomy, the composition of the auditory bullae, musculature and their almost opposable toe, and insectivore-like their chondrocranium, dentition and tarsus. Some zoologists have identified traits that link them to the marsupials, others to rodents, and still others to lagomorphs.

Some unique derived features

- The incisors are canine-like (fig. 1). They are particularly prominent and sharp on the lower jaw (fig. 2) and are used for preening. The upper jaw quadrants of tree shrews include two incisors (*I*), one canine (*C*), three premolars (*Pm*) and three molars (*M*); the lower jaw quadrants, three incisors, one canine, three premolars and three molars. Fig. 1 shows a lateral left view (1a) and ventral view (1b) of the upper dentition of a tree shrew. Fig. 2 outlines the dentition of the common tree shrew (*Tupaia glis*), showing the extreme development of the lower incisors and the atrophy of the canines. For comparison, the dentition of a carnivore (*Canis*) is shown in fig. 3. We note on this figure that the incisors are smaller than the canines.

- The canines are small and premolar-like (fig. 1 and 2).

- All the euarchontoglires have a hemochorial placenta except the scandentians and the strepsirrhinians. The tree shrews have an endotheliochorial placenta, where the uterine epithelium disappears and the uterine capillaries connect with the fetal tissue, without forming blood sinuses. With a hemochorial placenta, the uterine capillaries spread and form maternal blood pools that are in direct contact with the fetal tissue.

NOTE: we find endotheliochorial placentas by convergence in certain microchiropterans and carnivores, as well as in the tethytherians, the aardvark and the sloths.

Ecology

The scandentians are diurnal animals that can be either arboreal or terrestrial in nature. They live in the trees and thick underbrush of the tropical forests of Southeast Asia. They are omnivorous, feeding on both fruit and insects. They also sometimes kill mice or young rats. They often forage on the ground. The tree shrews typically live alone or in couples and defend their territory against congeners. They mark their territory on the ground by an odor produced by a glandular zone at the front of the body, and on the trees by drops of urine on the branches. Gestation lasts 43 to 56 days. The females give birth to 1 to 3 young that are reared in a moss nest found in the trees. The young reach adulthood at 3 months old and are able to reproduce by the time they are 4 months old. Females will be obviously pregnant with the next litter by the time the young are six weeks old. The longevity is roughly two years.

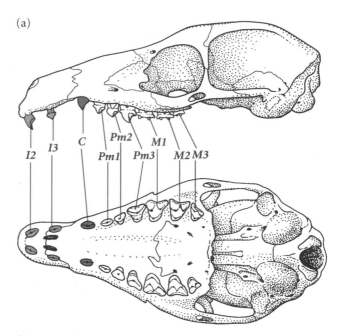

(a)

(b)

Figure 1

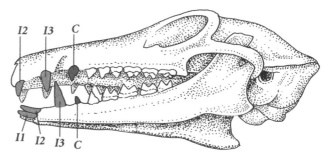

Figure 2

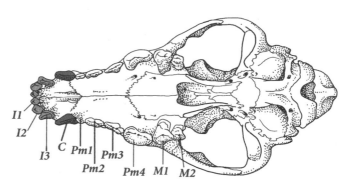

Figure 3

Number of species: 19.
Oldest known fossils: *Eodendrogale* from China dates from the Eocene (40 MYA) and may be a tree shrew. *Paleotupaia* from India dates from the Miocene (15 MYA).
Current distribution: India, Southeast Asia, Java, Sumatra, Borneo, Philippines.

Examples

Common tree shrew: *Tupaia glis;* northern tree shrew: *T. belangeri;* large tree shrew: *T. tana;* Madras tree shrew: *Anathana ellioti;* Philippine tree shrew: *Urogale everetii;* Bornean smooth-tailed tree shrew: *Dendrogale melanura;* pen-tailed tree shrew: *Ptilocercus lowii.*

➡ *see chap. 14, p. 478*

Primates (*sensu stricto*)

General Description

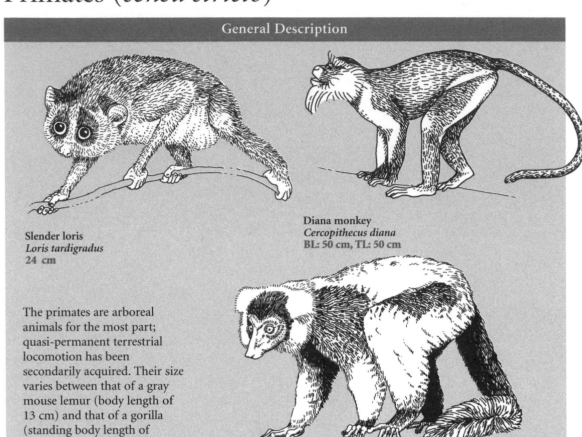

Slender loris
Loris tardigradus
24 cm

Diana monkey
Cercopithecus diana
BL: 50 cm, TL: 50 cm

Ruffed lemur
Lemur variegatus
BL: 55 cm, TL: 60 cm

Human
Homo sapiens
170 cm

The primates are arboreal animals for the most part; quasi-permanent terrestrial locomotion has been secondarily acquired. Their size varies between that of a gray mouse lemur (body length of 13 cm) and that of a gorilla (standing body length of 175 cm for a weight of between 135 and 275 kg). The neurocranium overwhelms the facial area in volume (the facial area being reduced). The orbits are oriented forward, enabling binocular vision. The form of the face is quite variable: in the strepsirrhini (lemuriformes and lorisiformes) the face is more or less elongate with a rhinarium and whiskers (vibrissae), whereas the haplorrhini (tarsiers and simiiformes, the true monkeys) have a somewhat flattened face and the whiskers and rhinarium are replaced by a nose. Sight is the dominant sense, as evident from the development of the occipital lobe of the brain. The neck is short and the limbs are generally slender, the anterior limbs often being longer than the posterior. The radius and ulna articulate freely, enabling the pronation and supination of the hands. The hands are prehensile with a thumb opposable to the other four fingers. The feet are also prehensile, except in the tarsiers (jumping animals) and humans (bipedal). The claws have been replaced by flattened nails, except for the second and third toes of the tarsiers and callitrichids (platyrrhini, simiiformes). The strepsirrhini have all retained a claw on the second toe, except for the particular case of the aye-aye; this species has flat nails only on the thumb and big toe. The sitting posture has freed the hands. The density and color of the hair is extremely variable, as is the length of the tail, which has completely disappeared in the hominoids. Several catarrhinian species have ischial callosities (buttock pads), whose temporary appearance and color can signal an individuals's reproductive status.

Some unique derived features

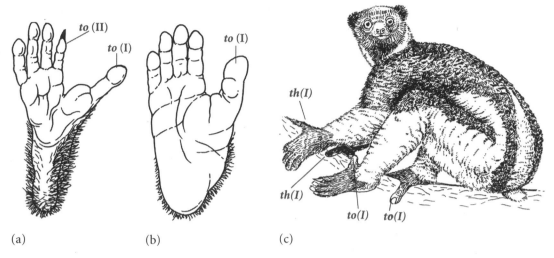

Figure 1

■ On both the feet (fig. 1) and hands (fig. 2), digit I is opposable to the others. This arrangement allows both the feet and hands to be prehensile. Fig. 1 shows the bottom of the feet of a lorisiform primate (galago or bushbaby, 1a) and a hominoid primate (chimpanzee, 1b). Fig. 1c shows the opposable nature of the first digit on the hands and feet of an indri. *th(I):* thumb, *to(I):* big toe.

NOTE: the tree shrews can spread out their first toe, but this toe is not opposable enough to be prehensile. The big toes of humans and tarsiers are not opposable, but these are exceptions that have been secondarily acquired as adaptations for locomotion (bipedalism in humans and vertical clinging and leaping in the tarsiers).

■ The fingers and toes have flat nails (fig. 2, *fna*), but certain species have retained claws (for example, the second toe of the galago, fig. 1a).

■ The orbits are oriented toward the front, enabling binocular vision. It should be noted that the orbits are still closed posteriorly by the postorbital bar, as in the tree shrews. However, even though the tree shrews show the start of the forward migration of the orbits, they are still more lateral than frontal in this group. An apical view of a chimpazee's skull (3a), along with vertical views of a pygmy marmoset's skull (3b) and a eulipotyphle's skull (hedgehog, 3c) are shown in fig. 3. Note the lateral position of the hedgehog's orbits, in contrast to the frontal position in the pygmy marmoset.

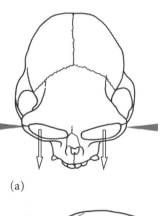

(a)

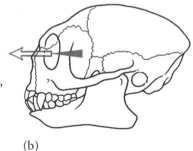

(b)

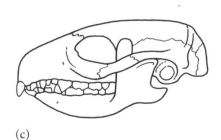

(c)

Figure 3

Figure 2

- In the β-globulin multigenic family, the η-globulin gene is inactivated and has become a pseudogene.

- The genome contains a particular class of short transposable elements ("SINEs" type) called the "Alu SINEs."

- There is a characteristic insertion of a long, LINE 1 type transposable element (more precisely, the L1MA4 type) between exons 40 and 41 of the type-I alpha-2 collagen protein gene (*COL1A2* gene).

- There is a characteristic insertion of a FLAM type (free left Alu monomers) transposable element between exons 5 and 6 of the *HBX2* gene.

- Molecular phylogenies based on the sequences of 19 nuclear genes and 3 mitochondrial genes regroup the primates. These genes include hemoglobin, myoglobin, alpha-crystallin, cytochrome c, and ribonuclease genes along with mitochondrial ribosomal genes. In addition, independent analyses of other nuclear (a fragment of *RAG1, c-myc* oncogene, growth hormone) and mitochondrial (*ND6*) genes support this clade.

Ecology

The primates inhabit the forests, savannas, steppes, treed plains and high mountains of warm countries, that is, in the intertropical zone where there are no harsh winters. There are, however, two exceptions: these involve certain Chinese and Japanese macaques that live through snowy winters and humans, which have colonized all terrestrial biotopes, even those above the Arctic Circle. The primates are never aquatic or marine animals. They are either omnivores or herbivore/frugivores; no exclusively carnivorous species are known. Certain species that feed exclusively on leaves, like the colobus monkeys, have developed specialized stomach compartments where anaerobic fermentation occurs for cellulose digestion. But for a few exceptions (orangutan), primates live in groups, have a well-developed social life and possess rich and intense communication skills. The natural enemies of primates include large carnivores, birds of prey and sometimes other primates. Certain species have been domesticated by humans, especially as circus animals and laboratory subjects. The ecological importance of one primate, *Homo sapiens,* cannot be ignored: this species alone is in the process of rapidly and profoundly changing the biosphere to a degree never before attained by another species. The human demographic explosion, in addition to our social and economic organization, has led to a rapid decline of the biodiversity, largely through the destruction of natural habitats.

Number of species: 182.
Oldest known fossils: *Purgatorius* from Montana dates from the start of the Paleocene (65 MYA) and was for a long time considered to be the oldest known primate. However, this species is closer to the dermopterans than to the primates. The oldest incontestable primates are the Ypresian adaptids *Donrusselia* (France) and *Cantius* (Belgium, France, Wyoming, 55 MYA).
Current distribution: worldwide (including humans). Excluding humans, the primates are found in the intertropical regions of all continents except Australia. They are therefore absent from North Amercia, Europe (except Gibraltar), northern Asia above the 40° parallel (except Japan, where the island of Honshu represents the most northern point where monkeys are found), and South Amercia below the Tropic of Capricorn.

Examples

Guinea baboon: *Papio papio;* ruffed lemur: *Lemur variegatus;* greater dwarf lemur: *Cheirogaleus major;* indri: *Indri indri;* aye-aye: *Daubentonia madagascariensis;* slender loris: *Loris tardigradus;* Senegal bushbaby: *Galago senegalensis;* Philippine tarsier: *Tarsius syrichta;* common marmoset: *Callithrix jacchus;* Goeldi's marmoset: *Callimico goeldii;* northern night monkey: *Aotus trivirgatus;* yellow baboon: *Papio cynocephalus;* Diana monkey: *Cercopithecus diana;* western red colobus: *Colobus badius;* black gibbon: *Hylobates concolor;* human: *Homo sapiens;* orangutan: *Pongo pygmaeus.*

Glires

A few representatives

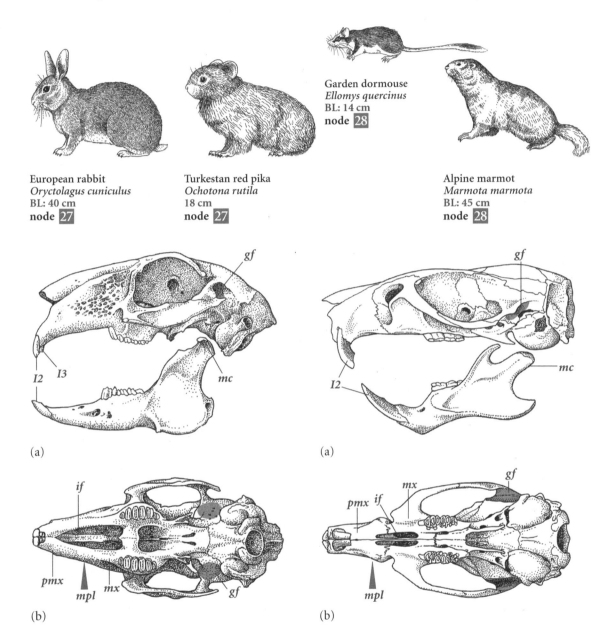

Garden dormouse
Ellomys quercinus
BL: 14 cm
node 28

European rabbit
Oryctolagus cuniculus
BL: 40 cm
node 27

Turkestan red pika
Ochotona rutila
18 cm
node 27

Alpine marmot
Marmota marmota
BL: 45 cm
node 28

(a)

gf

I2 *I3*

mc

(a)

gf

I2

mc

(b)

if

pmx

mpl *mx*

gf

(b)

mx *gf*

pmx *if*

mpl

Figure 1

Figure 2

Some unique derived features

■ The glenoid fossa, which receives the mandibular
condyle, is found well above the basicranium and is
no longer delimited posteriorly by the postglenoid
process. The fossa is elongated and opens
posteriorly. A lateral left view (1a) and a ventral
view (1b) of the skull of a European hare (*Lepus
europaeus*) is shown in fig. 1 (*mc:* mandibular
condyle). A lateral left view (2a) and ventral view
(2b) of a rat's skull (*Rattus rattus*) are also shown

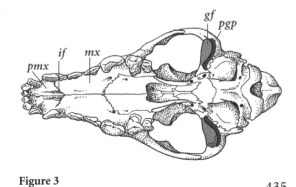

gf *pgp*

if *mx*

pmx

Figure 3

435

(fig. 2). On the lateral views, the location of the glenoid fossa (*gf*) is indicated. For comparison, a ventral view of a non-glires (*Canis*, fig. 3) is shown; note the transversal arrangement (and not elongated) of the glenoid fossa and how it is closed posteriorly by the the postglenoid process (*pgp*).

■ The incisive foramens (*if*), paired aperatures, are very developed and open posteriorly on the palate. We see them clearly in figs 1b and 2b on the hare and the rat, respectively, and their reduced size on the *Canis* (fig. 3).

■ The maxillary (*mx*) and premaxillary (*pmx*) bones contribute equally to the palate. We see these bones in the hare and the rat (fig. 1b and 2b, *mpl*: maxillary–premaxillary limit). The premaxillary contributes very little to the palate of *Canis* (fig. 3).

■ The dental formula is very special. The glires lose the first upper and lower incisors, the third lower incisors, the canines and the first premolars. The second incisors are also lost, but in their place powerful milk incisors form (*I2*) (fig. 1a and 2a).

■ The posterior process of the premaxillary is very long and connects dorsally to the frontal bone (*fr*); the maxillary no longer contacts the frontal bone along the front part of the orbit. Fig. 4 (*na*: nasal) shows apical views of a glires skull (jerboa, *Jaculus jaculus*, 4a) and a non-glires (*Canis*, 4b). There is no contact point between the maxillary and the frontal in front of the orbit in the first case, whereas in the second, the maxillary has a wide contact zone with the frontal in this area.

■ The nasal bone is very long (fig. 4).

■ The uterus is divided. Fig. 5 shows ventral views of part of the female reproductive system in a glires (5a: double uterus) and in a non-glires where the uterus is simple and median (human, 5b; *cv*: cervix, *ft*: fallopian tubes, *ut*: uterus, *vg*: vagina).

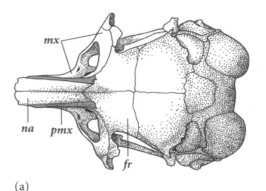

(a)

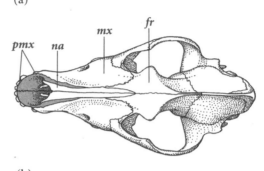

(b)

Figure 4

■ Molecular phylogenies based on the sequences of 19 nuclear genes and 3 mitochondrial genes unite the glires. These genes include hemoglobin, myoglobin, alpha-crystallin, cytochrome c, and ribonuclease genes along with mitochondrial ribosomal genes.

Number of species: 2,101.
Oldest known fossils: if the Zalambdalestids are considered to be lagomorphs, the group dates from the middle Cretaceous (*Zalambdalestes* and *Barunlestes* from Mongolia, 90 MYA). However, the oldest known incontestable lagomorphs are *Hsiuannania, Paranictops, Anictops* and *Cartictops* from China (Paleocene, 60 MYA) and are of the same age as the oldest known rodents, *Heomys* and *Mimotona* from China, *Acritoparamys* and *Paramys* from Wyoming, USA, and *Microparamys* and *Pseudoparamys* from Meudon, France (Paleocene, 60 MYA).
Current distribution: we find the glires everywhere except Australia, New Zealand, Antarctica and in circumpolar zones. However, humans have introduced this group to Australia and New Zealand.

Examples

■ LAGOMORPHA: European hare: *Lepus europaeus*; European rabbit: *Oryctolagus cuniculus*; pygmy rabbit: *Brachylagus idahoensis*; alpine pika: *Ochotona alpina*.
■ RODENTIA: Alpine marmot: *Marmota marmota*; red squirrel: *Sciurus vulgaris*; house mouse: *Mus musculus*; crested porcupine: *Hystrix cristata*; guinea pig: *Cavia aperea porcellus*; mara: *Dolichotis patagonum*.

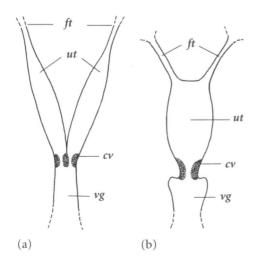

(a)   (b)

Figure 5

Lagomorpha

General Description

European rabbit
Oryctolagus cuniculus
BL: 40 cm, TL: 5 cm

Brown hare
Lepus capensis
BL: 60 cm, TL: 9 cm

Turkestan red pika
Ochotona rutila
18 cm

The lagomorphs (hares, rabbits and pikas) are small- to medium-sized placental mammals, with body lengths that range from 12 to 70 cm for weights of 100 g to 7 kg. The tail is very short and bushy. They are either digitigrade (walking on their toes) or plantigrade (walking on their heels) quadrupeds. The hindlegs can be long and powerful (hares and rabbits) or short (pikas). The forelimbs are never prehensile, as is sometimes the case for rodents. The nostrils are protected by a skin fold that can be closed off. Hearing is highly developed: the hares and rabbits have very elongated ear pinnae.

The two pairs of upper incisors (the I2 and I3 milk teeth) and the single pair of lower incisors (the I2 milk teeth) grow throughout the animal's life, are completely covered in enamel, and are separated from the cheek teeth by a large diastema. The canine teeth are absent. The jaws can make lateral movements owing to the action of specialized muscles. Indeed, the glenoid fossa is not closed off posteriorly or laterally. The condyles therefore have a wide breadth of movement, which is more often laterally than vertically. The intestinal caecum is very large and is used to digest cellulose. The fur is thick and soft and includes a wide variety of colors: white, gray, black, yellow, reddish and all shades of brown. In northern and alpine species, the fur changes color with the seasons, becoming white or light-colored in the winter.

Some unique derived features

- **Incisors:** the lagomorphs have two incisors on each quadrant of the upper jaw, the second milk incisor of continuous growth (*I2*) and the third vestigial incisor (*I3*). Fig. 1 shows a left lateral view of the skull of a European hare (*Lepus europaeus*) and fig. 2, a latero-ventral view of the premaxillary region of this same species to indicate the position of the vestigial incisors (*mx:* maxillary, *na:* nasal, *pmx:* premaxillary).

- The facial region of the maxillary bone is fenestrated (fig. 1, *fmx:* facial region of the maxillary bone).

- The palate is covered in small hairs.

- Caecotrophy: the lagomorphs all practice caecotrophy, which requires caecal fermentation and coprophagia.

- Molecular phylogenies based on the sequences of 19 nuclear genes and 3 mitochondrial genes unite the lagomorphs. These genes include hemoglobin, myoglobin, alpha-crystallin, cytochrome c, and ribonuclease genes along with mitochondrial ribosomal genes. In addition, independent analyses of other nuclear (a fragment of *RAG1, c-myc* oncogene, growth hormone) and mitochondrial (*ND6*) genes support this clade.

Ecology

The lagomorphs are found in all terrestrial biotopes and in all climates, including the most inhospitable (high alpine, polar zones and arid deserts). They are never aquatic or arboreal. They dig burrows where they take refuge and give birth to their young. They are agile on land and can run very fast. They are all strict herbivores. In addition to the normal, hard feces, the lagomorphs also produce soft, spherical pellets that they eat without chewing. This practice is called caecotrophy; a part of the food undergoes a second digestive process, a process similar to that occurring in ruminant animals. The soft pellets are formed in the caecum, where cellulose is digested by anaerobic bacteria. The caecum walls absorb the volatile fatty acids produced by the cellulose fermentation process, while the fecal pellets become enriched in vitamin B1. By eating these soft pellets, the lagomorph allows them a second passage through the digestive tube. In the colon, the vitamins are absorbed and the bacteria are digested; 80% of the bacteria are lysed and 85% of the nitrogen compounds cross the colon wall and move into the bloodstream.

Lagomorph demography, including that of the pikas, is legendary. The females are mated during gestation and become fertilized either just before parturition (superfetation in hares) or just after (in rabbits). In this way, one of the uterine horns can contain fetuses ready to be born while the other contains recently implanted embryos. Hares can have up to 4 litters per year and rabbits up to 7. However, the number of litters per year and the number of young produced per litter depends on the hierarchical status of the female and the local habitat conditions. The maximum number of young produced by a single wild rabbit seems to be around 30. This is one reason why each time humans introduce rabbits to predator-free regions, there is a population explosion resulting in an ecological catastrophe. When the Romans became familiar with rabbits in Spain, they introduced them to Italy and to several Mediterranean islands, including the Balearic Islands,

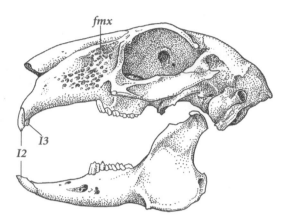

Figure 1

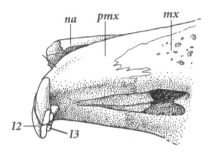

Figure 2

where, soon after, the inhabitants demanded the help of Caesar Augustus to combat the scourge. The same problem occurred in the fifteenth century on the small island of Porto Santo, north of Madeira Island; in the nineteenth century on the Kerguelen Islands and New Zealand; and in the twentieth century in Chile and Australia. Rabbits are appreciated both for hunting and for their meat. They have been bred by humans since Roman antiquity, but more intensely in the Western world for the last three centuries.

Examples

European hare: *Lepus europaeus;* brown hare: *L. capensis;* snowshoe hare: *L. americanus;* European rabbit: *Oryctolagus cuniculus;* eastern cottontail rabbit: *Sylvilagus floridanus;* pygmy rabbit: *Brachylagus idahoensis;* Sumatran rabbit: *Nesolagus netscheri;* riverine rabbit: *Bunolagus monticularis;* alpine pika: *Ochotona alpina.*

Number of species: 80.
Oldest known fossils: if the Zalambdalestids are considered to be lagomorphs, the group dates from the middle Cretaceous (*Zalambdalestes* and *Barunlestes* from Mongolia, 90 MYA). However, the oldest known incontestable lagomorphs are *Mimotona, Hsiuannania, Paranictops, Anictops* and *Cartictops* from China (Paleocene, 60 MYA)
Current distribution: we find lagomorphs everywhere, except Australia, New Zealand, Antarctica, Madagascar, southern South America and certain regions of Indonesia. However, humans have introduced rabbits and hares to all of these regions. Prior to the Quaternary glaciations, rabbits were found across western Europe. After the retreat of the last glaciers, they were found only in Spain and North Africa. They have been reintroduced to all areas of Europe by humans.

Rodentia

General Description

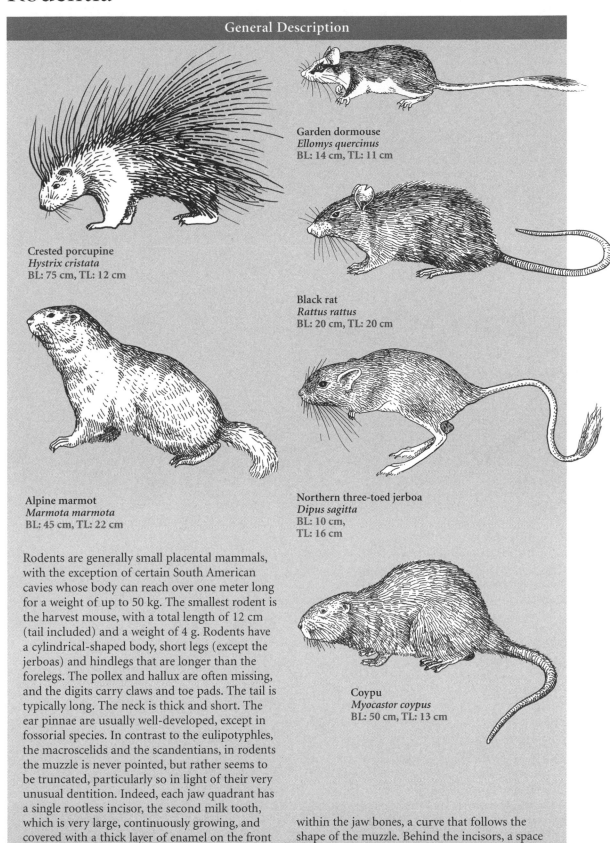

Crested porcupine
Hystrix cristata
BL: 75 cm, TL: 12 cm

Alpine marmot
Marmota marmota
BL: 45 cm, TL: 22 cm

Garden dormouse
Ellomys quercinus
BL: 14 cm, TL: 11 cm

Black rat
Rattus rattus
BL: 20 cm, TL: 20 cm

Northern three-toed jerboa
Dipus sagitta
BL: 10 cm,
TL: 16 cm

Coypu
Myocastor coypus
BL: 50 cm, TL: 13 cm

Rodents are generally small placental mammals, with the exception of certain South American cavies whose body can reach over one meter long for a weight of up to 50 kg. The smallest rodent is the harvest mouse, with a total length of 12 cm (tail included) and a weight of 4 g. Rodents have a cylindrical-shaped body, short legs (except the jerboas) and hindlegs that are longer than the forelegs. The pollex and hallux are often missing, and the digits carry claws and toe pads. The tail is typically long. The neck is thick and short. The ear pinnae are usually well-developed, except in fossorial species. In contrast to the eulipotyphles, the macroscelids and the scandentians, in rodents the muzzle is never pointed, but rather seems to be truncated, particularly so in light of their very unusual dentition. Indeed, each jaw quadrant has a single rootless incisor, the second milk tooth, which is very large, continuously growing, and covered with a thick layer of enamel on the front side only. These incisors run very deeply into the maxillary and dentary bones, forming a large arc within the jaw bones, a curve that follows the shape of the muzzle. Behind the incisors, a space with no teeth, called a diastema, separates these teeth from the cheek teeth (molars). The canines and certain premolars are absent. The zygomatic arches are strong. The glenoid fossa is elongated, permitting the jaw to move backwards and forwards, but not from side to side. The upper lip is usually split to form a "harelip." This lip forms, among other things, the labial lobes, lobes that lie laterally to the palate behind the upper incisors. These lobes form expandable cheek pouches that are used to transport food and that are found in many rodent species. The type of hair and its color are extremely variable from one species to another.

They have a complete or partial scaly layer on the tail and limbs. The hair follicules have sebaceous glands. There are several localized aggregations of cutaneous glands that form various glandular bodies (on the cheeks of the marmots, on the ears of the lemmings, on the hindlegs or groin of voles, in the anal region of porcupines, or at the oil and preputial glands in beavers).

The rodents are the largest group of mammals, both in terms of the number of species and the number of individuals living across the planet.

Some unique derived features

■ The rodents have only a single incisor per jaw quadrant: the second milk tooth, which has continuous growth (fig. 1, *I2*). Fig. 1 shows a lateral left view of a rat's skull (*Rattus rattus*).

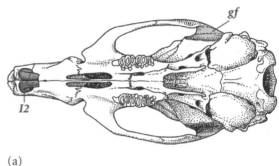

(a)

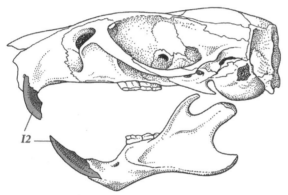

Figure 1

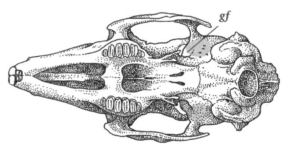

(b)

■ The glenoid fossa, which receives the mandibular condyle, is no longer limited posteriorly by the postglenoid process. In the rodents, it forms a deep rut that is elongated in an anterior-posterior direction, permitting the backward and forward movements of the condyle but no lateral movements. Fig. 2 shows ventral views of the skulls of a rat (2a, *Rattus rattus*), a hare (2b, *Lepus europaeus*) and a carnivore (2c, *Canis*). On the first, we see the elongated glenoid fossa (*gf*), on the second, the lack of a clearly delimited glenoid fossa, and on the third, a very differently shaped glenoid fossa, oriented transversally and closed posteriorly by the postglenoid process (*pgp*).

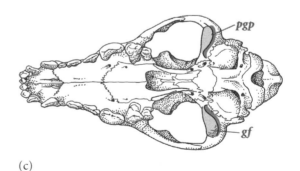

(c)

Figure 2

■ Molecular phylogenies based on the sequences of 19 nuclear genes and 3 mitochondrial genes unite the rodents. These genes include hemoglobin, myoglobin, alpha-crystallin, cytochrome c, and ribonuclease genes along with mitochondrial

ribosomal genes. In addition, independent analyses of other nuclear (a fragment of *RAG1*, *c-myc* oncogene, growth hormone) and mitochondrial (*ND6*) genes support this clade.

Ecology

The rodents are either herbivores or omnivores. They live in all environments, including the most arid. They include arboreal, terrestrial, semi-aquatic, fossorial, and even gliding species. Smell and touch (via tactile hairs) are the dominant senses. Their forelimbs enable them to walk, dig, hold food, hang and construct their nest. Their powerful incisors are formidable weapons that few materials can resist. They need to be constantly worn away, otherwise they become a danger for the individual.

The rodents are of considerable ecological importance, as much for their biomass in various ecosystems as for their role as primary consumers and prey for different species of squamates, birds and mammals. In many rodent species, we see rich and complex social structures and strong learning abilities. The family often makes up the basal unit for different groups within colonies. Each family group generally occupies a territory delimited by odorous signals, like urine or glandular secretions. In fact, very few rodents live solitary lives (e.g., squirrels, hamsters, dormice). After the battles and the nuptial parades, mating is generally short. In smaller species, the females are fertilized again immediately after the birth of the young. Gestation lasts from 16 days (in the golden hamster) to 5.5 months (in the capybara). Most species are nidicolous, meaning that the young are born naked, pink, and blind and are protected in a nest (murids such as rats and mice, squirrels, pocket gophers). In contrast, the young are born very well-developed in nidifugous species like guinea pigs, capybaras and coypus. Most rodents reproduce very rapidly. The number of young per litter range from 1 (in the paca) to 18 (in the hamster). A house mouse living indoors can produce young throughout the year, even in winter. Under these conditions, the number of young produced by a single female in one year can be over 80. The lifespan is typically less than two years in the small species. However, flying squirrels can live up to 13 years and Old World porcupines up to 18 years. In the sciurids and muroids, certain species undergo extreme demographic fluctuations, unmatched in other mammals. There are cyclic population explosions between which population minima are reached. For example, lemmings in Scandinavia undergo population explosions every 9 to 11 years. Several species (voles, American and Russian squirrels) have irregular demographic explosions. These events often result in massive levels of emigration (except in voles). The factors that control these demographic cycles remain unclear. The quantity of available resources, the effect of certain vegetative substances on ovarian activity, and the absence of disease epidemics have been put forward as possible explanations. However, population minima are reached even in the absence of disease epidemics and predators. This reduction has been suggested to be due to the physiological stress caused by overcrowding.

The relationship between humans and rodents became problematic as soon as humans became sedentary. Indeed, many rodents profit from human activities, particularly by attacking food reserves but also by using other types of materials, like wood, leather, canvas, textiles, paper and even cables. In addition, they are often disease reservoirs; the most famous and devastating disease in human history, the plague, is carried by rodents. More recently, humans have exploited rodents, not as a food resource, but for their fur (beaver, chinchilla, marmot, fat dormouse, etc.) and as laboratory animals (house mouse, rat, guinea pig, hamster). The annual number of rodents used in biomedical laboratories worldwide is in the order of tens of millions of individuals. The advances in human health made over the last 150 years is largely due to the use (and sacrifice) of rodents. The guinea pig (*Cavia porcellus*) was domesticated in the Andes starting in 4000 B.C.

Number of species: 2,021.
Oldest known fossils: *Heomys* from China, *Acritoparamys* and *Paramys* from Wyoming, USA, and *Microparamys* and *Pseudoparamys* from Meudon, France, all date from the Paleocene (60 MYA).
Current distribution: we find rodents everywhere, except Australia, New Zealand, and in circumpolar zones. However, humans have introduced them to Australia and New Zealand.

Examples

- SCIUROMORPHA: alpine marmot: *Marmota marmota;* red squirrel: *Sciurus vulgaris;* giant flying squirrel: *Petaurista grandis;* plains pocket gopher: *Geomys bursarius;* Eurasian beaver: *Castor fiber;* Pel's scaly-tailed squirrel: *Anomalurus peli.*
- MYOMORPHA: house mouse: *Mus musculus;* black rat: *Rattus rattus;* muskrat: *Ondatra zibethica;* black-bellied hamster: *Cricetus cricetus;* Norway lemming: *Lemmus lemmus;* common vole: *Microtus arvalis;* fat dormouse: *Glis glis;* lesser Egyptian jerboa: *Jaculus jaculus;* lesser mole rat: *Spalax leucodon.*
- HYSTRICOMORPHA: crested porcupine: *Hystrix cristata;* African mole rat: *Cryptomys hottentotus.*
- CAVIOMORPHA: coypu: *Myocastor coypus;* plains viscacha: *Lagostomus maximus;* chinchilla: *Chinchilla laniger;* guinea pig: *Cavia aperea porcellus;* mara: *Dolichotis patagonum;* capybara: *Hydrochoeris hydrochoeris;* paca: *Cuniculus paca;* agouti: *Dasyprocta aguti;* North American porcupine: *Erethizon dorsatum;* Brazilian porcupine: *Coendou prehensilis.*

Laurasiatheria

European hedgehog
Erinaceus europaeus
BL: 20 cm
node 30

European free-tailed bat
Tadarida teniotis
BL: 6 cm
node 32

European wild cat
Felis sylvestris
BL: 60 cm
node 36

Bottlenose dolphin
Tursiops truncatus
290 cm
node 46

Some unique derived features

■ There is a characteristic deletion of 10 base pairs in the 5' non-coding region of the *PLCB4* (phosphoinositide-specific phospholipase-C-beta 4) protein gene.

■ Molecular phylogenies based on the sequences of 19 nuclear genes and 3 mitochondrial genes unite the laurasiatheria. These genes include hemoglobin, myoglobin, alpha-crystallin, cytochrome c, and ribonuclease genes along with mitochondrial ribosomal genes. In addition, independent analyses of other nuclear (a different fragment of *RAG1*, *c-myc* oncogene, growth hormone) and mitochondrial (*ND6*) genes support this clade.

Number of species: 1,814.
Oldest known fossils: *Otlestes* from Uzbekistan and *Deccanolestes* from India (Cimolesta, *Ferae*) date from the middle Cretaceous (95 MYA).
Current distribution: worldwide.

Examples

■ EULIPOTYPHLES: European hedgehog: *Erinaceus europaeus*.
■ SCROTIFERA: red fox: *Vulpes vulpes*; giant pangolin: *Manis gigantea*; common long-eared bat: *Plecotus auritus*; mountain zebra: *Equus zebra*; Bactrian camel: *Camelus ferus*; desert warthog: *Phacochoerus aethiopicus*; elk: *Cervus elaphus*; bottlenose dolphin: *Tursiops truncatus*; hippopotamus: *Hippopotamus amphibius*.

Eulipotyphles

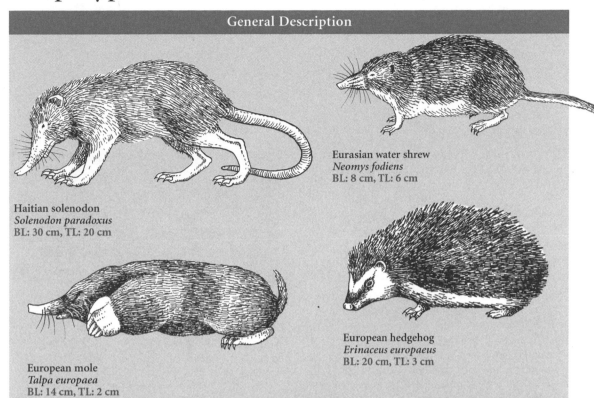

General Description

Eurasian water shrew
Neomys fodiens
BL: 8 cm, TL: 6 cm

Haitian solenodon
Solenodon paradoxus
BL: 30 cm, TL: 20 cm

European hedgehog
Erinaceus europaeus
BL: 20 cm, TL: 3 cm

European mole
Talpa europaea
BL: 14 cm, TL: 2 cm

The "insectivore" order that we find in classic zoology texts has always been unstable. At its maximum size, this order even included the macroscelids (elephant shrews) and the scandentians (tree shrews). However, regardless of its assemblage, it has always been difficult to find uniting characters for this order. Recent genomic approaches have shown that the "insectivores" were polyphyletic and that the holarctic families (solenodontidae, erinaceidae, talpidae, soricidae) were not related to the strictly African families (Afrosoricida: tenrecidae, potamogalidae, chrysochloridae). The eulipotyphles are small or very small animals: the body length varies from 3.5 cm (for the pygmy white-toothed shrew) to 45 cm (for the moonrat). The pygmy white-toothed shrew is in fact the smallest mammal, with an

adult mass of only 2 g. The fur is dense, and the legs are typically short with powerful claws. The eyes are small. Each of the four families has physical traits that differentiate it from the others. The solenodontids (solenodons) are the largest, with long legs and a long, stiff, scaly tail. The claws are large. The nose is elongated and the eyes are small. The soricids (shrews) are the smallest in size, but the largest in terms of species richness. The body is cylindrical with small legs, a proportionately large head, and a long tail. The nose is pointed, almost trunk-like. The ear pinnae are small, as are the eyes, which may sometimes be concealed under the fur. The fur is short, dense and soft, often dark-colored dorsally and light-colored ventrally. The erinaceids (hedgehogs) are easily recognized by their spiny backs. Their eyes and ears are well-developed and they have

a short tail. Specialized back muscles enable them to roll up into a ball. The talpids (moles) have a cylindrical body and a conical muzzle supported by the anterior nasal bone. The proportionately large skull is slightly flattened. The very small eyes are concealed by the dense, velvety fur. The ear pinnae have disappeared and the tail is short. The forelegs are short and strong, with five toes carrying particularly powerful claws. It is interesting to note that over their evolution the mammals have produced three mole-like groups: once in Africa with the chrysochlorids, once in the holarctic region with the talpids, and once in Australia with the notoryctid marsupials. The "hedgehog" form has also evolved twice: once in the holarctic region with the erinaceids and once in Madagascar with the hedgehog tenrecs (*Setifer* and *Echinops* genera).

443

Some unique derived features

■ Molecular phylogenies based on the sequences of 19 nuclear genes and 3 mitochondrial genes unite the eulipotyphles. These genes include hemoglobin, myoglobin, alpha-crystallin, cytochrome c, and ribonuclease genes along with mitochondrial ribosomal genes. In addition, independent analyses of other nuclear (a different fragment of *RAG1*, *c-myc* oncogene, growth hormone) and mitochondrial (*ND6*) genes support this clade.

Ecology

If it is difficult to find morphological similarities among the eulipotyphles, it is equally difficult to find similarities in their way of life. They are nevertheless all terrestrial and fossorial animals, even when they are amphibious (the desmans and several shrew species spend most of their time in rivers). They are present at all latitudes in forests, savannas, plains, steppes, deserts, tundra, bogs, mountains, with the majority of species living in rain forests. They dig temporary burrows, but also permanent and more sophisticated burrows. Certain species, such as the hedgehog, may even use burrows dug by other species. The eulipotyphles are insectivores or omnivores, with very high energetic needs: the smallest species must ingest more than their body weight in food daily. The senses of smell and hearing are dominant. They can be diurnal or nocturnal. They are typically solitary, territorial animals that are very aggressive with conspecifics. There are no common reproductive features that distinguish the eulipotyphles from other groups.

Number of species: 297.
Oldest known fossils: *Litolestes* from North America (Wyoming) dates from the upper Paleocene (50 MYA).
Current distribution: all continents except South America, Australia and Papua New Guinea, and Antarctica. The solenodontids are limited to Cuba and Haiti.

Examples

European hedgehog: *Erinaceus europaeus;* moonrat: *Echinosorex gymnurus;* lesser white-toothed shrew: *Crocidura suaveolens;* pygmy white-toothed shrew: *Suncus etruscus;* pygmy shrew: *Sorex minutus;* European water shrew: *Neomys fodiens;* armored shrew: *Scutisorex congicus;* Chinese shrew mole: *Uropsilus soricipes;* Pyrenean desman: *Galemys pyrenaicus;* European mole: *Talpa europaea;* eastern mole: *Scalopus aquaticus;* star-nosed mole: *Condylura cristata;* Haitian solenodon: *Solenodon paradoxus.*

Scrotifera

A few representatives

European free-tailed bat
Tadarida teniotis
BL: 6 cm
node 32

Mountain zebra
Equus zebra
BL: 240 cm
node 34

European wild cat
Felis sylvestris
BL: 60 cm
node 36

Bottlenose dolphin
Tursiops truncatus
290 cm
node 46

Some unique derived features

■ Molecular phylogenies based on the sequences of 19 nuclear genes and 3 mitochondrial genes unite the scrotifera. These genes include hemoglobin, myoglobin, alpha-crystallin, cytochrome c, and ribonuclease genes along with mitochondrial ribosomal genes. In addition, independent analyses of other nuclear (a different fragment of *RAG1*, *c-myc* oncogene, growth hormone) and mitochondrial (*ND6*) genes support this clade.

Number of species: 1,517.
Oldest known fossils: *Otlestes* from Uzbekistan and *Deccanolestes* from India (Cimolesta, *Ferae*) date from the middle Cretaceous (95 MYA).
Current distribution: worldwide.

Examples

■ CHIROPTERA: straw-colored fruit bat: *Eidolon helvum;* common long-eared bat: *Plecotus auritus.*
■ FEREUNGULATA: red fox: *Vulpes vulpes;* giant pangolin: *Manis gigantea;* mountain zebra: *Equus zebra;* Bactrian camel: *Camelus ferus;* desert warthog: *Phacochoerus aethiopicus;* elk: *Cervus elaphus;* hippopotamus: *Hippopotamus amphibius;* bottlenose dolphin: *Tursiops truncatus.*

Chiroptera

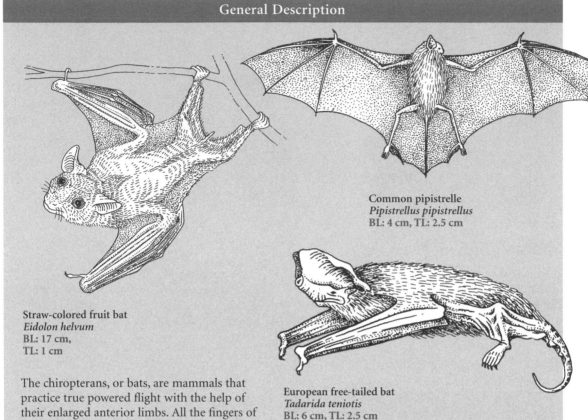

General Description

Straw-colored fruit bat
Eidolon helvum
BL: 17 cm,
TL: 1 cm

Common pipistrelle
Pipistrellus pipistrellus
BL: 4 cm, TL: 2.5 cm

European free-tailed bat
Tadarida teniotis
BL: 6 cm, TL: 2.5 cm

The chiropterans, or bats, are mammals that practice true powered flight with the help of their enlarged anterior limbs. All the fingers of the forelimbs (except the first) are extremely lengthened in the metacarpal region. A flight membrane extends between them (dactylopatagium) and is also attached to the body and all four limbs (plagiopatagium). This flight membrane also runs from the thumb to the neck (propatagium) and between the hindlegs (uropatagium). The uropatagium is also supported by a boney spur (calcar) that projects from the ankle (calcaneum bone). The thumb is not attached to the flight membrane and, like the toes, ends in a sharp claw. All the other fingers lack claws. The wing membrane is elastic and made up of both dermal and epidermal tissue layers. It attaches to the body high on the back and contains muscles, nerves and blood vessels. The sternum and pectoral girdle are powerful. The shoulder blades are well-developed and dorsal. The pelvis is small and the hindlegs are short. The body is covered in short, thick fur and the face has numerous vibrissae and large sebaceous-type skin glands. These glands are also present on the throat, shoulders, elbows and around the anus. There are no sweat glands. The ear pinnae are large and may sometimes have complex shapes. Skin folds or nasal folds sometimes develop around the nostrils. The complexity of the facial features and ears is linked to their use in echolocation and gives these animals their unique appearance. Despite the selective constraints linked to the ability to fly, we still find a large diversity of forms and colors within this group. The size varies from 3 to 42 cm in body length (excluding the tail), 18 to 142 cm in wing span, and 4 to 900 g in weight. The chiropterans are divided into two groups, the megachiroptera—large animals with elongated muzzles—and the microchiroptera—typically small with a flattened face, complex facial features, small eyes and large ears.

Some unique derived features

■ The patagium is a thin flight membrane largely supported by the elongated II, III, IV and V phalanges (fig. 1).

■ The cranial bones fuse and the sutures disappear at an early age.

Figure 1

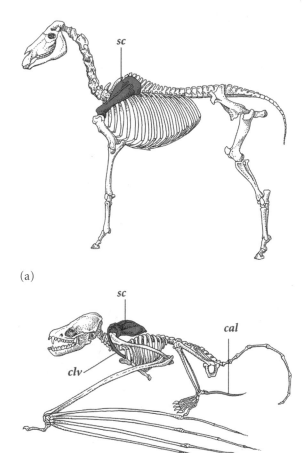

(a)

(b)

Figure 2

■ The pectoral girdle is enlarged and lies dorsal to the rib cage. Fig. 2 (*clv:* clavicle, *sc:* scapula or shoulder blade) depicts left lateral views of a horse (*Equus,* Perissodactyla) and a little brown bat (*Myotis,* Chiroptera). In the first, the pectoral girdle lies lateral to the rib cage, whereas in the second, it lies dorsally and is proportionately larger in size.

■ There is a characteristic process, cartilaginous or bony, that projects from the ankle and supports the posterior part of the patagium (uropatagium). This elongated process extends from the calcaneum (heel bone) and is called the calcar (*cal,* fig. 2b).

■ Molecular phylogenies based on the sequences of 19 nuclear genes and 3 mitochondrial genes bring together the microchiropterans and megachiropterans. These genes include hemoglobin, myoglobin, alpha-crystallin, cytochrome c, and ribonuclease genes along with mitochondrial ribosomal genes. In addition, independent analyses of other nuclear (a different fragment of *RAG1, c-myc* oncogene, growth hormone) and mitochondrial (*ND6*) genes support this clade.

Ecology

The bats are nocturnal animals that spend the daylight hours in shelters where they rest upside-down. These shelters can include trees, caves, rock crevices or human habitations. It is generally thought that the aerial habitat and the nutritional resources needed by flying vertebrates were largely exploited by the birds, most of which are diurnal. This left the night without avian competition for the development of the bats. Indeed, in the chiropterans we find a range of feeding regimes with avian equivalents: there are frugivores and nectarivores in the megachiropterans; and insectivores, piscivores, carnivores and frugivores in the microchiropterans. There is, however, one particularity in the chiropterans that is not found in birds: some species are exclusive blood-feeders, ecologically equivalent to mosquitoes, horseflies, ticks and leeches. The vampire bats have been at the origin of many myths about bats, highly exaggerated with respect to danger posed by these inoffensive species. Indeed, the amount of blood consumed by a vampire bat is minimal compared with the relative size of their sleeping prey (cattle, horses, goats, etc.). In the chiropterans, metabolism is very high and digestion occurs very quickly. The intestine is therefore short, and particularly so in insectivorous species.

Bats navigate and locate their prey in flight via echolocation. Using their well-developed larynx, they produce a series of short, ultrasonic sounds and are able to hear and interpret the reflections of these sounds. Perturbations in these echoes indicate the location of obstacles and prey items. Bats are social animals: solitary species are rare. Colony size varies from two to several hundred thousand (*Tadarida, Hipposideros, Miniopterus*). A single cave in Gabon harbors a colony of 600,000 individuals of the genus

Hipposideros. The layer of guano on the ground of this cave is over 2.5 m deep and the evening departure of individuals from the cave continues at a steady rate from 5:30 P.M. to 9:00 P.M. It has been shown that individuals hunting on the same territory belong to the same colony. Bats need their shelter for reproduction. They typically produce a single young, but some species can have two. In many species, the pregnant females leave the males and form large maternal colonies together (creches). When the mothers leave the colony to feed, the young remain and can be nursed by several females. However, in a few species, like the pipistrelles and the lump-nosed bats, the mother will only feed her own offspring. In temperate climates, bats hibernate. At the start of winter, individuals will put on fat reserves. During dormancy, the internal body temperature lowers to different degrees according to the species (up to 5.8°C in *Myotis*). If they are disturbed during hibernation, they will fly and quickly consume their fat reserves. Disturbance of these wintering sites therefore leads directly to the death of the individuals because they no longer have enough fat reserves to finish the season. In North America, certain species (genus *Lasiurus*) are truly migratory, spending the summer months in northern Canada and the winter months in the south of the United States.

Number of species: 925.
Oldest known fossils: *Icaronycteris* from France (Meudon), Belgium (Dormaal) and North America (Wyoming) dates from the lower Eocene (55 MYA). *Archaeonycteris* and *Palaeochiropteryx* from France (Rians) and Belgium (Dormaal) also date from this period.
Current distribution: equatorial, tropical and temperate regions of all continents. They are only absent above the two polar circles.

Examples

Egyptian fruit bat: *Rousettus aegyptiacus;* greater short-nosed fruit bat: *Cynopterus sphinx;* bulldog bat: *Noctilio labialis;* lesser horseshoe bat: *Rhinolophus hipposideros;* Sundevall's leaf-nosed bat: *Hipposideros caffer;* common vampire bat: *Desmodus rotundus;* greater spear-nosed bat: *Phyllostomus hastatus;* common long-eared bat: *Plecotus auritus;* common pipistrelle: *Pipistrellus pipistrellus;* mouse-eared bat: *Myotis myotis;* Brazilian free-tailed bat: *Tadarida brasiliensis.*

Fereungulata

A few representatives

Mountain zebra
Equus zebra
BL: 240 cm
node 34

European wild cat
Felis sylvestris
BL: 60 cm
node 36

Giant pangolin
Manis gigantea
BL: 80 cm
node 37

Bottlenose dolphin
Tursiops truncatus
290 cm
node 46

Some unique derived features

■ There is an epitheliochorial type placenta rather than the primitive endotheliochorial type. There has been a secondary reversion in the carnivores to either an endotheliochorial or hemochorial placenta and a convergence in the strepsirrhinian primates.

NOTE: In an epitheliochorial placenta (fig. 1a), the uterine epithelium is present and the uterine capillaries do not penetrate the placental tissue. In a hemochorial placenta (fig. 1b), the uterine epithelium disappears and the uterine capillaries spread and form lacunae where maternal blood is in contact with fetal tissue (the chorionic epithelium). In endotheliochorial placentas, the uterine epithelium disappears and the uterine capillaries infiltrate the fetal tissue, but without forming maternal lacunae.

■ Molecular phylogenies based on the sequences of 19 nuclear genes and 3 mitochondrial genes bring together the cetartiodactyla, *Ferae* and perissodactyla. These genes include hemoglobin, myoglobin, alpha-crystallin, cytochrome c, and ribonuclease genes along with mitochondrial ribosomal genes.

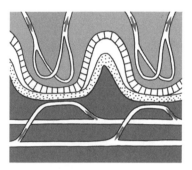

Figure 1a

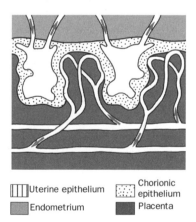

| | | Uterine epithelium | | Chorionic epithelium |
| | | Endometrium | | Placenta |

Figure 1b

Number of species: 592.
Oldest known fossils: *Otlestes* from Uzbekistan and *Deccanolestes* from India (Cimolesta, *Ferae*) date from the middle Cretaceous (95 MYA).
Current distribution: worldwide; in Australia and New Zealand they have been introduced by humans (dingo, cats, ruminants, etc.).

Examples

■ Perissodactyla: mountain zebra: *Equus zebra*.
■ Ferae: red fox: *Vulpes vulpes*; giant pangolin: *Manis gigantea*.
■ Cetartiodactyla: Bactrian camel: *Camelus ferus*; desert warthog: *Phacochoerus aethiopicus*; elk: *Cervus elaphus*; bottlenose dolphin: *Tursiops truncatus*; hippopotamus: *Hippopotamus amphibius*.

Perissodactyla

General Description

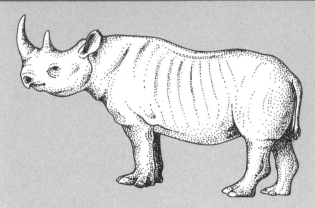

Black rhinoceros
Diceros bicornis
BL: 340 cm, TL: 70 cm

Malayan tapir
Tapirus indicus
BL: 220 cm, TL: 10 cm

Mountain zebra
Equus zebra
BL: 240 cm, TL: 80 cm

The perissodactyls include the tapirs, rhinos, horses, donkeys and zebras. They are placental mammals of medium to large size (body length ranging from 180 to 420 cm). They are excellent runners with unguligrade locomotion; they walk on the tips of their hooved toes. The median plane of symmetry passes through the third digit. This arrangement is primitive for mammals but is accompanied by a reduction in the number of lateral digits. On both the forefeet and hindfeet, the third digit is the longest, and in equids it is the only digit. The rhinos have three toes on both the fore and hindfeet, whereas the tapirs have three and four, respectively (with the third being the strongest). The muzzle is generally high and elongated with strong, undivided, prehensile lips. The ears and eyes are relatively small (with the exception of the ears of certain donkeys). The canine teeth are small or completely absent, whereas the cheek teeth have elongated ridges that run between the cusps. The stomach is simple and the intestinal caecum is very large: in horses, its maximum capacity can exceed 30 liters. Anaerobic fermentation of cellulose is carried out by the microbial flora of this chamber, analogous to the rumen of the ruminants. However, in contrast to other large herbivorous mammals, the perissodactyls do not ruminate. The rhinocerotids carry one or two horns in the nasal region; the fibrous make-up of these horns is completely different from that of the ruminant animals. The tapirs have a short, trunk-like muzzle. The type of hair and the color of these animals varies greatly from one species to another (black, gray, brown, and even black and white strips in the zebras). The equidae (horses) are typically taken as the representative examples of this group. However, this family has accumulated so many derived characters that the tapirs are, in fact, better model organisms.

Some unique derived features

■ The main supporting axis of the hind limbs passes through the third toe, with a reduction in the lateral toes. Paradoxically, this arrangement, known as the mesaxonic condition, is primitive for the perissodactyls because we also find it in other mammals. However, in this group, it is accompanied by a reduction in the number of lateral toes. The I and V digits have disappeared in all perissodactyls except the tapirs that have kept digit V on the forefeet. Only the third digit is found in the equids. We find a convergence in certain fossil ungulates, the litopterans of South Amercia (Tertiary). Fig. 1 shows anterior views of the left foot skeleton of two perissodactyls (1a: American tapir, 1b: horse) and a non-perissodactyl (1c: macaque). The main support axis is highlighted in color (*as*: astragalus, *cal*: calcaneus, *asj*: astragalus-navicular junction, *cbd*: cuboid, *cun*: cuneiform, *nav*: navicular bone).

■ Primitively, the astragalus is flat on the navicular surface. In this group, which corresponds to a restricted notion of "perissodactyls," this face is lightly curved, creating a slightly concave form. The contact between the astragalus and the navicular bone is

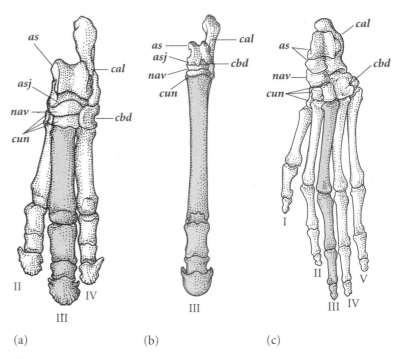

(a)   (b)   (c)

Figure 1

therefore described as being "saddle-shaped" (fig. 1a). Fig. 2 shows anterior views of the left astragalus of a perissodactyl (2a: tapir), a cetartiodactyl (2b: pig) and a primate (2c: human). We note that the navicular face of this bone is flat and concave in the tapir. However, this concave shape disappears in the equidae, whose navicular face is horizontal (see above, fig. 1b; *caf*: calcaneous face, *cbf*: cuboid face, *tbf*: tibial face, *naf*: navicular or scaphoid face).

Note: certain of the characters of the perissodactyls find no convergence outside the group, but do show homoplasy within. For example, we can mention the large contact zone between the lacrimal and nasal bones; this character is seen on the horse skull and in many fossil perissodactyls, but has been secondarily lost in the tapirs and modern rhinoceros. The navicular face of the astragalus is flat rather than "saddle-shaped" in the equidae.

■ Molecular phylogenies based on the sequences of 19 nuclear genes and 3 mitochondrial genes bring together the tapirs, horses and rhinoceros. These genes include hemoglobin, myoglobin, alpha-crystallin, cytochrome c, and ribonuclease genes along with mitochondrial ribosomal genes. In addition, independent analyses of other nuclear (a different fragment of *RAG1*, *c-myc* oncogene, growth hormone) and mitochondrial (*ND6*) genes support this clade.

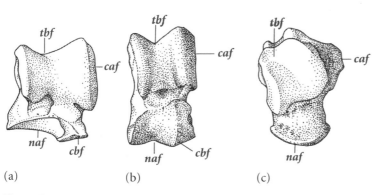

(a)   (b)   (c)

Figure 2

Ecology

The perissodactyls are exclusively terrestrial and herbivorous, leaf-eaters and grazers. They live in plains, high plateaus, steppes and semi-arid regions (equidae), deserts (onager), savannas (rhinoceros, zebras) and rain forests (certain rhinoceros, tapirs). The equidae are all social mammals, whereas the rhinoceros and tapirs are primarily solitary. However, some white rhinoceros females can temporarily live with a group of young animals. Only the equidae form herds directed by a dominant male. There is only a single young produced per litter; it is born fully formed and able to follow its mother. Sexual maturity occurs at 4 to 6 years old, with a longevity of between 20 to 30 years depending on the species.

The oldest domesticated perissodactyls seem to have been the donkey (*Equus asinus*), around 5000 years B.C. in the Nile valley, and the onager (*Equus hemionus onager*) during the same period in Mesopotamia. The domestic horse (*Equus przewalskii caballus*) has played an important role in human history as a beast of burden, means of transportation, companion and food source (except during the Christian era, when horse meat was taboo, a taboo that was lifted near the end of the nineteenth century). Its ancestors include three Eurasian sub-species. The first attempt to domesticate the horse probably took place 4000 years B.C. in Mesopotamia and China and then again in 3500 B.C. in southern Ukraine. Horses disappeared from North America during the Pleistocene. When Europeans debarked in the Americas, they brought domesticated English and Spanish horses with them. These animals did not waste any time returning to the wild state in the wide open plains of both South America (the wild horses of the Pampa) and North America (where the "mustang" comes from). The same scenario occurred in Australia, where the wild horses are called brumbies.

Number of species: 18.

Oldest known fossils: this declining group included a large diversity of forms during the Eocene and was found on all continents, except Australia, during the Quaternary. The oldest known fossil could be *Radinskya* from the upper Paleocene from China (55 MYA). The oldest perissodactyl *sensu stricto* is *Hyracotherium* from the lower Eocene from Europe and North America (53 MYA), but other families follow closely behind.

Current distribution: Africa, Middle East, central Asia, southern Asia to Borneo, northern half of South America and Central America. The equidae have been introduced to all continents by humans.

Examples

Brazilian tapir: *Tapirus terrestris;* Malayan tapir: *T. indicus;* Indian rhinoceros: *Rhinoceros unicornis;* black rhinoceros: *Diceros bicornis;* white rhinoceros: *Ceratotherium simum;* mountain zebra: *Equus zebra;* onager or Asiatic ass: *E. hemionus;* domestic horse: *E. przewalskii caballus.*

Ferae

A few representatives

Walrus
Odobenus rosmarus
310 cm
node 36

European wild cat
Felis sylvestris
BL: 60 cm
node 36

Brown bear
Ursus arctos
BL: 220 cm
node 36

Giant pangolin
Manis gigantea
BL: 80 cm
node 37

Some unique derived features

- *Osseum tenorium*: this is a bony plate in the skull that separates the cerebrum and the cerebellum. We find this plate by convergence in the tubulidentata.

- There is a characteristic deletion of 363 base pairs in the apolipoprotein B gene (*ApoB*).

- Molecular phylogenies based on the sequences of 19 nuclear genes and 3 mitochondrial genes group the carnivores and pholidota together. These genes include hemoglobin, myoglobin, alpha-crystallin, cytochrome c, and ribonuclease genes along with mitochondrial ribosomal genes. In addition, independent analyses of other nuclear (a different fragment of *RAG1*, *c-myc* oncogene, growth hormone) and mitochondrial (*ND6*) genes support this clade.

Number of species: 278.
Oldest known fossils: the *ferae* include the Cimolesta and Creodonta fossil groups. The oldest are *Otlestes* from Uzbekistan and *Deccanolestes* from India (Cimolesta) that date from the middle Cretaceous (95 MYA).
Current distribution: worldwide; in Australia and New Zealand they have been introduced by humans (Dingo, cats, etc.).

Examples

- CARNIVORA: red fox: *Vulpes vulpes*; brown bear: *Ursus arctos*; walrus: *Odobenus rosmarus*; European wild cat: *Felis sylvestris*.
- PHOLIDOTA: giant pangolin: *Manis gigantea*; long-tailed pangolin: *M. tetradactyla*.

Carnivora

General Description

Spotted hyena
Crocuta crocuta
BL: 155 cm, TL: 30 cm

Walrus
Odobenus rosmarus
310 cm

Red fox
Vulpes vulpes
BL: 65 cm, TL: 40 cm

Brown bear
Ursus arctos
BL: 220 cm, TL: 15 cm

Raccoon
Procyon lotor
BL: 50 cm, TL: 30 cm

The carnivores are, for the most part, flesh-eating mammals of medium to large size with highly developed sense organs. The forelimbs are pentadactyl and armed with sharp, powerful claws. The hind limbs generally have four toes. The incisors are sharp and always number three on each jaw quadrant. The canines form long, pointed fangs. The premolars are flattened from side to side and blade-like. The molars have numerous, slicing cusps. The zygomatic arches greatly depass the skull case on each side of the skull, leaving a large space for the passage of the powerful chewing muscles. The orbit is not completely closed posteriorly. The jaw can move vertically, but not horizontally. These animals are covered in fur. Overall, there are few sweat glands, except in very

localized zones (e.g., between the toes of the canids), but sebaceous glands are abundant. Indeed, the carnivores have a complex of glands in the anal/genital region that produces the distinct odor of each individual. Anal glands are invaginations of the skin that enclose sebaceous and sweat-like glands. Secretions from this region give off a very strong odor (skunk, civet, genet,

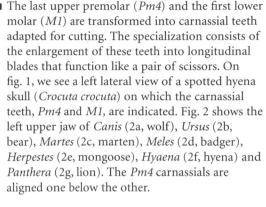

Some unique derived features

■ The last upper premolar (*Pm4*) and the first lower molar (*M1*) are transformed into carnassial teeth adapted for cutting. The specialization consists of the enlargement of these teeth into longitudinal blades that function like a pair of scissors. On fig. 1, we see a left lateral view of a spotted hyena skull (*Crocuta crocuta*) on which the carnassial teeth, *Pm4* and *M1*, are indicated. Fig. 2 shows the left upper jaw of *Canis* (2a, wolf), *Ursus* (2b, bear), *Martes* (2c, marten), *Meles* (2d, badger), *Herpestes* (2e, mongoose), *Hyaena* (2f, hyena) and *Panthera* (2g, lion). The *Pm4* carnassials are aligned one below the other.

Eurasian badger
Meles meles
BL: 70 cm, TL: 15 cm

Eurasian lynx
Lynx lynx
BL: 110 cm, TL: 18 cm

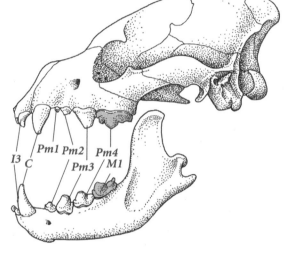

Figure 1

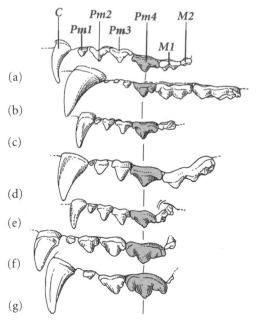

European wild cat
Felis sylvestris
BL: 60 cm, TL: 30 cm

etc.). The form and color of these animals can vary widely: we find animals from 13 cm long (least weasel, not including the tail) to up to 650 cm long and 3600 kg in weight (certain male elephant seals). This group includes wolves, bears, raccoons, weasels, otters, genets, hyenas, panthers, fur seals, true seals, walrus and a fossil group, the creodonts.

Figure 2

Ecology

The carnivores are flesh-eaters, scavengers, sometimes omnivores and exceptionally exclusive vegetarians (case of the giant panda). They are found in a diverse range of terrestrial biotopes, from polar zones (seals and polar bears) to tropical forests (jaguar), passing through the most arid deserts (fennec) and high altitudes (snow leopard). We also find them in the sea; indeed, the pinnipeds are semi-aquatic (walrus, seals, sea lions). Contrary to the cetaceans and sirens, the pinnipeds have not lost the ability to move and give birth on land. On every continent, the carnivores make up a major part of the secondary and tertiary terrestrial consumers, that is, the terminal link in the trophic food chain (except in Australia, where, if we do not include the carnivores introduced by humans over the last three centuries, only the dingo occupies this place next to the flesh-eating marsupials). Given the loss of energy that occurs during each step in the food chain, only a few carnivores can be supported in a single ecosystem. Carnivores can regulate herbivore populations and are indirect resources for scavengers (vultures, for example) and detritivores (many insects) that feed on the remains of the prey they kill. The disappearance of carnivores from an ecosystem can greatly modify all its components. Several examples demonstrate this effect. In 1907, 7000 deer were counted on the Kaibab Plateau in Arizona (USA). Nineteen years later, following a campaign to exterminate all predators, the number of deer had increased to 100,000 and 30,000 hectares of rangeland were severely overgrazed.

Except for the marine species, young carnivores are born very small, blind and unable to move much. It is assumed that this state is the result of the predatory way of life: females must still be able to hunt even during gestation and large young would greatly reduce their running abilities. Species that hunt prey larger than themselves are typically social animals whose organization and sophisticated modes of communication are unique to each species (wolf, African wild dog, lion, spotted hyena, etc.). The carnivores have provided us with our main domestic animals. The dog was likely domesticated several times; around 2400 B.C. in America, 7600 B.C. in Western Europe, 9500 B.C. in Persia and Palestine, 10,000 B.C. in central Europe and 13,000 B.C. in Siberia. The cat, being much less flexible, was domesticated around 3500 B.C. in Egypt, Libya, Pakistan and maybe also in Southeast Asia. The genet was much appreciated as a pet around the Mediterranean region starting in antiquity. The cheetah was domesticated around 3000 B.C. by the Sumerians, then in central Africa and from northern Africa to India.

Number of species: 271.

Oldest known fossils: the carnivores are part of a larger group, the *Ferae*, which also includes the Cimolesta and Creodonta fossil groups. The oldest fossils of the *Ferae* lineage are *Otlestes* from Uzbekistan and *Deccanolestes* from India (Cimolesta) that date from the middle Cretaceous (95 MYA). The oldest creodont is the oxyaenid *Tytthaena* from Wyoming (USA) that dates from the Paleocene (60 MYA). Finally, among the carnivores *sensu stricto*, the oldest species are the didymictid *Protictis* from North America, the miacid *Ulntacyon* and the viverravid *Simpsonictis*, both from Wyoming (Paleocene, 60 MYA).

Current distribution: worldwide, with the exception of Australia. However, certain sea lions (arctocephalinae) are found along the coasts of southern Australia. The dingo is a large wild Australian dog that was introduced by humans 6000 years ago. Over the last three centuries, humans have intensified the introduction of carnivores to Australia (dogs, cats), resulting in damage to the resident marsupial fauna.

Examples

Least weasel: *Mustela nivalis;* Eurasian badger: *Meles meles;* striped skunk: *Mephitis mephitis;* European otter: *Lutra lutra;* common raccoon: *Procyon lotor;* giant panda: *Ailuropoda melanoleuca;* brown bear: *Ursus arctos;* European genet: *Genetta genetta;* African palm civet: *Nandinia binotata;* Egyptian mongoose: *Herpestes ichneumon;* red fox: *Vulpes vulpes;* dog: *Canis familiaris;* wolf: *C. lupus;* Mediterranean monk seal: *Monachus monachus;* walrus: *Odobenus rosmarus;* southern elephant seal: *Mirounga leonina;* California sea lion: *Zalophus californianus;* striped hyena: *Hyena hyaena;* European wild cat: *Felis sylvestris;* lion: *Panthera leo;* cheetah: *Acinonyx jubatus.*

Pholidota

The pangolins (or scaly anteaters) are stocky mammals that vary from 75 to 150 cm long and can weigh up to 35 kg (giant pangolin). The long tail can make up 45–65% of the total body length. The skull is elongated without protrusions or zygomatic arches for the attachment of jaw muscles (fig. 1). The lower jaws are narrow and reduced to boney blades without teeth. The muzzle tapers down to a tube-like snout and houses a very long tongue. The eyes, ear pinnae and mouth are small. The dorsal surface is covered in large, brown epidermal scales that overlap like roof shingles. Sparse hairs lie between the scales. The tail is completely covered in scales. The ventral surface has light-colored hairs. The limbs are five-toed (pentadactyl), but are functionally three-toed (tridactyl). The claws are extremely powerful. The tongue is long, thin and very protractile. In large pangolin species, it can measure up to 40 cm long, or 18 cm long in the small, arboreal African species. In the resting position, the tongue is retracted into a sheath and goes as far back as the rib cage. The salivary glands are highly developed; the tongue is covered in thick saliva that is used to trap insects. The xiphoid process

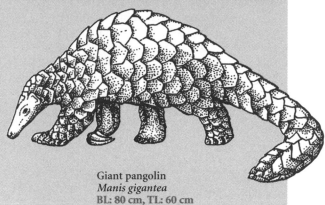

Giant pangolin
Manis gigantea
BL: 80 cm, TL: 60 cm

of the sternum has undergone complex modifications; it is elongated or widened (depending on the species) and is linked to multiple cartilaginous rods. The ensemble can continue ventrally from the outside the peritoneum to the pelvis. The ribbon-like xiphoid process of the African pangolins is even rolled up upon itself within the pelvic region, whereas that of the Asiatic pangolins is spade-shaped. These structures help support the tongue musculature.

Some unique derived features

- The pangolins lack teeth. Fig. 1 shows a left lateral view of the skull of *Manis tetradactyla*.

- The mandible is slender (fig. 1).

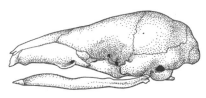

Figure 1

- The animal is covered in large, horny scales of epidermal origin. Fig. 2 depicts a longitudinal cut through 3 scales of *Manis tricuspis* (*der*: dermis, *sca*: scale, *hep*: horny epidermal layer, *mep*: mucosal epidermal layer, *dp*: dermal papilla).

- The internal lining of the stomach has a kcratinized epithelium in the anterior region and a muscular organ armed with hard teeth in the pyloric region. Indeed, as the pangolins do not chew their food in the mouth, the insects on which they feed must be ground up by the stomach. The insects, trapped in saliva, arrive in the stomach without being

chewed. They are engulfed in the mucus secreted by the large mucus glands of the stomach, which also contains grit and small stones. The insects are then ground up by the pyloric organ as the gastric glands release pepsin to degrade the proteins.

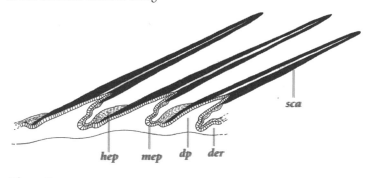

Figure 2

Ecology

The pangolins are primarily nocturnal animals that feed on ants and termites. They are either terrestrial or arboreal and live in the forests and treed savannas of the tropical regions of the Old World. The arboreal pangolins feed on tree termites. Terrestrial species can attack solid termite mounds as well as ant hills. The stomach has been transformed into a grinding organ. Rather than a mucosal surface, the anterior region has a thick, keratinized epithelium with "horny teeth" that are used to dilacerate ingested insects. To move rapidly, the pangolins are able to run on their hind legs. To protect themselves from large carnivores (leopards, tigers), they roll themselves into a ball. A considerable strength is required to unroll them; a man could not succeed. The African species give birth to a single young, whereas the Asiatic species produce from 1 to 3. In arboreal species, the young ride on their mother's back or hold on to the base of her tail, lying astride with their tail wrapped around hers. The young of terrestrial species are protected within a burrow.

Number of species: 7.
Oldest known fossils: *Eomanis* from Germany dates from the Eocene (40 MYA).
Current distribution: Africa (south of the Sahara), India, southern China, Southeast Asia, Java, Borneo, Sumatra.

Examples

Indian pangolin: *Manis crassicaudata*; giant pangolin: *Manis gigantea*; ground pangolin: *M. temmincki*; tree pangolin: *M. tricuspis*; Chinese pangolin: *M. pentadactyla*; long-tailed pangolin: *M. tetradactyla*; Malayan pangolin: *M. javanica*.

Cetartiodactyla

A few representatives

Bactrian camel
Camelus ferus
BL: 290 cm
node 39

Wild boar
Sus scrofa
BL: 140 cm
node 41

Blue wildebeest
Connochaetes taurinus
BL: 200 cm
node 43

Bottlenose dolphin
Tursiops truncatus
290 cm
node 46

Some unique derived features

■ The posterior limb axis passes between digits III and IV. This paraxonic arrangement in the certartiodactyls is associated with an even number of toes. Fig. 1 shows anterior views of the left foot of two certartiodactyls (1a: cow, *Bos taurus*, 1b: wild boar, *Sus scrofa*) and one of a non-certartiodactyl (1c: tapir, *Tapirus terrestris*) (*as:* astragalus, *cal:* calcaneus, *cbd:* cuboid, *cun:* cuneiform, *cun3:* third cuneiform, *mtt:* metatarsus; *mtt3:* third metatarsal, *nav:* navicular bone, *fi:* fibula, *ti:* tibula). The vertical line indicates the supporting axis.

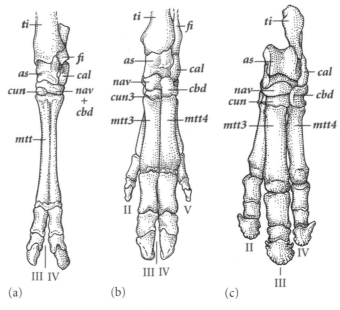

Figure 1

■ There is a double-trochleated astragalus, with a tibial trochlea (groove) proximally and a navicular trochlea distally. Note that this character is not found in extant cetaceans, as they have lost the astragalus (along with their hind limbs). Fig. 2 shows anterior views of the left astragalus of a cetartiodactyl (2a: pig) and a non-cetartiodactyl (2b: human); *caf:* calcaneous face, *cbf:* cuboid face, *naf:* navicular or scaphoid face, *tbf:* tibial face.

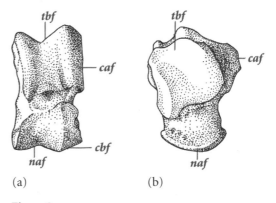

Figure 2

- In the ankle, the astragalus is in direct contact with the cuboid bone. The cuboid face of the astragalus is indicated in fig. 2a. Fig. 3 shows schematic anterior views of the left tarsus of a human (3a) and a cetartiodactyl, the pig (3b). In the pig, we see the contact between the astragalus and the cuboid, whereas in humans this contact is absent.

- The monophyly of the cetartiodactyls is supported by sequences of the mitochondrial cytochrome b gene and that of exon 7 of the β-casein gene.

- Molecular phylogenies based on the sequences of 19 nuclear genes and 3 mitochondrial genes bring together all of the cetartiodactyla. These genes include hemoglobin, myoglobin, alpha-crystallin, cytochrome c, and ribonuclease genes along with mitochondrial ribosomal genes. In addition, independent analyses of other nuclear (a different

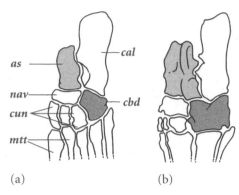

Figure 3

fragment of *RAG1*, *c-myc* oncogene, growth hormone) and mitochondrial (*ND6*) genes support this clade.

Number of species: 296.
Oldest known fossils: this lineage includes the fossil group Mesonychia. The oldest known mesonychians slightly predate the cetaceans and include *Yatanglestes* and *Dissacusium* from the lower Paleocene from China and *Ankalagan* from the same period from New Mexico (USA) (60 MYA). If we consider only extant groups, the oldest known fossils are the cetaceans *Ambulocetus*, *Pakicetus*, *Gandakasia* and *Ichthyolestes* from the lower Eocene from Pakistan (50 MYA). However, also dating from this period are fossil suids, *Cebochoerus* from Spain, *Khirtharia*, *Indohyus*, *Kunmunella* and *Raoella* from Asia, along with the dacrytherid tylopod *Cuisitherium* from France and the mixtotherid tylopod *Mixtotherium* from Spain, England and France.
Current distribution: worldwide, except Australia and New Zealand. They have been introduced to these regions by humans. All the seas and oceans.

Examples

- TYLOPODA: Bactrian camel: *Camelus ferus*.
- CLADE 40: lesser Malay mouse-deer: *Tragulus javanicus*; Siberian musk deer: *Moschus moschiferus*; elk: *Cervus elaphus*; giraffe: *Giraffa camelopardalis*; pronghorn: *Antilocapra americana*; Grant's gazelle: *Gazella granti*; domestic cow: *Bos taurus*; hippopotamus: *Hipopotamus amphibius*; blue whale: *Balaenoptera musculus*; sperm whale: *Physeter catodon*; killer whale: *Orcinus orca*; wild boar: *Sus scrofa*; babirusa: *Babyrousa babyrussa*.

Tylopoda

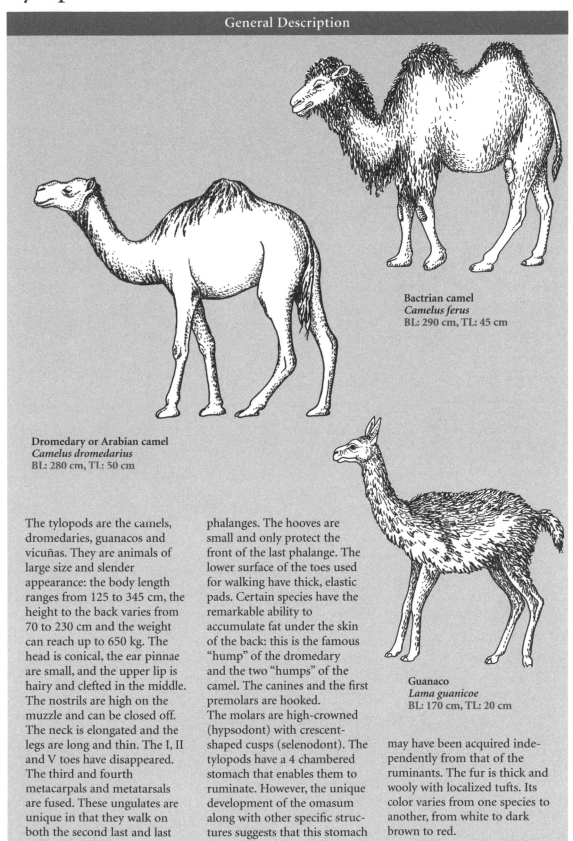

Bactrian camel
Camelus ferus
BL: 290 cm, TL: 45 cm

Dromedary or Arabian camel
Camelus dromedarius
BL: 280 cm, TL: 50 cm

Guanaco
Lama guanicoe
BL: 170 cm, TL: 20 cm

The tylopods are the camels, dromedaries, guanacos and vicuñas. They are animals of large size and slender appearance: the body length ranges from 125 to 345 cm, the height to the back varies from 70 to 230 cm and the weight can reach up to 650 kg. The head is conical, the ear pinnae are small, and the upper lip is hairy and clefted in the middle. The nostrils are high on the muzzle and can be closed off. The neck is elongated and the legs are long and thin. The I, II and V toes have disappeared. The third and fourth metacarpals and metatarsals are fused. These ungulates are unique in that they walk on both the second last and last phalanges. The hooves are small and only protect the front of the last phalange. The lower surface of the toes used for walking have thick, elastic pads. Certain species have the remarkable ability to accumulate fat under the skin of the back: this is the famous "hump" of the dromedary and the two "humps" of the camel. The canines and the first premolars are hooked. The molars are high-crowned (hypsodont) with crescent-shaped cusps (selenodont). The tylopods have a 4 chambered stomach that enables them to ruminate. However, the unique development of the omasum along with other specific structures suggests that this stomach may have been acquired independently from that of the ruminants. The fur is thick and wooly with localized tufts. Its color varies from one species to another, from white to dark brown to red.

Some unique derived features

- The tylopods walk on the lower surface of the last and second-last phalanges (*pha*). The hoof covers only the anterior surface of the last phalange. The sole of the foot has a thick elastic pad (*elp*). This contrasts with other terrestrial cetartiodactyls that walk on the tips of their hooves (*hf*), which completely surround the last phalange. Fig. 1 shows schematic diagrams of longitudinal sections of the feet of a tylopod (1a: llama; *ks*: keratinized sole) and a ruminant (1b: sheep).

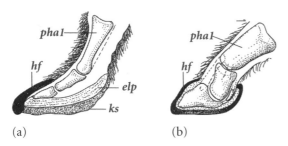

(a)   (b)

Figure 1

- The connection of the fourth toe to the metatarsal (*A4*) is lower than that of the third toe (*A3*). Anterior views of the left hind foot of a tylopod (2a, camel, where only the terminal end of the leg is shown) and a ruminant (2b, domestic cow, where the basipod (tarsal, *ta*), metapod (metatarsal, *mta*) and the acropod (phalanges, *pha*) are also included) are shown in fig. 2.

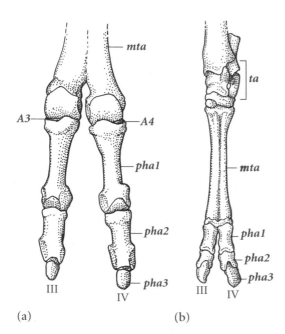

(a)   (b)

Figure 2

- Rumination: the tylopods ruminate, but this process of cellulose digestion utilizes different anatomical structures from those used by the ruminants. Indeed, this similarity is most certainly the result of a convergent acquisition. The omasum (*omas*) is very long, the lining almost completely lacks leaf-like folds, and it is not clearly distinct from the abomasum (*abo*) (in the ruminants, the lining of the omasum has a hundred or so broad longitudinal folds). The internal wall of the rumen (*rum*) lacks papillae (in contrast to that of the ruminants). Finally, the wall of the reticulum (*ret*) and part of the wall of the rumen have a dense network of deep pouches, aquiferous cells (*aqc*), that can store water. Fig. 3 shows diagrams of the stomach of a tylopod (3a) and a ruminant (3b) in order to illustrate the relative size of the different stomach compartments (*du*: duodenum, *omas*: omasum, *esg*: esophageal groove, *es*: esophagus).

NOTE: for more information on the function of the different stomach compartments, see 13.43.

- Diaphragm: the tylopods are the only mammals with ossifications of the diaphragm.

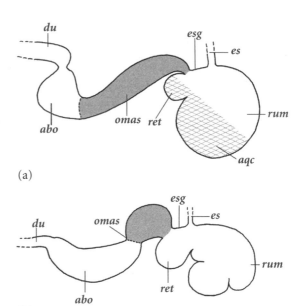

(a)

(b)

Figure 3

- The red blood cells of the tylopods are oval and flat (that is, they are not biconcave), of small size (the large diameter of the oval is 8 μm, the small one is 4 μm) and found in extremely high numbers (there are 14 million per mm³ in the llama and 19 million per mm³ in the Bactrian camel).

Ecology

The tylopods are exclusively terrestrial and herbivorous. They typically live in mixed groups in the driest and most inhospitable regions: the most arid deserts to dry valleys, steppes and high mountain plateaus. Groups are generally made up of females and young animals, with a single dominant male. Other adult males live solitarily. During the mating season, the male attempts to take over a harem and will confront other males in spectacular combats where each individual trys to knock over and bite the other. The gestation period is long (10 months in the vicuña, 12 to 14 months for the dromedary) and only a single young is born. The newborn will walk within the first three hours after birth. Sexual maturity is reached between 1 to 3 years old, depending on the species.

The Bactrian camel was domesticated around 3000 B.C. somewhere between Turkestan and Manchuria. However, domestication did not have much effect on this animal. It can survive in a remarkably wide range of temperatures (from + 50°C in summer to −25°C in winter). In contrast, the dromedary (or Arabian camel) only tolerates hot and dry climates. This species was domesticated around 2500 B.C. on the Arabian Peninsula. This animal holds the record for resistance to dehydration. A man will fall into a coma when his level of dehydration reaches between 5 and 10% of his body weight and, in the heat, a level of dehydration of 12% will cause death due to the loss of water from the blood. A dromedary can reach a dehydration level of 30% of its body weight without greatly affecting the water level of the blood. This can go down to as far as 40% if the animal is in good health. In fact, if an individual loses a quarter of its weight in water, the blood will only lose 10%. If the temperature does not go above 40°C and it can graze on vegetation, a dromedary can drink only once every two weeks and survive. However, when it does drink, its capacity to absorb water is enormous: 15 liters per minute and up to 120 to 140 liters. In addition, this animal efficiently accumulates heat. At night, its internal temperature is 34°C. It requires an entire morning, including the hottest hours of the day, to reach 40°C, the temperature at which it starts to sweat. With such characteristics, the dromedary has been named the "desert vessel" and has enabled humans to cross the Sahara. This species no longer exists in the wild, but some feral groups have returned to the wild state.

The vicuña and the guanaco are animals of dry plains (Gran Chaco plains, Patagonia), plateaus and the Andean Mountains. They are absent from hot, humid plains. It is for this reason that in equatorial regions we find them only at high altitudes. The vicuña was domesticated in Peru. The domestication of the llama and alpaca is very ancient and dates from before the Incas, approximately 4000 B.C. in the central Andes. The alpaca (*Lama guanicoë pacos*) is a cross between the vicuña and the guanaco and is exploited for its wool. The llama (*Lama guanicoë glama*) was domesticated for its meat and as a beast of burden.

Number of species: 6.
Oldest known fossils: the tylopods include the oreodontoids, the protoceratids, the oromerycids (all fossil groups) and the camelids. The oldest known fossil is a camelid, *Poebrodon*, from the middle Eocene (45 MYA) from North Amercia (Utah and California). However, a certain number of previously enigmatic fossil families are thought to be related to the tylopods: the cainotherids, anoplotherids, dacrytherids, mixtotherids and xiphodontids. In this case, the oldest known fossils in the tylopod lineage would be the dacrytherid *Cuisitherium* from France and the mixtotherid *Mixtotherium* from Spain, England and France, both dating from the lower Eocene (50 MYA).
Current distribution: the distribution is fragmented as a result of relatively recent migrations. The group appeared in North America during the Eocene. During the middle Miocene, it reached its maximum expansion. It is only during the late Miocene and Pilocene that this group moved into Europe, North Africa, and Asia and then during the Pleistocene that it reached South America. Currently, the group is absent from North America, but it remains in central Asia with the wild camels, from Turkestan to Tibet and including the Gobi Desert (Mongolia, China) The distribution of the Bactrian camel spreads from the west of the Caucasus Mountains to Turkey. The dromedary is found in North Africa, in the Sahara, on the Arabian Peninsula and in the Middle East, from Palestine and Turkey to the border of India. The guanacos and vicuñas live along the Andes mountain range, from the Peruvian Andes to Tierra del Fuego. In the past, the distribution of the South American tylopods included the large dry plains of the southern third of the continent, where these animals were hunted by humans.

Examples

Bactrian camel: *Camelus ferus*: dromedary or Arabian camel: *Camelus dromedarius;* guanaco: *Lama guanicoe glama;* vicuña: *L. vicugna.*

Unnamed clade

A few representatives

Wild boar
Sus scrofa
BL: 140 cm
node **41**

Blue wildebeest
Connochaetes taurinus
BL: 200 cm
node **43**

Hippopotamus
Hippopotamus amphibius
BL: 390 cm
node **45**

Bottlenose dolphin
Tursiops truncatus
290 cm
node **46**

Some unique derived features

■ Retrotransposed sequences: the swines, ruminants, hippopotamuses and cetaceans all possess repeated retrotransposed sequences in exactly the same place in the genome. These mobile genetic elements, called LINES (*long interspersed elements*), constitute a reliable phylogenetic marker when they are found in exactly the same position (to within one nucleotide) in the genome of different species. Indeed, the probability that these mobile genetic elements insert independently at exactly the same site twice is extremely small. This means that if two or more species have these elements in the same position, they were acquired during a single retrotransposition event that occurred in a common ancestor. The swines, ruminants, hippopotamuses and cetaceans all have repetitive ARE sequences (LINEs family) at the ino locus. Due to the mechanism of replication by which these sequences are integrated into a new locus, their presence in the genome is irreversible. This means that if the tylopods do not have this sequence, they never did, and that this insertion event took place in the common ancestor of the swines, ruminants, hippopotamuses and cetaceans.

■ Molecular phylogenies: the cetartiodactyls (excluding the tylopods) are united by the sequence structure of the mitochondrial cytochrome b gene and that of exon 7 and intron 7 of the nuclear β-casein gene. Molecular phylogenies based on the sequences of 19 nuclear genes and 3 mitochondrial genes unite the swines, ruminants, hippopotamuses and cetaceans. These genes include hemoglobin, myoglobin, alpha-crystallin, cytochrome c, and ribonuclease genes along with mitochondrial ribosomal genes. In addition, independent analyses of other nuclear (a different fragment of *RAG1*, *c-myc* oncogene, growth hormone) and mitochondrial (*ND6*) genes support this clade.

NOTE: older classifications separated the artiodactyla (terrestrial cetartiodactyls) into those with bunodont teeth (molars with rounded, hill-shaped cusps) and those with selenodont teeth (high-crowned molars with crescent-shaped cusps). The suina and hippopotamida were part of the first group, whereas the tylopoda and ruminantia were included in the second. Selenodont molars were certainly acquired several times over the course of cetartiodactyl evolution, at least once in the tylopods and a second time in the ruminants.

> **Number of species:** 290.
> **Oldest known fossils:** this lineage includes the fossil group Mesonychia. The oldest known mesonychians slightly predate the cetaceans and include *Yatanglestes* and *Dissacusium* from the lower Paleocene from China, and *Ankalagan* from the same period from New Mexico (USA) (60 MYA). If we consider only the extant groups, the oldest known fossils are the cetaceans *Ambulocetus*, *Pakicetus*, *Gandakasia* and *Ichthyolestes* from the lower Eocene from Pakistan (50 MYA). However, also dating from this period are fossil suids, *Cebochoerus* from Spain, and the raoellids *Khirtharia*, *Indohyus*, *Kunmunella* and *Raoella* from Asia.
> **Current distribution:** worldwide, except Australia and New Zealand. They have been introduced to these regions by humans. All the seas and oceans.

Examples

■ Suina: wild boar: *Sus scrofa*; babirusa: *Babyrousa babyrussa*.
■ Clade 42: lesser Malay mouse-deer: *Tragulus javanicus*; Siberian musk deer: *Moschus moschiferus*; elk: *Cervus elaphus*; giraffe: *Giraffa camelopardalis*; pronghorn: *Antilocapra americana*; Grant's gazelle: *Gazella granti*; domestic cow: *Bos taurus*; hippopotamus: *Hippopotamus amphibius*; blue whale: *Balaenoptera musculus*; sperm whale: *Physeter catodon*; killer whale: *Orcinus orca*.

Suina

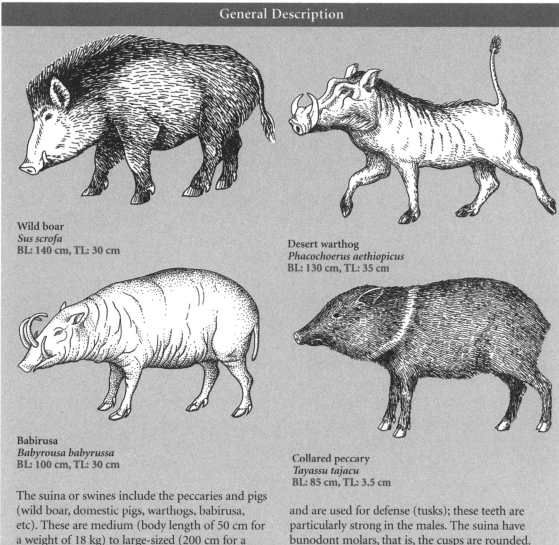

General Description

Wild boar
Sus scrofa
BL: 140 cm, TL: 30 cm

Desert warthog
Phacochoerus aethiopicus
BL: 130 cm, TL: 35 cm

Babirusa
Babyrousa babyrussa
BL: 100 cm, TL: 30 cm

Collared peccary
Tayassu tajacu
BL: 85 cm, TL: 3.5 cm

The suina or swines include the peccaries and pigs (wild boar, domestic pigs, warthogs, babirusa, etc.). These are medium (body length of 50 cm for a weight of 18 kg) to large-sized (200 cm for a weight of 300 kg or more) animals with either a stocky, cylindrical or barrel-shaped body, or a laterally flattened body (in peccaries). The legs, neck and tail are short. The legs are relatively slender and the neck is very thick. The head tapers into a conical, truncated muzzle with a snout and two terminal nostrils. The snout is supported by a cartilaginous disk that is characteristic of the group. The snout is hairless and mobile. The eyes are small and placed high on the sides of the head. The ear pinnae are medium-sized, pointed and sometimes hanging. In certain species, the face has large protuberances (as in the warthogs, for example). The canine teeth grow continuously and are used for defense (tusks); these teeth are particularly strong in the males. The suina have bunodont molars, that is, the cusps are rounded. In contrast to the tylopods, ruminants and hippopotamuses, the stomach is simple with a single compartment. There are 4 toes on each foot, but only the hoofs of phalanges III and IV support the body; the two other toes (II and V) only slightly touch the ground. A sustantial layer of fat lies under the thick skin. The skin has few sebaceous or sweat glands. The hairs or bristles are thick and heavy. There is often a ridge of hair along the back of the neck and back that stands erect when the animal is irritated. The color is variable, from black to grey, light brown or red. The color of the piglets differs from that of the adults; they are typically striped or spotted.

Some unique derived features

- Snout: the suina are characterized by the presence of a mobile snout at the tip of the muzzle. The snout is supported by two specialized prenasal bones attached to the nasal bones, the premaxillary bones and sometimes to the mesethmoid bone. The snout is moved by specialized muscles. Fig. 1 shows a left lateral view of the skull of a young pig and

465

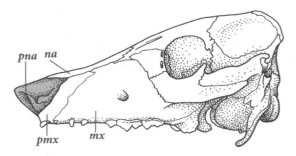

Figure 1

outlines the position of the prenasal bones (*mx*: maxillary bone, *na*: nasal bone, *pna*: prenasal bone, *pmx*: pre-maxillary bone). Fig. 2 presents an anterior view of a pig's head to show the position of the two nostrils on the snout disk.

■ Genes: the suina are also characterized by the structure of the gene sequences of the mitochondrial cytochrome b gene and by several nuclear genes: exon 1 and 2 for the protamine P1 gene, exon 7 and intron 7 of the β-casein gene, exon 4 of the κ-casein gene, exons 2 and 4 and introns 2 and 3 of the γ-fibrinogen gene.

■ Retrotransposed sequences: swines all possess repetitve retrotransposed sequences in exactly the same locations in the genome (see clade 40): the tayassuidae and suidae all have repetitive ARE sequences (LINEs family) in the gpi and pro sites in the genome.

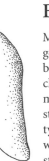

Figure 2

Ecology

Most species of swine are generalists and live in a variety of biotopes with a wide diversity of climates: plains, cultivated zones, marshes, alpine zones, forests, steppes, savannas. However, they typically need to be close to a water source (although this hasn't stopped the peccaries from living in semi-desert regions of Mexico and Texas). Although these animals are omnivores, the major proportion of food consumed is vegetative. They are social animals that live in pairs or small groups. Older males tend to be solitary outside the period of the rut. The quest for food can cause them to regroup into larger herds. They communicate using a diverse range of sounds and recognize one another by smell. The sense of sight is poorly developed.

The males can have spectacular combats during the period of the rut: they butt each other with their heads, snouts, and tusks. Before mating, the male will usually massage the female with his snout. Gestation lasts 4 to 5 months. The female digs a shallow cavity in the ground that she covers with vegetation to form a nest. A litter ranges from 2 to 14 young, born fully formed and able to walk very early. There is normally one litter per year, but two is possible if food resources are abundant. Warthogs sometimes protect their young in lairs abandoned by other animals (aardvark, for example). Our domestic pig orginates from none other than the Eurasian wild boar and was domesticated by the Thessalians around 9000 B.C.

Number of species: 19.
Oldest known fossils: the suina lineage includes the extant families tayassuidae (single genus *Tayassu*) and suidae (includes all other extant genera), and several fossil families. The oldest known swine fossils date from the lower Eocene (50 MYA) and include the cebochoerid *Cebochoerus* from Spain, and the raoellids *Khirtharia*, *Indohyus*, *Kunmunella* and *Raoella* from Asia. The oldest fossils among the extant families are the tayassuids *Perchoerus* and *Dollochoerus* from the upper Eocene (35 MYA) from South Dakota (USA) and France, respectively.
Current distribution: Central America, South America (except the Andes and the plains of Argentina), sub-Saharan Africa, Eurasia, Sunda Islands, Java, Sumatra, Borneo, Philippines, Sulawesi. Wild boars were introduced to North and South America, Australia, Tasmania, and New Zealand by humans. The domestic pig has been introduced worldwide.

Examples

Red river hog: *Potamochoerus porcus*; wild boar or domestic pig: *Sus scrofa*; Java pig: *S. verrucosus*; bearded pig: *S. barbatus*; desert warthog: *Phacochoerus aethiopicus*; giant forest hog: *Hylochoerus meinertzhageni*; babirusa: *Babyrousa babyrussa*; collared peccary: *Tayassu tajacu*; white-lipped peccary: *T. albirostris*.

Unnamed clade

A few representatives

Okapi
Okapia johnstoni
BL: 210 cm
node 43

Blue wildebeest
Connochaetes taurinus
BL: 200 cm
node 43

Hippopotamus
Hippopotamus amphibius
BL: 390 cm
node 45

Bottlenose dolphin
Tursiops truncatus
290 cm
node 46

Some unique derived features

- Nuclear gene sequences: this clade brings together the ruminants, hippopotamuses and cetaceans and is founded on the sequence structure of several nuclear genes: intron 7 and exon 7 of the β-casein gene, exon 4 of κ-casein, exons 2 and 4 and introns 2 and 3 of the γ-fibrinogen, and α-hemoglobin. In addition, molecular phylogenies based on the sequences of 19 nuclear genes and 3 mitochondrial genes bring together the ruminants, hippopotamuses and cetaceans. These genes include hemoglobin, myoglobin, alpha-cristallin, cytochrome c, and ribonuclease genes along with mitochondrial ribosomal genes. In addition, independent analyses of other nuclear (a different fragment of *RAG1*, *c-myc* oncogene, growth hormone) and mitochondrial (*ND6*) genes support this clade.

- Retrotransposed sequences: the ruminants, hippopotamuses and cetaceans all possess short retrotransposed sequences in exactly the same places in the genome. These mobile genetic elements, called SINEs (*short interspersed elements*) or LINEs (*long interspersed elements*), constitute reliable phylogenetic markers when they are found in exactly the same position (to within one nucleotide) in the genome of different species (see clade 40). The ruminants, hippopotamuses and cetaceans all have SINEs of the CHR1 family at four specific sites in the genome: aaa228, aaa792, Gm5, and HIP5.

Number of species: 271.
Oldest known fossils: this lineage includes the fossil group Mesonychia. The oldest known mesonychians slightly predate the cetaceans and include *Yatanglestes* and *Dissacusium* from the lower Paleocene from China, and *Ankalagan* from the same period from New Mexico (USA) (60 MYA). If we consider only the extant groups, the oldest known fossils are the cetaceans *Ambulocetus*, *Pakicetus*, *Gandakasia* and *Ichthyolestes* from the lower Eocene from Pakistan (50 MYA).
Current distribution: worldwide, except Australia and New Zealand. They have been introduced to these regions by humans. All the seas and oceans.

Examples

- RUMINANTIA: lesser Malay mouse-deer: *Tragulus javanicus*; Siberian musk deer: *Moschus moschiferus*; elk: *Cervus elaphus*; giraffe: *Giraffa camelopardalis*; pronghorn: *Antilocapra americana*; Grant's gazelle: *Gazella granti*; domestic cow: *Bos taurus*.
- CLADE 44: hippopotamus: *Hipopotamus amphibius*; blue whale: *Balaenoptera musculus*; sperm whale: *Physeter catodon*; killer whale: *Orcinus orca*.

467

Ruminantia

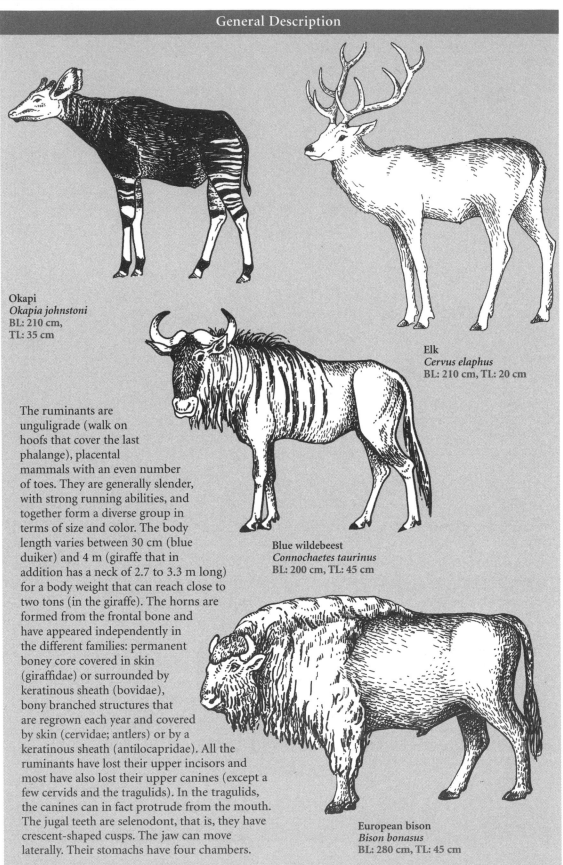

General Description

Okapi
Okapia johnstoni
BL: 210 cm,
TL: 35 cm

Elk
Cervus elaphus
BL: 210 cm, TL: 20 cm

Blue wildebeest
Connochaetes taurinus
BL: 200 cm, TL: 45 cm

European bison
Bison bonasus
BL: 280 cm, TL: 45 cm

The ruminants are unguligrade (walk on hoofs that cover the last phalange), placental mammals with an even number of toes. They are generally slender, with strong running abilities, and together form a diverse group in terms of size and color. The body length varies between 30 cm (blue duiker) and 4 m (giraffe that in addition has a neck of 2.7 to 3.3 m long) for a body weight that can reach close to two tons (in the giraffe). The horns are formed from the frontal bone and have appeared independently in the different families: permanent bony core covered in skin (giraffidae) or surrounded by keratinous sheath (bovidae), bony branched structures that are regrown each year and covered by skin (cervidae; antlers) or by a keratinous sheath (antilocapridae). All the ruminants have lost their upper incisors and most have also lost their upper canines (except a few cervids and the tragulids). In the tragulids, the canines can in fact protrude from the mouth. The jugal teeth are selenodont, that is, they have crescent-shaped cusps. The jaw can move laterally. Their stomachs have four chambers.

Some unique derived features

■ The ruminants are characterized by a stomach with four specialized compartments where anaerobic fermentation of cellulose occurs via a large number of resident unicellular organisms. The four chambers (fig. 1; *du*: duodenum, *esg*: esophagal groove, *es*: esophagus) have the following structure: the rumen (*rum*) with a mucosa covered in papillae; the reticulum (*ret*) with a lining containing a network of polygon-shaped muscular pockets; the omasum (*omas*), which is elongated with a hundred or so longitudinal folds in the mucosa; and finally the abomasum (*abo*), or true stomach, where gastric secretions prepare the food for passage into the small intestine (the other three compartments are in fact derived from the esophagus). In other herbivores capable of fermentation, the mucosae of the chambers do not have the same structure (see the tylopods, clade 39, for example).

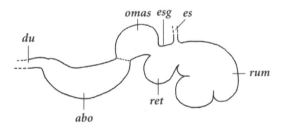

Figure 1

■ Rumination is a form of delayed mastication, that is, the animals regurgiate their food to chew it a second time. Fig. 2 shows the trajectory of a food bolus in the ruminant stomach. A mouth-full of poorly chewed vegetation (1) enters into the rumen (2) and into part of the reticulum (3), where it is partially broken down by the action of the resident bacteria. Indeed, as these animals do not have the enzymes required to break down cellulose, they rely on anaerobic fermentation carried out in the rumen by different bacteria, notably *Clostridium* and *Ruminococcus*. The food, or cud, is then regurgitated into the mouth for ruminating (4), that is, the cud is chewed again and mixed with saliva. It is then swallowed a second time and sent to the other two chambers, the omasum (5), where it is pressed and where water is removed, and the abomasum (6), where true digestion occurs by the release of gastic juices. As soon as fermentation of the food starts in the rumen, the animal recovers the products of the cellulose degradation in the blood, notably short-chained fatty acids that, when combined with bases, cross the rumen mucosa and move

into the veins. Water is recovered in the omasum. The bacteria and the remains of the cud are digested in the final compartments (abomasum and intestine, 6, 7) by proteolytic gastric enzymes. The liberated proteins are absorbed by the blood during passage through the very long intestine (7). Rumination is thought not only to be advantageous in terms of increased digestion but also to limit the amount of time that the animal spends feeding. During feeding the animal is more vulnerable to attack by predators. By delaying mastication, food can be fully processed under safer conditions. Rumination has evolved several times in the mammals: in the kangaroos, the ruminants and the tylopods. Bacterial fermentation of cellulose is used even more frequently by herbivorous mammals: caecal fermentation is found in the lagomorphs, hyracoids, and perissodactyls; stomacal fermentation appeared independently in the colobus monkeys (primates), the ruminants, the tylopods and the hippopotamuses.

■ Retrotransposed sequences: the ruminants have short retrotransposed elements (SINEs) of the CHR1 family in two specific sites in the genome: c21-352 and Pgha (see clade 42).

■ The monophyly of the ruminants is supported by the sequence structure of several nuclear genes: intron 7 and exon 7 of the β-casein gene, intron 1 and exons 1and 2 of the protamine P1 gene, exon 4 of κ-casein, exons 2 and 4 and introns 2 and 3 of the γ-fibrinogen gene.

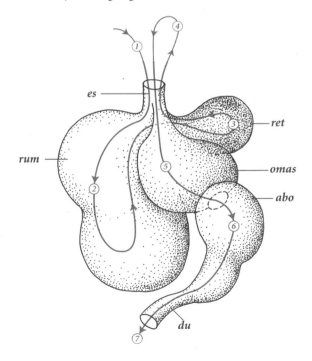

Figure 2

Ecology

The ruminants are all terrestrial and live in dense forests, prairies and savannas. They are herbivores, feeding on leaves, herbaceous plants and grasses. As primary consumers, they play a major role in the terrestrial trophic food chain. In many biotopes they are prey to diverse carnivores. This has selected for certain vigilance behaviors and the speed that is characteristic of all small and medium-sized ruminants. The young are born fully formed and are quickly ready to run. The horns are used as weapons, but more often in combats with other males than for defense against predators. They are considered to be structures that enable access to females, to a high social rank, and/or to the appropriation of a territory. Indeed, the ruminants are most often social animals that live in female herds with a dominant male. Bovines, ovines and caprines are of considerable economic importance because they have been the principal source of animal protein consumed by humans since domestication, more than 6000 years B.C. Indeed, around this date, several domestications occurred: the cow (*Bos primigenius*) in eastern Europe and the Middle East, the zebus (*Bos indicus*) in Syria and Pakistan, the sheep (*Ovis aries*) in the Middle East, and the goat (*Capra hircus*) in Iran. The reindeer (*Rangifer tarandus*) was domesticated around 3000 B.C. in northern Eurasia and since has served as the principal resource of humans living above the Arctic Circle.

Number of species: 191.
Oldest known fossils: most of the modern families date from the lower Miocene (23 MYA). The oldest known ruminant is the leptomerycid *Hendryomeryx* from the middle Eocene from Wyominig and Texas (USA) (40 MYA); the leptomerycidae are a fossil family of the paraphyletic *Tragulina* group, a group which includes extant tragulids. The amphimerycid *Amphimeryx* from Montmartre (France) dates from at least this period and may be even older (50 MYA). As the age of this fossil is questionable, we should also mention the members of another fossil family, the hypertragulids *Hypertragulus* and *Parvitragulus* from North America and *Miomeryx* from Asia, all dating from the upper Eocene (35 MYA).
Current distribution: worldwide, except Australia, Tasmania and New Zealand. They have been introduced to these regions by humans.

Examples

Lesser Malay mouse-deer: *Tragulus javanicus;* Siberian musk deer: *Moschus moschiferus;* elk: *Cervus elaphus;* giraffe: *Giraffa camelopardalis;* pronghorn: *Antilocapra americana;* blue duiker: *Cephalophus monticola;* royal antelope: *Neotragus pygmaeus;* greater kudu: *Tragelaphus strepsiceros;* yak: *Bos mutus;* domestic cow: *Bos taurus;* blue wildebeest: *Connochaetes taurinus;* sable antelope: *Hippotragus niger;* Grant's gazelle: *Gazella granti;* waterbuck: *Kobus ellipsiprymnus;* saiga antelope: *Saiga tatarica;* alpine ibex: *Capra ibex;* argali: *Ovis ammon;* musk ox: *Ovibos moschatus.*

Unnamed clade

A few representatives

Sperm whale
Physeter catodon
1400 cm
node 46

Hippopotamus
Hippopotamus amphibius
BL: 390 cm
node 45

Bottlenose dolphin
Tursiops truncatus
290 cm
node 46

Humpback whale
Megaptera novaeangliae
1180 cm
node 46

Some unique derived features

■ Retrotransposed sequences: the hippopotamuses and cetaceans have short retrotransposed elements (SINEs) of the CHR1 family in four specific sites in the genome: HIP24, KM14, HIP4, and AF (see clade 42).

■ Nuclear gene sequences: among extant fauna, the exclusive relatedness of the hippopotamuses and cetaceans is shown by the sequence structure of several nuclear genes: exon 7 of the β-casein gene, exon 4 of κ-casein, exons 2 and 4 and introns 2 and 3 of the γ-fibrinogen gene. In addition,

molecular phylogenies based on the sequences of 19 nuclear genes and 3 mitochondrial genes bring together the hippopotamuses and cetaceans. These genes include hemoglobin, myoglobin, alpha-crystallin, cytochrome c, and ribonuclease genes along with mitochondrial ribosomal genes. Furthermore, the independent analyses of other nuclear (a different fragment of *RAG1*, *c-myc* oncogene, growth hormone) and mitochondrial (*ND6*) genes also support this clade.

Number of species: 80.
Oldest known fossils: this lineage includes the fossil group Mesonychia. The oldest known mesonychians slightly predate the cetaceans and include *Yatanglestes* and *Dissacusium* from the lower Paleocene from China, and *Ankalagan* from the same period from New Mexico (USA) (60 MYA). If we consider only the extant groups, the oldest known fossils are the cetaceans *Ambulocetus*, *Pakicetus*, *Gandakasia* and *Ichthyolestes* from the lower Eocene from Pakistan (50 MYA).
Current distribution: intertropical sub-Saharan Africa. All the seas and oceans. Brackish estuary waters, but also certain rivers and lakes of Southeast Asia, tropical South America, northern North America, and Eurasia.

Examples

■ HIPPOPOTAMIDAE: hippopotamus: *Hipopotamus amphibius.*

■ CETACEA: bowhead whale: *Balaena mysticetus;* gray whale: *Eschrichtius robustus;* blue whale: *Balaenoptera musculus;* sperm whale: *Physeter catodon;* Cuvier's beaked whale: *Ziphius cavirostris;* Ganges river dolphin: *Platanista gangetica;* boto: *Inia geoffrensis;* La Plata dolphin: *Pontoporia blainvillei;* narwhal: *Monodon monoceros;* harbor porpoise: *Phocaena phocaena;* rough-toothed dolphin: *Steno bredanensis;* bottlenose dolphin: *Tursiops truncatus;* killer whale: *Orcinus orca.*

Hippopotamidae

General Description

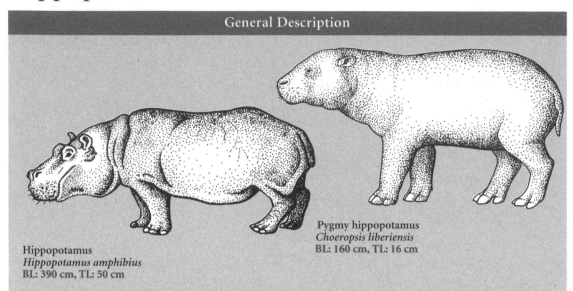

Hippopotamus
Hippopotamus amphibius
BL: 390 cm, TL: 50 cm

Pygmy hippopotamus
Choeropsis liberiensis
BL: 160 cm, TL: 16 cm

The hippopotamuses are heavy, thick animals with cylindrical bodies and barrel-shaped chests. The total body length varies from 150 cm for the pygmy hippopotamus to 450 cm for the hippopotamus, for a weight of 180 kg to 3200 kg respectively. The legs are pillar-like, and the neck and tail are short. The massive head carries a large mouth that can open very widely, small eyes and small ears. The sense organs (eyes, nostrils, ears) are aligned along the top of the head, a characteristic of amphibious tetrapods like the crocodiles. However, this pattern is weaker in the pygmy hippopotamuses. The nostrils can be closed by contraction. The canines are tusk-like and are particularly strong on the lower jaw. The canines and incisors grow continuously throughout the life of the animal. The stomach is large and divided into different compartments. It is the site of bacterial anaerobic fermentation, an arrangment also found independently in the ruminants and tylopods. Two diverticula flank the upper area of the stomach (caecums), followed by a compartmentized cylindrical middle region, and a bulgy lower region containing a highly folded mucosa. The lungs are long: a hippopotamus can hold its breath more than 5 minutes. There are four webbed toes on each foot. The skin is thick, rich in mucous glands and almost completely lacking hair. The little hair present is concentrated on the upper lip and the tip of the tail.
These animals are brownish in color, blackish-brown in the pygmy hippopotamus and slate brown in the hippopotamus with lighter more purple tones on the ventral surface and around the eyes and ears. The young are grayish-pink. The mucous glands secrete a thick reddish liquid that protects the skin against water and dehydration in the open air.

Some unique derived features

- The thorax is very long, enclosing not only the lungs but also almost all of the abdominal viscera. This is visible on the left lateral view of the hippopotamus skeleton shown in fig. 1.

- Incisors: the hippopotamuses have powerful, conical incisors and enormous canines that are triangular in cross-section (*C*, fig. 1). Both types of teeth grow

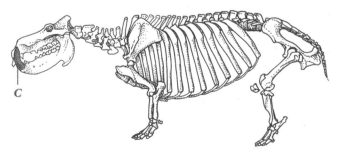

Figure 1

continuously throughout the lifetime of the individual.

- The muzzle is enlarged and the sense organs (nostrils, eyes, ears) lie high on the head (fig. 2: hippopotamus).

- Retrotransposed sequences: the hippopotamuses have a short retrotransposed element (SINEs) of the CHR2 family at a specific site in the genome called "HIP5" (see clade 42).

Figure 2

Ecology

The hippopotamuses inhabit lakes and rivers adjacent to grasslands. They are absent from the very arid desert regions of Namibia and South Africa. They are herbivores that consume grasses, aquatic shoreline plants, foliage and fruits. They feed in the grasslands during the night and return to the water before sunrise. Most of the day is spent in the water. In contrast to the pygmy hippopotamus, the hippopotamus is a social animal living in open habitats in groups of ten or so individuals. Males engage in violent combats during which they can inflict and incur serious injuries. These fights are undertaken in order to establish a harem, but also to retain the best resting places. The frequency of these combats increases with the density of individuals. Males mark their territory in the water and on the grasslands by scattering their excrement around using their tail. Mating takes place in the water. After 233 days of gestation, the birth of the single young takes place on land in a quiet spot prepared by the female. The young is nursed for 8 months. Females form "creches" where the young are protected when the mother is absent. Adult males remain close, each one dominating one area of the creche. Adult animals have almost no natural enemies but must defend their young against lions and crocodiles. The life span is approximately 50 years.

The pygmy hippopotamus lives a solitary existence in the marshes and bush surrounding freshwater bodies in undisturbed forests. Here, they defend a small territory against conspecifics. They are active at night. The principal enemy of this species is the panther. Gestation lasts 207 days and a single young is produced. The life span is around 30 years.

Number of species: 2.
Oldest known fossils: the hippopotamid lineage includes the fossil family Anthracotheridae. The oldest fossils of the lineage are several anthracotherid species from the middle Eocene (43 MYA): *Anthracokeryx* from Pakistan, *Slamotherium* from Thailand, *Probrachyodus, Brachyodus, Bakalovia,* and *Anthracosenex* from China, *Prominatherium* from the Balkans, *Ulausuodon* from Mongolia. The oldest hippopotamid *sensu stricto* is *Kenyapotamus* from Kenya and dating from the Miocene (15 MYA).
Current distribution: sub-Saharan and intertropical Africa.

Examples

Hippopotamus: *Hippopotamus amphibius;* pygmy hippopotamus: *Choeropsis liberiensis.*

Cetacea

General Description

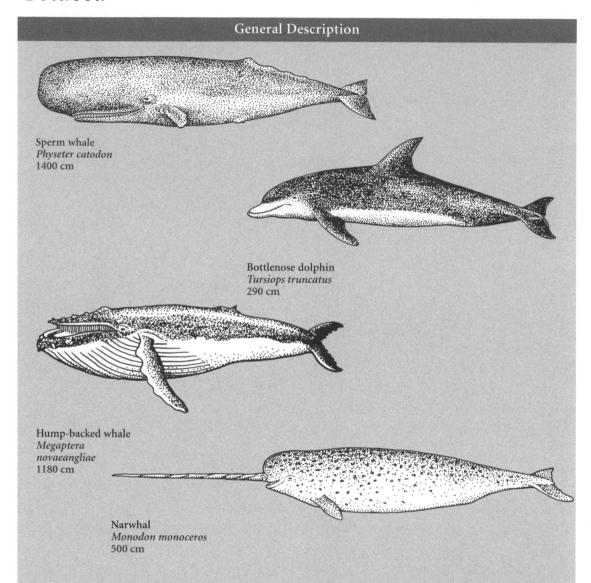

Sperm whale
Physeter catodon
1400 cm

Bottlenose dolphin
Tursiops truncatus
290 cm

Hump-backed whale
*Megaptera
novaeangliae*
1180 cm

Narwhal
Monodon monoceros
500 cm

The cetaceans are large placental mammals (1.25 to 33 m long and weighing 23 kg to 136 t) that differ greatly from other mammals by their entirely aquatic life style. Their "fish-like" appearance has accompanied a great elongation of the face, a posteriorly postioned cranium, a shortening of the neck, the transformation of the anterior members into swimming limbs often with greatly elongated phalanges, a vestigial pelvic region and vestigial or absent posterior limbs, and the development of a very powerful and horizontally enlarged tail. The lobes of the tail lack an underlyinig skeleton, as does the fold of conjunctive tissue that forms the dorsal fin. There is no organ (except maybe the heart) that has not been modified in form and function by the pelagic way of life. Each mammary gland has its own teat hidden within a pocket and the penis, lodgcd within a slit of the ventral surface, is exposed only to urinate and for mating. Given their fusiform shape, the body surface is relatively small for their size. Indeed, the skin of an elephant the size of a blue whale would cover 440 m², whereas that of the blue whale covers only half of this area. This morphology provides a hydrodynamic form and diminishes the surface area for temperature exchange with the sea water. Indeed, these homotherms tolerate a temperature difference with the surrounding water of 15 to 35°C. This situation is not comparable to that of a terrestrial animal as water conducts heat better than air. The outer keratinized layer, the epidermis and the dermis are all thin, but a thick sub-cutaneous layer of blubber serves as thermal isolation.

Hair is rare or totally absent: a few sensory hairs are still present on the muzzle of whales, and a hair is sometimes also present in the auditory canal. Integument glands are absent. In the pelvic region, only two small bones remain; these bones are no longer attached to the vertebral column, but are simply embedded in the muscle masses. Interestingly, a vestigial femur, a small bone connected to the vestige of the pelvis, has persisted in the balenopteridae (rorquals). In the right whale, another small bone, a vestigial tibia, extends from this femur. None of these vestigial parts are apparent externally. The skeleton and body form a rigid unit and only the anterior limbs and the powerful tail act as propulsory organs.

Classically, we divide this group into the toothed whales (Odontoceti) and the baleen whales (Mysticeti). The toothed whales have a homodont dentation (all teeth are identical) made up of milk teeth. The teeth are conical with a single root. They are used to capture prey, but not to chew them. Young whales are born without teeth. The baleen (or whalebone) is not a tooth formation, but rather continuously growing keratinized extensions of the epidermis. The internal edges are shredded and the extremities have long, stiff hairs. These structures lie transversally to the maxillary bones and are arranged in intervals of 6 to 10 mm. They are used to trap or filter plankton. The lower maxillary is very long and lacks an ascending process. The upper maxillary is elongated and is pushed upwards to lie above the frontal bone, the normal position for the nasal bone. Indeed, the nares, or blowholes, are found high on the head, with two in the mysticetes and one in the odontocetes. Cetaceans never inhale or exhale air by their mouths but rather by the blowhole, which can be closed off during immersion. The eyes are small and short-sighted outside the water. The ear pinnae have disappeared, the middle ear has been highly modified and the inner ear is not part of the skull wall. The petrotympanic bone is attached to the skull only via ligaments. In contrast to the rest of the skeleton, which is relatively soft, this bone is compact and very hard and remains long after the decomposition of the animal. Generally, it is the only thing that remains of whale carcasses found on the bottom of the sea.

Some unique derived features

■ The nares (blowhole, *bh*) lie posteriorly on the skull. Their position is shown on a left lateral view of the anterior end of the skeleton of *Eubalaena australis* (*car*: carpus, *ul*: ulna, *fr*: frontal, *h*: humerus, *j*: jugal, *mx*: maxillary, *na*: nasal, *or*: orbit, *pa*: parietal, *pmx*: premaxillary, *r*: radius, *sc*: scapula, *soc*: supraoccipital, *sq*: squamosal).

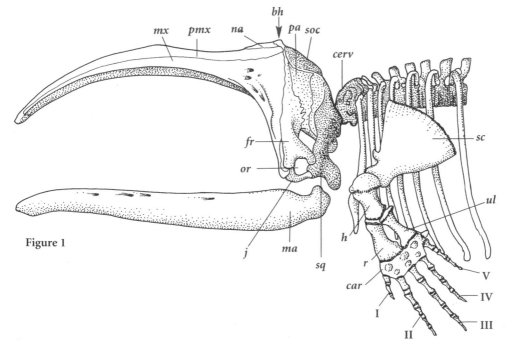

Figure 1

- The mandible (fig. 1; *ma*) has no ascending process.

- The cervical vertebrae are highly compressed (fig. 1; *cerv*).

- The anterior limbs have been transformed into swimming fins: the carpus and hand lie on the same plane. The humerus is short, and the radius and ulna are flat. The swimming fin is elongated by the addition of extra phalanges to digits II and III; we can sometimes count up to 14 phalanges. Fig. 2 shows the pectoral girdle and the left forelimb of a dolphin, where the extension of digits II and III by hyperphalangy is notable. This extension is even more pronounced on the forelimb of the killer whale (fig. 3a). In contrast, the fin whale's fin (fig. 3b) is elongated by the extension of the flattened radius and ulna (zeugopod) in addition to the hyperphalangy.

- The posterior limbs have disappeared.

- All integument glands have disappeared, except the mammary glands.

- The tail is enlarged to form a horizontal, bilobed caudal fin. The two lobes have no skeletal support.

- Retrotransposed sequences: the cetaceans have short retrotransposed elements (SINEs) of the

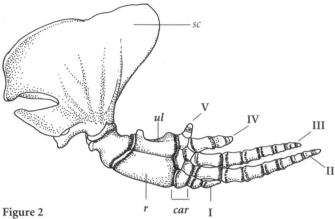

Figure 2

CHR2 family at three sites in the genome: Pm52, Pm72, and M11 (see clade 42).

- The monophyly of the cetaceans is also supported by the structure of mitochondrial gene sequences, such as that of the cytochrome b gene, the 12S and 16S ribosomal RNA sequences, and several nuclear gene sequences: intron 7 and exon 7 of the β-casein gene, exon 4 of κ-casein gene, exons 2 and 4 and introns 2 and 3 of the γ-fibrinogen gene, exons 1 and 2 of the protamine P1 gene, and α-hemoglobin gene.

Ecology

The cetaceans are exclusively marine or freshwater mammals. Some odontocetes (the sperm whale, for example) can dive to depths of up to 1000 m. The toothed whales feed on fish, cephalopods, crustaceans and even other mammals or seabirds. For example, a killer whale eats penguins, seals and harbor porpoise. Baleen whales feed only on plankton. Most cetaceans are social animals, but the number and composition of the groups vary with the species; from two breeding pairs and their young in the rorquals, to several thousand individuals in the dolphins. The pygmy sperm whale and the Ganges river dolphin live a solitary existence. The cetaceans are intelligent animals with sonar communication and a capacity to learn complex behaviors. They are able to emit a wide range of sounds. For communication, they use low- and medium-frequency sounds. In addition, the odontocetes use echolocation, that is, they are able to locate their prey using the echo from the high-frequency sounds that they emit. Spallanzani's famous experiment with bats was repeated with a bottlenose dolphin. This scholar hung bells from strings in a darkened room and released some bats; not a single bell rang. To test the dolphin, metal rods were suspended in the water; the dolphin was able to swim without touching them by emitting high-frequency clicks. The Ganges river dolphin is

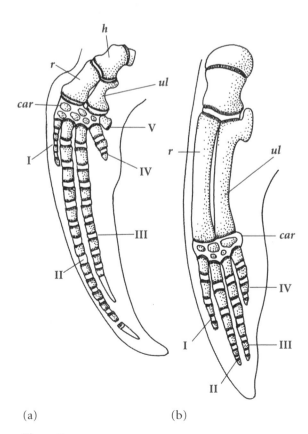

(a)   (b)

Figure 3

practically blind but can ably navigate in turbid waters and capture prey by emitting ultrasounds. The mysticetes emit much longer, low-frequency sounds. The humpback whale "sings" during the breeding season using a variety of modulated sounds.

After a nuptial parade, mating takes place in the water by either vertical (humpback whale, blue whale, sperm whale), or horizontal (dolphin, gray whale, sperm whale) copulation. Gestation lasts from nine months (dolphins) to 17 months (sperm whales). A female will produce only a single young during each reproductive cycle. Birth occurs close to the sea surface and the young is born tail first. It is then brought immediately to the surface to breath. The young will remain close to its mother for a long time, nursing for an average of one year in the odontocetes. Milk is actively expressed by the contraction of a localized muscle. As the cetaceans must go to the water's surface in order to breath, one may wonder how they sleep. In fact, it is thought that sleep is unihemispheric, that is, only one hemisphere of the brain sleeps at a time so that the animal is never completely asleep. Whales are thought to sleep an average of eight hours per day. The need to move continuously in the pelagic environment may explain the complete absence of REM sleep (rapid eye movement) in the cetaceans (sleep accompanied by a reduction in muscular function). The mysticetes carry out migrations for reproduction. They spend time in Arctic and Antarctic waters, where they feed on krill (*Euphausia*), and then move to tropical waters for giving birth. It is thought that the young, born without a blubber layer, could not tolerate the temperatures of the cold seas. After a few months of nursing the thick maternal milk, the young has built up a fat layer that helps it to face the lower temperatures. During the migration and their stay in the tropical waters, the mysticetes do not feed.

Around 1500 B.C., the inhabitants of Alaska already hunted whales. In Europe, written traces of whale hunts start in 890; the fat was the principal goal of these hunts. In the twentith century, and despite intensive hunting that brought many species to the brink of extinction, the contribution of cetaceans as a human food source has remained minimal. The products that we derive from these animals are instead rather superfluous (food for domestic animals, fertilizers, soaps, synthetic resins, glues, gelatins, etc.). The drastic reduction in population sizes has led to the strict regulation of whale hunting that we see today. The cetaceans have been domesticated very recently as circus animals (killer whales, beluga whales, dolphins) or military service animals (notably, the bottlenose dolphin, *Tursiops truncatus*).

Number of species: 78.
Oldest known fossils: the oldest known cetacean fossils in the strictest sense are the *Ambulocetus*, *Pakicetus*, *Gandakasia* and *Ichthyolestes* from the lower Eocene from Pakistan (50 MYA). The sister lineage to the hippopotamuses also includes the fossil group Mesonychia. The oldest known mesonychians slightly predate the cetaceans and include *Yatanglestes* and *Dissacusium* from the lower Paleocene from China and *Ankalagan* from the same period from New Mexico (USA) (60 MYA).
Current distribution: the cetaceans are present in all the world's seas and oceans. They live in brackish estuary waters but also certain rivers and lakes of Southeast Asia, tropical South America, northern North America, and Eurasia.

Examples

- MYSTICETI: bowhead whale: *Balaena mysticetus*; gray whale: *Eschrichtius robustus*; blue whale: *Balaenoptera musculus*.
- ODONTOCETI: sperm whale: *Physeter catodon*; Cuvier's beaked whale: *Ziphius cavirostris*; Ganges river dolphin: *Platanista gangetica*; boto: *Inia geoffrensis*; La Plata dolphin: *Pontoporia blainvillei*; narwhal: *Monodon monoceros*; harbor porpoise: *Phocaena phocaena*; short-beaked dolphin: *Delphinus delphis*; rough-toothed dolphin: *Steno bredanensis*; bottlenose dolphin: *Tursiops truncatus*; killer whale: *Orcinus orca*.

Chapter 14

Primates

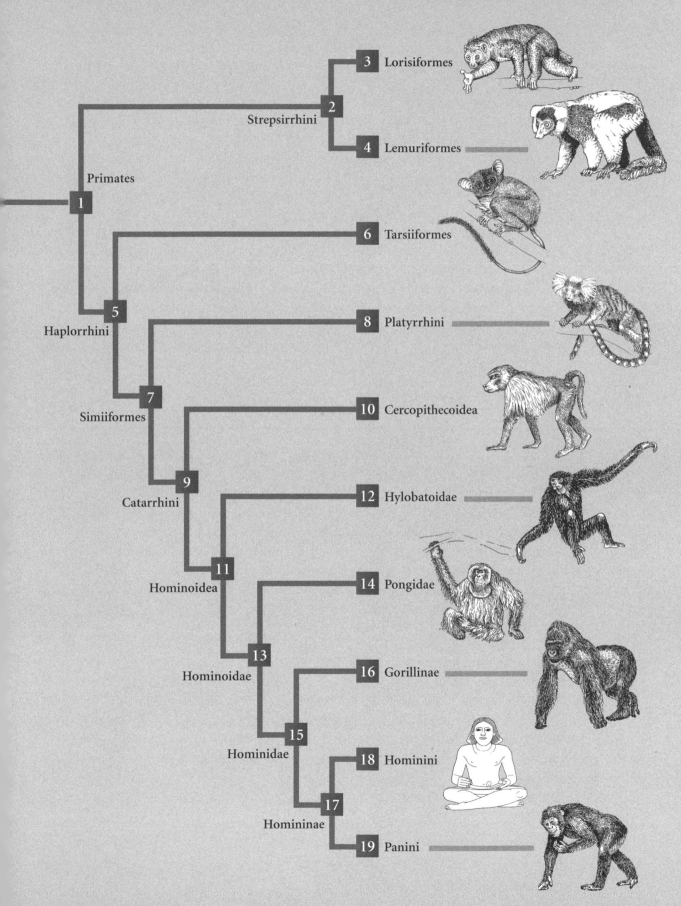

			3 Lorisiformes
		2 Strepsirrhini	
			4 Lemuriformes
1 Primates			
			6 Tarsiiformes
	5 Haplorrhini		8 Platyrrhini
		7 Simiiformes	
			10 Cercopithecoidea
		9 Catarrhini	12 Hylobatoidae
		11 Hominoidea	14 Pongidae
		13 Hominoidae	16 Gorillinae
		15 Hominidae	18 Hominini
		17 Homininae	19 Panini

Primate phylogeny has been widely studied, but within an arena of heated methodological disputes. Adopting a cladistic approach, that is, a classification based on sister-group relationships without considering grades, may seem easy when one works with green algae. Indeed, grades are readily forgotten when they are not associated with any anthropocentric ideas. This is not the case when one works with primates. It is among primatologists that we find the strongest resistance to strictly phylogenetic classifications and a tendency sometimes to rely on theological rather than scientific classifications, notably those inspired by teilhardian paleontology. It is still convenient for many well-known primatologists and paleoanthropologists to place our species in a group apart, a group whose only effect is to mask its true relationships to its sister species. For example, the pongidae, in the classic sense, is a grade that includes the orangutans, gorillas, and chimpanzees. Humans are part of a different group, despite their sister-group relationship to the chimpanzees. An objective classification would require that humans (*Homo sapiens*) and chimpanzees (*Pan troglodytes* and *P. paniscus*) be placed in the same family (the hominidae family), or even in the same subfamily (homininae) if gorillas are included in the hominidae. Given the genetic similarity of the two, both M. Goodman and E. Watson have even suggested placing humans and chimpanzees in the same genus, *Homo*. This would change the scientific name of the chimpanzee to *Homo troglodytes,* a proposition which would certainly not be widely accepted! One of our colleagues jokingly remarked that it would be much more amusing to change the scientific name of humans to *Pan sapiens*!

Another example of the persistence of grades in the primates is found in the prosimians, the non-simiiforme primates, which include the lemurs, lorisiformes and tarsiers. The tarsiers are in reality more closely related to the simian monkeys than to the lemurs. Nonetheless, the term *prosimian* continues to be used, a consequence not only of conservatism but also of pure methological error. This is the case for several French studies of primate cytogenetics. An impressive amount of data on the chromosome structure of this group has been gathered. The trees produced from the comparison of these structures actually agree with the one presented here, but only if we take into account the fact that they are incorrectly rooted. The tree topologies are the same, but the chromosomal tree is not presented from the right perspective. Indeed, the chromosomal tree is rooted on the simiiforme node, giving the impression that the prosimians are monophyletic. This is due to the methods used by the authors; rather than using only synapomorphies, they also include symplesiomorphies and do not use an outgroup. This impressive chromosomal data set therefore merits reanalysis using a true cladistic approach.

The primate tree presented in this book is robust; it is supported by numerous types of data. The one exception concerns the sister group of humans. Different results have followed one after the other: the gorillas, the chimpanzees, or the clade that includes the gorillas + the chimpanzees. The chimpanzees may even be paraphyletic, with the bonobo (*Pan paniscus*) more closely related to humans than to the chimpanzee (*Pan troglodytes*). Regardless of which hypothesis is actually correct, none calls into question the paraphyly of the classic pongidae group. The analysis of several different types of sequence data seems to favor a sister-group relationship between the genera *Pan* and *Homo*.

Strepsirrhini

A few representatives

Slender loris
Loris tardigradus
24 cm
node 3

Senegal galago
Galago senegalensis
BL: 16 cm
node 3

Ruffed lemur
Lemur variegatus
BL: 55 cm
node 4

Aye-aye
Daubentonia madagascariensis
BL: 40 cm
node 4

Some unique derived features

■ The front teeth form a dental comb (fig. 1), made up of four incisors (*I*) and two forward-angled canines (*C*). This comb is used to rip off vegetative material (e.g., tree gum) for feeding and to remove ectoparasites when grooming.

■ All euarchontoglires have a hemochorial placenta, except the scandentians and the strepsirrhines. The strepsirrhines have an epitheliochorial placenta, a rare placental type for eutherian mammals. In this type of placenta (fig. 2a), the uterine epithelium is present and the uterine capillaries do not directly penetrate the placental tissues. In a hemochorial placenta (fig. 2b), the uterine epithelium disappears and the uterine capillaries spread and form maternal blood lacunae that mix with the fetal tissue (with the chorionic epithelium).

NOTE: An epitheliochorial placenta is found by convergence in the fereungulata, a group that includes the cetartiodactyla, perissodactyla, pholidota and the carnivora (some members of the latter group have secondarily acquired another placental mode).

Number of species: 32.
Oldest known fossils: the fossil family adapidae is included in the strepsirrhini and is recorded from the Ypresian period (55 MYA) with *Donrussella* (France) and *Cantius* (Belgium, France, and Wyoming, USA).
Current distribution: sub-Saharan Africa (except South Africa), Madagascar, India, the mainland and islands of Southeast Asia.

Examples

LORISIFORMES: slender loris: *Loris tardigradus;* potto: *Perodicticus potto;* greater galago or bushbaby: *Galago crassicaudatus.*
LEMURIFORMES: ring-tailed lemur: *Lemur catta;* bamboo lemur: *Hapalemur griseus;* greater dwarf lemur: *Cheirogaleus major;* indri: *Indri indri;* aye-aye: *Daubentonia madagascariensis.*

C I

Figure 1

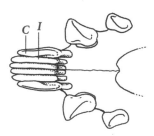

Figure 2a

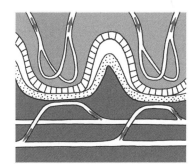

Uterine epithelium

Chorionic epithelium

Endometrium

Placenta

Figure 2b

481

Lorisiformes

General Description

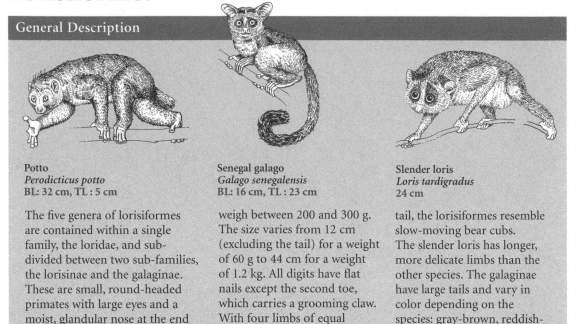

Potto
Perodicticus potto
BL: 32 cm, TL : 5 cm

Senegal galago
Galago senegalensis
BL: 16 cm, TL : 23 cm

Slender loris
Loris tardigradus
24 cm

The five genera of lorisiformes are contained within a single family, the loridae, and sub-divided between two sub-families, the lorisinae and the galaginae. These are small, round-headed primates with large eyes and a moist, glandular nose at the end of a pointed muzzle. Most species weigh between 200 and 300 g. The size varies from 12 cm (excluding the tail) for a weight of 60 g to 44 cm for a weight of 1.2 kg. All digits have flat nails except the second toe, which carries a grooming claw. With four limbs of equal length and a relatively short tail, the lorisiformes resemble slow-moving bear cubs. The slender loris has longer, more delicate limbs than the other species. The galaginae have large tails and vary in color depending on the species: gray-brown, reddish-brown, beige.

Some unique derived features

- The ectotympanic bone is fused at the opening of the tympanic bulla, but is not elongated into an external auditory canal. The ectotympanic bone (*ect*) is a ring which supports the tympanum. The tympanic (or auditory) bulla is a boney cavity of the basicranium delimited by the petrous bone (*pet*) in the primates. Fig. 1 shows cross-sections of the tympanic bulla as seen in the eulipotyphles (a), the lemuriformes (b), the lorisiformes and the platyrrhines (c), and the tarsiiformes and catarrhines (d). The lorisiformes and the platyrrhines both display a fusion of the ectotympanic and the petrous bones, but this similarity has been acquired by convergence. This fusion is also observed in the tarsiiformes and the catarrhines (1d), but here the ectotympanic bone is elongated and transformed into an external auditory canal.

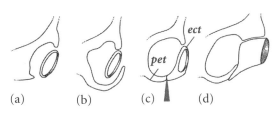

(a)   (b)   (c)   (d)

Figure 1

Ecology

The lorisiformes live in the trees of rain forests, dry deciduous forests, and savannas. They are the only strepsirrhines that are completely nocturnal. All lorids are omnivores feeding mainly on insects, fruits and tree gum. The two families can be clearly distinguished from one another by their mode of locomotion. The six galaginae species live in Africa and are agile climbers that use escape as a defense strategy; they are able to run in the trees and jump vertically from branch to branch. The four lorisinae species (2 in Asia, 2 in Africa) move slowly in a more chameleon-like manner and have lost the ability to jump. Their defense strategy is based on camouflage and immobility. The social structure varies among the different species.

Number of species: 10.
Oldest known fossils: a lorisiforme genus is recorded from Egypt and dates from the upper Eocene (37 MYA).
Current distribution: sub-Saharan Africa (except South Africa), India, the mainland and islands of Southeast Asia.

Examples

- LORISINAE: slender loris: *Loris tardigradus;* slow loris: *Nycticebus coucang;* Calabar angwantibo: *Arctocebus calabarensis;* potto: *Perodicticus potto.*
- GALAGINAE: greater galago or bushbaby: *Galago crassicaudatus;* western needle-clawed galago: *G. elegantulus.*

Lemuriformes

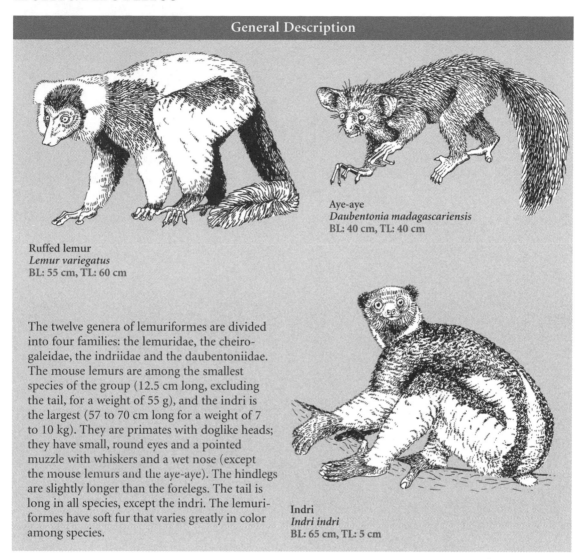

General Description

Ruffed lemur
Lemur variegatus
BL: 55 cm, TL: 60 cm

Aye-aye
Daubentonia madagascariensis
BL: 40 cm, TL: 40 cm

Indri
Indri indri
BL: 65 cm, TL: 5 cm

The twelve genera of lemuriformes are divided into four families: the lemuridae, the cheirogaleidae, the indriidae and the daubentoniidae. The mouse lemurs are among the smallest species of the group (12.5 cm long, excluding the tail, for a weight of 55 g), and the indri is the largest (57 to 70 cm long for a weight of 7 to 10 kg). They are primates with doglike heads; they have small, round eyes and a pointed muzzle with whiskers and a wet nose (except the mouse lemurs and the aye-aye). The hindlegs are slightly longer than the forelegs. The tail is long in all species, except the indri. The lemuriformes have soft fur that varies greatly in color among species.

Ecology

The lemuriformes are quadrumanous arboreal species that jump vertically from branch to branch (no brachiation). Some species may also be partially terrestrial. However, only the ring-tailed lemur spends more time on the ground than in the trees.

The lemuridae and the indriidae are diurnal vegetarians. The cheirogaleidae and daubentoniidae are nocturnal omnivores feeding on insects, fruit and leaves.

The social organization and hierachical structure varies greatly from one species to another. The aye-aye, a species currently at the brink of extinction, is without a doubt the strangest and most specialized primate. It is a solitary, nocturnal species that consumes fruit and insect larvae. It is unique among the primates in that it has powerful, forward-angled incisors, which it uses to dig into wood to uncover insect larvae. Using its very large ear pinnae, it first detects the presence of the larvae by sound. After opening the wood, it then extracts its prey using its highly adapted third finger (modified to be long and thin).

All of the lemuriforme species are threatened with extinction because of habitat destruction. Fourteen lemurid species have disappeared since humans settled in Madagascar 2000 years ago. Some of these species, the megaladapinae, were the size of orangutans.

Some unique derived features

■ The ectotympanic bone is suspended within the auditory (or tympanic) bulla. It is a ring that supports the tympanum. Fig. 1 shows cross-sections of the tympanic bulla in different groups: the eulipotyphles (a), the lemuriformes (b), the lorisiformes and the platyrrhines (c), and the tarsiiformes and catarrhines (d). The ectotympanic bone is suspended within the tympanic bulla in (b), whereas it is outside the bulla in (a) and fused to the petrous bone in (c) and (d).

Number of species: 22.
Oldest known fossils: fossils that resemble some of the indriidae date from the lower Pilocene (5.2 MYA). All the lemuriforme families are found in Madagascar in a sub-fossil state. If we accept that the fossil family adapidae should be placed among the lemuriformes, this group dates from the Ypresian period (55 MYA) with *Donrusselia* (France) and *Cantius* (Belgium, France, and Wyoming, USA).
Current distribution: Madagascar.

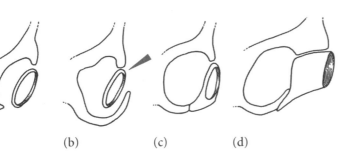

(a)   (b)   (c)   (d)

Figure 1

Examples

■ LEMURIDAE: ring-tailed lemur: *Lemur catta;* weasel sportive lemur: *Lepilemur mustelinus;* bamboo lemur: *Hapalemur griseus.*
■ CHEIROGALEIDAE: brown mouse lemur: *Microcebus rufus;* greater dwarf lemur: *Cheirogaleus major.*

■ INDRIIDAE: indri: *Indri indri;* avahi: *Avahi laniger;* Verreaux's sifaka: *Propithecus verreauxi.*
■ DAUBENTONIIDAE: aye-aye: *Daubentonia madagascariensis.*

Haplorrhini

A few representatives

Spectral tarsier
Tarsius spectrum
BL: 12 cm
node **6**

Common marmoset
Callithrix jacchus
BL: 20 cm
node **8**

Guinea baboon
Papio papio
BL: 70 cm
node **10**

Common chimpanzee
Pan troglodytes
90 cm
node **19**

Some unique derived features

- The rhinarium (naked nose pad) has disappeared and is replaced by a nose. The nasal regions of a lemurid (1a) and of two haplorrhines, a catarrhine such as a chimpanzee (1b) and a platyrrhine (1c), are shown in fig. 1.

- The whiskers on the muzzle have disappeared.

- In the β-globin multigenic family, there is a partial gene conversion of the δ-globin gene by the β-globin gene.

- Short transposable elements of the "Alu SINEs" type are inserted at characteristic sites in the genome: site C12, between exons 3 and 4 of the ATP synthase subunit β gene, site C7 of the zonadhesin gene, and site C9, between

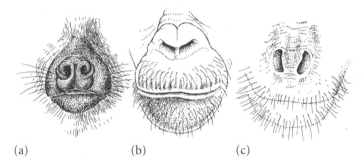

(a)   (b)   (c)

Figure 1

exons 4 and 5 of the alpha 1-microglobulin-bikunin gene.

- Molecular phylogenies based on the sequences of 19 nuclear genes and 3 mitochondrial genes bring together the haplorrhines. These genes include hemoglobin, myoglobin, alpha-crystallin, cytochrome c, and ribonuclease genes along with mitochondrial ribosomal genes. In addition, independent analyses of other nuclear (a different fragment of *RAG1*, *c-myc* oncogene, growth hormone) and mitochondrial (*ND6*) genes support this clade.

Number of species: 150.
Oldest known fossils: *Afrotarsius*, from Egypt, dates from the start of the Eocene. *Shoshonius* from Wyoming, USA, also dates from this period (55 MYA). If the fossil family omomyidae are included in the tarsiiformes, the haplorrhines date from the upper Paleocene with *Decoredon* from China and *Altiatlasius* from Morocco (57 MYA).
Current distribution: worldwide.

Examples

Philippine tarsier: *Tarsius syrichta;* common marmoset: *Callithrix jacchus;* yellow baboon: *Papio cynocephalus;* white-handed gibbon: *Hylobates lar;* orangutan: *Pongo pygmaeus;* gorilla: *Gorilla gorilla;* human: *Homo sapiens;* common chimpanzee: *Pan troglodytes.*

485

Tarsiiformes

General Description

The tarsiiformes are small arboreal primates with large round heads, wide eyes and large ears. Like owls, they can rotate their heads almost 180°. The body length (excluding the tail) varies from 9.5 to 14.5 cm, for respective body weights of between 105 and 135 g. The hindlegs are longer than the forelegs and are divided into three sections of equal length – thigh, calf and foot. The body length is only half that of hindlegs. Their long legs provide these animals with excellent jumping abilities. Different species are recognizable by the tufts of fur on the tail. The pelage is soft, gray-brown, reddish-brown or beige. The digits all carry flat nails except the second and third toes, which have grooming claws. The third finger of the western tarsier is almost as long at its arm (3 cm).

Spectral tarsier
Tarsius spectrum
BL: 12 cm, TL: 23 cm

Some unique derived features

- The orbit is enlarged (fig. 1).
- The ectotympanic bone is elongated to form the external auditory canal. Fig. 2 shows cross-sections of the tympanic (or auditory) bulla as found in the eulipotyphles (a), the lemuriformes (b), the lorisiformes and the platyrrhines (c), and the tarsiiformes and the catarrhiines (d). The ectotympanic bone has been elongated and transformed into the external auditory canal convergently in these last two groups.

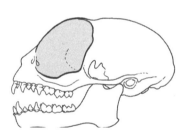

Figure 1

 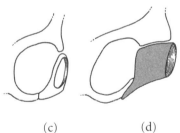

(a) (b) (c) (d)

Figure 2

Ecology

The tarsiers are arboreal species that live in tropical rain and scrub forests, groves, and plantations. They are active at twilight and through the night. They are insectivorous and carnivorous (lizards, snakes, bats, birds). They are excellent climbers-leapers and have enormous, forward-facing eyes—each eye weighs as much as the brain; these traits enable them to judge distances at night accurately so that they can jump and capture prey. They live in pairs or family groups. The life span varies from 8 to 12 years.

Examples

Philippine tarsier: *Tarsius syrichta;* western tarsier: *T. bancanus;* spectral tarsier: *T. spectrum.*

Simiiformes

A few representatives

Common marmoset
Callithrix jacchus
BL: 20 cm
node **8**

Guinea baboon
Papio papio
BL: 70 cm
node **10**

Siamang
Symphalangus syndactylus
85 cm
node **12**

Human
Homo sapiens
170 cm
node **18**

Some unique derived features

■ The orbit is closed posteriorly by bones of the skull. Lateral views of the skulls of a lemuriforme (1a) and a common marmoset (1b) are shown in fig. 1. The arrow indicates the posterior opening of the lemuriforme's orbit (ancestral state); the absence of an arrow in (1b) underlines that the orbit is closed posteriorly in the marmoset.

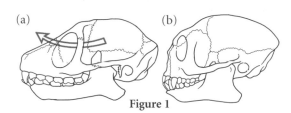

Figure 1

■ The cerebral cortex is highly folded. Fig. 2 shows the brains of a lorisiforme (loris, 2a) and a simiiforme (macaque, 2b).

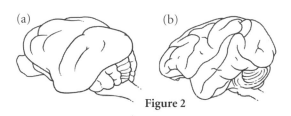

Figure 2

Number of species: 147.
Oldest known fossils: simians are recognized starting at the base of the Oligocene (34 MYA) from Egypt (*Oatrania*, *Parapithecus*, and *Apidium* from the fossil family Parapithecidae) and Hungary (*Pilopithecus* from the fossil family Pilopithecidae). The catarrhines, such as *Propilopithecus* and *Aegyptopithecus* from the fossil family Propilopithecidae, are also found in Egypt during this period.
Current distribution: worldwide.

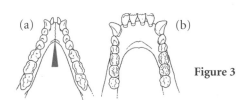

Figure 3

■ The mandibular symphysis has disappeared with the fusion of the two dentary bones, which form the mandible in mammals (fig. 3). These bones meet in the front and form a symphysis in mammals such as the tarsier (3a). This symphysis has disappeared in the simiiformes by their fusion into a single dentary bone (3b: chimpanzee).

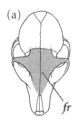

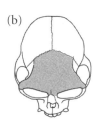

Figure 4

■ The right and left frontal bones have fused into a single frontal bone (*fr*, fig. 4). Fig. 4 shows the skull of a lemur (4a), where the suture between the two frontal bones is visible, and that of a chimpanzee (4b), where it has disappeared.

Examples

■ PLATYRRHINI: brown howler monkey: *Alouatta fusca*; common marmoset: *Callithrix jacchus*.
■ CATARRHINI: yellow baboon: *Papio cynocephalus*; western red colobus: *Procolobus badius*; orangutan: *Pongo pygmaeus*; gorilla: *Gorilla gorilla*; human: *Homo sapiens*; common chimpanzee: *Pan troglodytes*.

Platyrrhini

General Description

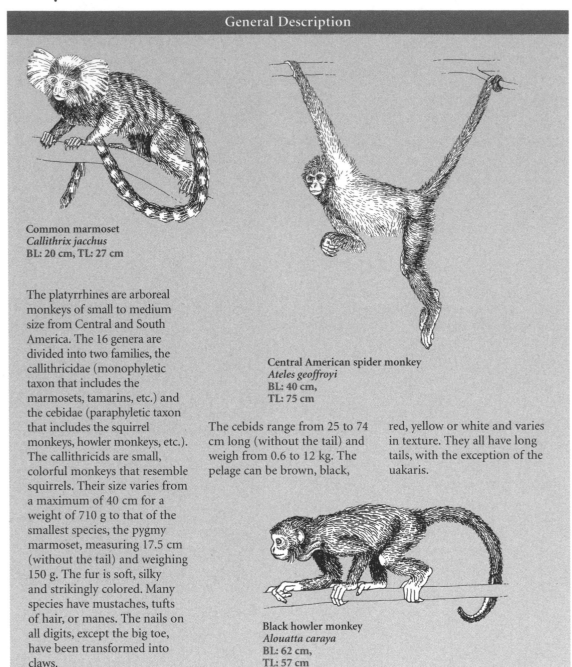

Common marmoset
Callithrix jacchus
BL: 20 cm, TL: 27 cm

Central American spider monkey
Ateles geoffroyi
BL: 40 cm,
TL: 75 cm

Black howler monkey
Alouatta caraya
BL: 62 cm,
TL: 57 cm

The platyrrhines are arboreal monkeys of small to medium size from Central and South America. The 16 genera are divided into two families, the callithricidae (monophyletic taxon that includes the marmosets, tamarins, etc.) and the cebidae (paraphyletic taxon that includes the squirrel monkeys, howler monkeys, etc.). The callithricids are small, colorful monkeys that resemble squirrels. Their size varies from a maximum of 40 cm for a weight of 710 g to that of the smallest species, the pygmy marmoset, measuring 17.5 cm (without the tail) and weighing 150 g. The fur is soft, silky and strikingly colored. Many species have mustaches, tufts of hair, or manes. The nails on all digits, except the big toe, have been transformed into claws.

The cebids range from 25 to 74 cm long (without the tail) and weigh from 0.6 to 12 kg. The pelage can be brown, black, red, yellow or white and varies in texture. They all have long tails, with the exception of the uakaris.

Some unique derived features

- In the platyrrhines, the nostrils (fig. 1) open laterally and are strongly separated (1b); in the catarrhines, they open downward and are separated by a thin septum (1a).

- There is a suture between the parietal (*pa*) and jugal (*j*) bones in the platyrrhines (2a: spider monkey). In the cercopithecidae (2b: guenon), the

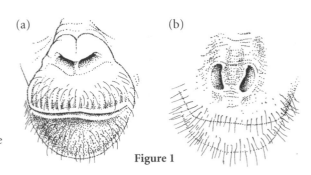

(a)   (b)

Figure 1

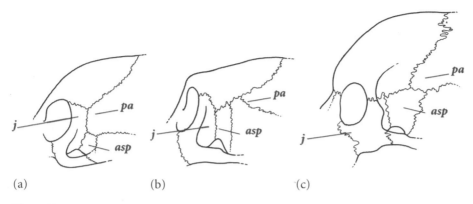

Figure 2

parietal bone does not touch either the jugal or the alisphenoid (*asp*). In hominoids (2c: human), there is a sphenoparietal suture.

■ The tail is prehensile (fig. 3).

■ The ectotympanic bone is fused to the opening of the tympanic (auditory) bulla, but is not elongated into an external auditory canal. Cross-sections of the tympanic bulla as observed on a eulipotyphle (4a), a lemuriforme (4b), a lorisiforme or playrrhine (4c), and a tarsiiforme or catarrhine (4d) are shown in

fig. 4. A fusion of the ectotympanic and the petrous bones has arisen convergently in both the platyrrhines and lorisiformes. This is also observed in the tarsiiformes and catarrhines (4d), except in these groups the ectotympanic bone has been transformed into an external auditory canal.

Figure 3

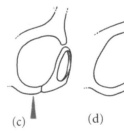

Figure 4

Ecology

The platyrrhines (or New World monkeys) live in tropical rain forests, gallery forests and savanna forest patches. Many of the cebids have a varied diet of fruit, leaves, insects, seeds, and small mammals. However, there are more specialized species as well, such as the leaf-eating mantled howler monkey or the insectivorous South American squirrel monkey. In the cebids, the social organization varies from individual monogamous couples to large polygamous groups. We find the only nocturnal simian monkey in this group, the northern night monkey. The life span ranges from 12 to 25 years in most cebids, but is unknown in natural populations for the callithricids. This latter group contains monogamous monkeys that live in social groups. They are insectivorous and specialize in the consumption of tree gum.

Number of species: 51.
Oldest known fossils: *Branisella* from Bolivia dates from the middle Oligocene (29 MYA). The cebidae are represented in Argentina by the genus *Tremacebus* starting in the upper Oligocene (25 MYA).
Current distribution: Central America and the northern half of South America.

Examples

Black spider monkey: *Ateles paniscus;* brown howler monkey: *Alouatta fusca;* red uakari: *Cacajao rubicundus;* northern night monkey: *Aotus trivirgatus;* dusky titi: *Callicebus moloch;* South American squirrel monkey: *Saimiri sciureus;* tufted capuchin: *Cebus apella;* emperor tamarin: *Saguinus imperator;* common marmoset: *Callithrix jacchus.*

Catarrhini

A few representatives

Guinea baboon
Papio papio
BL : 70 cm
node 10

Siamang
Symphalangus syndactylus
85 cm
node 12

Orangutan
Pongo pygmaeus
130 cm
node 14

Human
Homo sapiens
170 cm
node 18

Some unique derived features

■ The nostrils (fig. 1) open downwards and are separated by a thin septum in the catarrhines (1a), whereas they are laterally oriented and widely separated in the platyrrhines (1b).

■ The third premolar (*Pm3*) has disappeared (fig. 2). The ancestral dentition, that of the platyrrhines for example (2a), shows the following pattern from back to front: the third, second and first molars (*M*), the third, second and first premolars (*Pm*), then the canine (*C*) and the incisors (*I*). The dentition of the catarrhines (2b) is modified: the third, second and first molars are followed by the second and first premolars, and then the canines and incisors.

■ In the multigenic β-globin gene family, we see a duplication of the γ-globin gene.

Examples

■ CERCOPITHECOIDEA: yellow baboon: *Papio cynocephalus;* Japanese macaque: *Macaca fuscata;* hanuman langur: *Semnopithecus entellus;* western red colobus: *Procolobus badius.*

■ HOMINOIDEA: white-handed gibbon: *Hylobates lar;* orangutan: *Pongo pygmaeus;* gorilla: *Gorilla gorilla;* human: *Homo sapiens;* common chimpanzee: *Pan troglodytes.*

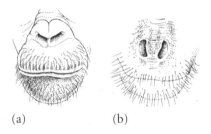

(a)   (b)

Figure 1

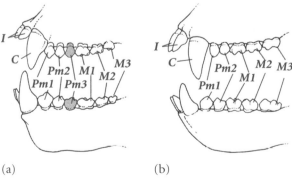

(a)   (b)

Figure 2

Number of species: 96.
Oldest known fossils: catarrhines from the fossil family Propilopithecidae, such as *Propilopithecus* and *Aegyptopithecus,* have been found in Egypt and date from the start of the Oligocene (34 MYA).
Current distribution: worldwide.

Cercopithecoidea

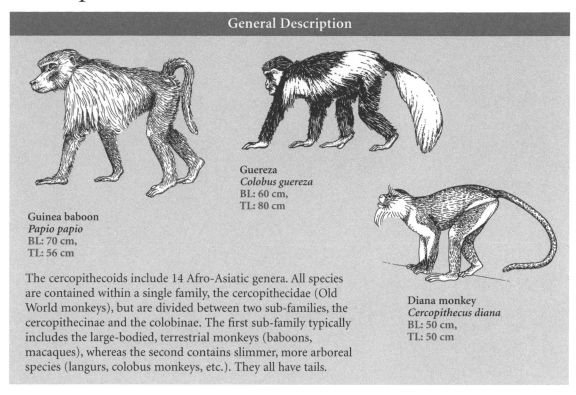

General Description

Guinea baboon
Papio papio
BL: 70 cm,
TL: 56 cm

Guereza
Colobus guereza
BL: 60 cm,
TL: 80 cm

Diana monkey
Cercopithecus diana
BL: 50 cm,
TL: 50 cm

The cercopithecoids include 14 Afro-Asiatic genera. All species are contained within a single family, the cercopithecidae (Old World monkeys), but are divided between two sub-families, the cercopithecinae and the colobinae. The first sub-family typically includes the large-bodied, terrestrial monkeys (baboons, macaques), whereas the second contains slimmer, more arboreal species (langurs, colobus monkeys, etc.). They all have tails.

Some unique derived features

- The molars are bilophodont, meaning there are two lophs, or ridges (fig. 1). On the right

(a)

(b)

Figure 1

upper molars of the gibbon, a hominoid, we see no real directional organization of the lophs (1a). The right upper molars of a cercopithecoid (1b), however, show a central longitudinal valley that follows the alignment axis of the molars and separates the two parallel lophs.

- The medial epicondyle at the distal extremity of the humerus has disappeared. Compare anterior views of the humerus

of a human (hominoid, fig. 2a) and a baboon (cercopithecoid, 2b). In the ancestral state, as seen in the human, the medial epicondyle forms a large, prominent eminence; this has disappeared in the baboon.

(a)   (b)

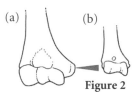

Figure 2

Ecology

The cercopithecoids live in a variety of different habitats: tropical forests and winter snow-covered mountains, but also in savannas and bushland. The diet is typically composed of fruit but can vary greatly among species; the genera *Colobus* and *Procolobus* are strictly leaf-eating species, whereas yellow baboons are partial carnivores and can hunt hares or young gazelles in social groups.

Number of species: 82.
Oldest known fossils: *Prohylobates* from Libya and Egypt dates from the lower Miocene (20 MYA).
Current distribution: sub-Saharan Africa and Maghreb, India, Southeast Asia, southern China, Sunda Islands, Japan.

Examples

Yellow baboon: *Papio cynocephalus;* gelada baboon: *Theropithecus gelada;* Barbary macaque: *Macaca sylvanus;* Japanese macaque: *Macaca fuscata;* African green monkey: *Cercopithecus aethiops;* red-capped mangabey: *Cercocebus torquatus;* talapoin monkey: *Miopithecus talapoin;* hanuman langur: *Semnopithecus entellus;* western red colobus: *Procolobus badius;* proboscis monkey: *Nasalis larvatus.*

Hominoidea

A few representatives

Siamang
Symphalangus syndactylus
85 cm
node 12

Orangutan
Pongo pygmaeus
130 cm
node 14

Human
Homo sapiens
170 cm
node 18

Common chimpanzee
Pan troglodytes
90 cm
node 19

Some unique derived features

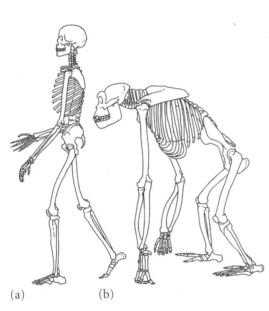
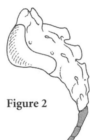

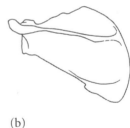

Figure 2

Figure 3

(a)

(b)

(a)   (b)

Figure 1

Figure 4

(a)   (b)

- Coccyx (or tailbone): There is no longer a true tail (fig. 1); the caudal vertebrae are reduced and have fused into a single coccyx (fig. 2).

- The arms are longer than the legs (fig. 1b), with the exception of humans (1a).

- We see the fusion of the scaphoid (*sca*) and os centrale (*osc*) in the hand skeleton. As seen in fig. 3, the scaphoid and os centrale are distinct bones in the lemurid (3a) but are fused in the human (3b).

- The scapula is dorsal and elongated along the anterior-posterior axis in hominoids (fig. 1 and 4a), whereas it is laterally stretched in non-hominoids like the macaque (4b).

Number of species: 14.
Oldest known fossils: the famous genus *Proconsul* from East Africa dates from the lower Miocene (23 MYA). *Dryopithecus* has been found in Europe and dates from this same period.
Current distribution: worldwide.

Examples

- HYLOBATOIDAE: white-handed gibbon: *Hylobates lar;* siamang: *Symphalangus syndactylus.*
- HOMINOIDAE: orangutan: *Pongo pygmaeus;* gorilla: *Gorilla gorilla;* human: *Homo sapiens;* common chimpanzee: *Pan troglodytes:* bonobo: *Pan paniscus.*

Hylobatoidae

General Description

The two hylobatoid genera, *Hylobates* and *Symphalangus* (generally used synonymously), form a single family, the hylobatidae. The gibbons are slender, tailless apes with extremely long arms. Most species measure between 45 and 65 cm in length for a weight that varies from 5.5 to 6.7 kg. The siamang is larger and can reach 75 to 90 cm in length and up to 10.5 kg in weight. In contrast to the other hominoids, the sexual dimorphism of this group does not involve body size. The pelage is dense. The color of the fur (black, brown, yellow or beige), the markings surrounding the face, and the

Siamang
Symphalangus syndactylus
85 cm

White-handed gibbon
Hylobates lar
85 cm

vocalizations are used to distinguish among the different species. The face is always gray but may be surrounded by a border of white hairs.

Some unique derived features

■ The hylobatoids use brachial locomotion, meaning they move by swinging with the arms from one hold to another in an alternating fashion, completely releasing one branch as they move to another. Brachiation requires certain specific anatomical characteristics of the skeleton and musculature that enable extremely ample arm rotations in all directions within a half circle (fig. 1).

■ The karyotype is unusual; there have been many more chromosomal rearrangements than we find in other primates.

Examples

White-handed gibbon: *Hylobates lar;* crested gibbon: *H. concolor;* pileated gibbon: *H. pileatus;* Borneo gibbon: *H. muelleri;* siamang: *Symphalangus syndactylus.*

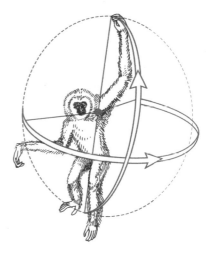

Figure 1

Number of species: 9.
Oldest known fossils: no fossilized gibbon has been found.
Current distribution: Southeast Asia, from east India to southern China, Malay Peninsula, Sumatra, western Java and Borneo.

Ecology

The gibbons live in the evergreen rainforests of Southeast Asia and in the monsoon forests of continental Asia. They are excellent brachiators that spend most of their time in the trees. However, as in the trees, their terrestrial bipedal locomotion is better than that of the great apes. They are monogamous, live in family groups and defend the family's territory. They feed primarily on fruit. Gibbons communicate using very sophisticated vocalizations. The life span is 25 to 30 years in natural populations. Deforestation has rendered their future uncertain.

Hominoidae

A few representatives

Orangutan
Pongo pygmaeus
130 cm
node 14

Gorilla
Gorilla gorilla
180 cm
node 16

Human
Homo sapiens
170 cm
node 18

Common chimpanzee
Pan troglodytes
90 cm
node 19

Some unique derived features

■ The cerebral cortex is highly folded (fig. 1). This is clearly seen if we compare the cortex of a cercopithecoid (macaque, 1a) and that of a human (1b).

(a)

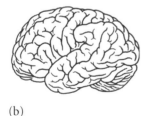

(b)

Figure 1

■ The maxillary and premaxillary bones lie closer together and delimit the incisive canal (arrow). Sagittal sections of the naso-maxillary region of a gibbon (2a), a gorilla (2b) and a chimpanzee (2c) are shown in fig. 2. In the gibbon, the maxillary (*mx*) is clearly separated from the premaxillary (*pmx*). In the two others, these bones are much closer and delimit a canal.

■ There is a frontal sinus (*fsi*). On a para-sagittal section of the facial region of an adult gorilla, we note that the frontal sinus is high on the head, just below the prominent brow ridges (*br*, fig. 3).

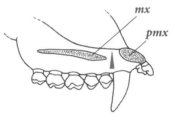

(a)

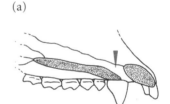

(b)

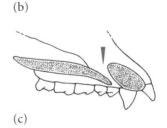

(c)

Figure 2

Number of species: 5.
Oldest known fossils: *Ramapithecus* and *Sivapithecus* from Europe and Africa date from the middle Miocene (14 MYA).
Current distribution: worldwide.

Examples

■ PONGIDAE: orangutan: *Pongo pygmaeus.*
■ HOMINIDAE: gorilla: *Gorilla gorilla*; human: *Homo sapiens*; common chimpanzee: *Pan troglodytes:* bonobo: *Pan paniscus.*

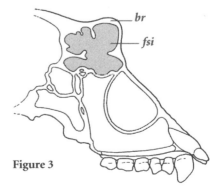

Figure 3

Pongidae

General Description

The pongidae are represented by a single extant species, the orangutan. It is a large, tailless ape; the males measure approximately 1.30 m in height for a weight of 75 kg. The forelimbs are much longer than the hindlimbs. The pelage is spare, long and rough. It is orange in young animals and becomes reddish brown in adults. The face is naked and gray, with pink around the eyes and on the muzzle in the young. The adult males have a well-developed double chin; the second chin is actually a throat sac used to resonate sound during vocalizations.

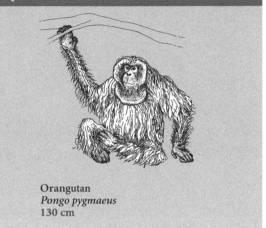

Orangutan
Pongo pygmaeus
130 cm

Some unique derived features

■ The orbits are higher than they are wide. Fig. 1 shows the skull of an orangutan (1a) and that of a chimpanzee (1b).

■ The premaxillary bone overlaps the maxillary bone. Sagittal sections of the naso-maxillary region of a gibbon (2a) and an orangutan (2b) are shown in fig. 2. In the gibbon, the premaxillary (*pmx*) follows the maxillary (*mx*) along the same plane, whereas in the orangutan this bone overlaps the maxillary dorsally.

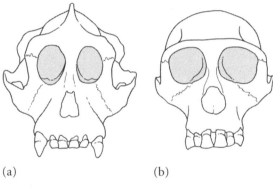

(a)   (b)

Figure 1

Ecology

Orangutans inhabit the tropical rain forests of plains and sub-mountainous regions and spend almost all of their time in the trees. With their very long arms and hook-like hands and feet, they are excellent brachiators; this species is the only completely arboreal hominoid. They are solitary, diurnal animals that feed principally on fruit (60% of the diet). The remainder of the diet consists mostly of leaves, although they also sometimes eat insects, eggs and small vertebrates. The life span is approximately 35 years in natural populations and 50 years in captivity. Orangutans are heavily protected; hunting and trade of this species is strictly forbidden. We can find natural populations within numerous wildlife reserves in Indonesia and Malaysia. However, the rain forests of Sumatra and Borneo are disappearing at an alarming rate, strongly threatening the survival of this animal.

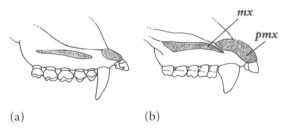

(a)   (b)

Figure 2

Number of species: 1.
Oldest known fossils: the most well known are *Ramapithecus* and *Sivapithecus* from Europe and Africa dating from the middle Miocene (14 MYA). Fossils of giant orangutans have been found in China and date from the Pleistocene.
Current distribution: Borneo and northern Sumatra.

Examples

Orangutan: *Pongo pygmaeus.*

Hominidae

A few representatives

Gorilla
Gorilla gorilla
180 cm
node 16

Human
Homo sapiens
170 cm
node 18

Common chimpanzee
Pan troglodytes
90 cm
node 19

Bonobo
Pan paniscus
75 cm
node 19

Some unique derived features

■ The fusion of the os centrale (*osc*) and the scaphoid (*sca*) bone in the hand skeleton is prenatal in the hominids, whereas it is postnatal in other hominoids. As seen in fig. 1, the scaphoid and os centrale are distinct bones in the lemurid (1a) and fused in the human (1b).

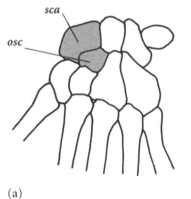

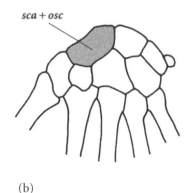

(a)

(b)

Figure 1

Number of species: 4.
Oldest known fossils: *Kenyapithecus* from Kenya dates from the Miocene (13 MYA) and is placed within the genus *Sivapithecus* by certain authors. *Pierolapithecus* from Spain also dates from this period. Fossils of *Graecopithecus* have been found in Greece and date from the late Miocene (10 MYA).
Current distribution: worldwide.

Examples

■ GORILLINAE: gorilla: *Gorilla gorilla*.
■ HOMININAE: human: *Homo sapiens;* common chimpanzee: *Pan troglodytes:* bonobo: *Pan paniscus.*

Gorillinae

General Description

The gorilla is the largest extant primate. It measures 1.25 to 1.80 m in height in its normal position (on all fours) and up to 2.3 m when standing upright. The arm span ranges from 2 to 2.75 m. The female weighs from 70 to 140 kg, whereas the male varies between 135 and 275 kg. These tailless animals are stocky with extremely strong limbs. The pelage is thick, black or gray-black with silver streaks on the backs of older males. The face, feet and hands are black and hairless. The ears are small and the brows and sagittal crest are pronounced.

Gorilla
Gorilla gorilla
180 cm

Some unique derived features

- Infraorbital foramen: this foramen lies far below the lower margin of the orbit in the gorillas (fig. 1a). In contrast, it is very close to the margin in humans (fig. 1b).

Ecology

The plains gorilla lives in the dense forests of western equatorial Africa, whereas the mountain gorilla inhabits the plains forests and mountain forests (up to 3500 m in altitude) in eastern Democratic Republic of Congo. Gorillas live in small social groups led by an older dominant male within a territory of 25 to 40 km². They are exclusive vegetarians, eating leaves, berries, fruit and bark. They are terrestrial during the day but climb trees at night to sleep. The life span is approximately 35 years in natural populations. The gorilla is officially a protected species, but the mountain gorilla remains in danger due to poaching and deforestation.

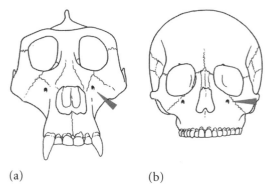

(a)   (b)

Figure 1

Number of species: 1.
Oldest known fossils: there are no known gorilla fossils.
Current distribution: western equatorial Africa, eastern Democratic Republic of Congo, Rwanda, Burundi and Uganda.

Examples

Western lowland gorilla: *Gorilla gorilla gorilla;* mountain gorilla: *G. g. beringei;* eastern lowland gorilla: *G. g. graueri.*

Homininae

A few representatives

Human
Homo sapiens
170 cm
node 18

Common chimpanzee
Pan troglodytes
90 cm
node 19

Bonobo
Pan paniscus
75 cm
node 19

Some unique derived features

■ Molecular phylogeny: the most reliable current arguments in favor of the homininae-paninae relatedness are the results of independent molecular phylogenetic studies. A recent review of this question analyzed 45 separate genes with a total of 46,855 nucleotides. More than half of these genes support a human-chimpanzee clade. The others are divided among two alternative groupings, those that support a human-gorilla clade and those that support a gorilla-chimpanzee clade. Several genetic mechanisms have been proposed to explain why the phylogeny of certain genes does not always

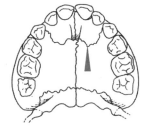

(a)   (b)

Figure 1

correspond to that of the species.

■ The incisive suture between the maxillary and premaxillary

disappears in the adult. Fig. 1 compares the palate of a child (fig. 1a) with that of an adult (fig. 1b).

Number of species: 3.
Oldest known fossils: *Sahelanthropus tchadensis* from northern Chad dates from the early Pilocene (7 MYA). *Orrorin tugenensis* is from Kenya and dates from the same period (6 MYA).
Current distribution: worldwide.

Examples

Human: *Homo sapiens;* common chimpanzee: *Pan troglodytes:* bonobo: *Pan paniscus.*

Hominini

General Description

Humans distinguish themselves from the rest of the extant fauna by their permanent vertical position. They are large primates. The females are smaller than the males. The lower limbs are longer than the upper limbs, and there is no tail (more exactly, the tail is reduced and part of the coccyx). The face, mandible, brows and ears are reduced in size, but the cranium is large. The hands are more square-shaped than in other primates. The foot has lost the opposable big toe. There is little hair except on the top of the head, under the arms and in the pubic region. The males have facial hair, a beard. The skin color varies from light pink to deep black, with all shades of brown in between.

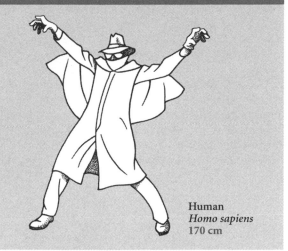

Human
Homo sapiens
170 cm

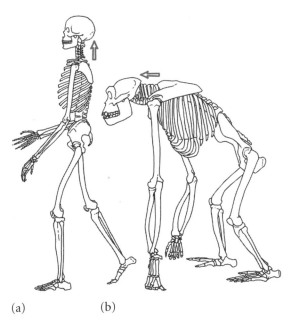

(a)   (b)

Figure 1

Some unique derived features

- Bipedalism has resulted in a movement of the occipital orifice (*foramen magnum*) to the underside the cranium (fig. 1a: human; 1b gorilla), an S-shaped spinal column, and the presence of buttocks.

- The pelvis (fig. 2) is short and wide in the hominines (2a), whereas it is long and thin in the other hominoids (2b: gorilla).

- The premaxillary bone (fig. 3) has a vertical orientation in hominines (3b), whereas it is oblique in chimpanzees (3a).

- The dentary arch (fig. 4) has a parabolic shape (4c). The primitive state is the V-shaped arch of the tarsiers (4a). The arch of the chimpanzee is U-shaped (4b).

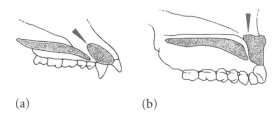

(a)   (b)

Figure 3

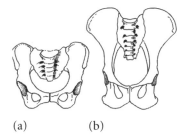

(a)   (b)

Figure 2

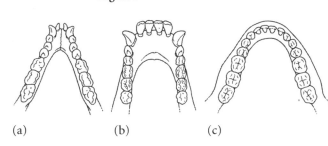

(a)   (b)   (c)

Figure 4

Ecology

Humans are super-predators that live in complex, hierarchical societies. Their ability to learn, communicate, reason and adapt are very well developed. This species possesses exceptional technical capabilities, partially because bipedalism has freed the hands. As most primates, humans are diurnal and omnivorous. The most frequent type of social union is that of two partners for life, but there are numerous variations to this general form. Human societies owe their cohesion both to the formation of the family unit and to the strongly developed sense of group identity maintained by rites, customs, beliefs and a hierarchical organization. Initially, human populations were made up of hunter-gatherers. However, this general lifestyle and certain biological features were dramatically altered by the emergence of agriculture 11,000 years ago, when humans became sedentary. The first cities were formed 8,000 years ago. Human longevity used to be from 35 to 40 years, but the sedentary lifestyle and human technical skills have prolonged this to 80 to 100 years. This species is currently in full expansion and gravely threatens all biodiversity, along with the climatic and ecological state of the planet.

Number of species: 1.
Oldest known fossils: if *Sahelanthropus tchadensis* is considered to be a hominine, it is the oldest (lower Pilocene from northern Chad, 7 MYA). Otherwise, the oldest hominine is *Orrorin tugenensis* from Kenya (Pilocene, 6 MYA).
Current distribution: worldwide.

Examples

Human: *Homo sapiens.*

Panini

General Description

The chimpanzee is a large, tailless primate with longer arms than legs. The body length ranges from 70 to 92 cm. The males may reach an upright height of 1.70 m and may weigh 50 kg. The females are smaller. The pelage is generally thick and relatively long and varies from dark brown to shiny black. However, the face, genital and anal regions, and palms and soles are hairless. The cheeks have a black beard. The mobile, protrusive lips are strong, and the brows and ears are large. The bonobo differs from the common chimpanzee by its smaller size, more slender build, red lips, external female sexual organs, behavior, vocalizations and mating.

Common chimpanzee
Pan troglodytes
90 cm

Bonobo
Pan paniscus
75 cm

Some unique derived features

- The η-globin pseudogene contains nucleotide signatures unique to the two species *Pan paniscus* and *Pan troglodytes*.

Number of species: 2.
Oldest known fossils: there are no known chimpanzee fossils. However, if *Sahelanthropus tchadensis* is considered to be a panine, it is the oldest (lower Pilocene from northern Chad, 7 MYA).
Current distribution: central and western Africa.

Ecology

Chimpanzees live in bands of 10 to 30 individuals in the dense forests, woodlands, plains and treed savannas of Africa. They are as agile on the ground as they are in trees. However, trees are mostly used at night for sleeping. Ninety percent of their diet is composed of fruit, but they also eat flowers, leaves, buds, bark and insects. In certain savanna regions, they may also sometimes kill and eat small mammals (the young of certain monkeys, antelopes, or bush pigs). In natural populations, the life span is around 40 years.

Examples

Common chimpanzee: *Pan troglodytes:* bonobo: *Pan paniscus.*

Chapter 15

ACTINOPTERYGII

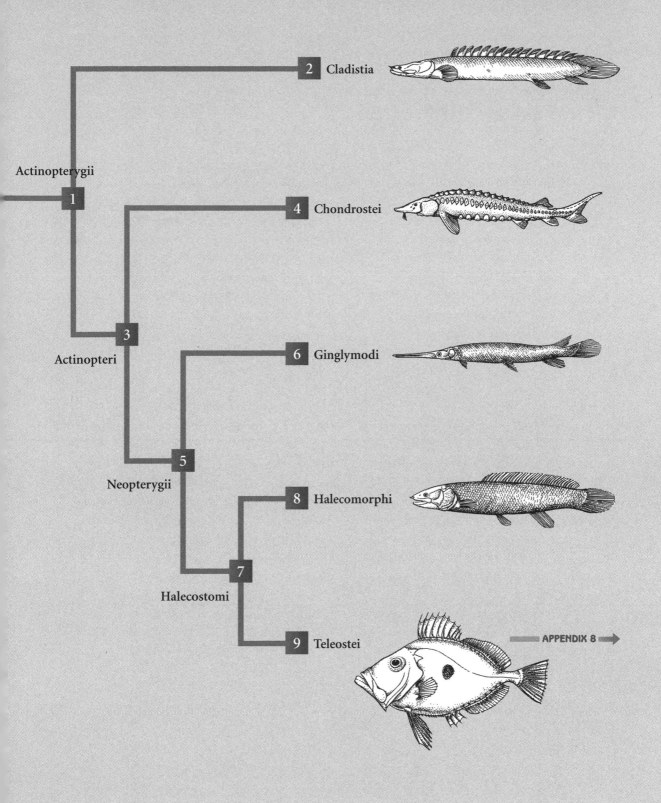

Actinopterygii

1

2 Cladistia

Actinopteri

3

4 Chondrostei

Neopterygii

5

6 Ginglymodi

Halecostomi

7

8 Halecomorphi

9 Teleostei

APPENDIX 8 ➡

ACTINOPTERYGII

The actinopterygians (or ray-finned fishes) bring together half of all living vertebrate species. This category includes four relic lineages and one enormous group, the teleostei, that alone makes up the quasi-totality of the actinopterygii and thus almost half of all craniates. The four other lineages are considered to be relics because their fossil diversity is higher than their current low, and therefore residual, diversity. Indeed, the actinopterygians had an initial "boom" in diversity during the beginning of the Carboniferous. A second "boom" then occurred during the Jurassic with the radiation of the teleost fishes.

The cladistians include the polypterids, lunged fishes that zoologists had difficulty classifying for a long time. They were sometimes placed with the sarcopterygians, typically within the paraphyletic group crossopterygii, but sometimes within a separate class, the brachiopterygii. These animals, considered to be one of the earliest lineages of actinopterygii, have been studied since the start of the 1950s and especially so since the introduction of cladistic analysis to ichthyology. The chondrosteans (paddlefish and sturgeons) have a stable phylogenetic position. These relic representatives of a formerly prosperous group are all currently threatened with extinction in their natural habitat. The ginglymodi (lepisosteidae) and the halecomorphi (only a single extant species: the bowfin *Amia calva*) were previously placed within the holostei. However, cladistic analyses showed this group to be paraphyletic. Ironically, although the holosteans had previously constituted a grade because of insufficient methodology, they might emerge again today as a group, but this time on a solid basis, as a result of both phylogenetic systematics (or cladistics) and the discovery of new fossils. Indeed, the discovery of new fossils has called into question the characters used to demonstrate the paraphyly of the holosteans, meaning they may well be monophyletic after all. Nonetheless, many recent cladistic analyses have produced contradictory or unresolved results. In addition, no robust and repeatable molecular phylogeny has yet been found. For these reasons, we have conserved the classic phylogenetic interpretation of this group. The phylogeny of the teleosteans was still almost entirely unknown at the start of the 1960s. The advent of phylogenetic systematics in ichthyology starting in 1967 has resulted in spectacular progress in our understanding of the diversity of this large clade, and this in only five or so years. The group has always been rich in characters, but before phylogenetic systematics there was no appropriate method for analyzing them. Today, there still remains much work to be done within crown teleosts, particularly within the acanthomorpha (a group of teleosts with spines in both the dorsal and anal fins), whose 14,000 species account for more than half of all teleosts.

Cladistia

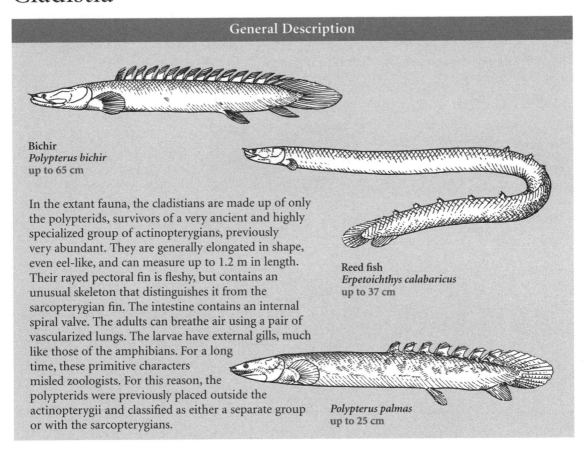

Bichir
Polypterus bichir
up to 65 cm

In the extant fauna, the cladistians are made up of only the polypterids, survivors of a very ancient and highly specialized group of actinopterygians, previously very abundant. They are generally elongated in shape, even eel-like, and can measure up to 1.2 m in length. Their rayed pectoral fin is fleshy, but contains an unusual skeleton that distinguishes it from the sarcopterygian fin. The intestine contains an internal spiral valve. The adults can breathe air using a pair of vascularized lungs. The larvae have external gills, much like those of the amphibians. For a long time, these primitive characters misled zoologists. For this reason, the polypterids were previously placed outside the actinopterygii and classified as either a separate group or with the sarcopterygians.

Reed fish
Erpetoichthys calabaricus
up to 37 cm

Polypterus palmas
up to 25 cm

Some unique derived features

■ The dorsal fin has independent rays; it is segmented into small "flags," the pinnules (fig. 1). Fig. 1a shows a lateral view of a pinnule (*sp:* spine; *le:* lepidotrich;

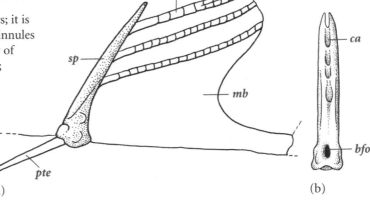

(a)

(b)

Figure 1

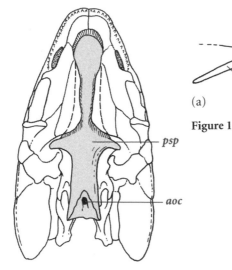

Figure 2

mb: membrane; *pte:* pterygiophore); fig. 1b shows a posterior view (*bfo:* basal foramen; *ca:* canal receiving the lepidotrichs).

■ The bone shaft that extends from the base of the cranium and joins the lower surface of the snout, called the parasphenoid, has a long posterior projection (fig. 2, *psp*) that encircles the aortic canal (*aoc*). In the early groups of fossil actinopterygians, the parasphenoid extended only to the otic region.

Ecology

The cladistians are predators in warm, freshwater bodies of Africa. They hunt at twilight using a lie-and-wait strategy, attacking any aquatic animal. They can live buried in silt or outside of the water during the dry season. They can also survive in muddy waters low in oxygen by using their lungs. The first polypterid known to Western science (*Polypterus bichir*) was discovered by Geoffroy Saint-Hilaire during Napoleon Bonaparte's expedition to Egypt.

Number of species: 10 (up to 15 according to certain authors).
Oldest known fossils: *Polypterus* from Bolivia dates from the upper Cretaceous (70 MYA).
Current distribution: central Africa and the Nile basin.

Example

Senegal bichir: *Polypterus senegalus;* Nile bichir: *P. bichir; P. palmas;* reed fish: *Erpetoichthys calabaricus.*

Actinopteri

A few representatives

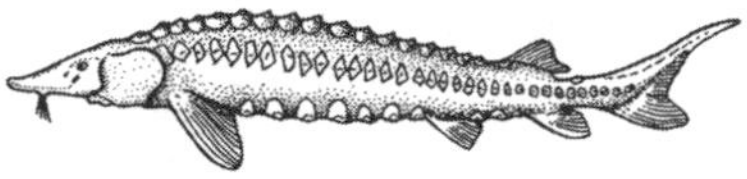

Longnose gar
*Lepisosteus
osseus*
180 cm
node 6

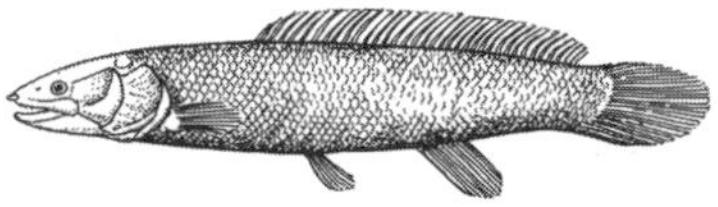

John dory
Zeus faber
up to 70 cm
node 9

European sturgeon
Acipenser sturio
up to 300 cm
node 4

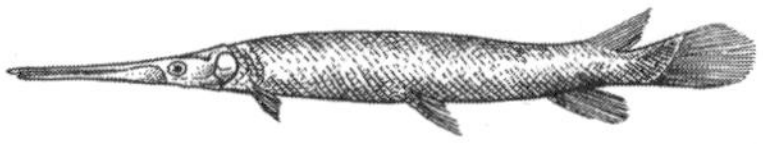

Bowfin
Amia calva
up to 87 cm
node 8

Some unique derived features

■ Air sacs: primitively, all the osteichthyans had air sacs, or lungs,
connected to the digestive
system. In the actinopterans,
the air sacs are connected to
the dorsal side of the digestive
tube. These sacs are lungs in
the primitive condition but
can be transformed into a gas-
filled flotation organ, the swim
bladder, in certain actinopteran
groups (the chondrosteans and
teleosteans). Fig. 1 shows four
different arrangements of the
air sacs in relation to the
digestive tube. In the actinopterans (fig. 1a), a swim bladder
or lung is connected to the

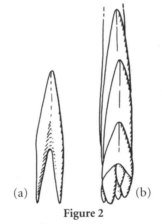

(a)   (b)

Figure 2

dorsal surface of the digestive
tube (in certain teleosteans, the
connection to the digestive
tube has been secondarily lost).
In the cladistians (fig. 1b), the
lungs connect ventrally and
run posteriorly and dorsally. In
the dipnoans, the lateral (or

dorsal in *Neocertodus*) lungs
connect ventrally (fig. 1c in
Protopterus). The lungs of the
tetrapods and the coelacanths
(fig. 1d) are also connected
ventrally to the digestive tube
but lie in a ventro-lateral
position.

■ Each fin has modified scales
along its anterior edge: the
fulcra. An isolated fulcrum is
shown in fig. 2a. The arrangement of the fulcra on the anterior bone of the dorsal fin
is also shown (anterior view,
fig. 2b).

■ A new bone is present in the
mandible, the supra-angular
(*san*). Fig. 3 shows a lingual view
of the mandible of *Amia calva*.

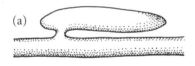
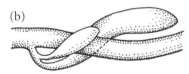
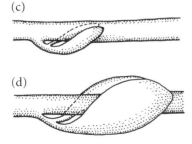

Figure 1

(a)

(b)

(c)

(d)

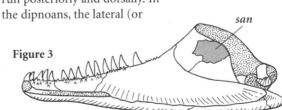

san

Figure 3

Number of species: 23,702.
Oldest known fossils:
Stegotrachelus, *Moythomasia*
and *Orvikuina* from Scotland,
Germany and Estonia, all from
the middle Devonian (380 MYA).
Current distribution:
worldwide.

Examples

■ CHONDROSTEI: Siberian sturgeon:
Acipenser baeri; *A. sturio*.
■ NEOPTERYGII: longnose gar:
Lepisosteus osseus; European eel:
Anguilla anguilla; northern pike:
Esox lucius; guppy: *Lebistes
reticulatus*; Atlantic bluefin tuna:
Thunnus thynnus; John dory: *Zeus
faber*.

Chondrostei

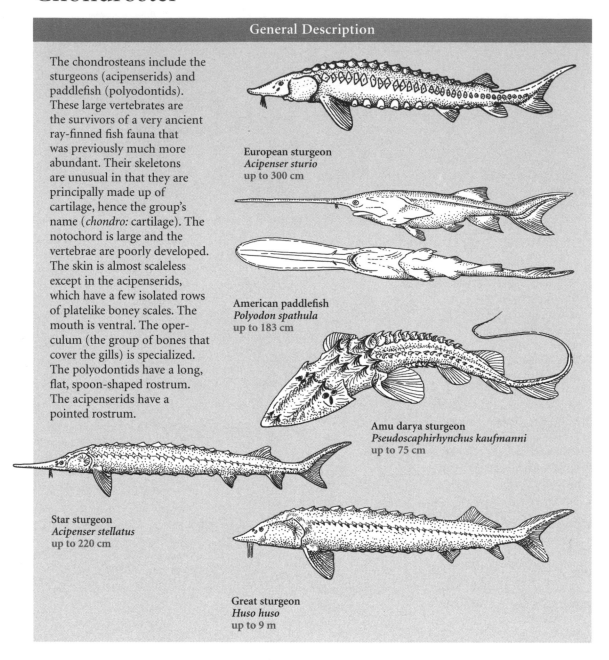

General Description

The chondrosteans include the sturgeons (acipenserids) and paddlefish (polyodontids). These large vertebrates are the survivors of a very ancient ray-finned fish fauna that was previously much more abundant. Their skeletons are unusual in that they are principally made up of cartilage, hence the group's name (*chondro:* cartilage). The notochord is large and the vertebrae are poorly developed. The skin is almost scaleless except in the acipenserids, which have a few isolated rows of platelike bony scales. The mouth is ventral. The operculum (the group of bones that cover the gills) is specialized. The polyodontids have a long, flat, spoon-shaped rostrum. The acipenserids have a pointed rostrum.

European sturgeon
Acipenser sturio
up to 300 cm

American paddlefish
Polyodon spathula
up to 183 cm

Amu darya sturgeon
Pseudoscaphirhynchus kaufmanni
up to 75 cm

Star sturgeon
Acipenser stellatus
up to 220 cm

Great sturgeon
Huso huso
up to 9 m

Some unique derived features

- There are sensory barbels (*bb*) in front of the mouth (*mo*, fig. 1, the head of *Scaphirhynchus platorynchus*).

- The chondrosteans have highly modified jaws that often lack teeth. Several bones of the upper jaw are fused together (maxillary, premaxillary and dermopalatine) and are not directly attached to the neurocranium. Lateral left views of the skulls of *Acipenser* (2a) and *Polyodon* (2b) are shown in fig. 2 (*ga:* gill arch; *ar:* articular; *ri:* ribs;

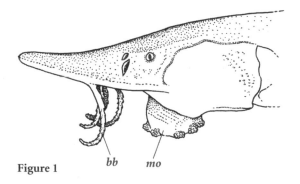

Figure 1

bb   *mo*

de: dentary; *hm:* hyomandibular; *ih:* interhyal; *paq:* palatoquadrate; *pmx:* pre-maxillary–maxillary; *nvr:* vertebrae fused to the neurocranium).

- The interhyal bone is hypertrophied. The hyoid arches (skeletal arches behind the jaws) of four boney fishes are shown in fig. 3 (*ach:* anterior ceratohyal; *pch:* posterior ceratohyal; *hh:* hypohyal): a sarcopterygian fossil, *Eusthenopteron* (3a), a sarcopterygian, *Latimeria* (3b), a chondrostean, *Acipenser* (3c), and an actinopteran, *Amia* (3d). In the chondrosteans, the interhyal is very large and plays the role of the symplectic bone (*sy*), meaning that it is involved in supporting the jaw and operculum.

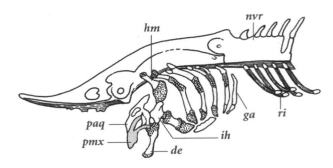

(a)

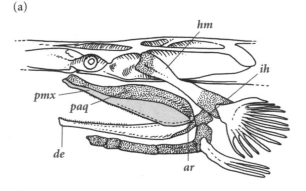

(b)

Figure 2

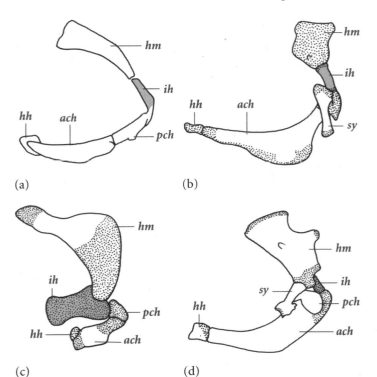

(a) (b)

(c) (d)

Figure 3

Examples

European sturgeon: *Acipenser sturio;* Siberian sturgeon: *A. baeri;* star sturgeon: *A. stellatus;* great sturgeon: *Huso huso;* American paddlefish: *Polyodon spathula;* Chinese paddlefish: *Psephurus gladius;* shovelnose sturgeon: *Scaphirhynchus platorynchus;* amu darya sturgeon: *Pseudoscaphirhynchus kaufmanni.*

Ecology

The chondrosteans include the world's largest freshwater fish, the great sturgeon (*Huso huso*). The largest individual captured of this species measured 8 meters long and weighed 1.3 tons. Certain chondrosteans spend part of their life cycle at sea. Some species, like the American paddlefish, feed on crustaceans and other small prey in the water column. Others, like the sturgeon or Chinese paddlefish, feed on fish and aquatic mollusks. All of the natural populations of these fishes are more or less threatened with extinction due to human activities and past over-fishing. Indeed, the meat and eggs of species from this group have been greatly appreciated for a long time; caviar comes from the eggs of *Acipenser gueldenstaedti, A. stellatus, A. baeri, A. transmontanus, Huso huso* and, for the last thirty years, also from *Polyodon spathula.*

Number of species: 26.
Oldest known fossils: *Errolichthys* from Madagascar dates from the lower Triassic (235 MYA) and may be the oldest chondrostean. Otherwise, it is *Chondrosteus* from England dating from the lower Jurassic (200 MYA).
Current distribution: North America and Eurasia.

Neopterygii

A few representatives

Longnose gar
Lepisosteus osseus
180 cm
node 6

Bowfin
Amia calva
up to 87 cm
node 8

John dory
Zeus faber
up to 70 cm
node 9

Elephant fish
Mormyrus rume
up to 87 cm
node 9

Some unique derived features

■ The dermal bones of the dorsal fin (the lepidotrichia) have an equal number of supporting bones, pterygiophores (or radials). The fin rays and their supporting bones are shown in fig. 1a on a teleost fish, *Lampanyctus leucopsarus*. A lateral view of an isolated ray of *Esox lucius* is illustrated in fig. 1b (*le*: lepidotrich; *dpg*: cartilaginous distal pterygiophore; *mpg*: median pterygiophore; *ppg*: proximal pterygiophore).

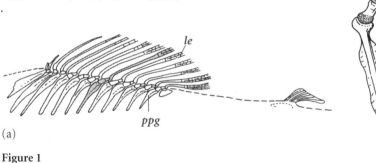

(a)

Figure 1

(b)

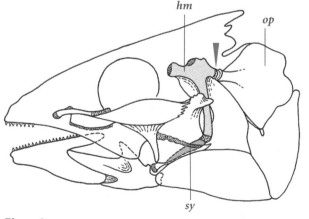

Figure 2

■ A new bone is involved in the articulation and support of the lower jaw and the operculum, the symplectic bone (*sy*). This bone is shown in fig. 2 on the mackerel.

■ Hyomandibular: the highest bone of the hyoidian arch that links the lower jaw to the skull, the hyomandibular (*hm*, fig. 2), has a branch that connects it to the opercular bone (*op*) of the operculum (see arrow on fig. 2).

Examples

■ GINGLYMODI: longnose gar: *Lepisosteus osseus*.
■ HALECOSTOMI: bowfin: *Amia calva*; European eel: *Anguilla anguilla*; northern pike: *Esox lucius*; guppy: *Lebistes reticulatus*; Atlantic bluefin tuna: *Thunnus thynnus*; John dory: *Zeus faber*.

Number of species: 23,676.
Oldest known fossils: *Elonichthys* is a palaeoniscid from the lower Carboniferous (355 MYA).
Current distribution: worldwide.

Ginglymodi

General Description

The ginglymodes are the gars, large freshwater vertebrates with an elongate body (up to 6 m in length). They are covered in a coat of thick diamond-shaped scales that articulate side-by-side and are covered in a layer of enamel (ganoin) that looks like porcelain and shines like ivory. Their heads carry a relatively long snout with numerous teeth and can resemble that of a crocodile. The dorsal and anal fins are symmetric, posteriorly angled, and directly opposite each other on the body. Vertebrates that resemble ginglymodes were very abundant during the Permian (250 MYA).

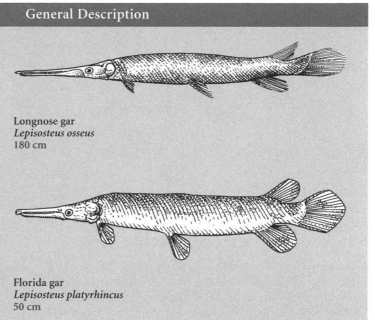

Longnose gar
Lepisosteus osseus
180 cm

Florida gar
Lepisosteus platyrhincus
50 cm

Some unique derived features

- The infra-orbital bones carry teeth (colored in fig. 1). However, this character is not found in all fossil species of the group.

- The vertebrae are opisthocoelous, meaning that they are convex on the anterior surface and concave on the posterior surface. Longitudinal sections of vertebral centra are shown in fig. 2, where one can see how an opisthocoelous vertebra (*ovr*) may develop

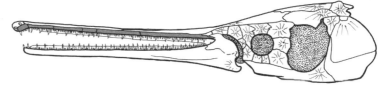

Figure 1

from an amphicoelous vertebra (*avr*) by the fusion of the intervertebral disk (*ivd*) to the anterior extremity.

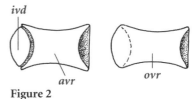

ivd

avr

ovr

Figure 2

Ecology

The ginglymodes are large freshwater predators with an ecology similar to that of the northern pike; they hide in aquatic vegetation and hunt fish using a lie-and-wait strategy. When these fishes find themselves in oxygen-poor waters, they are able to breathe air at the surface; the highly vascularized swim bladder plays the role of a lung. The longnose gar nests during the spring in slow-moving, shallow waters rich in vegetation. The rest of the year, they live in groups in deep water. It is here that they hibernate for the winter. In southern United States, the alligator gar is considered a "freshwater shark"; it is not good to eat and it can destroy fishnets when stealing the captured fish. For these reasons, this species is not popular among fishermen.

Number of species: 7.
Oldest known fossils: *Paralepisosteus* from western African dates from the lower Cretaceous (130 MYA).
Current distribution: North and Central America, from the Great Lakes to Panama.

Examples

Cuban alligator gar: *Atractosteus tristoechus;* longnose gar: *Lepisosteus osseus;* Florida gar: *L. platyrhincus.*

Halecostomi

A few representatives

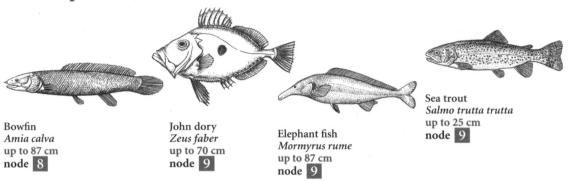

Bowfin
Amia calva
up to 87 cm
node 8

John dory
Zeus faber
up to 70 cm
node 9

Elephant fish
Mormyrus rume
up to 87 cm
node 9

Sea trout
Salmo trutta trutta
up to 25 cm
node 9

Some unique derived features

■ The maxillary bone is mobile on the upper jaw. The fixed position of the maxillary on the skull of a Senegal bichir, a cladistian, is shown in fig. 1a. A sketch of the boney head of a teleost fish, illustrating the mobility of the maxillary bone, is shown in fig. 1b.

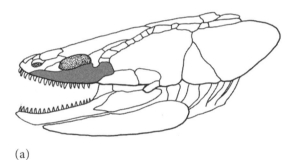

(a)

Figure 1

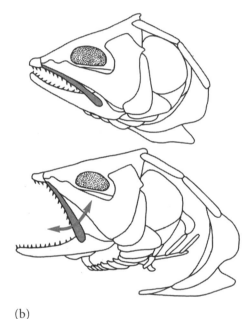

(b)

■ A new operculum bone, the interopercular bone (*iop*), appears. This bone is shown on the mackerel in fig. 2.

■ The quadratojugal, a bone from behind the upper jaw, has been lost. This bone has fused with another, the quadrate (*q*, fig. 2).

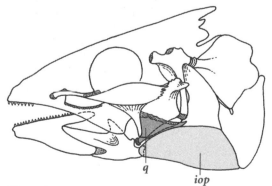

q

iop

Figure 2

Number of species: 23,669.
Oldest known fossils: *Acentrophorus* dates from the upper Permian (255 MYA).
Current distribution: worldwide.

Examples

■ HALECOMORPHI: bowfin: *Amia calva*.
■ TELEOSTEI: European eel: *Anguilla anguilla*; northern pike: *Esox lucius*; guppy: *Lebistes reticulatus*; Atlantic bluefin tuna: *Thunnus thynnus*; John dory: *Zeus faber*; dab: *Limanda limanda*.

Halecomorphi

General Description

Five halecomorph families inhabited the waters of the Mesozoic era. Only a single species of this group is still around today: the bowfin. This species has an elongate body and a long, spineless dorsal fin that undulates during slow movements. Indeed, each of the 42 to 53 rays of this fin can be moved by its own individual muscle. The head is armored in heavy, bony plates. The caudal fin is rounded. The swim bladder can play the role of a lung in certain conditions.

Bowfin
Amia calva
up to 87 cm

Some unique derived features

■ The symplectic and quadrate bones are both involved in the jaw articulation. Fig. 1 shows a partial view of the jaw articulation of *Amia* with the superficial dermal bones removed. The symplectic bone lies very low (*sy*) and participates with the quadrate bone (*q*) in the jaw articulation. For comparison, the placement of these bones is shown in fig. 2 on the mackerel, a recently evolved teleost fish; here, the symplectic bone lies behind the jaw articulation (*apa:* autopalatine; *Mc:* Meckel cartilage; *dpa:* dermopalatine; *ecp:* ectopterygoid, *enp:* endopterygoid; *hm:* hyomandibular; *mtp:* metapterygoid; *cop:* coronoid process).

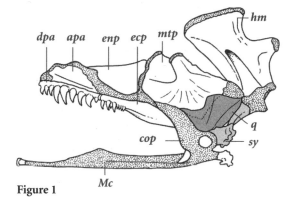

Figure 1

Ecology

The bowfin lives in stagnant fresh water. It is a voracious predator. The efficiency of the buccal capture is enhanced by the unique arrangement of the maxillary; this bone adopts a vertical position during the extremely rapid attack on its prey. The bowfin is able to breathe air using its swim bladder, which acts as a false lung. This species hibernates during the winter.

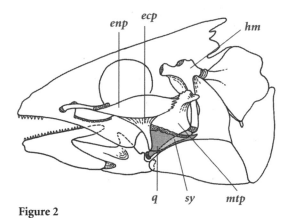

Figure 2

Number of species: 1.
Oldest known fossils: *Ospia* and *Broughia* date from the lower Triassic (245 MYA).
Current distribution: United States of America.

Examples

Bowfin: *Amia calva.*

see appendix 8, p. 527

Teleostei

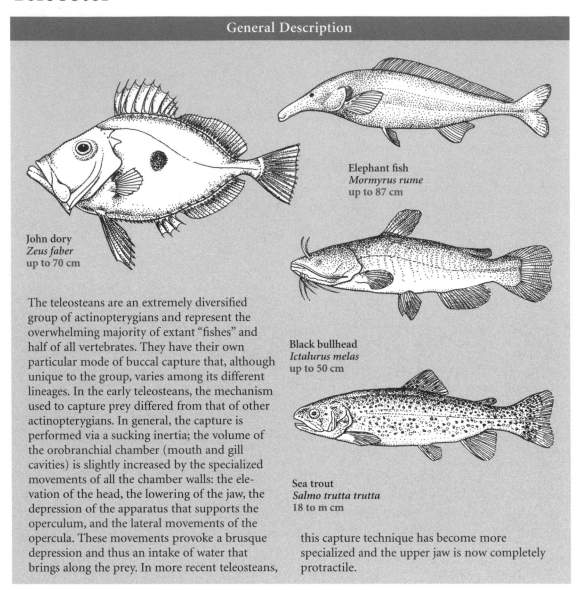

General Description

John dory
Zeus faber
up to 70 cm

Elephant fish
Mormyrus rume
up to 87 cm

Black bullhead
Ictalurus melas
up to 50 cm

Sea trout
Salmo trutta trutta
18 to m cm

The teleosteans are an extremely diversified group of actinopterygians and represent the overwhelming majority of extant "fishes" and half of all vertebrates. They have their own particular mode of buccal capture that, although unique to the group, varies among its different lineages. In the early teleosteans, the mechanism used to capture prey differed from that of other actinopterygians. In general, the capture is performed via a sucking inertia; the volume of the orobranchial chamber (mouth and gill cavities) is slightly increased by the specialized movements of all the chamber walls: the elevation of the head, the lowering of the jaw, the depression of the apparatus that supports the operculum, and the lateral movements of the opercula. These movements provoke a brusque depression and thus an intake of water that brings along the prey. In more recent teleosteans, this capture technique has become more specialized and the upper jaw is now completely protractile.

Some unique derived features

- The premaxillary bone of the upper jaw has become completely mobile. Figs. 1a and 1b show the mobile maxillary (*mx*) and immobile premaxillary (*pmx*) bones on a sketch of a non-teleost skull. The extreme mobility of the maxillary and premaxillary are then illustrated in fig. 1c on a carp skull.

- In the caudal skeleton, the ural neural arches separate from the vertebral bodies to form pairs of small bones, the uroneurals (*un1, un2, un3*, fig. 2c showing the caudal skeleton of a trout).

- Caudal fin: there is a visible symmetry in the caudal fin (fig. 2b: trout's tail). In other actinopterygians, the upper and lower lobes of the tail are not the same size: the caudal fin is asymmetric (fig. 2a: sturgeon's tail). The teleosteans acquired tail symmetry (fig. 2b) by the simultaneous development of the following bones:

the uroneurals (pairs of spines arranged on the caudal vertebrae, *un1–3*, fig. 2c) and the hypurals (flat bones on the lower lobe of the caudal fin, *hu1–6*, fig. 2c). The caudal fin remains fundamentally asymmetric because the vertebral column ends on the upper lobe.

- The deep asymmetry of the caudal fin is also due to the twisting of the caudal axis of the vertebral column at the first preural centrum (*pru1*, fig. 2c).

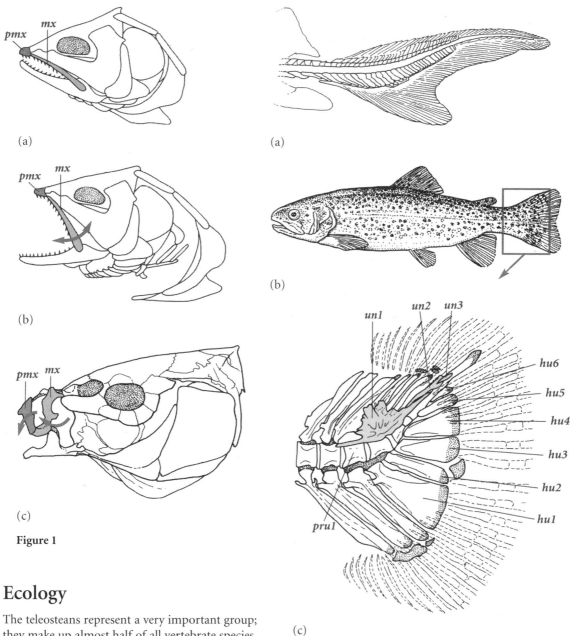

(a)

(b)

(c)

Figure 1

(a)

(b)

un1 un2 un3

hu6

hu5

hu4

hu3

hu2

hu1

pru1

(c)

Figure 2

Ecology

The teleosteans represent a very important group; they make up almost half of all vertebrate species. They have colonized all marine and freshwater habitats, from −11,000 m to +4,500 m and from hot springs (43°C) to cold polar waters (−1.8°C). They are highly diversified in terms of their morphology, ecology and behavior.

Number of species: 23,668.
Oldest known fossils: *Pholidophorus* dates from the lower Triassic (195 MYA).
Current distribution: worldwide.

Examples

Elephant fish: *Mormyrus rume*; European eel: *Anguilla anguilla*; Atlantic herring: *Clupea harengus*; tench: *Tinca tinca*; black bullhead: *Ictalurus melas*; northern pike: *Esox lucius*; Atlantic salmon: *Salmo salar*; sea trout: *Salmo trutta trutta*; guppy: *Lebistes reticulatus*; Atlantic bluefin tuna: *Thunnus thynnus*; John dory: *Zeus faber*; dab: *Limanda limanda*; perch: *Perca fluviatilis*.

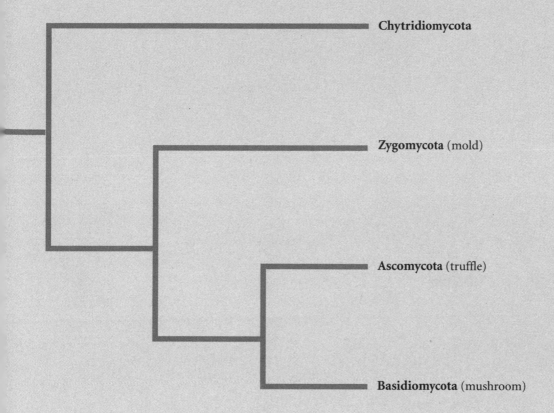

Chytridiomycota

Zygomycota (mold)

Ascomycota (truffle)

Basidiomycota (mushroom)

The list of orders is exhaustive. Some examples of the families found within each order are provided.

Amborellales (amborellaceae)

Nymphaeales (cabombaceae, nymphaeaceae)

Austrobaileyales (austrobaileyaceae, illiciaceae)

Chloranthales (chloranthaceae)

Ceratophyllales (ceratophyllaceae)

Magnoliales (annonaceae, magnoliaceae, myristicaceae)

Laurales (calycanthaceae, lauraceae, siparunaceae)

Canellales (canellaceae, winteraceae)

Piperales (aristolochiaceae, piperaceae, saururaceae)

Acorales (acoraceae)

Alismatales (alismataceae, araceae, butomaceae, hydrocharitaceae, juncaginaceae, posidoniaceae, potamogetonaceae, ruppiaceae, zosteraceae)

Miyoshiales (petrosaviaceae)

Dioscoreales (burmanniaceae, dioscoreaceae, nartheciaceae)

Pandanales (cyclanthaceae, pandanaceae, velloziaceae)

Liliales (colchicaceae, liliaceae, melanthiaceae, smilacaceae)

Asparagales (agapanthaceae, agavaceae, alliaceae, amaryllidaceae, asparagaceae, asphodelaceae, hemerocallidaceae, iridaceae, orchidaceae, ruscaceae)

Dasypogonales (dasypogonaceae)

Arecales (arecaceae)

Poales (bromeliaceae, cyperaceae, juncaceae, poaceae, sparganiaceae, typhaceae)

Commelinales (commelinaceae, haemodoraceae, philydraceae)

Zingiberales (cannaceae, heliconiaceae, musaceae, strelitziaceae, zingiberaceae)

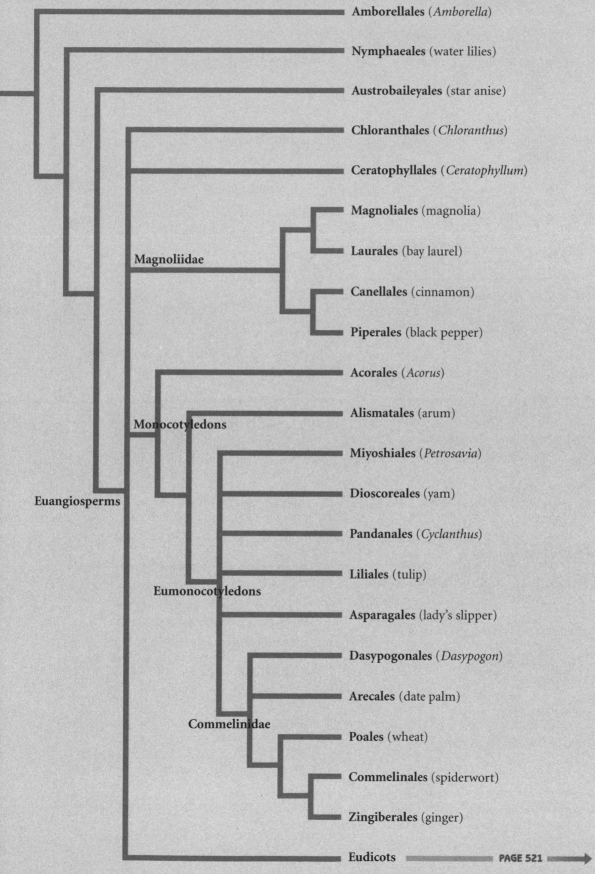

Amborellales (*Amborella*)

Nymphaeales (water lilies)

Austrobaileyales (star anise)

Chloranthales (*Chloranthus*)

Ceratophyllales (*Ceratophyllum*)

Magnoliales (magnolia)

Laurales (bay laurel)

Canellales (cinnamon)

Piperales (black pepper)

Acorales (*Acorus*)

Alismatales (arum)

Miyoshiales (*Petrosavia*)

Dioscoreales (yam)

Pandanales (*Cyclanthus*)

Liliales (tulip)

Asparagales (lady's slipper)

Dasypogonales (*Dasypogon*)

Arecales (date palm)

Poales (wheat)

Commelinales (spiderwort)

Zingiberales (ginger)

Eudicots PAGE 521

Magnoliidae

Monocotyledons

Euangiosperms

Eumonocotyledons

Commelinidae

The list of orders is exhaustive. Some examples of the families found within each order are provided.

Ranunculales (berberidaceae, papaveraceae, ranunculaceae)

Sabiales (sabiaceae)

Proteales (nelumbonaceae, platanaceae, proteaceae)

Buxales (buxaceae)

Trochodendrales (trochodendraceae)

Gunnerales (gunneraceae, myrothamnaceae)

Berberidopsidales (aextoxicaceae, berberidopsidaceae)

Santalales (loranthaceae, olacaceae, opliaceae, santalaceae)

Dilleniales (dilleniaceae)

Caryophyllales (amaranthaceae, cactaceae, caryophyllaceae, droseraceae, nepenthaceae, nyctaginaceae, plumbaginaceae, polygonaceae, tamaricaceae)

Saxifragales (crassulaceae, cynomoriaceae, daphniphyllaceae, grossulariaceae, iteaceae, paeoniaceae, saxifragaceae)

Vitales (vitaceae)

Crossosomatales (aphioiaceae, crossosomataceae, geissolomataceae)

Geraniales (geraniaceae, melianthaceae)

Myrtales (lythraceae, myrtaceae, onagraceae, vochysiaceae)

Zygophyllales (zygophyllaceae)

Celastrales (cclastraceae, parnassiaceae)

Malpighiales (euphorbiaceae, hypericaceae, linaceae, malpighiaceae, pandaceae, passifloraceae, phyllanthaceae, rhizophoraceae, salicaceae, violaceae)

Oxalidales (cephalotaceae, oxalidaceae)

Fabales (fabaceae, polygalaceae)

Rosales (cannabaceae, elaeagnaceae, moraceae, rhamnaceae, rosaceae, ulmaceae, urticaceae)

Cucurbitales (begoniaceae, cucurbitaceae)

Fagales (betulaceae, fagaceae, juglandaceae, myricaceae)

Brassicales (brassicaceae, caricaceae, resedaceae, tropaeolaceae)

Huertales (tapisciaceae)

Malvales (bixaceae, cistaceae, malvaceae, thymelaeaceae)

Sapindales (burseraceae, meliaceae, rutaceae, sapindaceae)

Cornales (cornaceae, hydrangeaceae, hydrostachyaceae)

Ericales (ebenaceae, ericaceae, primulaceae, sapotaceae, styracaceae, theaceae)

Garryales (garryaceae)

Gentianales (apocynaceae, gentianaceae, rubiaceae)

Lamiales (acanthaceae, bignoniaceae, calceolariaceae, gesneriaceae, lamiaceae, lentibulariaceae, oleaceae, orobanchaceae, paulowniaceae, plantaginaceae, scrophulariaceae, verbenaceae)

Solanales (convolvulaceae, hydroleaceae, solanaceae)

Aquifoliales (aquifoliaceae)

Apiales (apiaceae, araliaceae, pittosporaceae, torricelliaceae)

Asterales (asteraceae, calyceraceae, campanulaceae)

Dipsacales (caprifoliaceae, dipsacaceae, linnaeaceae, valerianaceae)

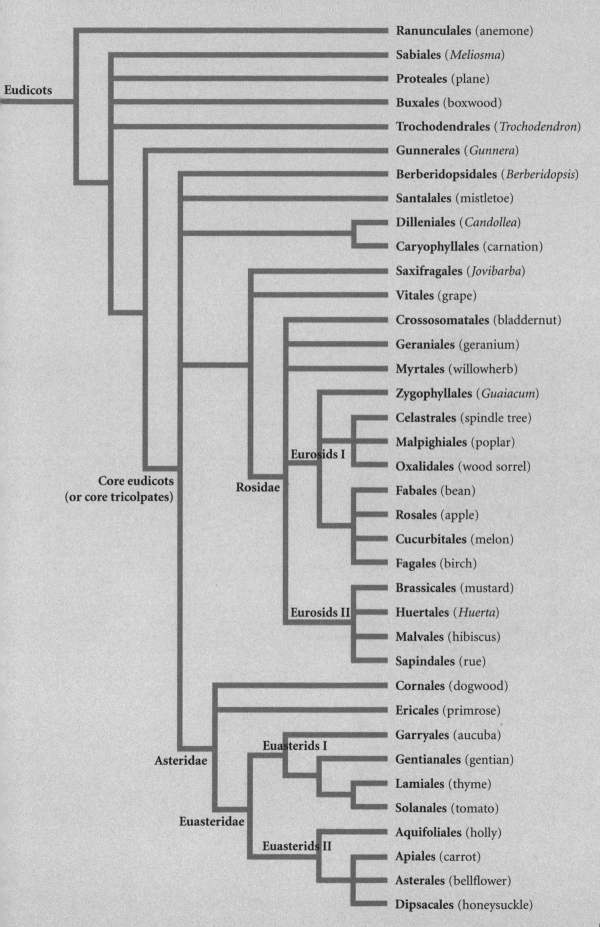

Eudicots

Ranunculales (anemone)
Sabiales (*Meliosma*)
Proteales (plane)
Buxales (boxwood)
Trochodendrales (*Trochodendron*)
Gunnerales (*Gunnera*)
Berberidopsidales (*Berberidopsis*)
Santalales (mistletoe)
Dilleniales (*Candollea*)
Caryophyllales (carnation)
Saxifragales (*Jovibarba*)
Vitales (grape)
Crossosomatales (bladdernut)
Geraniales (geranium)
Myrtales (willowherb)
Zygophyllales (*Guaiacum*)
Celastrales (spindle tree)
Malpighiales (poplar)
Oxalidales (wood sorrel)
Fabales (bean)
Rosales (apple)
Cucurbitales (melon)
Fagales (birch)
Brassicales (mustard)
Huertales (*Huerta*)
Malvales (hibiscus)
Sapindales (rue)
Cornales (dogwood)
Ericales (primrose)
Garryales (aucuba)
Gentianales (gentian)
Lamiales (thyme)
Solanales (tomato)
Aquifoliales (holly)
Apiales (carrot)
Asterales (bellflower)
Dipsacales (honeysuckle)

Core eudicots
(or core tricolpates)

Rosidae

Eurosids I

Eurosids II

Asteridae

Euasterids I

Euasteridae

Euasterids II

521

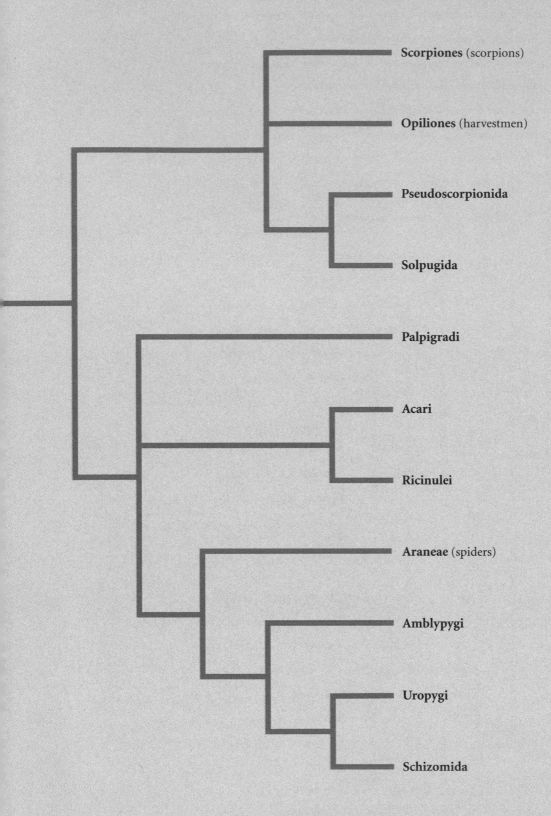

- Scorpiones (scorpions)
- Opiliones (harvestmen)
- Pseudoscorpionida
- Solpugida
- Palpigradi
- Acari
- Ricinulei
- Araneae (spiders)
- Amblypygi
- Uropygi
- Schizomida

Appendix 4: HEXAPODA

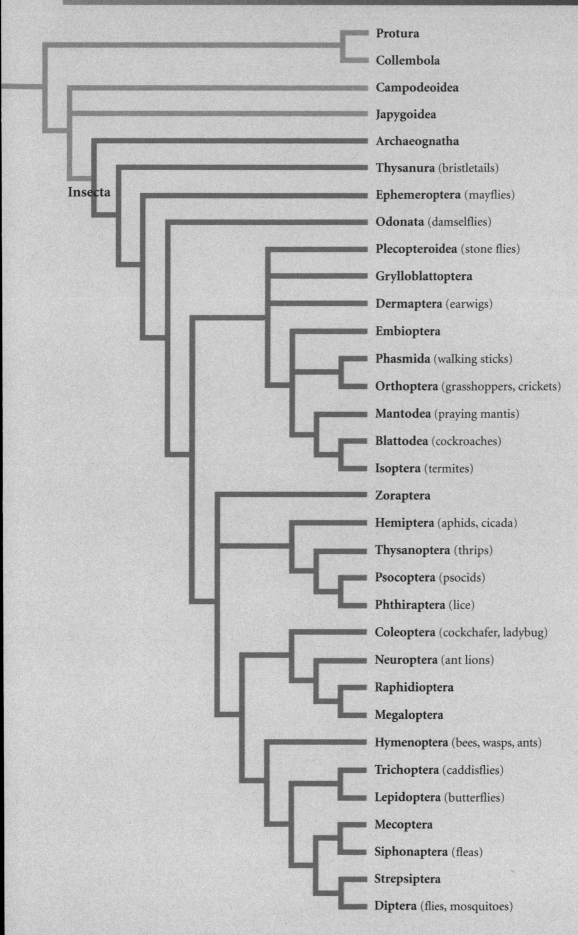

Protura

Collembola

Campodeoidea

Japygoidea

Archaeognatha

Thysanura (bristletails)

Ephemeroptera (mayflies)

Odonata (damselflies)

Plecopteroidea (stone flies)

Grylloblattoptera

Dermaptera (earwigs)

Embioptera

Phasmida (walking sticks)

Orthoptera (grasshoppers, crickets)

Mantodea (praying mantis)

Blattodea (cockroaches)

Isoptera (termites)

Zoraptera

Hemiptera (aphids, cicada)

Thysanoptera (thrips)

Psocoptera (psocids)

Phthiraptera (lice)

Coleoptera (cockchafer, ladybug)

Neuroptera (ant lions)

Raphidioptera

Megaloptera

Hymenoptera (bees, wasps, ants)

Trichoptera (caddisflies)

Lepidoptera (butterflies)

Mecoptera

Siphonaptera (fleas)

Strepsiptera

Diptera (flies, mosquitoes)

Insecta

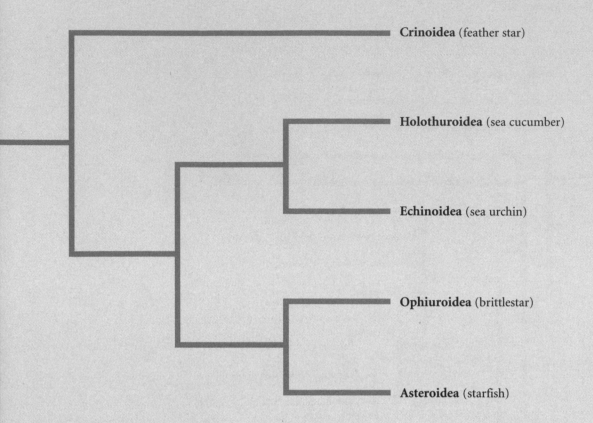

Crinoidea (feather star)

Holothuroidea (sea cucumber)

Echinoidea (sea urchin)

Ophiuroidea (brittlestar)

Asteroidea (starfish)

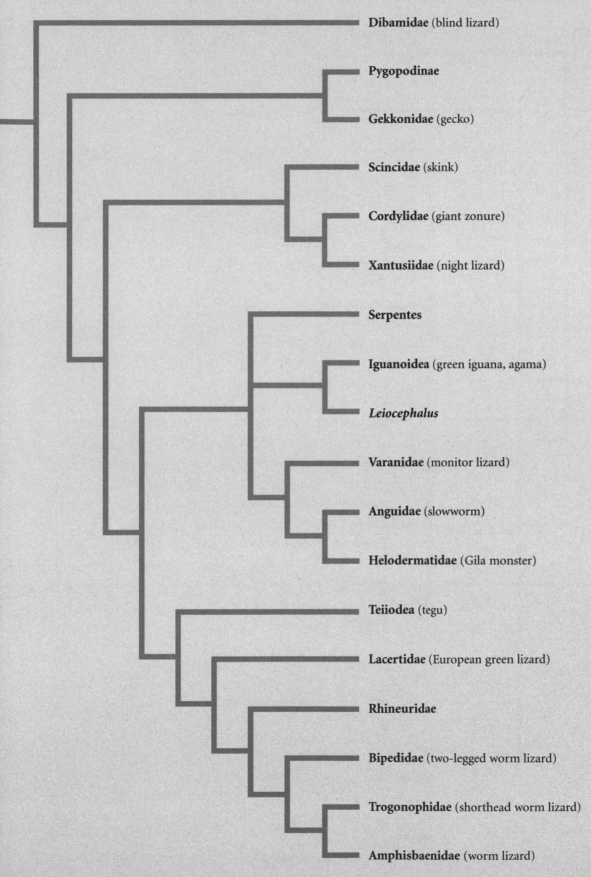

Dibamidae (blind lizard)

Pygopodinae

Gekkonidae (gecko)

Scincidae (skink)

Cordylidae (giant zonure)

Xantusiidae (night lizard)

Serpentes

Iguanoidea (green iguana, agama)

Leiocephalus

Varanidae (monitor lizard)

Anguidae (slowworm)

Helodermatidae (Gila monster)

Teiiodea (tegu)

Lacertidae (European green lizard)

Rhineuridae

Bipedidae (two-legged worm lizard)

Trogonophidae (shorthead worm lizard)

Amphisbaenidae (worm lizard)

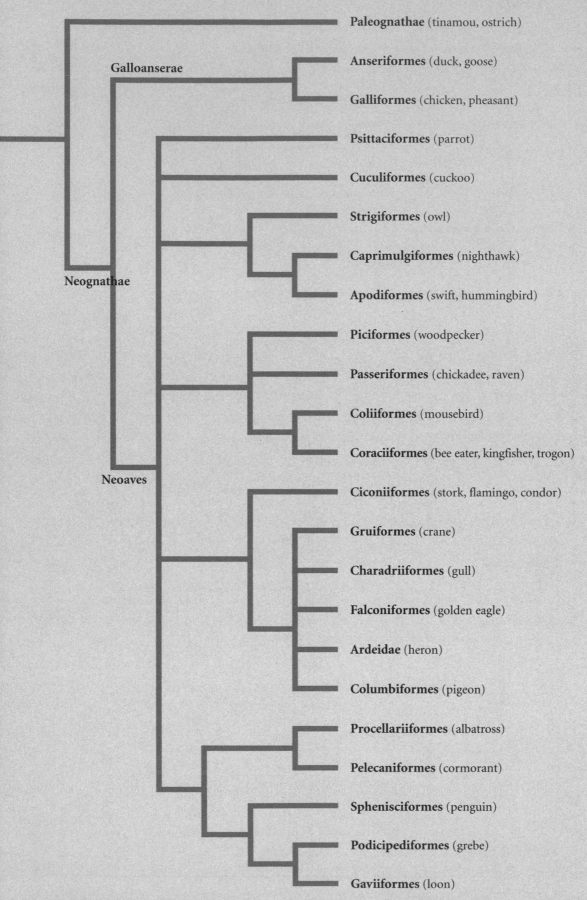

- Paleognathae (tinamou, ostrich)
- Galloanserae
  - Anseriformes (duck, goose)
  - Galliformes (chicken, pheasant)
- Neognathae
- Psittaciformes (parrot)
- Cuculiformes (cuckoo)
- Strigiformes (owl)
- Caprimulgiformes (nighthawk)
- Apodiformes (swift, hummingbird)
- Piciformes (woodpecker)
- Passeriformes (chickadee, raven)
- Coliiformes (mousebird)
- Coraciiformes (bee eater, kingfisher, trogon)
- Neoaves
- Ciconiiformes (stork, flamingo, condor)
- Gruiformes (crane)
- Charadriiformes (gull)
- Falconiformes (golden eagle)
- Ardeidae (heron)
- Columbiformes (pigeon)
- Procellariiformes (albatross)
- Pelecaniformes (cormorant)
- Sphenisciformes (penguin)
- Podicipediformes (grebe)
- Gaviiformes (loon)

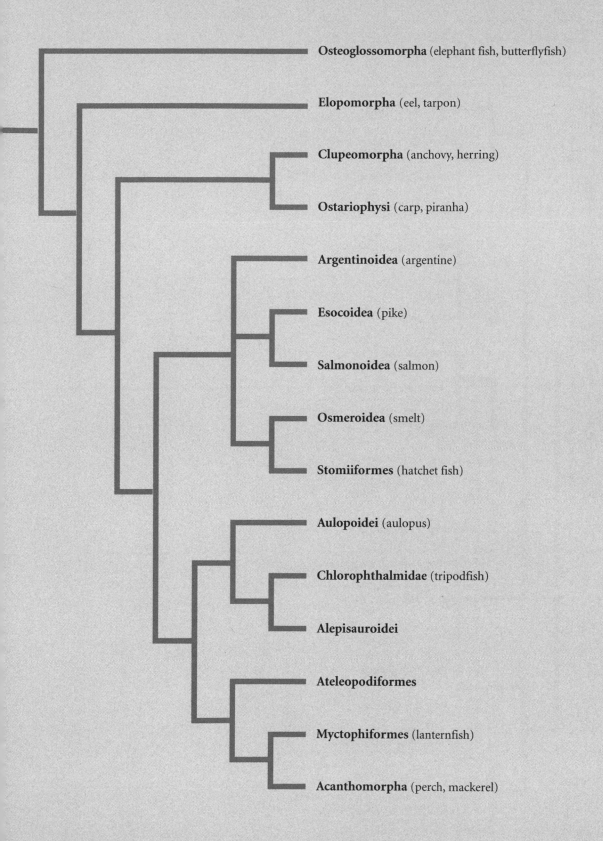

Osteoglossomorpha (elephant fish, butterflyfish)

Elopomorpha (eel, tarpon)

Clupeomorpha (anchovy, herring)

Ostariophysi (carp, piranha)

Argentinoidea (argentine)

Esocoidea (pike)

Salmonoidea (salmon)

Osmeroidea (smelt)

Stomiiformes (hatchet fish)

Aulopoidei (aulopus)

Chlorophthalmidae (tripodfish)

Alepisauroidei

Ateleopodiformes

Myctophiformes (lanternfish)

Acanthomorpha (perch, mackerel)

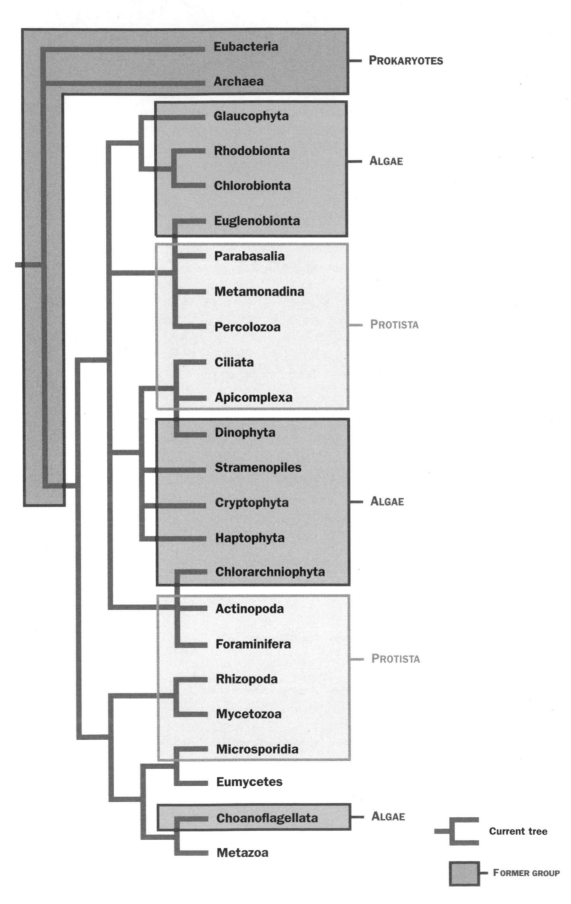

PROKARYOTES

- Eubacteria
- Archaea

ALGAE

- Glaucophyta
- Rhodobionta
- Chlorobionta
- Euglenobionta

PROTISTA

- Parabasalia
- Metamonadina
- Percolozoa
- Ciliata
- Apicomplexa

ALGAE

- Dinophyta
- Stramenopiles
- Cryptophyta
- Haptophyta
- Chlorarchniophyta

PROTISTA

- Actinopoda
- Foraminifera
- Rhizopoda
- Mycetozoa
- Microsporidia
- Eumycetes

ALGAE

- Choanoflagellata
- Metazoa

⊏ Current tree

▮ FORMER GROUP

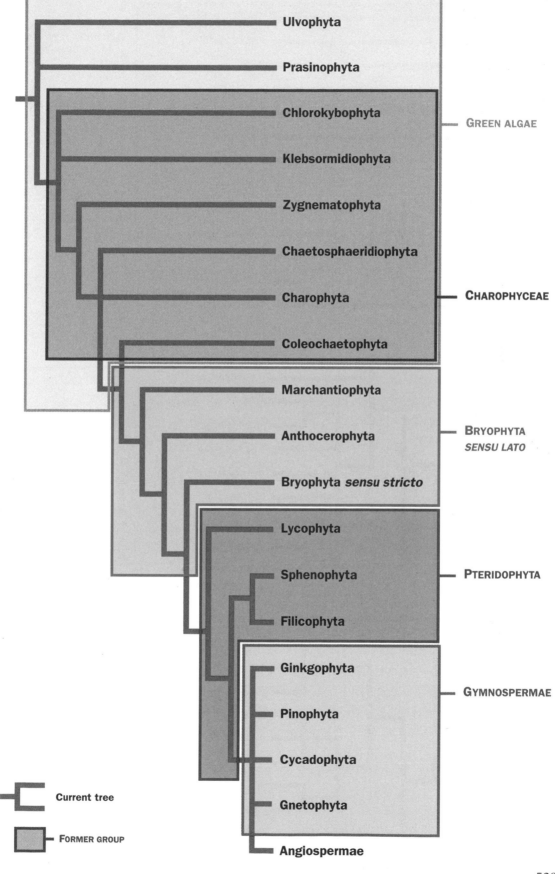

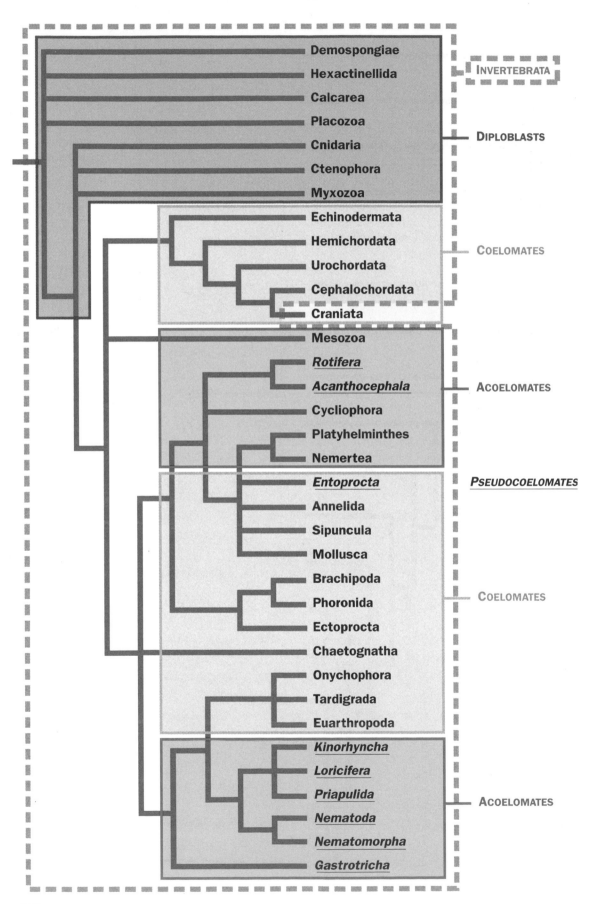

INVERTEBRATA

Demospongiae
Hexactinellida
Calcarea
Placozoa
Cnidaria — DIPLOBLASTS
Ctenophora
Myxozoa

Echinodermata
Hemichordata
Urochordata — COELOMATES
Cephalochordata
Craniata

Mesozoa
Rotifera
Acanthocephala — ACOELOMATES
Cycliophora
Platyhelminthes
Nemertea

Entoprocta — *PSEUDOCOELOMATES*
Annelida
Sipuncula
Mollusca
Brachipoda
Phoronida — COELOMATES
Ectoprocta
Chaetognatha
Onychophora
Tardigrada
Euarthropoda

Kinorhyncha
Loricifera
Priapulida — ACOELOMATES
Nematoda
Nematomorpha
Gastrotricha

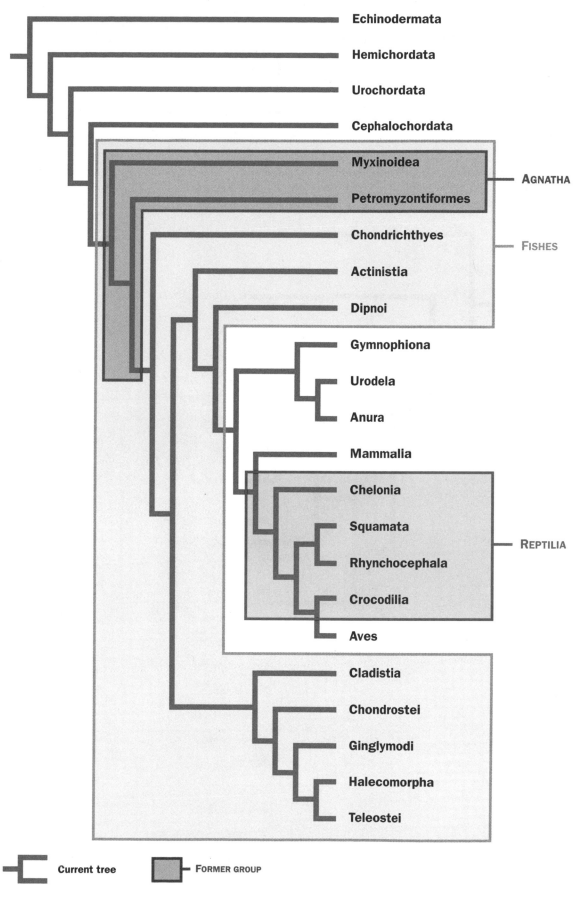

Echinodermata

Hemichordata

Urochordata

Cephalochordata

Myxinoidea

Petromyzontiformes

AGNATHA

Chondrichthyes

FISHES

Actinistia

Dipnoi

Gymnophiona

Urodela

Anura

Mammalia

Chelonia

Squamata

Rhynchocephala

Crocodilia

REPTILIA

Aves

Cladistia

Chondrostei

Ginglymodi

Halecomorpha

Teleostei

Current tree   FORMER GROUP

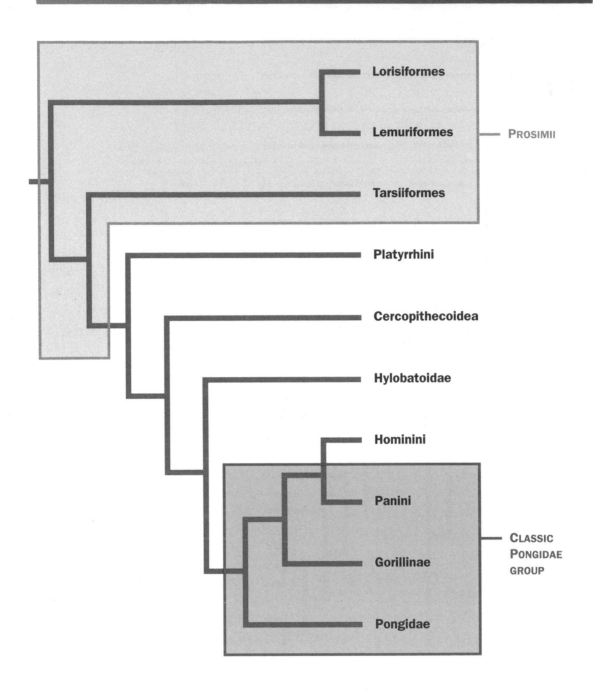

Lorisiformes

Lemuriformes

PROSIMII

Tarsiiformes

Platyrrhini

Cercopithecoidea

Hylobatoidae

Hominini

Panini

CLASSIC
PONGIDAE
GROUP

Gorillinae

Pongidae

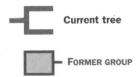

Current tree

FORMER GROUP

Sequenced Genomes

The following list presents the organisms for which the genome is either entirely sequenced or almost so, as of January 2006. For the Archaea and Eubacteria, genome sequencing of the listed species is complete, with a few rare exceptions. This is quite the opposite for the eukaryotes because, if we follow the exact definition of *complete*, even the genome of *Homo sapiens* is still not finished. The gaps are principally non-coding areas of the genome that would demand a large effort for a result that seems, at present, to hold little interest. In this context, we preferred to include organisms for which the genome sequence is complete (like *Saccharomyces cerevisiae*), but also those that are currently under study and for which all coding sequences will soon be finished (for example, the elephant *Loxodonta africana* or the chicken *Gallus gallus*). One will note that, given their obvious medical interest, many parasites have been sequenced, along with a wide variety of closely related species for some. Similarly, it is not surprising to find plants of agricultural interest—like wheat, soybean or rice—in the list, in addition to *Arabidopsis thaliana*. The number of organisms in each taxon (N) is also recalled.

Eubacteria (CHAPTER 2)

ε Proteobacteria (node **2** ; N: 232)

Campylobacter jejuni
Helicobacter hepaticus, H. pylori
Thimicrospira denitrificans
Wollinella succinogenes

δ Proteobacteria (node **3** ; N: 332)

Anaeromyxobacter dehalogenans
Bdellovibrio bacteriovorus
Desulfotalea psychrophila
Desulfovibrio desulfuricans, D. vulgaris
Desulfuromonas acetoxidans
Geobacter metallireducen, G. sulfurreducens
Pelobacter carbinolicus, P. propionicus

α Proteobacteria (node **4** ; N: 1,488)

Agrobacterium tumefaciens
Anaplasma marginale
Bartonella henselae, B. quintana
Bradyrhizobium japonicum
Brucella melitensis, B. suis
Caulobacter crescentus
Erlichia canis, E. ruminantium
Gluconobacter oxydans
Jannaschia sp.
Magnetococcus sp.
Magnetospirillum magnetotacticum
Mesorhizobium loti

Nitrobacter hamburgensis, N. winogradskyi
Novosphingobium aromaticivorans
Paracoccus denitrificans
Rhodobacter sphaeroides
Rhodospirillum rubrum
Rickettsia akari, R. conorii, R. prowazekii; R. sibirica, R. typhi
Silicibacter pomeroyi
Sinorhizobium meliloti
Sphingopyxis alaskensis
Wolbachia pipientis
Zymomonas mobilis

β Proteobacteria (node **5** ; N: 682)

Azoarcus sp.
Bordetella bronchiseptica, B. parapertussis, B. pertussis
Burkholderia cenocepacia, B. cepacia, B. mallei, B. pseudomallei, B. vietnamiensis, B. xenovarans
Chromobacterium violaceum
Cupriavidus metallidurans, C. necator
Dechloromonas aromatica
Methylobacillus flagellatus
Neisseria gonorrhoeae, N. meningitidis
Nitrosomonas europaea
Nitrospira multiformis
Polaromonas sp.
Ralstonia solanacearum
Rhodoferax ferrireducens
Rubrivivax gelatinosus
Thiobacillus denitrificans

γ Proteobacteria (node 6 ; N: 1,924)

Acinetobacter sp.
Azotobacter vinelandii
Buchnera aphidicola
Candidatus blochmannia floridanus
Chromohalobacter salexigens
Coxiella burnetii
Erwinia carotovora
Escherichia coli
Francisella tularensis
Haemophilus ducreyi, H. influenzae, H. somnus
Idiomarina loihiensis
Legionella pneumophila
Mannheimia succiniciproducens
Methylococcus capsulatus
Nitrosococcus oceani
Pasteurella multocida
Photobacterium profundum
Photorhabdus luminescens
Pseudomonas aeruginosa, P. fluorescens, P. putida,
*   P. syringae*
Psychrobacter cryopegella
Saccharophagus degradans
Salmonella enterica, S. typhimurium
Shewanella amazonensis, S. baltica, S. denitrificans,
*   S. frigidimarina, s. oneidensis, S. putrefaciens*
Shigella flexneri
Thiomicrospira crunogena
Vibrio cholerae, V. fischeri, V. parahaemolyticus,
*   V. vulnificus*
Wigglesworthia glossinidia
Xanthomonas axonopodis citri, X. campestris, X. oryzae
Xylella fastidiosa
Yersinia pestis, Y. pseudotuberculosis

Bacillus / Clostridium group (node 7 ; N: 2,513)

Bacillus anthracis, B. cereus, B. clausi, B. halodurans,
*   B. licheniformis, B. subtilis, B. thuringiensis*
Clostridium acetobutylicum, C. perfringens, C. tetani,
*   C. thermocellum*
Desulfitobacterium hafniense
Enterococcus faecalis, E. faecium
Exiguobacterium sp.
Geobacillus kaustophilus
Lactobacillus acidophilus, L. brevis, L. casei, L. delbrueckii
*   bulgaricus, L. gasseri, L. johnsonii, L. plantarum*
Lactococcus lactis lactis, L. lactis cremoris
Leuconostoc mesenteroides
Listeria innocua, L. monocytogenes
Mesoplasma florum
Moorella thermoacetica
Mycoplasma gallisepticum, M. genitalium, M.
*   hyopneumonae, M. mobile, M. mycoides,*
*   M. penetrans, M. pneumoniae, M. pulmonis*
Oceanobacillus iheyensis
Oenococcus oeni
Onion yellows phytoplasma
Pediococcus pentosaceus
Staphylococcus aureus, S. epidermidis
Streptococcus agalactiae, S. mutans, S. pneumoniae,
*   S. pyogenes, S. suis, S. thermophilus*
Symbiobacterium thermophilum

Thermoanaerobacter tengcongensis
Ureaplasma parvum

Actinobacteria (node 8 ; N: 2,134)

Arthrobacter sp.
Bifidobacterium longum
Brevibacterium linens
Corynebacterium diphtheriae, C. efficiens, C. glutamicum
Frankia sp.
Kineococcus radiotolerans
Leifsonia xyli
Mycobacterium bovis, M. leprae, M. avium
*   paratuberculosis, M. tuberculosis*
Nocardia farcinica
Propionibacterium acnes
Rubrobacter xylanophilus
Streptomyces avermitilis, S. coelicolor
Thermobifida fusca
Tropheryma whipplei

Cyanobacteria (node 9 ; N: 295)

Anabaena variabilis
Crocosphaera watsonii
Gloeobacter violaceus
Nostoc punctiforme
Prochlorococcus marinus
Synechococcus elongatus
Synechocystis sp.
Tricodesmium erythraeum

Thermus / Deinococcus group (node 10; N: 52)

Deinococcus geothermalis, D. radiodurans
Thermus thermophilus

Green sulfur bacteria (node 11; N: 28)

Chlorobium limicola, C. phaeobateroides, C. tepidum,
*   C. vibrioforme*
Chlorochromatium aggregatum
Pelodictyon luteolum, P. phaeoclathratiforme
Prosthecochloris aesturii

Flavobacteria (cytophagales) (node 12; N: 152)

Cytophaga hutchinsonii

Bacteroides (node 13; N: 181)

Bacteroides fragilis, B. thetaiotaomicron
Fusobacterium nucleatum
Porphyromonas gingivalis

Spirochaetes (node 14; N: 341)

Borrelia burgdorferi, B. garinii
Leptospira interrogans
Treponema denticola, T. pallidum

Chlamydiales (node 15; N: 107)

Chlamydia muridarum, C. pneumoniae, C. trachomatis
Chlamydophila caviae
Parachlamydia sp.

Planctomycetales (node 16; N: 82)

Rhodopirellula baltica

Green non-sulfur bacteria (node 17; N: 16)

Chloroflexus aurantiacus

Aquificales (node 18; N: 9)

Aquifex aeolicus

Thermotogales (node 19; N: 25)

Thermotoga maritima

Archaea (CHAPTER 3)

Thermoproteales (node 3; N: 16)

Pyrobaculum aerophilum

Desulfurococcales (igneococcales) (node 4; N: 14)

Aeropyrum pernix

Sulfolobales (node 5; N: 25)

Sulfolobus solfataricus, S. tokodaii

Thermococcales (node 7; N: 39)

Nanoarchaeum equitans (nanoarchaea), *Pyrococcus abyssi, P. furiosus, P. horikoshii, Thermococcus kodakaraensis*

Methanopyrales (node 9; N: 1)

Methanopyrus kandleri

Methanobacteriales (node 10; N: 109)

Methanothermobacter (synonym: *Methanobacterium*) *thermoautotrophicus*

Methanococcales (node 11; N: 20)

Methanocaldococcus (synonym: *Methanococcus*) *jannaschii, Methanococcus maripaludis*

Thermoplasmatales (node 12; N: 5)

Ferroplasma acidarmanus, Thermoplasma acidophilum, T. volcanium, Picrophilus torridus

Archaeoglobales (node 13; N: 6)

Archaeoglobus fulgidus

Halobacteriales (node 14; N: 105)

Haloarcula marismori, Halobacterium salinarum

Methanomicrobiales (node 15; N: 33)

Methanospirillum hungatei

Methanosarcinales (node 16; N: 36)

Methanococcoides burtonii, Methanosarcina acetivorans, M. barkeri, M. mazei

Eukaryotes (CHAPTER 4)

Rhodobionta (node 6; N: 5,500)

Cyanidioschyzon merolae
Galderia sulphuraria

Chlorobionta (node 7 and chaps. 5–6; N: 277,902)

Arabidopsis thaliana
Chlamydomonas reinhardtii
Lotus corniculatus var. *japonicus*
Lycopersicon esculentum
Medicago truncatula
Micromonas pusilla
Mesembryanthemum crystallinum
Oryza sativa
Ostreococcus tauri
Populus trichocarpa
Solanum tuberosum
Sorghum bicolor
Triticum aestivum
Volvox carteri

Ciliata (node 10; N: 8,000)

Paramecium tetraurelia
Sterkiella histriomuscorum (formerly: *Oxytricha trifallax*)
Tetrahymena thermophila

Apicomplexa (node 12; N: 5,000)

Babesia bigemine, B. bovis
Cryptosporidium parvum
Eimeria tenella
Neospora caninum
Plasmodium falciparum, P. knowlesi, P. berghei, P. yoelii
Perkinsus marinus
Sarcocystis neurona
Theileria annulata, T. parva
Toxoplasma gondii

Stramenopiles (node 13; N: 105,922)

Aureococcus anophagefferens
Phaeodactylum tricomutum
Phytophthora infestans
Thalassiosira pseudonana

Cryptophyta (node 14; N: 200)

Guillardia theta (nucleomorph)

Euglenobionta (node 21; N: 1,398)

Trypanosoma brucei, T. cruzi, T. congolense, T. vivax
Leishmania major, L. braziliensis, L. infantum, L. berghei, L. chabaudi, L. falciparum, L. gallinaceum, L. knowlesi, L. reichenowi, L. vivax

Parabasalia (node 22; N: 350)

Trichomonas vaginalis

Metamonadina (node 23; N: 300)

Giardia intestinalis
Spironucleus vortens

Percolozoa (node 24; N: 20)

Naegleria gruberi

Eumycetes (node 28; N: 100,000)

• Ascomycota
Ashbya gossypii
Aspergillus fumigatus, A. nidulans
Candida albicans, C. dubliniensis, C. glabrata
Cryptococcus neoformans

Debaryomyces hanserii
Eremothecium gossypii
Fusarium graminearum
Gibberella zeae
Kluyveromyces cerevisiae
Magnaporthe grisea
Neurospora crassa
Pneumocystis carinii
Saccharomyces cerevisiae
Schizosaccharomyces pombe
Yarrowia lipolytica

- **Basidiomycota**

Cryptococcus neoformans
Ustilago maydis

Microsporidia (node **29**; N: 800)

Encephalotizoon cuniculi

Metazoa (node **32** and **chaps. 7–15**; N: 1,211, 612)

- **Insecta**

Acyrthosiphon pisum
Aedes aegyptii
Anopheles gambiae
Apis mellifera
Bicyclus anynana
Bombyx mori
Culex pipiens
Daphnia pulex
Drosophila melanogaster
Glossina morsitans morsitans
Ixodes scapularis
Lutzomyia longipalpis
Nasonia vitripennis
Pediculus humanus
Rhodnius prolixus
Tribolium castaneum

- **Nematoda**

Brugia malayi
Caenorhabditis elegans

- **Platyhelminthes**

Echinococcus granulosus
Schistosoma mansoni
Schmidtea mediterranea

- **Chordata (except mammals)**

Ciona intestinalis, C. savignyi
Danio rerio
Gallus gallus
Leucoraja erinacea
Oikopleura dioica
Takifugu rubripes
Tetraodon nigroviridis
Xenopus tropicalis

- **Mammalia**

Atelerix albiventris
Bos taurus
Callithrix jacchus
Canis familiaris, C. latrans, C. lupus
Cavia porcellus
Dasypus novemcinctus
Echinops telfairi
Felis catus
Homo sapiens
Lemur catta
Loxodonta africana
Macaca mulatta
Monodelphis domestica
Mus musculus
Pan troglodytes
Rattus norvegicus

- **Mollusca**

Aplysia californica

Mycetozoa (node **35**; N: 532)

Dictyostelium discoideum
Entamoeba histolytica, E. invadens, E. moshkovskii,
    E. terrapinae
Physarum polycephalum

General Bibliography

BENTON M., *The Fossil Record II*. Chapman and Hall, London. 1993.

BENTON M., *Vertebrate Palaeontology*. 2nd ed. Chapman & Hall, London. 1997.

BRIGGS D. E. G. and CROWTHER P. R., *Palaeobiology: A Synthesis*. Blackwell Science, Oxford. 1990.

BOLD H. C. and WYNNE M. J., *Introduction to the Algae*. Prentice-Hall, Englewood Cliffs, New Jersey. 1978.

BRUSCA R. C. and BRUSCA G. J., *Invertebrates*. 2nd ed. Sinauer Associates Inc., Sunderland, Massachusetts. 2002.

CONROY G., *Primate Evolution*. W.W. Norton and Co., New York. 1990.

CORLISS J. O., *The Ciliated Protozoa*. 2nd ed. Pergamon Press, Oxford. 1979.

CUSSET G., *Botanique. Les embryophytes*. Masson, Paris. 1997.

DE PUYTORAC P., GRAIN J. and MIGNOT J. P., *Précis de Protistologie*. Boubée, Paris. 1987.

DEVILLERS C. and CLAIRAMBAULT P., *Précis de Zoologie. Vertébrés*. Masson, Paris. 1976.

DORIT R. L., WALKER W. F. and BARNES R. D., *Zoology*. Saunders College Publishing, Philadelphia. 1991.

ELDREDGE N. and STANLEY S., *Living Fossils*. Springer-Verlag, Berlin, New York. 1984.

FRITSCH F. E., *The Structure and Reproduction of the Algae*. 1961. 4th ed. Cambridge University Press, Cambridge. 1984.

GOODRICH E., *Studies on the Structure and Development of Vertebrates*. 2nd ed. Dover, New York. 1958.

GRAHAM L. E. *Origin of Land Plants*. John Wiley & Sons, New York. 1993.

GRASSÉ P. P., *Traité de Zoologie*. Masson, Paris.

GRZIMEK B., *Le monde animal*. Stauffacher SA, Zurich. 13 volumes. 1975.

HAYWARD P. J. and RYLAND J. S. *Handbook of the Marine Fauna of North-West Europe*. Oxford University Press, Oxford. 1998.

HENNIG W., *Taschenbuch des speziellen Zoologie. Wirbellose I und II*. Verlag Harri Deutsch. Thun, Frankfurt am Main. 1984.

HOLT J. G. and KRIEG N. R., *Bergey's Manual of Systematic Bacteriology*. Williams & Wilkins Company, Baltimore. 1984–1989.

HYMAN L. H., *The Invertebrates: Protozoa through Ctenophora*. McGraw-Hill Book Co., New York. 1940.

JANVIER P., *Early Vertebrates*. Oxford University Press, Oxford. 1996.

JOLLIE M., *Chordate Morphology*. Coll. "Reinhold books in the biological sciences." Chapman & Hall, London. 1963.

JUDD W. S. et al., *Plant Systematics: A Phylogenetic Approach*. Sinauer Associates Inc. Sunderland, Massachusetts. 1999.

KARDONG K., *Vertebrates*. 2nd ed. McGraw-Hill, Boston. 1997.

KENRICK P. and CRANE P. R., *The Origin and Early Diversification of Land Plants: A Cladistic Study*. Smithsonian Institution Press, Washington. 1997.

MACDONALD D. W., *Encyclopédie des animaux*. Solar, Paris. 1990.

MARGULIS L. et al., *Handbook of Protoctista*. Jones & Bartlett Publishers, Boston. 1990.

MARGULIS L., McKHANN H. I. and OLENDZENSKI L., *Illustrated Glossary of Protoctista*. Jones & Bartlett Publishers, Boston. 1990.

MEGLITSCH P. A. and SCHRAM F. R., *Invertebrate Zoology*. 3rd ed. Oxford University Press, Oxford. 1991.

NELSON J. S., *Fishes of the World*. 3rd ed. John Wiley & Sons, New York. 1994.

NIELSEN C., *Animal Evolution*. Oxford University Press, Oxford. 2001.

NOVAK R.M. and PARADISO J. L., *Mammals of the World*. 4th ed. Johns Hopkins Univ. Press, Baltimore, London. 2 vol. 1983.

PRESCOTT L. M., HARLEY J. P. and KLEIN D. A., *Microbiology*. 4th ed. McGraw-Hill, Boston. 1999.

REMANE A., STORCH V. and WELSCH U., *Systematische Zoologie*. Gustav Fischer Verlag, Stuttgart. 1980.

REVIERS B. DE, *Biologie et phylogénie des algues*, vols. 1 and 2. Belin, Paris. 2002–2003.

RUPPERT E. E. and BARNES R. D., *Invertebrate Zoology*. Saunders College Publishing, New York. 1993.

TAYLOR J., *Origin and Evolutionary Radiation of the Mollusca*. Oxford University Press, London. 1996.

VAN DEN HOECK C., MANN D. G. and JAHNS H. M., *Algae. An Introduction to Phycology*. Cambridge University Press, Cambridge, 1998.

Glossary

n: noun
pn: proper noun
adj: adjective

ALIGNMENT n. An operation that consists of arranging portions of similar sequences, one under the other, and inserting blanks in such a way as to minimize their differences (we can align genes of the same multi-gene family*, genes of different species, etc.). If the genes are homologous*, amino acid or nucleotide differences between the aligned sequences are signs of past mutations.

ANALOGY n. Similar characters that fulfill the same biological functions.

APOMORPHIC adj. In a series of character state transformations, this designates the derived state with reference to the ancestral state. One can also say apomorphic state or, by extension, apomorphic character or derived character.

AUTAPOMORPHIC adj. When comparing characters among sister groups*, this adjective refers to the derived (or apomorphic*) state of a character that is unique to one of the two groups. By extension, one speaks of autapomophic characters or an autapomorphy.

CATEGORY n. In zoological and botanical nomenclature, a category is a hierarchical level or rank. For example, species, genus, tribe, family, order, class, and phylum are all different categories. To use imagery, a category is like a box and the taxon* is the content of this box. The categories are hierarchical such that each is part of the higher-ranking categories and incorporates all of the lower-ranking categories.

CHARACTER n. A set of attributes for which we can formulate at least one hypothesis of homology* (primary homology).

CHARACTER MATRIX n. A two-way table that generally has a series of species (or taxa) in rows and a series of characters in columns. The coded character states for each taxon are found within the cells of the table. A cladistic analysis of characters gives rise to this type of matrix with the character states generally coded as 0, 1, 2, etc.

CLADE n. From the Greek word that signifies branch. A clade is a strictly monophyletic taxon*, meaning that it contains the ancestor and all its descendants.

CLADISTIC 1. adj. Pertaining to a phylogenetic classification that must contain only monophyletic taxa or clades*. **2.** n. A synonym of *cladism*. A method of analyzing characters that attempts to determine the evolutionary sequence of transformations, that is, to identify their plesiomorphic*

(primitive) and apomorphic* (derived) states (Hennig's* method, 1966). Also see polarization*.

CLADOGRAM n. A diagram or "tree" that expresses a hypothesis about the phylogenetic relationships (i.e., sister-group relationships) among numerous taxa* and results from a cladistic* analysis. Each branching point, or node, is defined by one or more synapomorphies*. A cladogram therefore represents the distribution of apomorphic* characters for a given set of taxa*. A cladogram is also a nested classification.

CLASSIFICATION (BIOLOGICAL) n. The construction of a meaningful link among objects that has the form of a hierarchical system. In biology, this is the arrangement of living organisms into hierarchical groups according to a set of pre-defined rules (typologic, phylogenetic*, ecological, etc.).

CONGRUENT adj. Equivalent, or in agreement. More specifically, two phylogenetic trees that show the same topology—that is, the same relationships among taxa*—are said to be congruent.

CONVERGENCE n. Similar attributes that have appeared independently in different taxa* and thus have not been inherited from a common ancestor. A convergence that appears in closely related taxa is called a parallelism*. A convergence is a type of homoplasy*.

DELETION n. A type of mutation* in DNA that results in the disappearance of one or several base pairs. By extension, a deletion in a protein corresponds to the loss of one or several amino acids. Also see insertion*.

DENDROGRAM n. Comes from the Greek words for "tree" and "drawing." A diagram that expresses the links among taxa using a series of branches. Cladograms*, phenograms* and phylograms* are different types of dendrograms that are constructed according to their own set of rules.

DISTANCE (METHOD) n. Rather than determining the sequence of evolutionary transformations individually for each character*, this analysis measures the overall degree of difference between two character assemblages (i.e., between two taxa*) using a unique continuous variable, the "distance." There is therefore a distance value for each pair of taxa in the analysis. The distances are arranged in a matrix. The tree produced from these distances is called a phenogram* because it is based on the global similarity* among taxa*.

DISTANCE MATRIX n. A two-way table with the same series of species (or taxa) in both the columns and rows. The distance* (a number) that separates a pair of species is found in the corresponding cell of the table. A distance matrix therefore contains a distance for each possible pair of species (or taxa).

DUPLICATION n. A process by which a gene is copied and transposed to another location in the genome. Genes that are related in this way form multi-gene families*.

EQUIPARSIMONIOUS adj. Pertaining to two or more models or theories that require the same number of hypotheses. In systematics, several equiparsimonious trees involve the same number of evolutionary steps (hypotheses of mutation* or character* transformations).

GRADE n. **1.** The general degree of organization reached by a group at a given period of time. **2.** A group of living organisms that show a state of progress or evolutionary perfection (as defined by Huxley, 1957–58). In contrast to a clade*, a grade can be paraphyletic*.

GRADIST n. Evolutionary systematist who classes organisms by grades* as well as by clades* (not to be confused with gradualist*).

GRADUALIST n. Biologist who thinks that organismal evolution occurs gradually over time (there are no accelerations or stases).

HENNIG, WILLI pn. German entomologist (1913–1976) and founder of phylogenetic systematics*. The three major principles of Hennig's method are as follows:
—Studying the phylogeny starts with searching for sister groups* (and not the ancestor-descendant relationships). From an operational perspective, this starts with the polarization* of the characters*.
—The presence of a synapomorphic* character must be interpreted as an indication of a close relationship.
—The classification follows the constructed phylogeny.

HOMOLOGY (HOMOLOGOUS CHARACTERS) n. **1.** Two structures are homologous if, in different living organisms, they show the same fundamental organization and the same essential connections with neighboring organs, despite variations in the appearance of these structures (Owen, 1843). Historically, this notion comes from Geoffroy Saint-Hilaire (1818). This definition is one of positional homology and after being initially detected, can be used to determine the relationships among organisms. In our times, we also call this "primary homology" because it corresponds to a hypothesis of homology formulated at the beginning of a phylogenetic analysis. **2.** Similarity among organs, or parts of organs, in one or many species when we can presume that this similarity was inherited from a common ancestor. This definition is also referred to as "homology by common ancestry" or "secondary homology." As common ancestry can be identified on a tree, a secondary homology among characters is shown on the most parsimonious tree at the end of the analysis.

Remark: according to Hennig, different attributes associated with transformation steps from the same original attribute are homologous. This definition equally applies to molecular characters. Therefore, two similar attributes are assumed *a priori* to be homologous (primary homology), unless shown otherwise (Hennig's principle that we apply when aligning nucleotide or protein sequences).

HOMOPLASY n. A similarity among one or several species, organs, organ parts, DNA or protein sequences when we can presume or see on a tree that this correspondence is not due to an inheritance from a common ancestor. We distinguish among several types of homoplasy: convergence*, parallelism*, and reversion*.

Remark: for molecular characters, homoplasy is not often detectable *a priori*, but rather is revealed by the most parsimonious tree. For morphological characters, detailed analyses of the characters and of their primary homology sometimes enables one to detect homoplasies before the tree is constructed. As in the first case, homoplasies that are not initially detected will be revealed by the most parsimonious tree.

INGROUP n. A set of taxa for which we would like to determine their phylogenetic relationships. This group opposes the outgroup*, a group known *a priori* to lie outside the taxa of interest (or ingroup).

INSERTION n. A type of mutation* in DNA that results in the introduction of one or more additional base pairs to the sequence*. By extension, an insertion in a protein adds one or several amino acids* to the sequence. Also see deletion*.

LEAF n. A terminal taxon in a tree ("branch tip").

MOLECULAR CLOCK n. According to this hypothesis, molecules of the same functional class evolve regularly over time and at a similar rate among different lineages. In this way, the observed molecular divergence currently found between homologous sequences in different species can be used to estimate the time lapsed since their divergence from a common ancestor (or the time of divergence*).

MONOPHYLETIC GROUP n. In phylogenetic systematics*, a group that includes an ancestral species and all of its descendants. A monophyletic group is defined by at least one synapomorphy*. We also refer to a monophyletic group as a clade*.

MORPHOCLINE n. **1.** A sequence of character transformations within an evolutionary series (also see polarization*), going from the most "primitive" (plesiomorphic*) to the most recently derived (apomorphic*) state. **2.** More generally, an ordered scale of character states in which each state can be derived from the one that precedes it.

MULTIFURCATION (POLYTOMY) n. A node* in a tree that connects at least three terminal or internal branches. A multifurcation signals that the phylogenetic relationships of the groups (branches) involved could not be resolved.

MULTI-GENE FAMILY n. A group of genes produced by duplications*. They are therefore related, homologous genes found within the same genome.

MUTATION n. Designates any type of alteration to the genome that is not repaired. It is typically a replication error in a DNA sequence. By extension, we speak of protein "mutations," which are, in fact, the result of an alteration to the DNA sequence that codes for the protein.

Note: this term is most often used to designate the mutation process rather than the mutated state (the result of the mutation).

NEIGHBOR-JOINING n. One method of tree construction, that uses a distance matrix* invented by Saitou and Nei in 1987. When used to construct a tree, this distance method groups species together in such as way as to minimize the total tree length*.

NEUTRAL ALLELE n. An allele that has no effect on the fitness of the genotype. Othewise stated, an allele that carries no advantage or disadvantage for the organism with respect to natural selection.

NODE n. The meeting point of three branches or branch segments in a tree. If the tree is rooted, a node is formed by an upstream root segment and two downstream daughter branches or sister groups. If the classification is phylogenetic, that is, based on a cladogram*, a node represents a taxon (or group) that includes all the taxa lying downstream to the node. A node is defined by apomorphies or apomorphic characters*. These characters represent the only (fragmentary) knowledge that we might have about the last common ancestor of the sister taxa of the node. A node that contains several daughter branches is a multifurcation* indicating that the phylogenetic relationships among the different branches (taxa) could not be resolved.

OUTGROUP n. A group that is known *a priori* to lie outside the taxa of interest (those for which we are trying to determine the evolutionary relationships). The common ancestor of the ingroup and outgroup is usually the oldest of the tree.

PARALLELISM n. A similarity that has appeared independently in different closely related taxa*: the same apomophic* state evolves several times from the same ancestral character in different taxa. A parallelism is a special case of convergence*.

PARAPHYLETIC GROUP n. In phylogenetic systematics, a group that includes an ancestral species and only part of its descendants. A paraphyletic group is defined by at least one symplesiomorphy* or the absence of a character. Grades* are generally paraphyletic groups.

PARSIMONY n. A requirement of rational reasoning. Among several scenarios or theories, one chooses the one implying the fewest assumptions, or fewest *ad hoc* hypotheses. In systematics, the term designates a method of phylogeny construction that, given all possible dendrograms, retains the one with the lowest number of required evolutionary events (character state changes).

PHENETICS n. **1.** A synonym for numerical taxonomy*, a method whereby taxa* are identified and ranked according to their global similarity*. The evolutionary significance of the characters, or initially detected transformations, are not considered. The characters themselves are not polarized, meaning that they are not coded into apomorphic* or plesiomorphic* states. The differences between the ensemble of characters of two taxa are measured using a continuous variable (global similarity) that is used to determine the distance* between the taxa. We transcribe the distance between each taxon pair into a matrix. A classification is then constructed based on this data. **2.** A classification* constructed using numerical taxonomy.

PHENOGRAM n. A dendrogram* produced by numerical taxonomy or phenetics*, where the relationships among taxa represent their degree of global similarity*.

PHYLOGENY n. **1.** The historical pattern of descent of organized beings. **2.** A tree that illustrates the genesis of phyla. **3.** In the modern sense, a tree that illustrates sister-group relationships, or "who is most closely related to whom" or the degree of evolutionary relatedness*. **4.** In a more restrictive sense, a tree that provides secondary

homologies. This term is also used for a series of genes within a genome that are derived from the same ancestral gene.

PHYLOGRAM n. A representation of the evolutionary relationships (dendrogram) that expresses the branching pattern and the degree of divergence associated with each branch (taxon). For a long time, a phylogram was the traditional way to represent a phylogenetic tree.

PLESIOMORPHIC adj. Pertaining to the ancestral state of a character. Plesiomorphic is synonomous with the adjectives ancestral and primitive.

POLARIZATION n. An operation that consists of determining the evolutionary sequence of character transformations from the plesiomorphic* state to the apomorphic state(s)*. The operation is achieved using a certain number of criteria (principally three) proposed by Hennig (1966).

POLYPHYLETIC GROUP n. In phylogenetic systematics*, a group that includes a certain number of species or taxa*, but not the common ancestor to these species. In other words, a polyphyletic group contains two or more ancestral species. It is defined by at least one homoplasy*.

POSITION n. See site*.

PSEUDOGENE n. A non-functional gene that generally belongs to a multi-gene family*.

RELATEDNESS (EVOLUTIONARY) n. This indicates that two species or taxa* have an exclusive common ancestor. This ancestor is largely unknown except for the apomorphic characters carried by the studied taxa. Evolutionary relatedness contrasts with the relationship between an ancestor and its descendants, where a known taxon is the assumed ancestor of another. This type of relationship is equivalent to saying that the ancestral taxon is paraphyletic*.

Remark: Common ancestors are abstractions. As life's history is already past, we can never know the real ancestral species of a clade, in the genetic sense of the word. We can deduce the state of certain characters of an ancestral species only by examining the characteristics of the descendants. This also applies to fossil taxa. All organisms we classify using phylogenetics (either fossil or extant) are treated as terminal taxa.

REVERSION n. A derived character state that has returned to what seems to be the primitive (plesiomorphic*) state. More generally, in a series of character transformations (from the primitive to the derived state(s)), a reversion is a return to the morphological or molecular state of the preceding stage. A reversion is a special case of homoplasy*.

ROOT n. The branch segment upstream of the most inclusive ingroup* node: when the outgroup contains only a single species, the root is the position of this species in the tree. The root can be considered as a reference point for interpreting the characters: the character states of the outgroup are the plesiomorphic* states and the character states that differ from it are the apomorphic* states.

Remark 1: in order to readily compare two trees, one must root them on the same species or taxon.

Remark 2: when there are several outgroups, the upstream branch segment of the most important node is not always the separation point between the outgroup and ingroup.

ROOTED (TREE) adj. Refers to a tree (or dendrogram) that carries a root*.

SIMILARITY (GLOBAL) n. The overall resemblance between two taxa across all characters typically measured using a single, continuous variable, the distance*. More generally, this is the overall resemblance between two taxa analyzed without the polarization* of the character* transformations.

SISTER GROUPS n. Monophyletic* groups (or clades*) that have a unique common ancestor and, together with this ancestor, form a higher-ranking monophyletic* group.

SITE n. In a block of aligned sequences, a site corresponds to a vertical series of nucleotides or amino acids belonging to different species and considered to be homologous* by the alignment*. Each site is a molecular "character" when we use sequences to construct phylogenies. The nature of the amino acid or nucleotide is the character state. This definition also holds for a block of aligned sequences of paralogous genes.

SUBSTITUTION n. A type of mutation* in a sequence that consists of replacing one or more amino acids or nucleotides by different ones.

SYMPLESIOMORPHIC (CHARACTER) adj. Refers to a plesiomorphic* character state that is present in two or more taxa*.

SYNAPOMORPHIC (CHARACTER) adj. Refers to an apomorphic* character state shared by two or more taxa*.

SYSTEMATICS n. **1.** The science of biological classification and of the evolutionary relationships among organisms. **2.** The scientific study of the diversity and relationships among different organisms (according to Simpson, this means their phylogenetic relationships).

SYSTEMATICS (EVOLUTIONARY) n. A school of systematics* that classifies grades* and clades* and whose reference texts are those of Simpson and Mayr. Evolutionary classification attempts to express both the phylogenetic relationships and the degree of divergence among organisms and is supposed to reflect both adaptive jumps and degrees of complexity.

SYSTEMATICS (PHENETIC) n. A school of systematics* that classifies organisms according to their global similarity (see phenetics*) and whose reference texts are those of Sokal and Sneath. Phenetic trees express the degree of dissimilarity among organisms.

SYSTEMATICS (PHYLOGENETIC) n. See cladistics* and Hennig*. A school of systematics founded by Hennig* (1913–1976) that classifies only clades*. A phylogenetic classification expresses only the phylogenetic relationships among organisms determined using a cladistic analysis*.

TAXA n. Plural form of taxon*.

TAXON n. A group of recognized and defined organisms found within the different categories* of a hierarchical biological classification. In other words, a taxon is the actual content of a category. Example: the wolf, *Canis lupus*, is a taxon at the species level (category: species); the canidae (dog, wolf, fox) make up a taxon at the family level (category: family).

TAXONOMY n. **1.** The science of the rules for the classification of living organisms. **2.** The science of the theory and practice of classification (Mayr). From a practical point of view, taxonomy includes the recognition and identification of living organisms and their organization in a classification.

TIME OF DIVERGENCE (OR TIME OF SEPARATION OF TWO TAXA*) n. Date of the last common ancestor of two taxa. More precisely, and in other words, the date at which the ancestral species common to two taxa lost its genetic integrity and underwent speciation.

TREE LENGTH n. **1.** In a tree constructed using a distance method* (phenogram), this is the sum of the branch lengths, the branch lengths being proportional to the percent difference between two taxa. **2.** In a tree constructed using a parsimony method, this is the total number of evolutionary events (or steps) required by the tree (throughout all branches).

UPGMA (*UNWEIGHTED PAIR GROUP METHOD USING ARITHMETIC AVERAGE*) n. A phenetic* method of tree construction using a distance matrix*. It consists of first grouping those species or taxa that are most similar. The distances between each of these species and the other taxa are then averaged to obtain the distance between the new group (formed by the two species) and each of the other taxa. Among these new distances, the species (or taxon) with the shortest distance to the new group is then added to the tree at the midpoint between the two initial taxa. This forms a new group for which the distances to all remaining taxa can then be calculated using averages. Using the successive averages and clustering in an iterative manner, all the species (or taxa) are integrated one by one into the tree. If this method is used on molecular data, it requires a constant molecular clock*, that is, it assumes that two sister taxa have accumulated the same quantity of evolutionary changes since their last common ancestor. If this is not true (i.e., if the molecule does not have a constant rate of evolution in the taxa of interest), the results can be erroneous.

Index of Common Names

S

Index of Latin Names

The fossil names cited as "oldest known fossils" are not indexed.